Introduction to Protein Science

INTRODUCTION TO
PROTEIN SCIENCE

ARCHITECTURE, FUNCTION, AND GENOMICS

THIRD EDITION

ARTHUR M. LESK

The Pennsylvania State University

OXFORD
UNIVERSITY PRESS

OXFORD
UNIVERSITY PRESS

Great Clarendon Street, Oxford, OX2 6DP,
United Kingdom

Oxford University Press is a department of the University of Oxford.
It furthers the University's objective of excellence in research, scholarship,
and education by publishing worldwide. Oxford is a registered trade mark of
Oxford University Press in the UK and in certain other countries

First edition 2004
Second edition 2010
Impression: 1

1007560629

Published in the United States of America by Oxford University Press
198 Madison Avenue, New York, NY 10016, United States of America

British Library Cataloguing in Publication Data
Data available

Library of Congress Control Number: 2015942562

ISBN 978–0–19–871684–6

Printed in Italy by L.E.G.O. S.p.A.

Dedicated to the memory of Max Ferdinand Perutz

— Mozart, K. 584

CONTENTS

PREFACE TO THE FIRST EDITION

This book completes a trilogy, joining *Introduction to protein architecture: the structural biology of proteins*, and *Introduction to bioinformatics*. The other two books are more specialized—*Bioinformatics to sequences*, and *Architecture to structures*. This one is intended as an introduction to the other two. It also complements them, with a focus on protein function, including the integration and regulation of function: this book is as much about the logic of life as about its chemistry. These three components—sequence, structure, and function—define the significance of proteins in biology.

Sydney Brenner once stated the questions to ask about any living system:

1. How does it work?
2. How does its structure form?
3. How did it evolve?

Protein science has appealed to many scientific disciplines to address these problems:

- Parts of protein science have developed out of *chemistry*, including physical methods such as spectroscopy, kinetics, and techniques of structure determination; and organic/biochemical methods for working out the mechanisms of enzymatic catalysis. Much of classical protein chemistry and enzymology studied purified molecules in isolation. These studies revealed detailed information about the structures and activities of individual molecules.

- Parts of protein science have developed out of *molecular and cellular biology*. The challenge is to coordinate our knowledge of the properties and functions of isolated molecules, into an understanding of the biological context and integration of protein activity.

- High-throughput methods of genomics and proteomics provide very large amounts of data about sequences and expression patterns. We know the full sequences of the DNA of many organisms. From these genome sequences we can infer, to a large extent but certainly not perfectly, the amino acid sequences of an organism's full complement of proteins. Proteomics tells us how the expression patterns of these proteins vary. We have pieces of a jigsaw puzzle that extends in both space and time. How can we fit the pieces together to understand the complex and delicate instrument that is the living cell?

- Parts of protein science have developed out of *evolutionary biology*. Evolution takes place on the molecular scale, exploring variations in sequences, protein structures and functions, and patterns of expression. Much of what we see in living organisms depends on, and reveals, historical events no longer observable directly.

- Parts of protein science have emerged from *computing*. Computer science is a flourishing field with the goal of making the most effective use of information technology hardware. Computers are essential for storing and analysing the very large amounts of data that have been coming onstream. Databanks archive and organize access to sequences of genomes; to amino acid sequences, structures, expression patterns, and interaction networks of proteins; to metabolic pathways; to the molecular correlates of diseases; and to the scientific literature. Distribution of the information requires facilities of computer networks and the World Wide Web. Computer programs provide access to the data, and tools for their analysis. *Bioinformatics* is a hybrid of biology, chemistry, physics, and computer science. Its goal is to bring the data to bear on biological problems, including applications to clinical medicine, agriculture, and technology.

Some readers may wonder why nucleic acids receive substantial attention in a book on protein science. The answer is that they belong here. It is impossible to make sense of the biology of proteins without discussion of the ribosome, and of control of gene expression, just to pick two examples. Nucleic acids provide the logical framework of the stage on which proteins act.

Whole-genome sequencing projects have transformed molecular biology. Before, we were limited to studying selected available examples. The new results lay bare the choices that Nature has made.

The three-dimensional counterpart of genome-sequence determination, the provision of *structures* of all the proteins in an organism, is known as *structural genomics*.

When we know the sequences and structures of all macromolecules in an *E. coli* cell, we shall have a complete but *static* knowledge of the components of a living object. The next step will be to learn about how the components are assembled and the dynamics of how their individual activities are integrated.

The most profound significance of proteins lies in their collective properties as a class of biomolecules. These properties include all their potential, as well as their actual, characteristics: proteins have an underlying chemical unity, they have the ability to organize themselves in three dimensions; and the system that produces them can create inheritable structural variations, conferring the ability to evolve. Recent developments in genomics, proteomics, and structural genomics permit study of the entire complement of proteins of an organism and their coordinated activities, and give glimpses of the *proteosphere*—the entire spectrum of the proteins in Nature.

Proteins are where the action is

Proteins are fascinating molecular devices. They play a variety of roles: there are structural proteins (molecules of the cytoskeleton, epidermal keratin, viral coat proteins); catalytic proteins (enzymes); transport and storage proteins (haemoglobin, myoglobin, ferritin); regulatory proteins, (including hormones, many kinases and phosphatases, and proteins that control gene expression); and proteins of the immune system and the immunoglobulin superfamily (including antibodies, and proteins involved in cell–cell recognition and signalling).

Amino acid sequences are now available for over 1 million proteins, and detailed atomic structures for over 24 000. These data support a number of fascinating and useful scientific endeavours, including:

1. *Interpretations of the mechanisms of function of individual proteins.* The catalytic activities of enzymes can be explained in terms of physical-organic chemistry, on the basis of the interactions of substrates with residues in the active site of the protein.

2. *Patterns of molecular evolution.* For many families of proteins, we know dozens or even hundreds of amino acid sequences, and many structures. We can observe the structural and functional roles of the sets of residues that are strongly conserved. We can describe the consequences of mutations, insertions and deletions in the amino acid sequence that perturb protein conformation, and identify constraints that selection imposes on the structure to preserve function. When proteins evolve with changes in function, these constraints on the structure are relaxed—or rather, replaced by alternative constraints—and the sequences and structures change more radically.

On a broader scale, *comparative genomics* traces the changes and distribution of proteins and regulatory sequences in different species, and similarities and differences in genome organization.

3. *The amino acid sequences of proteins dictate their three-dimensional structures, and their folding pathways.*

(a) Under physiological conditions of solvent and temperature, proteins fold spontaneously to an active native state. Although we cannot yet confidently predict the conformation of a novel protein structure from its amino acid sequence in all cases, many principles of protein structure and folding are now understood.

(b) Relationships between the evolutionary divergence of amino acid sequence and structure in protein families do make possible the reliable prediction of protein structure from amino acid sequence, whenever we know the sequence and structure of one or more sufficiently closely related proteins. As we come closer to having a structure of every type that appears in Nature, then the nearest relative of known structure will provide a reasonable quantitative model for almost all unknown proteins. In the future, this will give us access to structures of the proteins encoded in any genome.

(c) The amino acid sequence of a protein must not only preferentially stabilize the native state, it must contain a road map telling the protein how to get there, starting from the many diverse conformations that comprise the unfolded state. A combination of physicochemical measurements and studies of the effects of mutations on the kinetics of folding and unfolding are illuminating the process of folding.

4. *Mapping the logic and the mechanisms of integration and control of life processes.*

(a) We can measure and compare patterns of biochemical activity in different tissues, or under different growth conditions. Comparisons of protein expression patterns, or of genotypes embedded in the DNA, using microarrays, have direct clinical applications. Many diseases involve failures in control mechanisms. Therefore differences in expression patterns in normal and disease states permit precise diagnosis of disease, producing a more accurate prognosis and choice of treatment. Prediction, from details of genotypes, of enhanced susceptibility to disease may suggest preventative measures, including alteration of lifestyle.

(b) We can trace pathways of information flow in regulatory networks. Proteins receive and transmit signals, often using conformational changes to control activity. Proteins 'talk' to one another, and also to DNA. DNA–protein interactions control gene expression.

(c) We can study patterns in time as well as in space. The cell cycle, and the unfolding of developmental programmes, are coordinated with changes in protein expression. Development of the nervous systems of higher organisms presents great challenges to diversity plus accuracy in molecular signalling and recognition.

Advances in protein science have spawned the biotechnology industry:

5. *Protein engineering*. It is now possible to design and test modifications of known proteins, and *to* design novel ones. Applications include:

(a) Attempts to enhance thermostability, for example by introducing disulphide or salt bridges, or by optimizing the choice of amino acids that pack the protein interior.

(b) Clinical applications, for instance therapeutic antibodies. Rats can raise antibodies against human tumours. Transfer of the active site from a rat antibody to a human antibody framework can produce a molecule that retains therapeutic activity in humans but reduces the side effects arising from patients' immunological response against the antigenic rat protein.

(c) Modifying antibodies to give them catalytic activity. Two features of enzymes are: the ability to bind substrate specifically, and the juxtaposition of bound substrate with residues that catalyse a chemical change. Antibodies can provide the binding and discrimination; the challenge to the chemist is to introduce the catalytic function.

(d) Modifications to probe mechanisms of function or folding, such as the identification of properties of folding intermediates by the effects of mutations on stability and folding kinetics.

6. *Drug discovery*. There are many proteins specific to pathogens that we want to deactivate. Knowing the structure of the AIDS proteinase, or the neuraminidase of influenza virus, it should be possible to design molecules that will bind tightly and specifically to an essential site on these molecules, to interfere with their function. Knowing the structures of receptors it should be possible to design agonists and antagonists of the signals to which they respond.

These and other topics comprise a new scientific speciality, *protein science*. By a combination of approaches and methods, gathered from many other fields of science, we are gaining insight into the principles governing this fascinating class of molecules. Let us apply this knowledge in medicine, agriculture, and technology to improve the lot of mankind.

I thank F. Arnold, M. Bashton, C. Chothia, J. Clarke, T.J. Dafforn, A. Doherty, D.S. Eisenberg, I. Fearnley, J.R. Fresco, A. Friday, M. Gait, E. Gherardi, W.B. Gratzer, R. Henderson, O. Herzberg, T.J.P. Hubbard, J. Janin, T. Kiefhaber, G. Kleywegt, P.A. Lawrence, E.L. Lesk, V.E. Lesk, V.I. Lesk, L. Liljas, L. Lo Conte, D.J. Lomas, J. Löwe, P.A. Lyons, A.J. McCoy, J.D. Mollon, S. Moran, J. Moult, A.G. Murzin, D. Neuhaus, M. Oliveberg, V. Ramakrishnan, R. Read, G.D. Rose, B. Rost, K. Scott, L. Serpell, R. Staden, R. Tregear, M. Vendruscolo, C. Vogel, G. Vriend, A.G. Weeds, J.C. Whisstock, A. Yonath, and L. You.

PREFACE TO THE SECOND EDITION

This new edition of *Introduction to protein science: structure, function, and genomics* seeks to retain the successful features of the first edition, and to expand the coverage of the subject. This includes a fuller treatment of methods of structure determination, of enzymatic catalysis, and of protein evolution. There have been significant advances in protein science since the first edition, both experimental and computational. In presenting these advances, I emphasize how, for many topics, the combination of theory and experiment has proved to be a very powerful and effective partnership.

In the first edition, Introduction to protein science was described as a companion volume to Introduction to protein architecture: the structural biology of proteins, and Introduction to bioinformatics. Some of Architecture has been integrated into this volume. A more recent book, Introduction to genomics, is another companion volume.

Proteins claim the attention of many readers, spanning a range of interests in the fields of biology and medicine, chemistry and physics. In using this book as a text for a course, instructors will find ample opportunity to tailor the material to their students' interests and needs, especially through choice of exercises, problems, and weblems.

PREFACE TO THE THIRD EDITION

Since the last edition of this book, protein science has advanced along many lines of research already active, and initiated new ones.

A landmark in structure determination was the growth of the world-wide Protein Data Bank to reach 100 000 entries, in May 2014. Electron microscopy of macromolecular structures and complexes will very soon achieve atomic resolution.

On the computational side, molecular dynamics has come a long way from the early days when Jeremy Knowles could dismiss it as 'an expensive way to denature proteins'. This is no longer true. Thanks to growth in computer power, molecular dynamics is no longer so expensive; thanks to the availability of mature force fields and the ability to include explicit solvent, it no longer denatures proteins. A confirmation of the maturity of the field was the award of the 2013 Nobel Prize in chemistry to molecular-dynamics pioneers Martin Karplus, Michael Levitt, and Arieh Warshel.

In addition to naturally-occurring proteins, artificial ones can be created by simulated 'evolution' in the laboratory, or even designed by computer programs. It is possible by genetic engineering to express modified proteins, for applications or for research. The monoclonal antibody enterprise, creating molecules for diagnostics and therapy, is currently a multibillion dollar per year industry.

Research in molecular evolution can in many cases suggest the sequences of proteins in the absence of experimental data. Thus, it may not be possible to revivify entire extinct organisms, but when analysis of phylogenetic trees gives us a confident statement of their sequences, their proteins can be synthesized, and tested for their purported functions. This relatively new field is called palaeomolecular biology.

Effective application of our knowledge requires powerful information-retrieval facilities. Not only has there been continued and accelerating growth of the traditional information resources about proteins—data banks of amino acid sequences and three-dimensional structures—these results are becoming ever more intimately integrated into more general data collections. These include databanks of protein function, coordinated into databanks of metabolic pathways; and data dealing with expression patterns, across different species and within the developmental programs of individual organisms. Proteomics and even genomics make use of data on proteins to create integrated descriptions of processes of life.

The new edition of this book reflects these changes. There is a general migration of material from the text to the web. In particular, instead of stereo pictures in the book as aids to the three-dimensional perception of structures, the associated website contains movies. (Many widely-available computer programs can generate additional pictures. Facility in their use is a necessary skill for study and research in this field.) Emphasis on a wider variety of databanks is another appropriate response to recent developments.

The 'weblems'—problems to be solved by appeal to websites, a prominent feature of earlier editions, have been expanded and now do not appear in the printed text but only on the companion website. They remain an integral part of the book.

Even more than in previous editions, there is an emphasis on applications. These may be thought of as both illustrations of the general principles that remain the backbone of the exposition, and also as exemplifying the contributions to society that the progress in fundamental research has made possible.

Plan of the book

Chapter 1: A general introduction to the subject, setting out the topics developed in the rest of the book, emphasizing that amino acid sequences of proteins determine their three-dimensional structures and functions.

Chapter 2: Detailed description of the structural chemistry of proteins and their constituent amino acids; survey of varieties of proteins.

Chapter 3: Protein structure determination, integrating experimental and computational methods.

Chapter 4: Protein bioinformatics: databases of sequences, structures and the results of other experiments, and tools for information retrieval and analysis.

Chapter 5: Enzyme catalysis and mechanisms, and the use of classical and modern tools to illuminate mechanisms of enzyme activity.

Chapter 6: Protein interactions, including functional aspects of complex formation.

Chapter 7: Protein evolution: by diverging in sequence and structure, proteins can retain, modify, or profoundly alter their function.

Chapter 8: Mechanism of the folding process by which proteins achieve their native states.

Chapter 9: Proteomics and Systems biology: analysis of proteins within their natural context, and description of the regulatory networks by which cells control and integrate the activities of their constituent molecules.

Learning aids throughout the book include boxes with a purple background containing summaries of key concepts. Interpolated short comments on green backgrounds are the equivalent of marginal notes; larger boxes with green backgrounds contain description of applications, many but not all clinical.

The list of people to thank has grown. In addition to those mentioned in the first edition, I acknowledge with gratitude Drs P. Aloy, A. Andreeva, P.C. Babbitt, H. and F. Bernstein, E.H. Egelman, J. Finer-Moore, J.R. Fresco, M.E. Glassner, K. Henrick, D.T. Jones, J.T.J. Lecomte, C. Praul, N. Seeman, J. Sussman, V.E. Womble, and E.B. Ziff for reading specific sections and answering questions, and many other colleagues and publishers for courtesy in allowing me to reproduce illustrations.

CHAPTER 1

Introduction

LEARNING GOALS

- *Understanding the biological context of protein science,* including the very wide variety of functions that different proteins achieve.

- *Recognizing that the structures of proteins are implicit in the amino acid sequences,* through a spontaneous folding process. Knowing the meaning and significance of protein denaturation and renaturation.

- *Appreciating the significance of the 'central dogma' describing information transfer and its expression,* in its original statement by Francis Crick, and subsequent extensions.

- *Knowing how DNA sequences code for amino acid sequences.* In nature, 64 triplets of nucleotides, or codons, encode 20 amino acids (occasionally 22), plus Stop signals.

- *Being able to sort out the relationships among gene sequences, amino acid sequences, protein structures, and protein functions.* Gene sequences encipher amino acid sequences. Amino acid sequences dictate the folding of proteins. The sequences and structures of proteins determine their functions. The evolution of genome sequences is the effect of the action of natural selection on protein function, and of genetic drift.

- *Recognizing the significance of examples of protein disorder* as replacing classical ideas of unique rigid native states.

- *Being able to define all the terms in Table 1.4.*

- *Appreciating the nature of mutations,* and their possible effects on protein structure, function, and regulation, with potential consequences for health and disease.

- *Recognizing the prevalence of alternative splicing of eukaryotic genes,* amplifying the genome into a proteome of enhanced richness.

- *Understanding the role of the ribosome,* as the protein-synthesizing machine that connects mRNA sequences transcribed from genomes, with amino acid sequences of proteins.

- *Knowing the basic levels of description of protein structure:* primary, secondary, tertiary, and quaternary structure.

- *Understanding the nature of the information available from spectroscopic measurements on proteins:* absorbance, circular dichroism (CD) and its relationship to secondary structure content, fluorescence and fluorescence resonance energy transfer (FRET).

- *Recognizing the importance of computing in protein science:* for archiving, curation, and distribution of information in databanks; for computational analysis of protein sequences, structures, and functions, and in particular for graphical representations of protein structures.

Proteins in their biological context

Proteins are a family of biological macromolecules that provide a variety of three-dimensional structures exquisitely shaped for their many different individual functions. They include:

- *structural proteins:* including the keratins of our hair and the outermost layer of our skin, proteins of the cytoskeleton, and the coats of viruses;

- *enzymes:* that catalyse the reactions of metabolism, and the replication and transcription of DNA;

- *antibodies:* that recognize and repel invading pathogens and inactivate toxins;

- *regulatory proteins:* including those that control transcription of genes;

- *sensors:* that detect and implement signals generated within our bodies and those impingent on us from the outside world;

- *transporters and pumps:* that control traffic into and out of cells and organelles;

- *transducers:* motor proteins that convert chemical to mechanical energy, such as the contractile proteins of muscle contraction, the kinesins and dyneins of intracellular transport, and ATP synthases that convert chemiosmotic energy generated by respiratory or photosynthetic electron transport chains, to the high-energy chemical bond of ATP.

Proteins create the structures required for their panoply of functions by variations on a common underlying chemical scheme. Proteins are linear polymers, all containing the same polypeptide backbone. Along this backbone a variable sequence of sidechains distinguishes different proteins. The sidechain at any position is almost always one of a canonical set of 20. These common features of the chemical structures of proteins make possible a common synthetic mechanism: ribosomes assemble the great variety of proteins under the direction of different messenger RNA (mRNA) sequences.

A sidechain is a set of atoms attached to each residue of the protein backbone.

Both the DNA sequences of genes and the amino acid sequences of proteins are one-dimensional. How then do proteins achieve their three-dimensional biologically active states? The three-dimensional structures of proteins are inherent in their amino acid sequences: *natural proteins fold spontaneously to form individual 'native' structures* (see Fig. 1.1). How is this achieved? The polypeptide chain is flexible, and can assume many different spatial conformations. Each conformation brings into proximity different constellations of sidechains, which interact with one another and with solvent. One of these conformations creates a set of interactions with a much greater stabilization than all other conformations. At equilibrium, the protein adopts this native state.*

- Rutherford famously said that 'All science is either physics or stamp collecting'. I reply that the study of protein structure combines the best elements of both.

  For what we see in biology today is the result of a combination of constraints imposed by physics and historical accident. T. Dobzhansky wrote that 'Nothing in biology makes sense except in the light of evolution'. Without even for a moment denying this, I would insist on adding thermodynamics to the list of things, except in the light of which, nothing in biology makes sense.

  The native structures of proteins represent a physicochemical compromise. Thermodynamics tells us that systems at constant temperature and pressure will seek equilibrium states of low energy and high entropy. Think of entropy as a measure of conformational freedom. In the denatured state, the entropy is high because there is a great multiplicity of conformations. In the native state, the entropy is low because the molecules adopt a unique conformation. Thus conformational entropy greatly favours the denatured state, and is a severe impediment to protein folding. For proteins to adopt native states, they must achieve substantial inter-residue attractive interactions to lower the energy, and/or find compensating ways to raise the entropy of the native state. Burying of aliphatic groups in the interiors of compact native states, achieves precisely this increase in entropy of the native state. (This 'hydrophobic effect' is the same as observed in the

*Definitions of bold-face terms appear in the glossary at the end of the book.

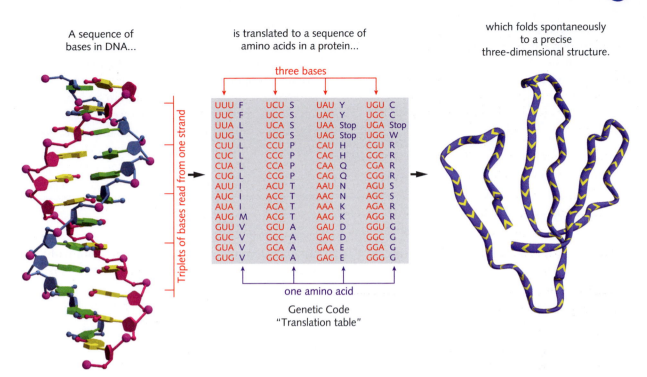

A sequence of bases in DNA...

is translated to a sequence of amino acids in a protein...

which folds spontaneously to a precise three-dimensional structure.

Figure 1.1 Expression of gene sequences as three-dimensional structures of proteins. The three-dimensional structure is implicit in the amino acid sequence. Logically, gene sequences are one-dimensional. Amino acid sequences are also one-dimensional. The spontaneous folding of proteins to form their native states is the point at which Nature makes the giant leap from the one-dimensional world of genetic and protein sequences to the three-dimensional world we inhabit. There is a paradox: the translation of DNA sequences to amino acid sequences is very simple to describe logically; it is specified by the genetic code. The folding of the polypeptide chain into a precise three-dimensional structure is very difficult to describe logically. However, translation requires the immensely complicated machinery of the ribosome, tRNAs, and associated molecules; but protein folding occurs spontaneously.

spontaneous separation of salad dressing into aqueous and oil phases, which is also entropy-driven.)

These thermodynamic considerations, and their implications about the features of native states of proteins, apply to the many different protein structures observed—this is *why* evolution has had the freedom to explore the many different folding patterns we see. To make sense of them we can afford to neglect neither evolution nor physics.

The great range of protein structures and functions depends on the variety in the properties of the sidechains. The 20 natural sidechains vary in size and in other physicochemical properties. Some are polar or charged. These can participate in **hydrogen bonding** and **electrostatic interactions**, with other residues and with solvent. The charged atoms occur at or near the ends of the relatively long and flexible sidechains; the atoms proximal to the backbone are non-polar.

Two sidechains with positive and negative charge can approach each other in space to form a **salt bridge**.

Other sidechains are non-polar. Large **aliphatic** sidechains have an unfavourable interaction with water, called the **hydrophobic effect**. Clustering of hydrophobic sidechains in the interiors of proteins provides a thermodynamic driving force for protein folding.

- Proteins are linear polymers with a constant polypeptide-chain backbone and one of a canonical set of 20 side chains attached to each residue. The unique amino acid sequence of each protein determines its three-dimensional structure.

It should be emphasized that the property of folding spontaneously to a unique native structure is not expected for any polypeptide with a random amino acid sequence. Amino acid sequences of natural proteins are selected by evolution for their individual equilibrium structures, and for the capacity to fold efficiently into them.

> The total number of 100-residue amino acid sequences possible is 20^{100}, a number so large as to imply that the proteins we see are a minute fraction of those possible.

It is the combination of a common synthetic machinery—the gene sequence dictating the amino acid sequence—with **spontaneous folding** to the native three-dimensional structure, that underlies the mechanism of molecular evolution. Mutations—inheritable changes in genome sequences—create variation in protein structures and functions. Selection acts on protein function.

The amino acids

Amino acids are the raw material of proteins. Each amino acid has the structure of a central carbon atom, to which are attached: a hydrogen atom, an amino group ($-NH_3^+$), a carboxylate group ($-COO^-$) and a sidechain (see Fig. 1.2(a)). Within each protein, the mainchain forms **peptide bonds** between the COO^- group of one amino acid and the NH_3^+ group of another (see Fig. 1.2(b)). Every protein has a free amino terminus at one end and a free carboxy terminus at the other (except for a few cyclic polypeptides).

Natural proteins are formed from a basic repertoire of 20 **standard amino acids** (see Table 1.1). Some proteins do contain amino acids outside the usual set of 20. The unusual amino acids selenocysteine and pyrrolysine are introduced *during* translation, by extensions of the normal genetic code. (In the laboratory, other extensions of the genetic code to incorporate unnatural amino acids are possible.) In contrast, **collagen** contains **hydroxyproline** and hydroxylysine produced by enzymatic action as **post-translational modifications** after the protein is synthesized.

In addition to amino acids, ions, small organic ligands and even water molecules are integral parts of many protein structures. For instance, haem groups are an essential component of haemoglobin and cytochrome *c*.

> • A roster of 20 amino acids is standard. Proteins achieve even greater chemical variety by: (1) binding cofactors, (2) post-translational modifications, and (3) incorporating additional amino acids (rare).

Proteins are polymers of amino acids containing a constant **mainchain** or **backbone** of repeating units, with a variable **sidechain** attached to each:

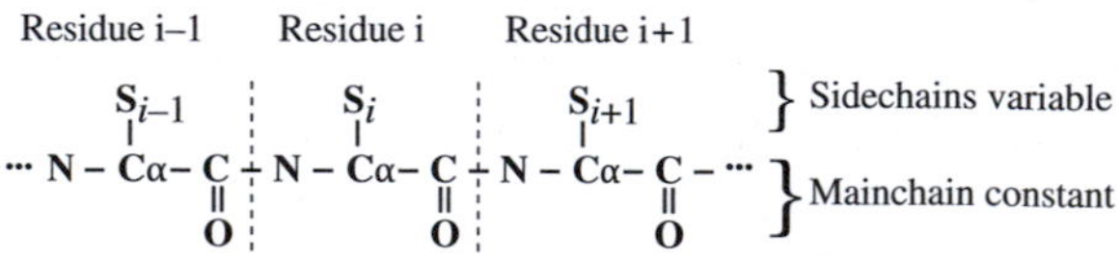

Think of strings of Christmas tree lights: the wire is like the repetitive backbone. The order of colours of the bulbs is variable. (However, the order of the bulb colours does not control the structure of the wire!) Each protein has a unique sequence of sidechains that determines its individual characteristics: The amino acid sequence of a protein is a text written in a 20-letter alphabet.

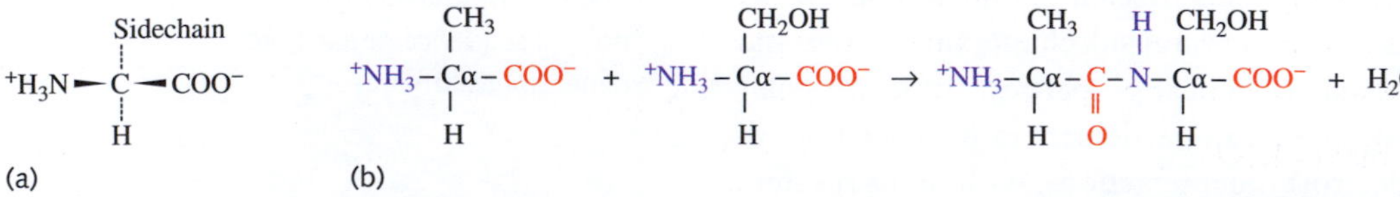

Figure 1.2 (a) General structure of an amino acid. There are 20 canonical amino acids, each with a different sidechain. For example: for alanine, sidechain = CH_3; for serine, sidechain = CH_2OH. The central carbon is designated the Cα. Successive atoms out along the sidechain are Cβ, Cγ or Oγ, etc. (b) Formation of the peptide bond between two amino acids: alanine + serine $\rightarrow$ alanylserine + H_2O.

Table 1.1 The 20 standard amino acids in proteins

Non-polar amino acids							
G	glycine	A	alanine	P	proline	V	valine
I	isoleucine	L	leucine	F	phenylalanine	M	methionine
Polar amino acids							
S	serine	C	cysteine	T	threonine	N	asparagine
Q	glutamine	Y	tyrosine	W	tryptophan		
Charged amino acids							
D	aspartic acid	E	glutamic acid	K	lysine	R	arginine
H	histidine						

Other classifications of amino acids can also be useful. For instance, histidine, phenylalanine, tyrosine, and tryptophan are aromatic, and are observed to play special structural roles in membrane proteins.

Amino acid names are frequently abbreviated to their first three letters—for instance Gly for glycine—except for isoleucine, asparagine, glutamine and tryptophan, which are abbreviated to Ile, Asn, Gln, and Trp, respectively. The rare amino acid selenocysteine has the three-letter abbreviation Sec and the one-letter code U. The even rarer amino acid pyrrolysine has the three-letter abbreviation Pyl and the one-letter code O.

It is conventional to write one-letter abbreviations of nucleotides in lower case and those of amino acids in upper case. Thus, atg = adenine-thymine-guanine and ATG = alanine-threonine-glycine.

Amino acid sequences are always stated in order from the N-terminal to the C-terminal. This is also the order in which ribosomes synthesize proteins: ribosomes add amino acids to the free carboxy terminus of the growing chain.

Dogmas—central and peripheral

In 1958, as the basic paradigm of protein synthesis emerged, Francis Crick brought the ideas together into a statement that he called the central dogma of molecular biology: DNA makes RNA makes protein (see Fig. 1.3). Crick emphasized that the pathway of information flow is unidirectional: from DNA, the archival genetic material in most organisms → by transcription to messenger RNA and then → by translation into the specific amino acid sequences of proteins. Reverse transcriptases of retroviruses were not known then.

Most people would now add another step: amino acid sequence determines protein structure. Most proteins adopt specific three-dimensional conformations dictated by their amino acid sequences alone.

Native protein structures are the primary agents of biochemical activity. Just as protein sequence determines protein structure, protein structure determines protein function. Perhaps a quick statement of the dogma should now read:

> DNA makes RNA
>   makes amino acid sequence
>   makes protein structure
>   makes protein function

(See Box 1.1)

The one-way flow of information requires Darwinian natural selection at the protein level to be a mechanism of evolution—as opposed to direct

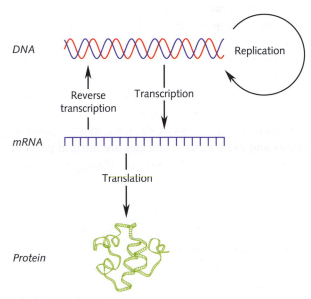

Figure 1.3 The central dogma of molecular biology. Reverse transcriptases, discovered after F.H.C. Crick originally enunciated the central dogma, are an exception to the one-directional flow of information.

> ## BOX 1.1   An expanded version of the central dogma
>
> - DNA is the archival genetic material in most organisms; some viruses have RNA genomes.
> - Gene sequences from DNA are transcribed into messenger RNA. In eukaryotes, exons are spliced together, often in alternative ways, to form mature messenger RNA. Multiple splicing patterns can generate proteins with different amino acid sequences from the same gene.
> - Messenger RNAs are translated into the polypeptide chains of proteins, by synthesis on ribosomes.
> - Amino acid sequences of proteins determine three-dimensional protein structures by a spontaneous folding process.
> - Protein structures determine protein functions.
> - Mutations change gene sequences, sometimes altering amino acid sequences, protein structures, and protein functions. Mutations in non-coding regions may affect splicing or regulation.
> - Mutations create variation, providing raw material on which evolution acts.
> - Control mechanisms govern different rates of transcription and translation, altering the repertoire and inventory of proteins active in different environments, at different stages of cell cycles and life histories, and in different cells and tissues of higher organisms.
> - Selection acts on function—including control mechanisms—effecting changes in the distributions of genome sequences within populations.
> - Non-selective genomic change—genetic drift—is also important, especially in small populations.

selection of DNA sequence changes. Reverse transcriptases aside, the only way to change individual DNA sequences is by mutation, which generates variation but does not itself move a population to a better adapted state. The **ribosome** makes possible the connection between *heritable genetic information*—in the form of nucleic acid—and the *agents of biochemical activity*—primarily proteins. Selection acts on the functions produced by the proteins, and RNAs, encoded in the genome, and on regulatory mechanisms.

Alternatively, some genetic change is non-selective: neutral evolution, or genetic drift. Drift is especially important in small populations. The results of both selection and drift are inherited alterations in genomes—including both changes in sequence and changes in **allele** frequencies. These changes modify the nature and expression patterns of the proteins and RNAs in the individuals of a population.

> - DNA → RNA
>   → amino acid sequence
>   → protein structure
>   → protein function.
>
> Selective pressure acts on protein function (including the function of individual proteins, and on regulation of transcription rates) to control differential survival and transmission of DNA sequences within populations. (Genetic drift can also affect gene frequencies, especially in small populations.)

The relationship between amino acid sequence and protein structure is robust

The rules relating protein sequence to structure can tolerate many but not all mutations. This is of the utmost significance for evolution, because it permits the exploration of sequence variations. A protein structure that required a unique amino acid sequence—so that any mutation would destroy it—would not be observed in nature because evolution could never find it. (If any mutation would destroy the structure, then it could not have any precursor.) If such a protein were designed and its gene artificially synthesized and released into nature, it would be unstable to mutation.

If we compare corresponding proteins from related species, we find that the sequences have diverged. However, in many instances the structures and even the functions have stayed very much the same. As an example, histidine-containing phosphocarrier proteins from *Escherichia coli* and *Streptococcus faecalis* have only 38% identical residues in an **alignment** of their sequences. However, their structures retain the same general spatial course of the mainchain (see Fig. 1.4).

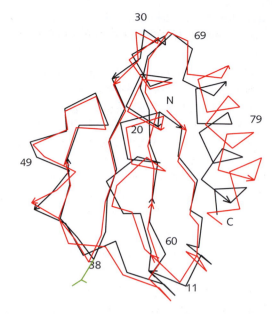

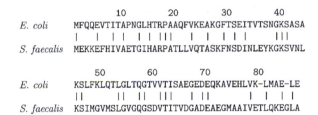

```
                10        20        30        40
E. coli    MFQQEVTITAPNGLHTRPAAQFVKEAKGFTSEITVTSNGKSASA
           |   | |   | | |  ||  |   |   |   |    |||
S. faecalis MEKKEFHIVAETGIHARPATLLVQTASKFNSDINLEYKGKSVNL

                50        60        70        80
E. coli    KSLFKLQTLGLTQGTVVTISAEGEDEQKAVEHLVK-LMAE-LE
           ||    || ||  |||   | ||        |  | | |
S. faecalis KSIMGVMSLGVGQGSDVTITVDGADEAEGMAAIVETLQKEGLA
```

Figure 1.4 Superposition of structures of histidine-containing phosphocarrier proteins from *Escherichia coli* [1POH] (black) and *Streptococcus faecalis* [1PTF] (red). Although over 60% of the residues have mutated, the folding pattern is completely intact. What residue in *E. coli* corresponds to the green sidechain in the *S. faecalis* structure?

Disorder in proteins

Crystal structures portray protein structures in fixed native conformations. There is no choice, really: any mobile structures or substructures are invisible in the electron-density maps derivable from X-ray diffraction. But it made for a very simple and attractive picture: the fully-disordered **denatured state** *v.* the **native state** with a unique conformation. This was in accord with Fischer's authoritative 'lock-and-key' model of enzyme-substrate specificity; ignored for a century (by almost everyone) was the unquestioned fact that locks contain mechanisms that move when keys are inserted and turned.

And yet there was evidence that this picture was overdrawn. There was suspicion that crystal structures might be statues, capturing one pose of a dynamic and active subject.

Thus, crystal structures of proteins in different states of ligation, or in different crystallization conditions, gave different structures. Oxy- and deoxy-haemoglobin were the classic example, but even so the implication was still not disorder, but merely that

haemoglobin had two conformations each of which was fixed. This also applied to the idea of 'induced fit'—conformational changes in enzymes associated with substrate binding.

Especially interesting examples of multiple conformers appear in icosahedral viruses. Most viruses assemble capsids from multiple copies of a unique coat protein. In many cases the coat proteins have the same sequence but adopt different conformations. (*Why* they cannot all have the same conformation is a very interesting story, which we shall discuss in Chapter 6.) Nevertheless, again the response was, 'Yes... but ...'; after all viruses were admittedly very clever. Moreover, this was also an example of non-uniqueness of the native state, but not disorder.

Many examples of proteins with multiple conformations are now known.

More to the point, there were many examples of regions that were missing from crystal structures because they lacked the fixed structure, reproduced exactly in every unit cell of the crystal, that was required for X-ray structure determination. Even these were resolutely ignored. Indeed in some cases these regions were excised from the protein in order to improve the crystals. Often they were dismissed as artefacts arising from crystal-packing interactions.

Crystallographers dismissed the greater apparent flexibility of NMR structures as simple underdetermination. That is, they insisted that ideally the conformations were fixed and unique, but NMR spectroscopists simply could not measure enough data to determine them.

Conversely, it is undeniable that X-ray crystallography selects for unique and rigid structures. Disordered proteins are unlikely to crystallize at all. If they do, crystal-packing interactions are forcing them to adopt unique conformations. These might not even correspond to the most highly-populated state when free in solution.

Gradually, the paradigm began to shift. The statues came to life. (A common theme in theatre: think Pygmalion, Hermione in The Winter's Tale, the Commendatore in Don Giovanni.) What was responsible? Two factors: (1) The evidence for genuine disorder, including but not limited to NMR studies, grew to the point where it could not simply be shrugged off. (2) The discovery of more and more exceptions to the lock-and-key model, in the form of molecules that showed less specificity in their binding and interactions.

One set of examples contrasts primary and secondary antibodies. Upon exposure to a pathogen, perhaps a new strain of flu virus, our immune systems contain antibodies that have enough flexibility to bind to epitopes from the pathogen, albeit relatively weakly. These are the primary antibodies. A process of internal evolution—mutation plus selection—modifies the sequences of these proteins to produce a **secondary response** of antibodies

with greater **affinity** and specificity. (This process takes about a week, which accounts for the clinical duration typical of bouts of a cold or of flu.) The lower specificity of the primary antibodies requires flexibility; the secondary antibodies sacrifice flexibility in order to achieve enhanced specificity.

The idea that has emerged is that of a *spectrum* of protein flexibility (see Fig. 1.5). At one end is the classic unique native state, with a rigid binding site and high specificity. The lock-and-key model applies. Proceeding in the direction of greater flexibility, one can encounter states in which ligation induces a fixed conformation in a more-disordered unligated state. (Induced fit is an example.) This is compatible with ligand specificity, but also with conformational selection in which different ligands induce different bound-state conformations. The unligated state can be close to the bound conformation, but, alternatively, could be disordered.

In most of these cases the picture is of flexibility (and possible promiscuity of potential ligands) of the unbound state but rigidity (and possible multiple rigid conformations) of the ligated state (see Box 1.2).

Yet another generalization is the 'fuzzy complex' (see Fig. 1.5(iii)) in which even the conformation of the ligated state is not rigid. Such a complex may pay a price—in terms of affinity—relative to a rigid complex, but functional reasons may require a molecule capable of less specific and weaker binding. An example is the complex between the cyclin-dependent kinase inhibitor Sic1 (stoichiometric inhibitor of Cdk1-Clb) and the SCF subunit of Cdc4 (Cdc4 = cell-division control protein-4; SCF = a ubiquitin ligase (E3) complex: the Skp1-Cdc53/CUL-F Box receptor E3 complex).

(The reader should keep this figure in mind when encountering complexes in succeeding chapters, and ask where they fit into Mannige's scheme.)

BOX 1.2   The distinction between flexibility and disorder

Flexibility and disorder are related concepts, but it is useful to distinguish them. Regions in proteins can show two, or more, stable mainchain conformations. If this set of conformations is discrete and small, let us call such regions *flexible*. In other cases, regions in a protein may be entirely *disordered*—there appear to be no preferred conform-

ations, at least no stable ones. They are just 'flapping in the breeze'.

A disordered region can become flexible; for instance, it may adopt different fixed conformations, when the protein binds two different ligands. But a flexible region may not necessarily be disordered, short of complete denaturation of the protein.

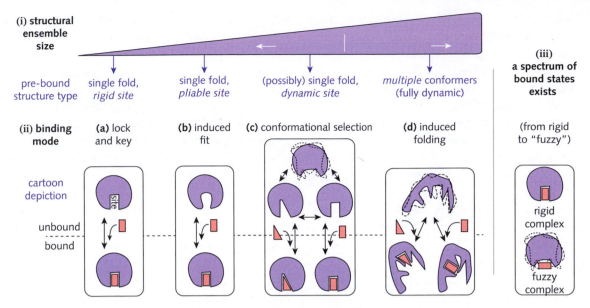

Figure 1.5 (i) The spectrum of protein native conformations, from the classical unique, rigid fold (left), through states of increasing flexibility and dynamics (right). The increase in ensemble size indicates that the greater the flexibility, the larger the number of individual microscopic conformations that the system can adopt. (ii) The corresponding nature of ligation, from the classical 'lock-and-key' model (left), to ligation-induced selection of conformational state (right), allowing for the possibility of reduced specificity of binding and ligand-dependent conformations. (iii) The alternatives of rigid ('lock-and-key') binding and 'fuzzy' binding, which gives up our idea of fixed complementarity between protein and ligand. Although this figure is drawn in terms of a large protein binding a relatively small ligand, similar considerations can apply to protein-protein interactions.

From: Mannige, R.V. (2014). Dynamic new world: refining our view of protein structure, function and evolution. Proteomes. 2, 128–153. Reproduced via the creative commons attribution 3.0 unported license.

Figure 1.6 shows an example of a disordered region that becomes ordered upon binding. In this case, the structure induced depends on the ligand.

Disordered regions play a role in **zymogen** activation. Digestive enzymes, such as trypsin and chymotrypsin, are synthesized in an inactive form containing disordered regions. Cleavage of an N-terminal oligopeptide by enterokinase causes a conformational change in which interactions with the new N-terminus fix the active structure of the catalytic site.

Databases of **intrinsically-disordered proteins** (IDP) include:

- DisProt (www.disprot.org)
- MobiDP (http://mobidb.bio.unipd.it/)
- Gallery of images of intrinsically-disordered proteins (http://iimcb.genesilico.pl/metadisorder/protein-disorder_intrinsically_unstructured_proteins_gallery_images.html)

- An attractive set of movies of macromolecular movements (http://www.molmovdb.org/)

Prediction of disordered regions is one of the current challenges of the **CASP** (Critical Assessment of Structure Prediction) programmes. The highest-ranked group in CASP10 (2012), that of D. Jones of University College London, scored an accuracy, measured by a combination of correctly and incorrectly predicted residues, of 72.3%.

The actual formula is:

$$\text{accuracy} = 0.5 \times \left[\frac{\text{TP}}{\text{TP} + \text{FN}} + \frac{\text{TN}}{\text{TN} + \text{FP}} \right]$$

where TP = true positive = residue correctly predicted to be disordered, TN = true negative = residue correctly predicted to be ordered, FP = false positive = residue incorrectly predicted to be disordered, and FN = false negative = residue incorrectly predicted to be ordered.

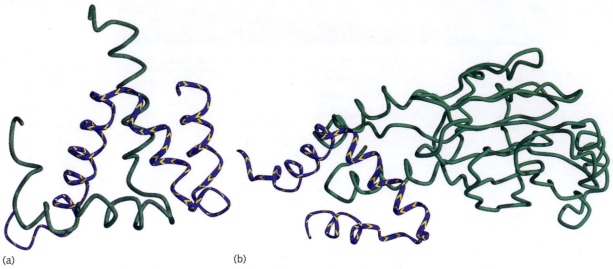

(a)   (b)

Figure 1.6 The NCBD (IbID) domain of the creb-binding protein (CBP) is disordered in the unligated state. It forms different structures in complex with (left) the ACTR domain of p160 and (right) with IRF3. In both pictures, the common NCBD domain is shown in blue. The C-terminal region of the NCBD domain, including two of the three helices, has approximately the same structure in both complexes.

Regulation

Even the extended central dogma is deceptively simple. We understand fairly well the mechanisms of many biochemical processes, in isolation. What is required for life is a set of mechanisms to integrate and control these activities.

Life is an integrated set of molecular activities that, even in an unchanging environment, involves regulation at the cellular level at different stages of the cell cycle, and also mechanisms for unfolding programmes of development, differentiation, and ageing in higher organisms. In any cell at any time, there must be an appropriate distribution of the traffic through metabolic pathways to maintain overall harmony. For instance, unless successive steps in a reaction sequence have matched 'throughputs', some intermediates will accumulate. As a cell goes through the stages of its cycle, certain processes must be turned on and off and others adjusted. For instance, cyclin-dependent regulation of protein activity is essential for the control of cell replication and division. Changes in the environment, especially those embodying threats, must elicit a response.

In a multicellular organism, different amounts of different proteins must be synthesized at different stages of life and in different tissues. For example, in fruit flies, genes involved in eye differentiation and vision are expressed only in the adult. Most humans produce a succession of five types of haemoglobin during their lifetime: three embryonic forms, produced in the yolk sac for 5–6 weeks after conception; a foetal form, produced in the liver from 5 to 6 weeks after conception until after birth; and an adult form afterwards. The pre-adult forms have properties that facilitate oxygen transfer from mother to child *in utero*.

Regulation must be effectively integrated across the genome. The process as a whole must therefore be robust; that is, stable to perturbations, but flexible enough to deal with changing environments. In principle, the source of robustness is redundancy. In cells, regulatory interactions form complex networks, such that an ultimate 'decision' about the level of expression of a gene is the resultant of combinations of stimulatory and repressive 'inputs'.

The genetic code

The genetic code is the translation table between nucleic acid sequences of genes and amino acid sequences of proteins (see Table 1.2). The unit of coding is a triplet of nucleotides, called a **codon**. Four nucleotides—adenine, thymine (or, equivalently, uracil), guanine and cytosine—can form a total of 64 possible triplets, or codons.

There is an excess of codons over amino acids: 64 codons correspond to only 20 amino acids, plus Stop signals. Most amino acids are translated from more than one codon. Five of the amino acids are encoded by blocks of four codons, which have the same nucleotides in the first two positions and all four possibilities in the third. Nine are encoded by blocks of two codons, with the same nucleotides in the first two positions and two nucleotides in the third position. Three amino acids are encoded by more than four codons, one amino acid by only three codons, and two amino acids by only one codon. The variability in the third position is related to the base pairing between the codon in mRNA and the **anticodon** in the corresponding tRNA. The pairing is more tolerant at the third position, according to the **'wobble' hypothesis**.

The genetic code shown in Table 1.2 is the standard version used by nuclear DNA of **eukaryotes** and most **prokaryotes**. Other versions appear in mitochondria, chloroplasts, and, sporadically, in a few species. These alternative genetic codes show only a few differences from the standard one. For example, in the yeast mitochondrial genetic code ctg is translated to Thr instead of Leu. In most other cases one or two of the usual Stop codons are translated into amino acids.

A mutation is a change in a genomic nucleotide sequence. Some, but not all **single-nucleotide polymorphisms (SNPs)**, or base substitutions, in coding regions produce changes in amino acid sequences, which can alter protein structure and function. Many substitutions at the third positions of codons are silent, leaving the protein invariant. SNPs, even in noncoding regions, can affect splice sites or regulatory sequences. They are also useful as genetic markers. As such they have important applications: for example, the identification of disease genes; and in anthropology, for detection of ancestry and tracing of migrations.

Table 1.2 The standard genetic code

ttt	Phe	tct	Ser	tat	Tyr	tgt	Cys
ttc	Phe	tcc	Ser	tac	Tyr	tgc	Cys
tta	Leu	tca	Ser	taa	STOP	tga	STOP
ttg	Leu	tcg	Ser	tag	STOP	tgg	Trp
ctt	Leu	cct	Pro	cat	His	cgt	Arg
ctc	Leu	ccc	Pro	cac	His	cgc	Arg
cta	Leu	cca	Pro	caa	Gln	cga	Arg
ctg	Leu	ccg	Pro	cag	Gln	cgg	Arg
att	Ile	act	Thr	aat	Asn	agt	Ser
atc	Ile	acc	Thr	aac	Asn	agc	Ser
ata	Ile	aca	Thr	aaa	Lys	aga	Arg
atg	Met	acg	Thr	aag	Lys	agg	Arg
gtt	Val	gct	Ala	gat	Asp	ggt	Gly
gtc	Val	gcc	Ala	gac	Asp	ggc	Gly
gta	Val	gca	Ala	gaa	Glu	gga	Gly
gtg	Val	gcg	Ala	gag	Glu	ggg	Gly

The layout of the genetic code—the assignments of codons to amino acids—is such that many substitutions that change the amino acid make conservative changes in physicochemical class and volume (see Chapter 2). For instance, a single-base substitution in one of the four codons for Val can change a Val to any of the following: Phe, Leu, Ile, Met, Ala, Asp, Glu, or Gly. Of these, Phe, Leu, Ile, Met and Ala are physicochemically similar to Val and are regarded as conservative substitutions. These changes might have only a minor effect on a protein, unless they were to occur at a particularly sensitive site. Of the other amino acids that can be produced from Val by single-base substitutions, Asp and Glu are charged, and Gly is a special residue, with only a hydrogen atom as its sidechain. Changing a Val to any of these would not be considered a conservative substitution. Such a mutation would threaten more serious consequences.

Single-nucleotide changes are either **transitions** = purine to purine (a → g or g → a) or pyrimidine to pyrimidine (t → c or c → t), or **transversions** = purine to pyrimidine (a or g → t or c), or pyrimidine to purine (t or c → a or g). Transitions are more common mutations than transversions. A single transition could change a Val to an Ile, Met, or Ala. These substitutions are *all* conservative, with respect to their expected effect on a protein structure. Such relationships within the genetic code help to make proteins robust to mutation.

> Many single-nucleotide changes in protein-coding regions are either silent, or produce only conservative changes. The layout of the genetic code gives proteins a degree of robustness to mutation.

With life so dependent on proteins, there is ample opportunity for things to go wrong

Some general types of problems include:

- *A mutation may lead to an absent or dysfunctional protein* (see Box 1.3). If a normal allele on the partner chromosome provides adequate healthy protein, the abnormal phenotype may be a recessive trait. Mutations on the X chromosome may lead to sex-linked abnormalities such as some forms of colour blindness.

- Some missing or dysfunctional proteins can be supplied, such as insulin for diabetes, Factor VIII or IX for the most common forms of haemophilia, and human growth hormone for children who produce inadequate amounts. Attempts to replace damaged genes using viruses as vectors, or DNA nanoparticles, or plasmid DNA encapsulated in liposomes, are in the experimental stage. Use of a lentivirus to insert a normal gene into blood stem cells has recently shown success in treating X-linked adrenoleukodystrophy.

- *Cutting a metabolic pathway* by knocking out an enzyme that catalyses one of the steps may lead to (a) loss of product of that step, and of downstream steps, (b) accumulation of precursor.

  - Phenylketonuria is a hereditary disease resulting from loss or dysfunction of phenylalanine hydroxylase, the enzyme that converts phenylalanine to tyrosine (see Box 1.3).

  - Some enzyme deficiencies can be accommodated with changes in lifestyle. Some show complex interactions with other diseases, with the result that some deleterious mutations have not been eliminated from populations by selection because they carry some compensating advantages. For example, the genes for sickle-cell anaemia, and for glucose-6-phosphate dehydrogenase deficiency confer *resistance to malaria. Glucose-6-phosphate dehydrogenase* (G6PDH) is the most common enzyme deficiency, affecting over 400 million people world-wide. It is a recessive X-linked genetic defect, affecting up to 10% of populations in which mutations are common (see Box 1.4).

  - Mutation in a targeting signal may result in mislocalization of a functional protein.

  - In some cases, species-wide loss of biosynthetic enzymes contributes to the list of essential nutrients. For instance, whereas most animals can synthesize vitamin C, we must provide it in our diet (see Box 2.8).

  - Loss of some proteins is surprisingly innocuous. Mice lacking myoglobin thrive, and even show athletic performance comparable to normal mice.

- *Dysfunction of a regulatory protein or receptor* can disorganize the operation of a pathway even if all components of the pathway regulated are normal. Some abnormal regulatory proteins cannot be activated at all, others are constitutively activated and cannot be shut off. The effects include:

  - *Physiological defects:* A number of diseases are associated with mutations in G protein-coupled receptors (see Box 2.9). Some mutations in opsins are associated with colour blindness. Certain mutations in the common G protein target of olfactory receptors lead to loss of sense of smell.

BOX 1.3   Proteins in health and disease

Mutations and molecular disease

Evolution has largely optimized proteins for their roles in healthy organisms. Therefore, most mutations causing amino acid sequence changes are deleterious, impairing protein function and threatening to produce disease.

- The first identification of a mutation as the cause of a disease was the discovery by Linus Pauling and colleagues in 1949 that **sickle-cell anaemia** *is the result of a mutation in haemoglobin.*

  Haemoglobin contains four polypeptide chains, two α-chains and two β-chains. The sickle-cell mutation changes residue 6 of the β-chain from a charged sidechain, glutamic acid, to a non-polar one, valine. This creates a 'sticky patch' on the surface of the molecule. As a result, the mutant haemoglobin forms polymers within the **erythrocyte**, in the unligated or deoxy state (in the deoxy form, typical of venous blood, haemoglobin does not bind oxygen). To flow through small capillaries, erythrocytes must be deformable, as their typical size, 7.8 mm (in humans), is larger than the diameter of small capillaries. The formation of the polymers has a rigidifying effect on the erythrocyte, impeding its flow and blocking capillaries. In the traffic jam building up behind a plugged capillary, arriving red cells release their oxygen to surrounding hypoxic tissues, become deoxygenated, and thereby aggravate the problem.

- *Some diseases involve single genes encoding metabolic enzymes.* Phenylketonuria (PKU) is a genetic disease caused by deficiency in phenylalanine hydroxylase, the enzyme that converts phenylalanine to tyrosine. If untreated, phenylalanine accumulates in the blood, to toxic levels. The high levels of phenylalanine cause developmental defects, including mental retardation. The disease cannot be cured, but symptoms can be avoided by lifestyle control: a phenylalanine-free diet. Screening of newborns for PKU is legally required in the United States and many other countries.

PKU is an example of how understanding molecular mechanisms of a disease can, for some conditions, help restore and preserve health through suggested changes in lifestyle and/or medical treatment.

- *Many other mutations in proteins have clinical consequences.* In some cases, treatment is possible to mitigate the effect. For example, regular injections of the normal forms of clotting factors VIII or IX prevent the symptoms of the most common forms of haemophilia. In other cases, detection of a recessive mutation in phenotypically normal 'carriers' permits genetic counselling of prospective parents.

  An example of a risk factor detectable at the genetic level involves α_1-**antitrypsin**, a protein that normally functions to inhibit elastase in the alveoli of the lung. People homozygous for the Z mutant of α_1-antitrypsin (342Glu→Lys) express only a dysfunctional protein. They are at risk of emphysema because of damage to the lungs from endogenous elastase unchecked by normal inhibitory activity, and also of liver disease because of accumulation of a polymeric form of α_1-antitrypsin in hepatocytes where it is synthesized. The combination of a genome homozygous for Z α_1-antitrypsin, and smoking cigarettes, is a guarantee of death from emphysema at a young age. The disease is brought on by a *combination* of genetic and environmental factors. This is a brutally simple illustration of J. Stern's maxim, 'Genetics loads the gun and environment pulls the trigger'.

- *Some diseases involve multiple genes, and complex interactions with the environment.* Over 100 genes are implicated in susceptibility to asthma. Several environmental factors also affect risk: life history of challenge to the immune system plays an important role. Thus, breast feeding, and bouts of common childhood diseases, appear to have protective effects. Occupational exposure to allergens and toxins contributes to the development of the disease, as well as triggering attacks in asthma sufferers.

- *Developmental defects:* Several types are traceable to mutations in hormone receptors. For instance, Laron syndrome, a phenotype including diminished stature, arises from a mutation in the human growth hormone receptor. Administration of exogenous growth hormone does not restore normal growth.

- *Cancer is a disease of genomic instability.* Several proteins act as tumour suppressors, with mutations correlating with increased risk of

BOX 1.4   Proteins in health and disease

Glucose-6-phosphate deficiency, food taboos, folk medicine, pharmacogenomics, and mosquito-breeding seasons

Glucose-6-phosphate dehydrogenase (G6PDH) catalyses the reaction:

glucose-6-phosphate + NADP →
   6-phosphogluconate + NADPH,

the first step in the pentose phosphate shunt.

This reaction produces the reduced glutathione needed to dispose of hydrogen peroxide (H_2O_2). It is particularly important in red blood cells, which, lacking nuclei and mitochondria, are metabolically impoverished, and have no alternative mechanism for detoxifying H_2O_2. Without active G6PDH, build-up of H_2O_2 will oxidize and denature haemoglobin, causing destruction of red blood cells, leading to a condition called haemolytic anaemia.

Eating fava beans, especially if uncooked, can induce anaemic episodes in people deficient in G6PDH. The danger of eating fava beans has been recognized since antiquity, and has been associated with food taboos and preparation techniques designed to reduce toxicity. Pythagoras, for example, banned eating of fava beans in his school. We now know that fava beans contain the compounds vicine and convicine, metabolized in the intestine to isouramil and divicine, which react with oxygen to produce hydrogen peroxide, subjecting cells to oxidative stress.

Other chemicals, including certain drugs, present the same danger to G6PDH-deficient people. During World War I, some patients were observed to suffer dangerous side effects of the antimalarial drug primaquine. Many drugs, including sulphonamides, are now contraindicated for G6PDH-deficient patients, as is the taking of large doses of vitamin C. The observation of variations in effectiveness and toxicity of different drugs in different people has developed into the new field of **pharmacogenomics**, the tailoring of drug treatments to the genotype of the individual patient.

Why have dysfunctional G6PDH genes remained at such a high level in the population? Why does primaquine produce haemolytic anaemia in G6PDH-deficient patients, and does this have a relationship to its antimalarial activity? And why have fava beans continued to be grown if non-toxic alternatives are available?

The malarial parasite invades the red blood cell of its host, and competes metabolically with normal activity. Primaquine and related drugs, such as **chloroquine**, subject the red blood cells to oxidative stress. Cells stressed by *both* parasite and drug are the most vulnerable, and if they die they take the parasite down with them. Because consumption of fava beans subjects cells to oxidative stress, they also provide an antimalarial effect, recognized in folk medicine. Indeed, fava beans have some effect against malaria even for people with normal G6PDH activity; those with abnormal G6PDH have a greater advantage, until the maturing *Plasmodium* produces its own G6PDH.

The link with malaria is the likely explanation of the persistence of the gene in the population and the fava bean in agriculture. A final clue appears in the calendar: there is good overlap between the fava bean harvest period and the peak *Anopheles* breeding season.

developing cancer. Some of these proteins act to repair DNA damage. Other mutations lead to cancer by interfering with regulatory functions directly. Mutations in Ras appear in 30% of human cancers. A cell containing p21 Ras trapped in a constitutively active state is continuously triggered to proliferate.

- *Protein aggregation.* Many diseases are associated with the formation of insoluble aggregates, usually of *misfold*ed proteins (see Chapter 6). These include classical amyloidoses, **Alzheimer's** and **Huntington's** diseases, **aggregates** of misfolded serpins, and **prion diseases**. Polymerization of insulin creates problems in production, storage and delivery for diabetes therapy.

- *Impairment or loss of defence to infection by pathogenic organisms.* Part of the ongoing wars between species. To avoid the necessity of developing *ad hoc* defences against every new threat, vertebrates developed a generalized system to detect and repel any foreign invader—the immune system. It works amazingly well—except in the case of AIDS, which attacks the immune system itself.

Genome sequences

A genome is a specification of a potential life. The genome of an organism is the sequence of its main genetic material, usually one or more molecules of DNA. Organelles—mitochondria and chloroplasts—contribute small amounts of additional genetic material. We now know the full genomes of almost 25000 prokaryotes, many viruses, organelles and plasmids, and almost 3500 eukaryotes (see Table 1.3), representing all major categories of living things (see http://www.genomesonline.org). Many other full-genome projects are in progress.

The ENCODE project (encyclopedia of DNA elements) has the goal of determining the functions of all significant regions of the human genome, including coding and regulatory regions. ENCODE started with a pilot project: for a selected 1% of the human genome (about 30 Mb), the corresponding regions in 30 vertebrate genomes were sequenced. These data illuminate one another. ENCODE applied a variety of experimental and computational techniques, including comparative genomics.

With the success of the pilot phase, and the lessons learnt, ENCODE scaled up its efforts to the full human genome. The basic goal has been to explode the *static* sequence of the human genome into a description of the functions of component parts, and the dynamic interactions among them.

The scaled-up ENCODE project has produced some unexpected results. For example, it was found that *most* of the DNA in the human genome is transcribed into RNAs that have some function, not limited to protein coding. Formerly it was believed that much of the human genome is 'junk'. Although function cannot be ascribed to all regions of the genome, we now know at least most are not transcriptionally silent.

ModENCODE is an extension, to the systematic comparison of functional elements in the human, fruit fly, and *C. elegans* genomes.

Table 1.3 Some completed eukaryotic nuclear genomes

Type of organism	Species	Genome size (10⁶ base pairs)
Primitive microsporidian	*Encephalitozoon cuniculi*	2.5
Fungi	*Saccharomyces cerevisiae* (Baker's yeast)	12.1
	Schizosaccharomyces pombe	13.8
	Neurospora crassa	40
Nematode worm	*Caenorhabditis elegans*	100
Fruit fly	*Drosophila melanogaster*	180
Mosquito	*Anopheles gambiae*	278
Malarial parasite (carried by *A. gambiae*)	*Plasmodium falciparum*	22.8
Thale cress	*Arabidopsis thaliana*	116.8
Rice	*Oryza sativa*	400
Chicken	*Gallus gallus*	1200
Platypus	*Ornithorhynchus anatinus*	2000
Mouse	*Mus musculus*	3454
Dog	*Canis familiaris*	2500
Human	*Homo sapiens*	3165

Gene sequence determines amino acid sequence

A gene is a segment, of DNA or viral genomic RNA, that encodes the sequence of a protein or structural RNA. In prokaryotes, genes for proteins are contiguous sequences of the DNA. In eukaryotes, genes may be split into exons (expressed regions) and introns (intervening regions). Introns are transcribed but not translated: following synthesis of a full-length messenger RNA from the gene—introns plus exons—the introns are spliced out to produce a mature messenger RNA to be translated by the ribosome. The splicing is carried out in the nucleus by a large ribonucleoprotein particle called the spliceosome.

In many cases, alternative splicing combinations produce more than one mature message from the same region of the DNA. Each of these messages produces a different protein. Alternative splicing patterns thereby give genes an extra dimension of versatility. They also make it more difficult to interpret and annotate a genome sequence! It is estimated that about half the genes in the human genome show alternative splicing.

- In bacteria, gene sequences in DNA correspond directly to mRNA sequences. Many eukaryotic genes contain introns that are spliced out to form mature mRNA for translation. Variability in splicing gives the DNA → protein correspondence an additional dimension of complexity.

Some genes in organelles—mitochondria and chloroplasts—contain introns. Because spliceosomes appear only in the nucleus, intron-containing genes in organelles are self-splicing. The transcribed RNA molecules have the required catalytic activity to excise the intron (including the catalytic site for splicing) and link the exons.

There has been considerable traffic of genes from mitochondria and chloroplasts to nuclei. (Why not in the other direction?) Once in the nuclear genome, in some cases genes originally continuous in the organelle can acquire introns (see Box 1.5).

The generation of the very great number and variety of antigen-binding segments of antibodies by the immune system arises also from the combinatorial splicing of genetic segments, but this is done at the DNA level, during differentiation of B cells.

Another kind of splicing is done at the protein level. Inteins are proteins that autocatalyse the excision of a stretch of residues within their sequences.

Usually, a protein is a faithful translation of a messenger RNA sequence. However, there are exceptions. In some cases, one genomic DNA sequence encodes parts of more than one polypeptide chain. For example, in *E. coli* one gene codes for both the τ and γ subunits of DNA polymerase III. Translation of the entire gene forms the τ subunit. The γ subunit corresponds approximately to the N-terminal two-thirds of the τ subunit. A frameshift on the ribosome leads to premature chain termination 50% of the time, yielding a 1:1 expression ratio of τ and γ subunits.

Another mechanism that breaks the fidelity of mRNA → protein translation is the enzymatic modification of certain bases between transcription and translation, a process known as mRNA editing. In the human intestine, two isoforms of apolipoprotein B appear. To produce one of these forms, mRNA editing alters one nucleotide, changing 'c' to 'u'. This turns a Gln codon (caa) into a Stop codon (uaa), truncating the altered isoform to half the size of the longer form. The longer form is the translation of the full unedited mRNA.

Post-translational modifications alter many proteins, often with important effects on function. Some of these are additions of a variety of chemical groups, for example glycosylation. Others involve excision of peptides: (1) Some proteolytic enzymes are synthesized as inactive *zymogens*. Cleavage of a 16-residue N-terminal peptide from trypsinogen results in a conformational change to active trypsin. (2) Active insulin contains two polypeptide chains, of 21 and 30 amino acids. The precursor proinsulin is a single polypeptide chain from which excision of an internal peptide produces the mature protein. The proinsulin stage appears to be essential for proper folding—see Chapter 8.

BOX 1.5   **An example of alternative splicing in rice (*Oryza sativa*) illuminates gene transfer from the mitochondrion to the nucleus**

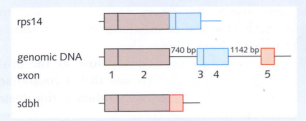

A region of chromosome 8 in the nuclear genome of rice contains five exons (centre strip), encoding two *mitochondrial* proteins: rps14 (ribosomal protein 14 of the small subunit) and sdhB (the B subunit of succinate dehydrogenase). rps14 is the translation of exons 1, 2, 3 and 4. sdhB is the translation of exons 1, 2 and 5.

These proteins are synthesized in the cytoplasm and transported into the mitochondria. It is likely that these genes were originally in the mitochondrial genome. They illustrate the traffic between organelle and nuclear genomes. A region in the rice mitochondrial genome similar to rps14 is translated, but has become non-functional as a result of four single-nucleotide deletions that destroy the reading frame. Certain other higher plants have functional mitochondrial genes for rps14 (broadbean, oilseed rape), whereas others resemble rice in containing non-functional mitochondrial rps14 genes but functional nuclear genes encoding the mitochondrial protein (potato, *Arabidopsis*).

Compared to other rps14 genes, the protein encoded by the rice nuclear gene for mitochondrial rps14 has an N-terminal extension derived from the sdhB exons. Incorporation of the additional exons provides the protein with a targeting signal directing the protein to mitochondria. This region of the protein is cleaved off in the mitochondria.

Genes similar to sdhB, in contrast, have not been observed in plant mitochondrial genomes. It is likely, therefore, that the move of the sdhB gene from the mitochondrial → nuclear genome is an old event, and the move of the rsp14 gene a relatively recent one.

Protein synthesis: the ribosome is the point of contact between genes and proteins—it is the fulcrum of genomics

Where do proteins come from? Ribosomes are subcellular particles that synthesize proteins. Ribosomes are large nucleoprotein particles, approximately 250 Å in diameter. They contain two subunits that dissociate reversibly upon lowering the Mg^{2+} concentration. Originally, the masses and sizes of particles were measured by their sedimentation constants in the ultracentrifuge. (The Svedberg, S, is a unit in which sedimentation coefficients are measured. It is named after the inventor of the ultracentrifuge. Other things being equal—notably molecular shape—the larger the sedimentation coefficient the heavier the particle.) The nomenclature survives: the complete *E. coli* ribosome is called the 70 S particle; it dissociates into 30 S and 50 S subunits. (Sedimentation constants are not additive!)

We now know their sizes precisely: the full *E. coli* ribosome has a relative molecular mass of 2.5×10^6 and the two subunits 1.65×10^6 and 0.85×10^6 (relative molecular mass, or r.m.m.—colloquially, 'molecular weight'—expresses mass on a scale in which the most common isotope of carbon has mass of exactly 12.)

The cytoplasmic ribosomes of eukaryotes are somewhat larger, and vary in size from species to species.

Ribosomes contain proteins and RNA. In prokaryotic ribosomes, the large subunit contains two molecules of RNA, one with about 2900 bases and another with 120 bases. Bases, not base pairs: the RNAs are single-stranded, although they can form intramolecular stretches of double helix. The large subunit also contains 31 different proteins. The small subunit contains one RNA molecule of 1500 bases, and 21 proteins. In eukaryotic cytoplasmic ribosomes, the large subunit contains three RNA molecules of approximately 4700, 156 and 120 bases; and 50 proteins, most of which are present in only one copy. The small subunit contains a 1900-base RNA molecule and 32 proteins.

Cells take ribosomes very seriously. An *E. coli* cell contains about 15000 ribosomes, comprising approximately 25% of the cell mass! Conversely, protein synthesis is so vital a process that ribosomes are the targets of several effective antibiotics, including streptomycin and tetracycline.

Ribosomes were implicated in protein synthesis very early on

It was observed in the 1950s that incorporated radioactive amino acids first appeared in protein associated with ribosomes. Preparations of purified ribosomes capable of cell-free protein synthesis proved the function. Recognition that, in eukaryotes, genes were in the nucleus and ribosomes in the cytoplasm suggested the idea of *messenger RNA* to link them. The inability to detect specific and selective amino acid–nucleic acid interactions suggested the **adaptor hypothesis**: a special form of RNA that would interact with both protein and nucleic acid and achieve the physical association of messenger RNA with specific amino acids. **Transfer RNAs** play precisely this role.

These ideas and the cell-free protein synthesis systems came together to allow deciphering of the genetic code. The use of synthetic RNAs of known sequence (or known simple compositions) as artificial messengers, and observation of which amino acids were taken up into the polypeptides synthesized, permitted assignment of codons to amino acids.

A general picture of protein synthesis emerged with the metaphor of the ribosome as a 'tape reader'. A ribosome 'reads' codons sequentially from messenger RNA. To process each codon, an appropriate transfer RNA, charged with a specific amino acid, brings its amino acid into proximity to the growing polypeptide chain on the ribosome. The peptide bond is synthesized enzymatically; then the transfer RNA is released, and the 'tape advanced' to the next codon.

One worrying portent: despite extensive investigation, no ribosomal protein that catalysed peptide bond synthesis was ever found.

Structural studies of ribosomes by X-ray crystallography and electron microscopy

A major triumph of structural biology in the new century has been the determination of the structures of ribosomal subunits and intact ribosomes to high resolution, achieved through a combination of X-ray crystallography and electron microscopy. These structures are the culmination of two decades of work since the first diffracting crystals were obtained in 1983.

The results came as a shock.

The cores of the subunits are nucleic acid, with the proteins mostly peripheral (see Fig. 1.7). The subunit interface is also nucleic acid. Indeed, the enzymatic activity resides in the RNA: the ribosome is a **ribozyme**.

Discovery of the ribosome structure was also an archaeological find of major significance. There is now consensus that before the current roster of biological molecules was established, there was a world of life based on RNA. Certainly the activity of the RNA in ribosomes is consistent with this. One puzzle about the differences between the RNA world and our own is how the transition was achieved. The ribosome provides a crucial link.

• Many people believe that an 'RNA world' preceded the development of proteins. The ribosome is a **ribozyme**—that is, it is the RNA in ribosomes that is catalytically active. The ribosome is a link to the RNA world.

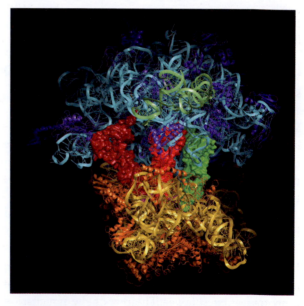

Figure 1.7 Structure of bacterial ribosome. Ribosomal RNAs are shown in turquoise, green, and yellow; proteins in purple and orange. Three tRNAs are bound between the 50S (top) and 30S (bottom) subunits. (Green = A-site tRNA; red = P-site tRNA; brown = E-site tRNA.) These sites correspond to the three positions of the tRNA during a cycle of amino addition during protein synthesis: A = acceptor site for attachment of incoming tRNA, which is then translocated to the P-site = peptidyl elongation site, the aminoacyl tRNA in which is attached to the growing polypeptide chain. The E-site = exit site is transiently occupied by deacylated tRNA before release.

Courtesy of Dr V. Ramakrishnan, MRC Laboratory of Molecular Biology.

Protein stability, denaturation, aggregation, and turnover

Native structures of most proteins are only marginally stable, and achieve stability only within narrow ranges of conditions of solvent and temperature. Overstep these boundaries, and proteins denature: they lose their definite compact structure, and take up states with disorder in the backbone conformation and few if any specific interactions among residues: the pieces of the jigsaw puzzle have been pulled apart and deformed. Homeostatic mechanisms that keep the internal environment of our bodies at a relatively constant temperature, salt concentration, and acidity, are essential to health if only because proteins lose the effectiveness of their function if any of these conditions vary beyond limits, which are in many cases quite narrow.

The native structure can be broken up, by heating or by high concentrations of certain chemicals such as urea—a process called **denaturation**. Denaturation destroys the three-dimensional conformation, but leaves the polypeptide chain intact.

If the original conditions are restored the protein will spontaneously **renature**, reforming its original structure and recovering its biological activity.

Formation of the native state is a *global* property of the protein. In most cases, the entire protein (or at least a large part) is necessary. This is because many of the stabilizing interactions involve parts of the protein that are very distant in the polypeptide chain, but brought into spatial proximity by the folding.

The integrity of the native state is essential for biological activity. The residues forming the **active site** of an enzyme may be distant in the sequence, but juxtaposed in the native structure. Typically, an active site may involve only 10% of the residues of a protein or even fewer. The rest of the structure is required as scaffolding to create the exquisite spatial relationships among the active residues. (Still more may be involved: (1) many proteins may provide more than a *fixed* scaffolding. In many cases, conformational changes are an essential part of the mechanism of activity. (2) Some residues in active sites are in strained conformations that play a role in catalysis. The scaffolding 'pays for' the energy required to allow these residues to adopt unfavourable conformations.)

In the very crowded regime inside cells, the smooth process of protein folding as observed in dilute solution is threatened by aggregation. Protein aggregates are associated with several diseases, including Alzheimer's, Huntington's, and the prion diseases, the **spongiform encephalopathies**. Misfolded or partially folded proteins are particularly prone to aggregation. Many mutant proteins show increased tendency to aggregate (see Chapter 6).

Cells defend themselves against aggregation of misfolded molecules with **chaperone** proteins. (The metaphor was suggested by the idea that chaperones inhibit the formation of improper alliances.) Chaperones unfold misfolded proteins and give them another chance to fold properly. But the activity of chaperones does not contradict the principle that all information necessary to determine the native structure of a protein is inherent in the amino acid sequence. Chaperones do not themselves contain any information about the correct structure of any particular protein. The fact that they can act to catalyse the folding of many different proteins proves this.

> • Misfolded proteins are prone to aggregation. Protein aggregates are implicated in many diseases, including Alzheimer's disease and spongiform encepathalopathies.

Protein turnover

Cells contain, or secrete, many different types of proteinases. Many function to digest nutrients, or for defence; others are involved in apoptosis (programmed cell death). A complex macromolecular particle, the proteasome, coordinates **protein turnover**, the controlled degradation of proteins:

• Cells must destroy normal functional proteins that have reached the end of their natural lifetimes. Some proteins, for example glycolytic enzymes, are long-lived. Others, such as those that initiate DNA recognition or cell division, must have their activities turned off after their signals have been received and acted upon.

- The N-terminal amino acid determines the half-life of a protein. Proteins beginning with Arg or Lys have short half-lives (2–3 min), and those beginning with Gly, Val or Met have long half-lives (> 1 day). Defective regulation of protein turnover is associated with several diseases. For example, the half-life of the proto-oncogene ornithine decarboxylase differs between normal tissues and tumours.

- Misfolded proteins must be removed. They are potentially dangerous—especially those present in high concentrations, which threaten to form aggregates. If a cell is unsuccessful at rescuing misfolded proteins with chaperones, degradation is the necessary alternative.

- The immune system uses proteolytically generated oligopeptides (~8–10 residues), presented by MHC proteins, to recognize foreign proteins (see Chapter 6).

A fundamental problem with an intracellular proteinase is keeping it away from the many proteins that the cell does *not* want to see degraded. Rogue proteases would quickly prove fatal to cells. The solution is to sequester the proteinase activity inside a large macromolecular assembly called the **proteasome**, with carefully controlled access to a narrow channel leading to the interior.

Proteasomes or related structures appear in **archaea,** bacteria, and eukaryotes. A typical human cell contains 30000 of them, distributed between the nucleus and cytoplasm.

The proteasome is a barrel-shaped structure, 15 nm high and 11 nm in diameter. The complete proteasome particle contains a 20S proteasome core complex + one or two copies of a regulatory complex of comparable size forming a cap at both ends (see Fig. 1.8). In archaea, the core complex contains multiple copies of two homologous proteins α and β, assembled into a stack of four sevenfold symmetric rings: α_7, β_7, β_7, α_7. The proteolytic activity resides on the inside of the central rings. The catalytic residue is the N-terminal threonine of the β subunits. In eukaryotes, the α and β subunits have diverged to form 14 independent proteins. Not all the β subunits are proteolytically active. In higher organisms, the cap structure and some of the β subunits can vary under the control of the cytokine interferon-γ during activation of the immune response.

To target a protein for degradation, the cell attaches to the condemned protein a linear chain of ubiquitins. Ubiquitin is a small, 76-residue protein. An enzyme attaches its C-terminus to a lysine of the original protein, or to the preceding ubiquitin in the chain. This is an unusual example of a peptide bond between a mainchain and a sidechain, creating a branched polymer. At the proteasome, a cap domain removes the ubiquitin adducts and unfolds the protein. The unfolded protein is threaded through a narrow channel into the central chamber of the proteasome, where hydrolysis occurs.

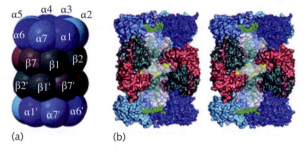

(a) (b)

Figure 1.8 The schematic structure of the eukaryotic proteasome. (a) Side view. Active sites are located at the N-termini of subunits β1, β2 and β5. (b) Cutaway view, in stereo, showing the sequestering of the active sites within a central chamber. This prevents the protease activity from promiscuous chopping up of proteins the cell needs to preserve. Entry to the catalytic chamber is through an opening in the centre of the ring of α subunits (shown in green.)

From: Rechsteiner, M. & Hill, C.P. (2005). Mobilizing the proteolytic machine: cell biological roles of proteasome activators and inhibitors. Trends Cell Biol. 15, 27–33.

Description of protein structures

The Danish protein chemist K.U. Linderstrøm-Lang described three levels of protein structure: the **primary, secondary,** and **tertiary structures.** For proteins composed of more than one subunit, J.D. Bernal added the level of **quaternary structure,** the composition and assembly of the monomers (see Table 1.4).

Table 1.4 Basic vocabulary of protein structure

Polypeptide chain	Linear polymer of amino acids.
Peptide bond	The covalent bond linking successive amino acids in the polypeptide chain.
Mainchain	Atoms of the repetitive concatenation of peptide groups. $\cdots N - C\alpha - (C=O)N - C\alpha - (C=O)\cdots$
Sidechains	Sets of atoms attached to each $C\alpha$ of the mainchain. Most sidechains in most proteins are chosen from a canonical set of 20.
Primary structure	The amino acid sequence of the protein and intra- and interchain cross-links.
Hydrogen bond	A weak interaction between two neighbouring polar atoms, mediated by a hydrogen atom.
Secondary structure	Substructures common to many proteins, compatible with mainchain conformations free of interatomic collisions, and stabilized by hydrogen bonds between mainchain atoms. Secondary structures are compatible with all amino acids, except that a proline necessarily disrupts the hydrogen-bonding pattern (see Chapter 2).
α-helix	Type of secondary structure in which the chain winds into a helix, with hydrogen bonds between residues separated by four positions in the sequence (see Fig. 1.9(a)).
β-sheet	Another type of secondary structure, in which sections of mainchain interact by lateral hydrogen bonding (see Figs 1.9(b) and (c)).
β-bulge	A common irregularity in β-sheets is the β-bulge, in which one or two (usually) residues in an edge strand of a sheet loop out of the strand, not participating in the regular hydrogen-bonding pattern of the sheet.
Folding pattern	Layout of the chain as a curve through space.
Tertiary structure	The spatial assembly of the helices and sheets, and the pattern of interactions between them. (Folding pattern and tertiary structure are nearly synonymous terms.)
Quaternary structure	The assembly of multisubunit proteins from two or more monomers.
Native state	The biologically active form of a globular protein, compact and low energy. Under suitable conditions proteins form native states spontaneously.
Intrinsically-disordered protein (IDP)	A protein which under physiological conditions does not adopt a unique native state, or which has one or more flexible regions, not characterized by unique conformations.
Denaturant	A chemical that tends to disrupt the native state of a protein.
Denatured state	Non-compact, structurally heterogeneous state formed by proteins under conditions of high temperature, or high concentrations of denaturant. Denaturation destroys secondary, tertiary, and quaternary structure, but leaves the polypeptide chains intact.
Post-translational modification	Chemical change in a protein, after its creation by the normal protein-synthesizing machinery.
Disulphide bridge	Sulphur–sulphur bond between two cysteine sidechains. A simple example of a post-translational modification.

- Major levels of description of protein structures are: primary, secondary, tertiary, and quaternary structure. What are the major contributors to the stabilization of each?

Primary structure

The amino acid sequence of a protein, plus intra- and interchain covalent cross-links if any, defines the primary structure.

Secondary structure: helices and sheets are favourable conformations of the chain that recur in many proteins

Because the polypeptide chain is flexible, any primary structure is compatible with a very large number of spatial conformations of the mainchain and sidechains. The mainchain conformation defines the secondary structure. Some conformations are preferred: α-helices and β-sheets recur in many proteins (see Fig. 1.9, Box 1.6). They (1) keep the mainchain in an unstrained conformation, and (2) satisfy the

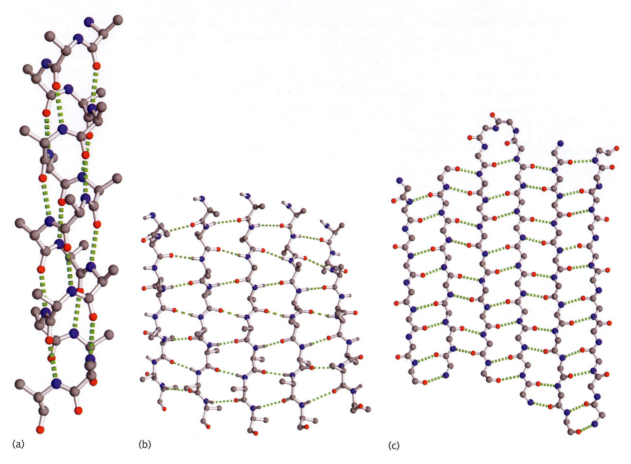

(a)   (b)   (c)

Figure 1.9 α-helices and β-sheets are two 'pre-fabricated' structures that appear in many proteins.
(a) α-helix. (b) Parallel β-sheet. (c) Antiparallel β-sheet. Note the different hydrogen-bonding patterns in
parallel and antiparallel β-sheets.

hydrogen-bonding potential of the mainchain N–H and C=O groups.

α-helices are formed from a single consecutive set of residues in the amino acid sequence (see Fig. 1.9(a)). They are a *local* structure of the polypeptide chain; that is, they form from a set of residues consecutive in the sequence. The hydrogen-bonding pattern links the C=O group of residue i to the H–N group of residue $i+4$. (3_{10} helices, hydrogen-bonding pattern $i-i+3$; and π helices, $i-i+5$, are rare.) From N-terminus to C-terminus defines the *direction* of a polypeptide chain. Thus residue $i+4$ is farther from the N-terminus of the chain than residue i.

β-sheets form by the lateral interactions of several independent sets of residues (Figs 1.9(b) and (c)). They can bring together sections of the chain separated

BOX 1.6   **Helices and sheets were predicted before they were discovered**

Helices and sheets are like 'lego' pieces, standard units of structure of which many proteins are built and that can be assembled in different ways. They were predicted by Linus Pauling and Robert Corey in 1950 from model building. Upon reading their report, Max Perutz went into his laboratory and photographed the X-ray diffraction pattern of a horse hair. The keratin in the hair contained the predicted helices! Later, he and his colleagues found α-helices to be the basis of the structures of myoglobin and haemoglobin, the first globular protein structures determined.

widely in the amino acid sequence. Figure 1.9(b) shows an ideal β-sheet, with all strands parallel. The strands are not fully extended, but accordion-pleated. The sheet is not flat, but each strand is rotated from its neighbours so that the sheet appears twisted in propeller fashion. Each of the three central strands has two neighbouring strands in the sheet. The two edge strands have only one neighbour in the sheet.

Figure 1.9(c) shows an antiparallel **β-sheet**, from bacteriochlorophyll *a* protein, in which every strand points in a direction opposite to its neighbours. β-sheets can be parallel, antiparallel, or mixed, with respect to the relative directions of the strands. In Fig. 1.9(c), the central pair of adjacent strands is connected through a short **loop**, making a **β-hairpin**.

The strands of some sheets form a closed structure, with no edge strands, called a **β-barrel**. The strands lie roughly on the surface of a cylinder.

A common irregularity in β-sheets is the **β-bulge**, in which one or two (usually) residues in an edge strand of a sheet loop out. The bulged residues do not participate in the regular hydrogen-bonding pattern of the sheet.

Tertiary and quaternary structure

Different proteins combine helices and sheets in different ways, to create different spatial arrangements of the chain, and different patterns of interaction between the helices and sheets. This is called the *tertiary structure,* or **folding pattern** (see Chapter 2). It is in their folding patterns that proteins show how versatile the polypeptide chain can be in forming structures adapted for different roles in living systems.

Many proteins contain more than one subunit, or monomer. These may be multiple copies of the same polypeptide chain, or combinations of different polypeptide chains. The assembly of the subunits is called the *quaternary structure.*

Progress in understanding protein structure has depended crucially on tools to visualize these complicated structures (see Box 1.7).

Folding patterns in native proteins—themes and variations

It is very interesting to classify the different types of structural patterns observed in native proteins, and to try to correlate different structure types with different functions (see Chapter 2). It has been estimated that only about 1000–2000 different folding patterns will be observed in the majority of proteins in nature. When we know this repertoire almost completely, we shall be in a better position to appreciate the set of observed proteins in terms of variations on a fixed set of themes.

One or two thousand folding patterns is much less than the number of proteins of known structure. Therefore many proteins must share the same folding pattern. This may come about in two ways:

1. *Evolutionary divergence.* When proteins evolve, they tend to maintain their folding pattern. We can therefore recognize families of proteins that share a folding pattern.

2. *Building up complex structures from different combinations of simpler ones,* in several ways:

   a. Create a larger, but still single, compact unit by synthesizing a longer polypeptide chain.

   b. Form oligomers by combining several monomers: sperm-whale myoglobin is a monomer, the globin from the ark clam *Scapharca inaequivalvis* is a dimer formed by two identical monomers, and human haemoglobin is a tetramer (see Fig. 1.11). Haemoglobin is an oligomer formed from non-identical monomers.

      Multisubunit proteins can contain very large numbers of subunits; for example, ribulose-1,5-bisphosphate carboxylase/oxygenase contains 16 subunits. Pyruvate dehydrogenase contains over 100 (see Fig. 1.12). Viral capsids may contain thousands of monomers.

   c. A kind of compromise between these two approaches is to build up a protein from small compact units, called **domains**, formed from successive regions of a single long polypeptide chain. It is common for structures seen as monomeric proteins to appear as domains within longer ones. Such proteins are called *modular* proteins. One effective mechanism of evolution is to recombine **modules** in different orders (see next section). Immunoglobulins illustrate *both* concatenation of domains and oligomer formation (see Chapter 6).

BOX 1.7

Molecular graphics—visualization of protein structures

Chemistry is the most visual of the physical sciences. Appreciation of the spatial features of molecular structures began with the discovery of the tetrahedral carbon atom by Paterno, van't Hoff and Le Bel. Pasteur's discovery in 1850 that yeast could metabolize only one mirror image form of tartaric acid brought three-dimensional structure into biochemistry. Emil Fischer published his famous 'lock-and-key' picture of the mechanism of enzyme specificity in 1894.

Determinations of protein structures have now provided full **atomic resolution**. What Pasteur and Fischer appreciated as general principles, we can now interpret in precise detail.

With so many known protein structures, it is impossible to build physical models, or to draw pictures of all of them by hand. (The late Irving Geis produced superb pictures, but of course of only a few structures. Among museum artists, the works of Salvador Dali and Ben Shahn include representations of molecules.)

Proteins are so large, and there are so many to examine, that the methods of computer graphics are required to display them. Molecular graphics converts the coordinates of

a structure to an image that can be printed on paper, or displayed on a screen. Displaying the molecule on a screen, allows use of a mouse, or other devices such as dials, to give the observer the ability to rotate the structure, to zoom in on a selected region, or to alter the conformation interactively. The most 'literal' representations of molecules focus either on the topology—showing individual bonds as line segments—or on the space-filling aspects—showing individual atoms as spheres. 'Ball-and-stick' representations are a compromise.

For molecules as large as proteins, however, pictures showing every bond and/or every atom are too complicated to be intelligible. It has been necessary to devise simplified, 'cartoon' representations of proteins. Many pictures of proteins show a smooth curve or ribbon along the course of the main chain. Many of the figures in this book are examples of this. Frequently, two common structural patterns—α-helices and strands of β-sheet—are further simplified to cylinders and broad arrows, respectively.

Figure 1.10 compares three representations of the same protein molecule. It is essential, in practical work, to be able

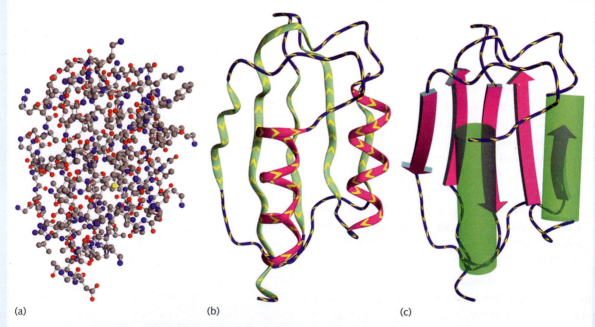

(a)   (b)   (c)

Figure 1.10 Proteins are sufficiently complex structures that it has been necessary to develop specialized tools to present them. This figure shows a relatively small protein, **acylphosphatase**, at three different degrees of simplification. (a) Complete skeletal model. (b) The course of the chain is represented by a smooth interpolated curve, the chevrons indicating the direction of the chain. (c) Schematic diagram, in which cylinders represent helices and arrows represent strands of sheet. The solid objects in the picture are represented as 'translucent'.

to compose a picture from a variety of representations. A program must permit the user to be able to select different levels of detail for different parts of a molecule. It may be desirable to show an active site in full atomic detail. The rest of the molecule may appear not at all, or in reduced detail.

A crucial advantage of computer graphics over physical models is that it is possible to *superpose* structures (see Fig. 1.4). Only in this way can one see structural differences easily.

Very high quality yet inexpensive computer graphics have become widely available because of applications in the entertainment industry. Software can generate photorealistic scenes, and hardware is fast enough for interactive animation. Biochemists have taken full advantage of these developments. (Of course, what it means to talk about a 'realistic' picture of proteins and nucleic acids is unclear.*)

Other aids to perspicuity in representing the spatial relationships within complex objects include: (1) the use of stereo, (2) the representation of features as transparent, translucent or opaque—these can be demonstrated in a 'still' picture (see Fig. 1.10)—and (3) the 'kinetic depth effect': displaying a structure in continuous rotation enhances depth perception. Many movies of molecules on the web offer examples.

There are a number of molecular graphics software packages in common use in the protein community. Readers should become adept in the use of at least one of these:

jV	http://www.pdbj.org/jV/TOP.html
RasMol	http://rasmol.org/
Jmol	http://jmol.sourceforge.net/
PyMol	http://www.pymol.org
ccp4mg	http://www.ccp4.ac.uk/MG/

General websites containing relevant links are:

http://www.molvis.org

http://www.rcsb.org/pdb/static.do?p = software/ software_links/molecular_graphics.html

http://jena3d.fli-leibniz.de

* Many centuries ago, a Chinese artist painting for the King of Ch'i was asked by the King which was harder to draw: dogs and horses, or demons and goblins. 'Demons and goblins are much easier, Your Majesty,' the artist replied, 'for everyone knows what dogs and horses look like.'

Modular proteins, and 'mixing and matching' as a mechanism of evolution

Many proteins are single domains: individual compact units typically about 100 residues in length. The combination of domains into a **modular protein**, as in immunoglobulins, or in proteins involved in blood coagulation, is an important way of creating complex proteins from simpler components (see Fig. 1.13).

- Similarity = shared features. Homology = descent from a common ancestor. It is rare that homology can be observed directly; usually homology is inferred from a sufficiently high degree of similarity (see Box 1.8).

In prokaryotes, approximately 2/3 of the proteins contain more than one domain. The several thousand proteins in a typical bacterium such as *E. coli* are made up of combinations of about 400 domains with different folding patterns. In eukaryotes, approximately 3/4 of the proteins are multidomain proteins, combinations of about 600–700 domains with different folding patterns. Many of these individual domains are common to bacteria, archaea, and eukaryotes, but some are—as far as we know now—specific to each group.

Sometimes separate genes for proteins in prokaryotes become fused into a single gene for a modular protein in eukaryotes. As an example of such **domain fusion**, five separate enzymes in *E. coli*, which catalyse successive steps in the pathway of biosynthesis of aromatic amino acids, correspond to five regions of a single protein in the fungus *Aspergillus nidulans*. In rarer cases, separate proteins in eukaryotes correspond to a single fusion protein in prokaryotes.

Combination of domains in modular proteins does not occur at random. A few domains combine with many different partners, but most combine with only one or a few. Conversely, many domain combinations are common, and others rare. If domains exert their function independently, they can appear in different orders in different proteins. In other cases, the order of the domains is important for function. Tandem repeats of the same domain occur frequently, created by gene duplication and subsequent divergence. A giant muscle protein, titin, contains about

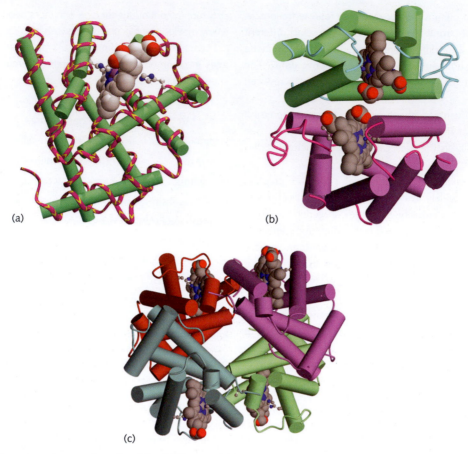

Figure 1.11 Proteins from the globin family assemble different combinations of oligomers. (a) Monomer: sperm-whale myoglobin [1MBO]. (b) Dimer of two identical subunits: ark clam globin [4SDH]. (c) Mixed tetramer: human haemoglobin, containing two α-chains and two β-chains.

BOX 1.8   Similarity and homology

The distinction between **similarity** and **homology** is subtle and important.

- *Similarity* is the observation or measurement of resemblance and difference, independent of the source of the resemblance.

- *Homology* means, specifically, that the sequences and the organisms in which they occur are descended from a common ancestor. The implication is that the similarities are shared ancestral characteristics.

Similarity of sequences or structures is observable *now*, and involves no historical hypotheses. In contrast, assertions of homology are statements about historical events that are almost always unobservable. Homology must be an *inference* from observations of similarity. Only in a few special cases is homology directly observable; for instance in pedigrees of human families carrying Huntington's disease; or in laboratory populations of animals, plants, or microorganisms; or in clinical studies that follow sequence changes in viral infections in individual patients.

As proteins evolve, their amino acid sequences diverge, and the corresponding structures also diverge. Structures tend to change more conservatively than sequences. Proteins that are very similar in sequence, structure, and function may be presumed to be homologous. For proteins with similar folding patterns, but for which no substantial similarities appear in the sequences, it can be difficult to decide whether they are homologues or not.

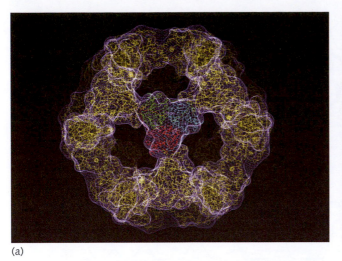

(a)

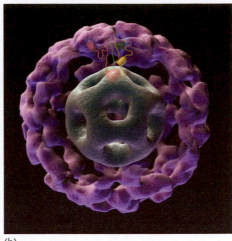

(b)

Figure 1.12 Pyruvate dehydrogenase catalyses the transfer of an acetyl group from pyruvate, via thiamine pyrophosphate and dihydroxylipoamide, to coenzyme A. The structure of the complete enzyme comprises three component enzymes, assembled in a complex containing over 100 subunits, arranged in concentric shells, about 300 Å in diameter. Mobile domains on the inner shell can access several active sites on the inner surface of the outer shell. (a) The core of the complex is an icosahedral array of 60 copies of one of the enzymes, E2. This figure shows a contoured electron-density map for the icosahedral E2 catalytic core of *B. stearothermophilus* pyruvate dehydrogenase, determined by averaging images of the particles obtained by cryo-electron microscopy (see Chapter 3). Copies of an X-ray structure of the subunits (yellow, cyan, red, and green) are superimposed on the density. The red, cyan, and green subunits are related by a threefold axis of symmetry. (b) Representation of the structure and mechanism of pyruvate dehydrogenase. The figure shows the inner icosahedral core domain (green)—the structure shown in (a)—containing the E2 subunits, and approximately half the outer shell, containing the E1 subunits. The outer shell is complete in the structure but cut away here to expose the inside of the particle. The outer diameter of the outer shell is 460 Å. The outer diameter of the inner shell is 225 Å. The width of the gap between the shells is 90 Å. Mobile regions of the E2 core extend into the space between the shells, and can make contact with active sites of several *both* E1 and E2 subunits.

Figure courtesy of Dr P B Rosenthal, and Drs J Milne, R Perham, and S Subramaniam. See: Milne, J. L. *et al.* (2002) Molecular architecture and mechanism of an icosahedral pyruvate dehydrogenase complex: a multifunctional catalytic machine. EMBO J. 21, 5587–5598.

300 modules including repeats of immunoglobulin and fibronectin type-3 domains.

Protein evolution thus appears to be going on at two levels. By variation of individual residues, mutations are exploring the immediate neighbourhoods of particular sequences, structures, and functions. Larger-scale evolutionary variation involves trying out different combinations of domains.

How do proteins develop new functions?

Evolution has pushed the limits in its exploration of sequence–structure–function relationships. Observed mechanisms of protein evolution that produce altered or novel functions include: (1) divergence, (2) recruitment, and (3) 'mixing and matching' of domains, or modular evolution. Divergence involves progressive localized changes in sequence and structure, leading initially to changes in specificity and ultimately to changes in the nature of the reaction catalysed. In recruitment, one protein is adapted, sometimes without any change at all, for a second function. Modular evolution involves large-scale structural changes. Individual domains may retain their function in a new context, or their function

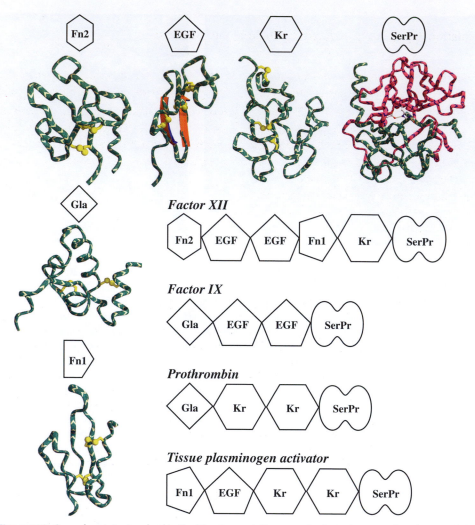

Figure 1.13 Several proteins involved in the blood coagulation cascade show structures that share modules. The composition and the order of the modules are not preserved. Each module is a relatively small compact unit in its own right. The serine proteinases (SerPr) contain two halves with structural similarities, which arose by gene duplication and divergence, but are never seen separately.

may be modified by their new surroundings, or they may participate in entirely different processes (see Box 1.9). We shall discuss evolution of novel functions in Chapter 7.

One consequence of the robustness of protein structure to mutation is a maintenance of structure in spite of divergence of sequences during evolution. But, although similar sequences determine similar structures, the converse is not true—proteins with very different sequences can have similar structures. Indeed, there are many cases of proteins with very

similar structures with *no* easily recognizable relationship between the sequences.

Even less reliable is the reasoning from sequence to function, or even from sequence + structure to function. Many proteins with similar sequences and structures have similar functions, but many do not, and conversely many functions can be carried out by proteins with unrelated sequences and structures.

Evolution of function is discussed further in Chapter 7.

BOX 1.9   Relationships among sequence, structure, and function

- *Similar sequences can be relied on to produce similar protein structures,* with divergence in structure increasing progressively with the divergence in sequence.

- Conversely, *similar structures are often found with very different sequences.* In many cases the relationships in a family of proteins can be detected *only* in the structures, the sequences having diverged beyond the point of our being able to detect the underlying common features.

- *Similar sequences and structures often produce proteins with similar functions,* but exceptions abound.

- Conversely, *similar functions are often carried out by non-homologous proteins with dissimilar structures;* examples include the many different families of proteinases, sugar kinases, and lysyl-tRNA synthetases.

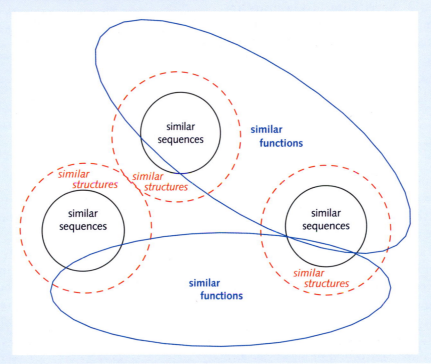

The study of proteins: in the laboratory, in the cell, in the computer

Early investigations of proteins emerged from two streams: biochemistry and polymer chemistry. The two key concepts were that proteins were very large molecules, and that they nevertheless had a definite three-dimensional structure. The success of X-ray structure determinations of proteins, and later NMR spectroscopy and cryo-electron microscopy, changed the field forever. Polymer chemists had used techniques, such as centrifugation, intrinsic viscosity, and light scattering, to derive the general sizes and shapes of proteins, modelled as ellipsoids. Spectroscopists devised methods to characterize protein secondary structure, and interactions between residues within a protein, and between proteins. But high-resolution structure determination gave us coordinates of each of the thousands of atoms in a protein structure. To a very large extent, structure determinations have superseded the older methods. For a forceful statement of this point of view, see W.B. Gratzer (1970) *Those woolly cotton effects*. Nature, 227, 94–95.

All these methods, including structure determination, report properties of isolated systems. This requires that not only the proteins, but the state studied, are stable. Study of the dynamics of conformational change or of interactions requires different approaches. An overriding development is the great interest in measuring properties of proteins within living cells. A number of spectroscopic methods support such investigations. A well-known example is the use of green fluorescent protein, coupled to a molecule of interest, to indicate tissue or subcellular location.

Computational studies of proteins have become another heavy industry in molecular biology. The award of the 2013 Nobel Prize in Chemistry for molecular dynamics emphasizes its importance. Computers proved essential for converting the experimental data of X-ray crystallography, NMR spectroscopy, and electron microscopy to structures,

and, more generally, for interpretation of a variety of experimental data and for simulations. Molecular graphics, as we have seen, is essential to allow us to examine the harvest of known structures.

In succeeding sections, we describe spectroscopic methods for characterizing proteins, and the contributions of computational methods to protein science. Structure determination merits a chapter of its own.

- Methods for study of protein structure include—separately *and in combination*—measurements of properties of proteins *in vivo* and in dilute solutions; structure determinations, mostly by X-ray crystallography, NMR spectroscopy and electron microscopy; and computational methods.

Spectroscopic methods of characterizing proteins in solution

The interaction of molecules with light provides information about molecular structure and dynamics (see Box 1.10). Absorbance, fluorescence, and circular dichroism are useful probes of protein conformation. Fluorescent tags, including fluorescent antibodies, are useful in mapping intracellular locations of molecules. Fluorescent *in-situ* hybridization (FISH) locates gene sequences on chromosomes. Fluorescence recovery after photobleaching (FRAP) measures the regain of fluorescence at a particular cellular site after localized destruction of tags by a laser beam. The motion of replacement molecules into the bleached area reveals their mobility.

For many purposes, spectroscopic methods have been superseded by atomic-resolution structure determinations, which contain much more detailed information.

However, spectroscopic measurements survive, at least as 'niche' methods, because they can be performed in real time, permitting observation of kinetics of reactions or folding; and they provide sensitive information about molecular structure, dynamics, and interactions. Circular dichroism permits determination of secondary structure content of proteins in solution, and is also in common use to monitor protein unfolding and refolding transitions. Fluorescence resonance energy transfer (FRET) gives

information about the structures of macromolecular complexes too large to be treated easily by crystallographic methods. Electron microscopy is taking over these investigations.

All spectroscopic experiments involve the temporary exchange of energy between light and matter; excitation involves only a 'short-term loan' of energy. An excited molecule will quickly release its energy and return to the ground state. Possible paths of de-excitation include:

- *Conversion of excitation energy to heat.* This is an efficient and fast process, occurring within about 10^{-12} s. As a result, some of the energy of a molecule that has absorbed visible or ultraviolet light is lost to warm up the surroundings.

- *Fluorescence* is efficient loss of excitation energy by re-emission of radiation. Typically, fluorescence occurs within about 10^{-8}–10^{-6} s. Because of the loss of some excitation energy to heat, the re-emitted light must be at a lower energy—that is, higher wavelength—than the exciting radiation. This displacement of the wavelengths of absorption and fluorescence is useful, experimentally. It makes it possible to filter out the exciting light, facilitating the detection of a relatively weak fluorescent signal. The quantum yield is a measure of the fluorescence

BOX 1.10

Light and molecules

- Atoms and molecules have energy levels. In dilute gases these energy levels are discrete and well separated. In condensed phases, interactions between molecules perturb the states, giving a continuous spectrum of energies.

- Absorption of light raises a molecule from its lowest energy level, or **ground state**, to a higher energy level, an **excited state**.

- For absorption of light of frequency v, the excitation energy is $\Delta E = hv$, in which Planck's constant $h = 6.626176 \times 10^{-34}\, Js^{-1}$. A packet of light energy is called a *photon*. Light of frequency v can be regarded as a beam of particles, each of energy hv.

- The frequency v and wavelength λ of *any* wave are related by the formula

$$v \times \lambda = \text{the speed of propagation of the wave}$$

The speed of light in a vacuum is a constant, independent of the frequency: $c = 2.99792458 \times 10^8\, m\, s^{-1}$. For light in a vacuum $v\lambda = c$.

- Frequency and wavelength are therefore equivalent variables with which to describe light. Most spectroscopic measurements in protein science are recorded in terms of wavelength rather than frequency.

- Different wavelengths of visible light correspond to different colours. Light of a wavelength of about 510 nm appears green. This corresponds to a frequency of 5.878×10^{14} Hz, a photon energy of 3.89×10^{-19} J, equivalent to 234.6 kJ mol^{-1}.

- The **visible region** of the electromagnetic spectrum comprises wavelengths in the range 780–380 nm, corresponding to photon energies in the range 2.55×10^{-19}–5.23×10^{-19} J, equivalent to 153.37–314.81 kJ mol^{-1}. (Note that the *lower* the wavelength the *higher* the energy, because of the relationship $E = hv = hc / \lambda$.). This energy of a photon of visible light is less than the energy of typical chemical bonds. This is why ordinary visible light rarely produces photochemical reactions, including not causing severe sunburn. The **near-ultraviolet** comprises wavelengths in the range 300–400 nm, corresponding to photon energies of 5.22×10^{-19}–9.93×10^{-19} J, equivalent to 314.81–598.13 kJ mol^{-1}. The **far-ultraviolet** lies from 100–200 nm, corresponding to photon energies in the range 9.93×10^{-19}–1.99×10^{-18} J, equivalent to 598.13–1196.27 kJ mol^{-1}.

- Many substances appear coloured, implying that their interaction with light varies with the frequency. If illuminated with white light (a mixture of different wavelengths), they differentially absorb some of the wavelengths, transmitting or reflecting light enriched in some colour. In many cases, a specific group within a molecule is responsible for interaction with light. This group is called a **chromophore.** For instance, the chromophore in haemoglobin is the haem-complexed iron. Groups that absorb and fluoresce only outside the visible region do not appear coloured but are called chromophores nevertheless.

intensity: it is the ratio of the number of photons emitted to the number of photons absorbed.

- *Phosphorescence* is *inefficient* loss of excitation energy by re-emission of radiation. Phosphorescence may last as long as 10^2–10^4 s. The reason for the delay is an intrinsically slow process involving a change in the orientation of an electron spin.

- *Energy transfer.* Under certain circumstances, an excited molecule can transfer its energy to another molecule (which then in turn has to decide what to do with it). If energy transfer leads to radiationless de-excitation, we say that the fluorescence of the originally excited chromophore has been *quenched*. Alternatively, fluorescence

may be observed from the molecule to which the excitation energy was transferred (Fig. 1.14). One requirement for efficient energy transfer is that the donor and acceptor molecules be nearby—within about 20–70 Å. Another requirement is that the donor and acceptor have the correct relative excitation energies. If the acceptor molecule can *absorb* light at the wavelength at which the donor molecule would *fluoresce*, then the transfer would balance in energy, and be more probable. The spectroscopist's criterion is that there should be good overlap between the fluorescence spectrum of the donor and the absorption spectrum of the acceptor.

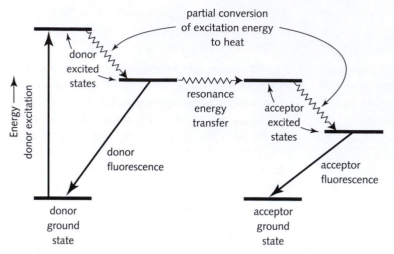

Figure 1.14 Energy-level diagram showing photochemical processes in fluorescence resonance energy transfer (FRET). Initial excitation of a donor molecule is followed by a fast radiationless de-excitation with release of energy as heat, leaving the donor in a low-lying excited state. Next, the donor may fluoresce to its own ground state, or transfer the excitation energy to the acceptor, which then converts some of *its* excitation energy to heat, leading to observation of fluorescence from the acceptor. The excitation energy transfer is most probable between states of equal energy. 'Molecular excitation is like skiing: going up is much less complicated than getting back down.'

- *Fluorescence resonance energy transfer (FRET)* takes advantage of the requirement for proximity to map out neighbouring groups in large macromolecular complexes; either statically or to detect conformational changes. The reasoning is that if excitation of a donor molecule is followed by fluorescence from an acceptor molecule, the two chromophores must be nearby in space.

- *Photochemical reaction.* Excited states are intrinsically more reactive than ground states, in part because they have more energy to spare. Some photochemical reactions are irreversible, and can damage proteins and other substances. (A molecule that loses its fluorescence after photochemical damage is said to be *bleached*.) In other cases—notably photosynthesis and vision—systems have evolved to capture excitation energy safely, channel it, and to store it as chemical energy (in photosynthesis) or to trigger a nerve impulse (in vision), and then to restore the initial state (see Box 1.11).

Absorbance and fluorescence of proteins

Absorption of light by proteins (see Table 1.5) involves energy levels of:

- *The peptide bond,* which has an intense transition at 190 nm, in the far-UV, and weak absorption at 210–220 nm. Sidechain carboxyl and amide groups of Asp, Glu, Asn, Gln; and Arg and His sidechains, also absorb in this region.

BOX 1.11   Natural FRET in vision and photosynthesis

- Insect visual pigments contain a chromophore, 3-hydroxyretinal, bound to a protein, opsin. Excitation of the chromophore triggers a conformational change in the protein. In flies, a second, auxiliary, chromophore, **3-hydroxyretinol**, is hydrogen bonded to the opsin, but not mechanically linked to trigger the protein conformational change directly. The absorption maximum of 3-hydroxyretinol is ~ 350 nm. The absorption maximum of 3-hydroxyretinal is 500 nm. The auxiliary pigment acts as a sensitizer, transferring its excitation energy to the main chromophore. As a result, the spectral response of the fly photoreceptor is broadened and its efficiency enhanced.

- Chloroplasts contain complex arrangements of accessory pigments that channel excitation energy to special sites of photochemical reaction. Some of these **accessory pigments** are chlorophylls, some are bound to proteins similar to globins, and some are similar to visual pigments.

- Conversely, and very unusually, the deep-sea dragon fish (*Malacosteus niger*) uses a molecule related to chlorophyll as a sensitizing pigment for vision.

Table 1.5 Chromophores in proteins

Chromophore	Wavelength of absorption maximum	Colour
NADH	260 nm and 340 nm	UV
NAD	260 nm	far-UV
Bacteriorhodopsin	498 nm	purple
Haem (oxy)	940 nm	red
Haem (deoxy)	550 nm	blue-purple
Cu^{2+}	600 nm	blue
Green fluorescent protein	395 nm	blue
Green fluorescent protein	475 nm	blue
Trp sidechain	280 nm	far-UV
Tyr sidechain	274 nm	far-UV
Phe sidechain	257 nm	far-UV
Peptide bond	210–220 nm	far-UV
Peptide bond	190 nm	far-UV

- *Sidechains of aromatic amino acids* absorb at longer wavelengths in the near-UV. Phe has an absorption band at 257 nm, Tyr at 274 nm, and Trp at 280 nm. The wavelengths of these transitions vary with the environment of the sidechain, including pH and solvent polarity. Ionization shifts the tyrosine transition to 295 nm. Shifts in absorption bands arising from changes in the polarity of the environment are sensitive probes of conformation. They can be applied to distinguish exposed from buried aromatic sidechains in a protein structure.

- *Ligands* that absorb in the visible region are responsible for most proteins that appear coloured. Examples include the red haem group of globins, and the blue complexed copper ion of **plastocyanin**. Nucleotides and related compounds such as NADH and NAD absorb strongly in the near-UV at 260 nm. In addition, NADH but not NAD absorbs at 340 nm; this provides the classic method for following the kinetics of NAD-linked enzymatic reactions.

- The *green fluorescent protein (GFP)* from the jellyfish *Aequorea victoria*, and its homologues, is a special case. GFP absorbs in the blue ($\lambda_{max} = 395$ nm and 475 nm), and fluoresces in the green ($\lambda_{max} = 509$ nm) (see Fig. 1.15). The chromophore of GFP is formed as a post-translational modification of a Ser-Tyr-Gly tripeptide within the protein. The transformation is autocatalysed, not requiring any additional enzyme. However, Ser-Tyr-Gly sequences in many other proteins do not form chromophores. Moreover, denatured GFP is not fluorescent. The special spectral properties depend on *both* the chemical modification, and the environment of the altered amino acids within the three-dimensional structure of the protein. The sensitivity of the spectral properties to the environment of the chromophore has led to exploration of natural variants and engineered mutants that fluoresce at different wavelengths.

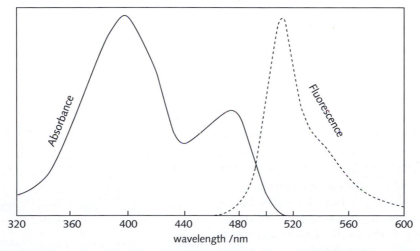

Figure 1.15 Absorption (solid line) and fluorescence (broken line) spectra of green fluorescent protein (GFP) from the jellyfish *Aequorea victoria*. Note the shift of the fluorescence spectrum to higher wavelength (lower energy) relative to absorbance.

Fluorescence is sensitive to the environment and dynamics of the chromophore

The intensity of fluorescence depends on the competition between fluorescence and alternative paths of de-excitation, such as excitation energy transfer. Acceptor molecules in the vicinity of a chromophore can quench its fluorescence. This provides a technique for determining solvent exposure of aromatic sidechains or ligands.

Interaction of a molecule with light may depend on the relative orientation of the molecule and the plane of polarization of the light. If a chromophore is fixed in orientation, there will be a correlation of the polarization of its fluorescence and that of the exciting light. If the chromophore is mobile within a structure, the degree of polarization of the fluorescence depends on how much time the chromophore has available to reorientate itself during the lifetime of its excited state. Therefore, measurements of **depolarization of fluorescence** reveal the mobility of the chromophore. (Rates of overall molecular tumbling, which also reorientate chromophores, are usually slower than fluorescence lifetimes.)

Spectroscopic measurements take advantage of the speed and time resolution with which data can be collected. For example, it has been possible to determine the rate of loss of mobility of tryptophan sidechains during protein folding.

Fluorescence resonance energy transfer (FRET)

Fluorescence resonance energy transfer (FRET) is a method for probing protein structure based on the observation that excitation energy transfer occurs only between chromophores nearby in the structure. It has been used to study the conformation and dynamics of proteins and nucleic acids, including static structure, conformational changes, folding pathways, and macromolecular interactions.

Two chromophores are required, one to receive the initial excitation, and the other as the destination of the energy transfer (see Fig. 1.14). The observable effect is a decrease of the fluorescence from the donor and increase in the fluorescence from the acceptor. There is a strong distance dependence, varying as $1/[1+(R/R_0)^6]$, where R is the distance between the chromophores, and R_0 is the distance at which 50%

of the energy is transferred. For this reason, FRET is often called a molecular ruler. There is also an orientation effect on the efficiency of energy transfer.

Circular dichroism

Circular dichroism (CD) measures the asymmetry of a chromophore or its environment. When polarized light passes through an asymmetric medium, it emerges with the plane of polarization reorientated. This is called optical rotation. (It was through measurements of the sign of the optical rotation of tartaric acid solutions that allowed Pasteur to correlate the macroscopic asymmetry of tartaric acid crystals with the microscopic asymmetry of the molecules themselves.) In addition, there is some loss of polarization, called **ellipticity**, the effect measured as circular dichroism.

The near-UV circular dichroism of a native protein is much stronger in the native state than in the denatured state. In the native state the peptide groups are in fixed and asymmetric surroundings, whereas in the denatured state the environment is random and fluctuating.

The most common applications of CD spectra of proteins are (1) monitoring unfolding and refolding transitions (discussed in Chapter 6), and (2) determining the amounts of helix and sheet in a protein structure.

Placing peptide groups in an asymmetric environment, for instance in an α-helix, gives rise to

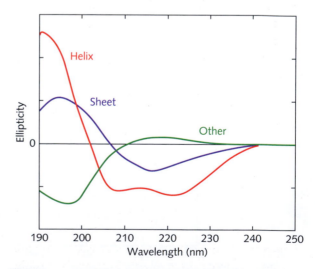

Figure 1.16 Contributions of regions of helix, sheet, and other conformations to the circular dichroism (CD) of a globular protein. A measured spectrum of a protein of unknown secondary structure content could be analysed as a weighted sum of these curves.

BOX 1.12   Conformational changes in the prion protein

Transmissible spongiform encephalopathies are a group of diseases causing central nervous system degeneration in humans and animals. They include **kuru** and **Creutzfeld–Jakob Disease (CJD)** in humans, **scrapie** in sheep and goats, and **bovine spongiform encephalopathy (BSE)** (colloquially, **'mad cow disease'**). These diseases tend to be species specific, and to have long incubation periods. A variant CJD with a much shorter incubation period in humans is believed to have arisen by a jumping of BSE across the species barrier.

The infectious agent of these diseases appears to be a prion protein that changes from a normal state (PrP^C = prion protein–cellular) to a disease one (PrP^{Sc} = prion protein–scrapie). The disease form catalyses the $PrP^S \rightarrow PrP^{Sc}$ transformation, converting additional protein to PrP^{Sc}, leading to fibrillar aggregates of PrP^{Sc}.

Circular dichroism measurements show that the prion protein changes from a predominantly α-helical conformation in the PrP^C state, to a predominantly β-structure in the PrP^{Sc} state (see Fig. 1.17).

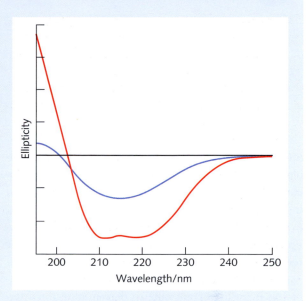

Figure 1.17 The near-UV CD spectra of the normal form of prion protein, PrP^C (blue), and the disease form, PrP^{Sc} (red). From these curves one can recognize a change in secondary structure in the prion protein from predominantly α-helical to predominantly β-structure. (Compare Fig. 1.16.)

See: Jackson, G.S. *et al.* (1999). Reversible conversion of monomeric human prion protein between native and fibrilogenic conformations. Science, **283**, 1935–1937.

a CD signal sensitive to the secondary structure. This effect arises from the interaction of the peptide groups. A β-sheet, although intrinsically less asymmetric than an α-helix, also gives rise to a CD signal.

The CD spectrum of a protein is the sum of contributions from its individual secondary structures. By measuring CD spectra of a set of proteins with known secondary structure contents, it is possible to solve for the spectra of α-helices and β-sheets (Fig. 1.16). The CD spectrum of a protein of unknown structure can then be fitted to a sum of contributions from helix, sheet, and a residual signal from other parts of the structure (see Box 1.12).

- Near-UV circular dichroism is sensitive to the content and type of secondary structure of a protein solution.

Protein expression patterns in space and time: proteomics

An integrated understanding of the roles of proteins in living systems can only emerge from analysis of the high-throughput, whole-organism data of **proteomics**. Here, we discuss some of the background and methods that provide these data.

Subcellular localization

Proteins have adapted to a variety of working environments (see Box 1.13). Some are secreted, and function outside the cell. Others are deployed in specific intracellular compartments or organelles, or are integrated into membranes. Differences in the physico-chemical properties of these environments are reflected in the properties of proteins. For instance, **membrane proteins** have surfaces designed to interact with the lipid bilayer. Disulphide bridges cannot form in the reducing environment of the cytosol of bacterial cells.

It is useful to know where a protein is active. Location is an important clue to determining the interaction networks in which a protein participates. It is thereby an aid to assigning function. Looking towards clinical applications, extracellular proteins provide more accessible drug targets. However, some proteins are active in more than one place; indeed, they may even have different functions at different sites.

Proteins are directed to their destinations by **sorting signals**. Short sequences, usually at or near the N-termini, serve as a kind of 'postcode' (or 'zip code') specifying the destination. For instance, the N-terminal sequence KDEL specifies proteins retained in the endoplasmic reticulum. Proteins destined for the nucleus have a nuclear localization signal containing a stretch of 4–8 positively charged residues. We have seen that the transfer of a mitochondrial gene to the nucleus in higher plants required the gene to pick up a signal sequence to direct the product to the mitochondrion.

Signal sequences specifying 'delivery addresses'—i.e., postcodes—can interact with receptors on a target organelle directly, or with a carrier protein complex. In eukaryotes, proteins destined for the *secretory pathway*—some but not all of which are exported from the cell—bind to a *signal recognition particle* even before synthesis is complete. These proteins pass through the endoplasmic reticulum and the Golgi apparatus; secondary signals then direct them to lysosomes, to the plasma membrane, or to discharge from the cell (**exocytosis**). Proteins targeted to the nucleus, or to intracellular organelles such as the mitochondria, chloroplasts, or peroxisomes, are translocated after they are synthesized. Proteins lacking targeting sequences remain in the cytoplasm.

Related to the intracellular protein distribution systems are mechanisms for uptake of proteins from outside the cell. Some viruses take advantage of this machinery to gain entry.

The transcriptome

In principle, for proteomics we should like the ability to measure the identities and amounts of all the proteins in a cell or in each subcellular compartment. However, methods based on nucleic-acid hybridization and sequencing make it much easier to analyse the **transcriptome**; that is, to identify and measure distributions of RNAs. For a long time, **microarrays** were the best available method. Now, direct sequencing of the DNA produced by reverse-transcribing the RNA molecule in a sample is taking over. This method is called **RNAseq**.

The transcriptome is of course very interesting and important in its own right. The RNAs in cells have many important functions other than as messengers, including but not limited to control over protein expression patterns. But in a book on proteins it is probably appropriate to emphasize that the transcriptome reveals the proteome. With some reservations, it must be admitted. The correlation between concentrations of an mRNA and the concentration of the corresponding protein is not perfect, because of different rates of translation and different mRNA and protein lifetimes. In RNAseq, bias in the yields of the reverse transcriptase is an additional confounding factor.

Proteins in health and disease

Applications of transcriptomics

- *Identifying genetic individuality in tissues or organisms.* In humans and animals, this permits correlation of genotype with susceptibility to disease. In bacteria, this permits identifying mechanisms of development of drug resistance by pathogens.

- *Investigating cellular states and processes:* Patterns of expression that change with cellular state or growth conditions can give clues to the mechanisms of processes such as sporulation, or the change from **aerobic** to **anaerobic** metabolism.

- *Diagnosis of disease:* Testing for the presence of mutations can confirm the diagnosis of a suspected genetic disease, including detection of a late-onset condition such as Huntington's disease. Detection of carriers can help in counselling prospective parents.

- *Genetic warning signs:* Some diseases are not determined entirely and irrevocably by genotype, but the probability of their development is correlated with genes or their expression patterns. A person aware of an enhanced risk of developing a condition can in some cases improve his or her prospects by adjustments in lifestyle, or even prophylactic surgery.

- *Drug selection:* Detection of genetic factors that govern responses to drugs, which in some patients render treatment ineffective and in others cause unusual serious adverse reactions.

- *Specialized diagnosis of disease:* Different types of leukaemia and other cancers can be identified by different patterns of gene expression. Knowing the exact type of the disease is important for prognosis, and for selecting optimal treatment.

- *Target selection for drug design:* Proteins showing enhanced transcription in particular disease states might be candidates for attempts at pharmacological intervention.

- *Pathogen resistance:* Comparisons of genotypes or expression patterns, between bacterial strains susceptible and resistant to an antibiotic, point to the proteins involved in the mechanism of resistance.

- *Following temporal variations in protein expression* permits timing the course of (1) responses to pathogen infection, (2) responses to environmental change, (3) changes during the cell cycle, and (4) stages of development.

Nevertheless, both methods are very widely used in research and applications. We introduce the topic here and develop it Chapter 6.

DNA microarrays

Microarrays analyse: (1) the RNAs in a cell, to reveal the transcription patterns of genes for proteins and for non-protein-coding RNAs; or (2) genomic DNAs, to reveal absent or mutated genes (see Box 1.14).

1. For an integrated characterization of cellular activity, we want to determine what proteins are present, where, and in what amounts.

   To determine the expression pattern of all of a cell's genes, it is necessary to measure the relative amounts of many different mRNAs. Hybridization is an accurate and sensitive way to detect whether a particular sequence is present. The key to high-throughput analysis is to run many hybridization experiments in parallel.

2. Knowing the human genome sequence can help identification of genes associated with propensity to diseases. Some diseases, such as cystic fibrosis, are associated with single genes. For these, isolating a region by classical genetic mapping usually leads to identifying the gene. Other diseases depend on interactions among many genes, with environmental factors as complications. Study of such multifactorial diseases requires the ability to determine and analyse patterns of expression of genes.

DNA microarrays, or DNA chips, are devices for checking a sample *simultaneously* for the presence of many sequences. DNA microarrays can be used (1) to determine expression patterns of different proteins by detection of mRNAs; or (2) for genotyping, by detection of different variant gene sequences, including but not limited to single-nucleotide polymorphisms (SNPs). It is possible to measure simple presence or absence, or to quantitate relative abundance (see Fig. 1.18).

> • DNA microarrays analyse the RNAs in a cell, to reveal the expression patterns of proteins, or genomic DNAs, to reveal absent or mutated genes.

Mass spectrometry

Mass spectrometry is a physical technique that characterizes molecules by measurements of the masses of their ions. Applications to molecular biology include:

- Rapid identification of the components of a complex mixture of proteins.
- Sequencing of proteins and nucleic acids.
- Analysis of post-translational modifications, or substitutions relative to an expected sequence.
- Measuring extents of hydrogen–deuterium exchange, to reveal the solvent exposure of individual sites. This provides information about static conformation, dynamics, and interactions.

Investigations of large-scale expression patterns of proteins—proteomics—require methods that give high throughput rates as well as accuracy and precision. Mass spectrometry achieves this, which has stimulated its development into a mature technology in widespread use.

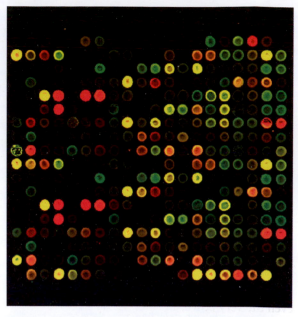

Figure 1.18 Comparison of gene expression patterns in liver (red) and brain (green). The liver RNA is tagged with a red **fluorophore**, the brain RNA with a green one, then both are exposed to the array. Red spots correspond to genes active in the liver but not in the brain. Green spots correspond to genes active in the brain but not in the liver. Yellow spots correspond to genes active in both brain and liver.

Courtesy Dr P A Lyons.

> • Proteomics requires high-throughput analysis of the protein content of cells. Mass spectrometry is an important contributor to these data. It identifies proteins by fragmenting them and measuring the charge-to-mass ratios of the fragments.

Computing in protein science

Computing has made major contributions to protein science, including:

- computer–instrumental partnerships in the laboratory;
- molecular graphics: so important it brooked no delay (see Box 1.7);
- simulations, including molecular dynamics;
- bioinformatics—the creation, coordination, and applications of large-scale databases.

Computer–instrument partnerships in the laboratory

One theme of our discussions has been the trend towards high-throughput experimental methods, in molecular biology in general and protein science in particular. Computers are essential for control, capture, and reduction of the data. Large-scale sequencing projects such as the human genome project are entirely computer controlled. Even studies of individual molecules often generate enough data to require computer data acquisition methods. X-ray crystallography is a good example; in a protein structure determination the intensities of hundreds of thousands of reflections may be measured.

Simulations, including molecular dynamics

Proteins obey the laws of physics. Why not apply these laws to predict the properties of proteins? (After all, this worked fine for the solar system.) **Molecular dynamics** is just such an attempt. The atoms of the proteins, and of the solvent in which they are embedded, are the particles. Newton's laws allow prediction of the trajectories, knowing the forces the particles exert on one another. One problem is that the forces are known only approximately, and therefore the results can at best be only approximate. Another is that even the very powerful computers now available cannot quite manage to simulate the motion of a protein long enough to see it fold up from a denatured state. (With the dedication of ever-more-powerful computers, this is changing—see Chapter 3).

• Physics tells us that if we know the force on a particle, we can predict how it will move. If, as in a protein, we have many particles, but we know the forces on all of them, we can predict how all the particles will move. In principle we should be able to start with an unfolded protein and follow its trajectory to the folded state. The problems are that there are very large number of particles, and that the forces are constantly changing, because they depend on the instantaneous positions of the particles. Despite substantial progress, the computations remain challenging.

However, with guidance from experimental data—from X-ray diffraction or NMR spectroscopy—molecular dynamics does a fine job of finding a suitable model consistent with the data. Such simulations now contribute to experimental protein structure determinations.

Not all simulations treat full-atom representations by Newtonian mechanics. Other computational approaches to understanding and predicting protein structure permit (1) other ways of applying experimental information, and (2) testing ideas about protein folding using simplified models and force fields. These ideas include the following.

Knowledge-based methods of protein structure prediction

Databases contain very large amounts of information about amino acid sequences and protein structures. They are applicable to predict structures of closely related proteins, by **homology modelling**. Homology modelling is now a reliable technique in very widespread use. Should it not be possible to apply knowledge-based methods to predicting protein structure from amino acid sequence, even if the target structure is an entirely novel fold? One crucial idea is the extent to which *local* sequence patterns determine *local* structural patterns—for instance, is there a signal in a stretch of residues that implies that this region will fold into an α-helix? It is possible to do a fairly good job of predicting the structures of local regions, or at least to narrow the possibilities. Of course one then faces the problem of putting the regions together into a complete structure. Another approach is to use patterns of mutation in a set of evolutionarily-related proteins to infer which residue positions are close together in space, and derive the structure from this information. These methods are under active development, and have achieved some impressive successes.

Simplified simulations

A *lattice model* represents the structure of a protein as a connected set of points distributed at discrete and regular positions in space, with simplified interaction rules for calculating energies of different conformations. For small lattice polymers, it is possible to enumerate every possible configuration. There is then no chance that a computer program failed to find the optimum because the simulation could not be carried out for long enough times.

The simplest lattice model is the two-dimensional hydrophobic-polar (HP) model, with the assumptions:

1. Positions of residues are restricted to points on a square lattice in a plane.

2. Each residue can be of one or two types: hydrophobic or polar—H or P.

3. The interaction energy is measured only between residues in contact: a neighbouring H–H pair contributes a constant attractive energy $-E$, and all other neighbours make no contribution to the conformational energy.

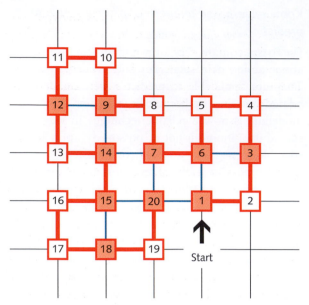

Figure 1.19 A two-dimensional lattice model of a protein. There are two types of residues, hydrophobic (H) shown in filled red squares, and polar (P) shown in open squares. The 'protein' contains 20 residues, with the 'sequence' HPHPPHHPHPPHPHHPPHPH. Light blue lines show H–H contacts between residues not consecutive in the sequence. In this conformation there are nine such H–H contacts. Can the reader find a lower-energy conformation of this system?

Figure 1.19 shows a two-dimensional lattice 'protein' in a low-energy conformation.

Another intensively studied lattice model represents a 27-residue polymer on a three-dimensional $3 \times 3 \times 3$ lattice. There are exactly 57 704 compact structures. It is very comforting to do statistical-mechanical calculations on systems with small, discrete sets of states, for which it is possible to consider every conformation explicitly. It's really too bad that they're not real proteins!

Bioinformatics

Bioinformatics is a hybrid of molecular biology and computer science. Computing is absolutely essential in managing the flood of data that the high-throughput methods are generating. A major goal of bioinformatics is to map out the world of sequences, structures, and functions of biological molecules, and the relations among them. Using these maps, molecular biologists travel in time—back, to deduce events in evolutionary history, and forward to greater controlled modifications of

biological systems. The methods of bioinformatics support applications to medicine, agriculture, and technology.

At the basis of the enterprise are the large-scale databases available on the world-wide web, and **algorithms** and software with which to interpret the data. All readers will have used the web: for reference material, for access to databases in molecular biology, for checking out personal information about individuals—friends or colleagues or celebrities—or just for browsing. The web provides a complete global village, containing the equivalent of library, post office, shops, and schools. It is not merely the number of entries in the web but the density of the connections between them—their **reticulation**—that makes it such a powerful tool. The web has already transformed the world. Its professional and social implications continue to be revolutionary.

- Bioinformatics is a hybrid of molecular biology and computer science. A major goal of bioinformatics is to curate biological data, impose a logical structure on them, and provide effective methods for **information retrieval**.

Introduction to databanks for protein science

The world-wide web contains many compilations of information in the field of molecular biology. Some are general and comprehensive databanks of sequence, structure, function, or bibliography; others are specialized 'boutique' collections. Still others are primarily indices of other websites. Indeed, given the crucial role of the reticulation of information sources, the utility of a site depends on the quality of the links it contains as well as the data it presents. Each year the January issue of the journal *Nucleic Acids Research* contains contributions from most of the important database projects. A companion issue collects descriptions of web-based software.

Major primary data curation projects in molecular biology archive and annotate nucleic acid sequences, protein sequences, and macromolecular structures. Many individuals or groups select and recombine data focused on particular topics, and provide links affording steamlined access to information about the topic of interest.

Archival databanks of amino acid sequences

The archiving of amino acid sequences of proteins, now determinined almost exclusively from translation of gene sequences, is being carried out by the United Protein Database (UniProt), a merger of the databases SWISS-PROT, The Protein Information Resource (PIR), and Translated EMBL (TrEMBL = translations of genes identified within DNA sequences in the EMBL Nucleotide Sequence Database). Other databanks depend on these archives for curation and preservation of data, but select, reformat, and re-present portions of the information. For example, several such derived databanks classify proteins into families based on the similarities of their sequences.

Archival databanks of protein structures

The world-wide Protein Data Bank (wwPDB) is the collection of publicly available structures of proteins, nucleic acids, protein-nucleic acid complexes, and other biological macromolecules. Related archives include the Nucleic Acid Data Bank, containing oligo- and polynucleotide structures, a database of protein–DNA complexes, and the BioMagRes Data Bank, containing NMR structures. The Cambridge (U.K.) Crystallographic Data Centre, and the PubChem database at the U.S. National Center for Biotechnology Information, archive structures of small organic molecules.

Currently (September 1, 2015) the wwPDB contains 111 558 sets of coordinates. There is a certain amount of duplication. In some cases there have been several determinations of the structure of a protein, crystallized under different conditions, or re-solved at higher resolution. In some cases the structure of a protein has been determined in different states of ligation; for example, an enzyme with no ligand, and the same protein binding an inhibitor.

In other cases the structures of very closely related molecules have been determined. This makes it difficult to state precisely how many unique protein structures the databank contains, but this number may be estimated at 1000–2000. Several sites present classifications of protein structures (see Chapter 2).

Submitted coordinates are subjected to stereochemical checks, and translated into a standard entry format. Entries contain technical information about the structure determination, some structural analysis such as assignments of helices and sheets, references to papers describing the structure, and the atomic coordinates.

The wwPDB website (http://www.wwpdb.org) permits retrieval of entries in computer-readable form. These and other sites provide search tools for identifying entries that match stated specifications, including molecule name, species of origin, depositor name, experimental technique, etc.

Information-retrieval tools

One function of archival databanks is conservation; protecting information against loss or corruption. However, conservation is not enough: databanks without effective access become data graveyards. Effective access requires computer programs that allow users to identify and retrieve data according to versatile sets of criteria. It is the responsibility of the database designers to organize the data with an adequate internal logical structure to make writing such programs possible.

In addition to scanning databanks for items of interest, facilities available on the web provide a wide range of computational tools for data analysis. They take the form of *web servers*—sites that allow the user to select input and launch calculations. Some of these are straightforward operations, such as the calculation of the molecular weight of a protein. Some, at the other extreme, are at the cutting edge of research—examples are the servers that attempt to predict the three-dimensional structure of a protein from its amino acid sequence.

If the computation is a fast one, the results may be returned by the browser 'on the fly.' If the computation is slow, the program may send the results by e-mail. The old model of 'install programs on your computer and download the data on which to run them' has given way to world-wide distributed computer facilities. Only connect.

Web access to the scientific literature

We are in an era of transition to paper-free publishing. The U.S. National Library of Medicine supports a bibliographical database of the biomedical literature: http://www.ncbi.nlm.nih.gov/PubMed. Scientific publications never appear solely on paper. (Well, maybe hardly ever, but I don't know of any.) A journal may post its table of contents, a table of contents together with abstracts of articles, or full articles.

Although most commercial publishers are moving towards web access, they and their customers are still feeling their way while a new economic *modus vivendi* evolves. Some journals place their issues on the web immediately upon publication. Some place their issues on the web after a delay—to encourage the purchase of paper copies. Others make available electronically only their tables of contents, or only summaries of their articles and require payment for full access. Moreover, although the medium of distribution has changed, the underlying 'newsstand' economic model—payment at the point of access to the information—has remained the same.

A radical alternative, **PubMed Central,** has created a central repository for literature in the life sciences. Negotiations with publishers, to allay their fears of loss of subscription income, involved agreements that the publishers would retain their customary copyright assignments for commercial use of the material published, and that there would be a delay between the publication of a journal on paper and free electronic access to its contents.

A new organization, the **Public Library of Science,** has the goal of making the scientific (including medical) literature publicly and freely accessible. A non-profit organization, the Public Library of Science, has received support from foundations, for its efforts in distributing literature published by others, and to start its own publications, which will permit exploration of different relationships—including but not limited to economic ones—between authors, publishers, and readers.

Searching the Web for information is a development of the classic library model in which you take action to select and harvest material. This material may be either a specific and known target of your visit to the library, or identified while browsing. The converse mode is that the repository—database or publisher—takes the initiative to send you information. You can submit a list of items of interest containing either topics—*i.e.,* keywords—or citations of publications, or database entries. When new information that is related to your choices emerges, an alert will appear in your e-mail. Most readers will be familiar with such alerts in on-line marketing techniques. Some people describe the difference between the two approaches as 'pull' or 'push' alternatives.

● USEFUL WEBSITES

The following list contains a number of websites that will be of general utility in pursuing the study of protein science from the text onto the internet. Of course electronic links are more convenient than printed ones and the reader is referred to the companion site www.oxfordtextbooks.co.uk/orc/leskprotein3de/ for classified lists of all links that appear in the book.

The Protein Data Bank   http://www.rcsb.org

The Protein Data Bank in Europe   http://www.ebi.ac.uk/pdbe/

Home page of ReLiBase, a system for analysing receptor-ligand complexes in the Protein Data Bank   http://www.ccdc.cam.ac.uk/Solutions/FreeSoftware/Pages/Relibase.aspx

Collection of on-line analysis tools, including database searches   www.expasy.org

Index to websites in molecular biology, including specialized databases   http://www.lib.berkeley.edu/BIOS/molebio.html

Site about electronic scholarly publishing, with emphasis on genetics   http://www.esp.org/

● RECOMMENDED READING

Alberts, B. (1998). The cell as a collection of protein machines: preparing the next generation of molecular biologists. *Cell,* **92,** 291–4. Thoughtful assessment of the field, albeit from some time ago, by an individual at the centre of the developments.

Babu, M.M., Kriwacki, R.W. and Pappu, R.V. (2012). Versatility from protein disorder. *Science* **337**, 1460–1, and Latysheva, N.S., Flock, T., Weatheritt, R.J., Chavali, S. and Babu M.M. (2015). How do disordered regions achieve comparable functions to structured domains? *Protein Sci.* **24**, 909–22. The role of flexibility and disorder in protein function or functions.

Berman, H.M., Goodsell, D.R. and Bourne, P.E. (2002). Protein structures: from famine to feast. *Am. Sci.*, **90**, 350–9. General description of the development of protein structure determination and the archiving, curation, and distribution of the results.

Branden, C.-I. and Tooze, J. (1998). *Introduction to protein structure*, 2nd edn. Garland Publishing Co., New York. A fine introductory textbook.

Brenner, S. (2003). Nobel lecture. Nature's gift to science. *Biosci. Rep.*, **23**, 225–37. Brenner's profound insights are always worth studying.

Dunker, A.K., Oldfield, C.J., Meng, J., Romero, P., Yang, J.Y., Walton Chen, J., Vacic, V., Obradovic, Z. and Uversky, V.N. (2008). The unfoldomics decade: an update on intrinsically disordered proteins. *BMC Genomics* **9**(Suppl 2):S1. Review of studies and interpretations of protein disorder.

Janin, J. and Sternberg, M.J.E. (2013). Protein flexibility, not disorder, is intrinsic to molecular recognition. F1000 *Biol Rep.* **5**, 2. A discussion by two of the leading experts in protein science, of the distinction between flexibility and disorder, and their role in creating ligand-binding sites.

Lesk, A.M., Bernstein, H.J. and Bernstein, F.C. (2008). Molecular graphics in structural biology. In: *Computational structural biology, methods and applications*, M. Peitsch, and T. Schwede, eds. World Scientific Publishing, Singapore, pp. 729–70. A review of molecular graphics and its applications.

Liljas, A., Liljas, L., Piskur, J., Lindblom, G., Nissen, P. and Kjeldgaard, M. (2009). *Textbook of structural biology*, World Scientific Publishing, Singapore. Another fine Scandinavian textbook, of which Carl-Ivar Brändén would be proud. Covers the material contained in this book, plus related aspects of the general biological context of protein science. Written by a team of leaders of the field.

Mannige, R.V. (2014). Dynamic new world: refining our view of protein structure, function and evolution. *Proteomes* **2**, 128–53. A review and clear exposition of the nature of the flexibility of protein structure and the implications for complex formation.

Richards, F.M. (1991). The protein folding problem. *Sci. Am.*, **264(1)**, 54–7, 60–3. A general introductory-level article by one of the founders of protein science, responsible for classic work introducing the computational approach to the subject, in addition to carrying out several classic experiments. (Prof. Richards died during the writing of the 2nd edition of this book, on January 11, 2009).

● EXERCISES AND PROBLEMS; FOR WEBLEMS SEE COMPANION WEBSITE

Three types of activities go with each chapter. Exercises are short and straightforward applications of material in the text. Problems also depend only on information contained in the text, but require lengthier answers or in some cases calculations. The third category, 'weblems', require access to the world-wide web, preferably through a graphical interface. Weblems are designed to give readers practice with the tools required for further study and research. In this edition, they appear on the companion website only.

Exercise 1.1 (a) To what other amino acids (or Stop signal) can an Ala be changed by a single nucleotide substitution? (b) To what other amino acids (or Stop signal) can a Ser be changed by a single nucleotide substitution? (c) To what other amino acids (or Stop signal) can a Lys be changed by a single transversion? A transversion is a substitution of purine by a pyrimidine (a or g ↔ t or c), or a pyrimidine by a purine (t or c ↔ a or g).

Exercise 1.2 What is the minimum number of nucleotide substitutions required to change a codon for Phe to a codon for Lys?

Exercise 1.3 What physicochemical class of amino acids tends to correspond to codons with a pyrimidine (t or c) in the second position?

Exercise 1.4 Which amino acids are encoded by six codons? Which by four codons? Which by three codons? Which by two codons? Which by one codon?

Exercise 1.5 What codons can you make from Ile codons by (a) any SNP (single-nucleotide polymorphism)? (b) by a single transversion? (c) by a single transition? Which of these amino acid changes are conservative, in terms of their change in the physicochemical properties, including volume, of the sidechain?

Exercise 1.6 (a) Which would you expect to be the least likely to create a dysfunctional protein? (1) Only mutations that do not change the amino acid sequence; (2) deletion of one base pair within the coding sequence of the gene; (3) deletion of three base pairs within the coding sequence of the gene. (b) Which would be the most likely to create a dysfunctional protein?

Exercise 1.7 Write the amino acid sequence of the pentapeptide glutamic acid–leucine–valine–isoleucine–serine in (a) three-letter code, (b) one-letter code.

Exercise 1.8 The sickle-cell mutation in haemoglobin changes residue 6 of the β-chain from a charged sidechain, glutamic acid, to a non-polar one, valine. The codon that appears at this position in the mutated form is likely to be either of which two of the four valine codons?

Exercise 1.9 (a) Why is gene transfer from mitochondria → nucleus observed but not nucleus → mitochondria? (b) Suppose an essential gene in the mitochondria is copied into the nucleus and integrated into the nuclear DNA. What would be required for the mitochondrial gene to become silent?

Exercise 1.10 (a) What is the frequency of a photon of wavelength 260 nm? (b) What is the energy of a photon of wavelength 260 nm? (c) What is the energy of a photon of wavelength 509 nm?

Exercise 1.11 Which of the following would increase the danger of a protein aggregation disease? (a) Absence or dysfunction of a chaperone protein. (b) Mutation in a codon that specifies an amino acid to a Stop codon, creating a truncated protein. (c) Mutation in a codon that specifies a Stop signal to a codon that specifies an amino acid, creating an extended protein. (d) Dysfunction in ubiquitin ligase.

Exercise 1.12 Which of the following would be expected to follow simple Mendelian inheritance as a recessive trait? (a) A mutation causing dysfunction in an essential metabolic enzyme encoded by a gene on human chromosome 10. (b) A mutation causing a gain of function in a protein.

Exercise 1.13 In H.G. Wells' 1898 novel, *The War of the Worlds,* invaders from Mars are overcome by disease. Assuming that life on Mars developed independently of life on Earth, why is it unlikely that the Martians would have died of viral infections?

Problem 1.1 There are $4^3 = 64$ possible single-point mutations in codons. (a) How many of them, if they occur within coding regions, leave the amino acid sequence unchanged? (b) How many possible transitions—purine to purine (a → g or g → a) or pyrimidine to pyrimidine (t → c or c → t) substitutions—are there? How many of them, if they occur within coding regions, leave the amino acid sequence unchanged? (b) How many possible transversions—purine to pyrimidine (a or g → t or c), or pyrimidine to purine (t or c → a or g) substitutions—are there? How many of them, if they occur within coding regions, leave the amino acid sequence unchanged?

Problem 1.2 On a photocopy of Fig. 1.4, showing the superposition of the histidine-containing phosphocarrier proteins of *E. coli* and *S. faecalis,* indicate with a highlighter the worst-fitting regions. Show these regions on a photocopy of the sequence alignment that appear with Fig. 1.4. Is there any correlation between the parts of the sequences with the fewest local matches and the regions of insertions and deletions?

Problem 1.3 For purposes of defining conservative mutations, amino acids can be divided into a larger number of classes than those appearing in Table 1.1, by taking into account sidechain volume and sign of the charged residues. Consider the following classes:

Small non-polar: Gly, Ala

Hydrophobic: Val, Leu, Ile, Phe, Pro, Met, Trp

Polar: Ser, Cys, Thr, Asn, Gln, Tyr, (His)

Positively charged: Asp, Glu, (His)

Negatively charged: Lys, Arg

Define a conservative mutation as one that interconverts amino acids within the same class. Analyse the sequence alignment of the histidine-containing phosphocarrier proteins from *E. coli* and *S. faecalis* (after Fig. 1.4). What percentage of the residue positions contain either an unchanged amino acid or a conservative mutation?

Problem 1.4 Draw a diagram similar to Fig. 1.9(b) and (c) showing the hydrogen-bonding pattern in the β-sheets in acylphosphatase (see Fig. 1.10). Assume for purposes of the drawing that each strand is six residues long. Acylphosphatase contains what is called a mixed β-sheet; i.e., the strands are not all either parallel nor are adjacent strands always antiparallel.

Problem 1.5 An infant is born with a mutation that destabilizes a protein, to the extent that the protein is largely unfolded at 40°C. The protein has a high intracellular concentration, and there is danger that it will form aggregates. What clinical advice might your offer the parents of this child, knowing that babies often 'spike' high fevers when ill?

LEARNING GOALS

- *Understanding the basic chemical structure of proteins:* the distinction between the repetitive backbone common to all proteins, and the variable sequence of sidechains.

- *Learning the structures and properties of the 20 natural sidechains:* atom nomenclature, size, polarity, charge, titratable groups.

- *Defining the degrees of freedom that specify different conformations of the polypeptide chain:* definitions and allowed values of conformational angles, the Sasisekharan–Ramakrishnan– Ramachandran plot, rotamers and rotamer libraries.

- *Knowing the properties of the native states of proteins:* compact, stable states of unique conformation, required for biological function.

- *Recognizing that many proteins have disordered regions.* Most proteins can form a relatively loosely-packed state called the 'molten globule'.

- *Categorizing the energetic factors stabilizing native states of proteins:* covalent bonds, including disulphide bridges; hydrogen bonding; the hydrophobic effect and accessible surface area; the coordination of metal ions; Van der Waals forces and the dense packing of protein interiors.

- *Recognizing that most globular proteins are only marginally stable:* about 20–60 kJ mol^{-1}.

- *Knowing the distinctions between:*

  primary structure—the amino acid sequence of the protein and other covalent bonds

  secondary structure—the formation of helices and sheets from particular regions of a protein

  tertiary structure—the way the helices and sheets are organized and interact in space

  quaternary structure—the formation of multisubunit proteins from individual polypeptide chains.

- *Being able to define supersecondary structures, domains, and modular proteins,* and to relate these terms to the classical hierarchy of primary, secondary, tertiary, and quaternary structure.

- *Developing proficiency at recognizing, in a drawing of a protein structure,* helices, parallel, antiparallel, and mixed β-sheets, β-barrels, supersecondary structures, and domains.

- *Recognizing the effectiveness of building up proteins from independent modules,* as a means of generating a variety of structures and functions.

- *Appreciating some of the great range of protein structures and functions.* Understanding the purpose of comparative analysis of protein folding patterns. Knowing about projects that provide complete structural classifications of proteins, and gaining familiarity with the associated websites.

- *Becoming familiar with different general types of proteins,* and the interdependence of structure and function.
- Recognizing that although proteins fold to native states, *conformational change* can be an essential aspect of the mechanism of protein function.
- *Appreciating the importance of regulation* to integrate the actions of the many different proteins in organized cellular activity.

Introduction

In this chapter we look into the structures of proteins in more detail. Building on the basic structural and physical chemistry of amino acids and protein conformations, we examine the features that stabilize native states. A survey of the types of protein structures that have been observed demonstrates convincingly that protein architecture is a fascinating topic in itself. Also of compelling interest is how structure illuminates function.

Proteins combine the unity of a common and repetitive mainchain with the variety available from a sequence of sidechains individual to each protein. The sequence of sidechains both: (1) determines the three-dimensional structure of the protein, including its overall folding pattern in general and the structure of the active site in particular, and (2) provides the chemistry required for function: a negative charge here, a hydrogen-bond donor there.

The 20 standard amino acids contain groups of several physicochemical types and are adequate for many but not all purposes. In addition to amino acids, many proteins bind prosthetic groups—ions or small organic molecules—that stabilize or modify the structure, and/or participate in catalysis. Some proteins undergo post-translational modification, such as glycosylation, phosphorylation and many others. Post-translational modifications can extend the functional repertoire of components of a protein, beyond those supported by the chemical properties of the 20 amino acids themselves. They also serve regulatory purposes: For instance, phosphorylation and dephosphorylation can switch proteins between active and inactive states.

Structures of the amino acids

The sidechains provide proteins with a useful variety of chemical properties. Box 2.2 and Fig. 2.1 show the sidechains, both as chemical structure diagrams, and in full-atomic representations—a sphere at the position of each non-hydrogen atom. These two representations emphasize the dual importance of the structure, polarity, and charges of the sidechains, on the one hand; and the space-filling aspects—important for the packing of protein interiors—on the other. The figure shows all non-hydrogen atoms only for glycine and alanine; for the other 18 only the sidechain atoms appear.

Glycine ($^+H_3NCH_2COO^-$) has a second hydrogen instead of a larger sidechain. All the other amino acids have sidechains containing between 4 and 18 atoms. All amino acids except glycine contain a chiral Cα. Only one of the two mirror-image forms appears in natural proteins, with a few exceptions that arise as post-translational modifications. (Which sidechain has another chiral carbon?)

Greek letters label the atoms of the sidechains, in progression out from the Cα atom (see Box 2.1).

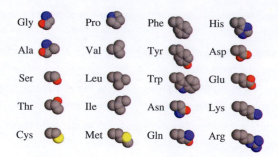

Figure 2.1 The 20 natural amino acids. White = carbon, red = nitrogen, blue = oxygen, yellow = sulphur. This is the cast of characters, which play all the different roles in different proteins. The sidechains vary in their physicochemical properties: size, hydrogen-bonding potential, and charge.

- *Some sidechains are electrically neutral:* Because of the thermodynamically unfavourable interaction of hydrocarbons with water, residues containing large aliphatic neutral sidechains are called '**hydrophobic**' residues.

- *Some sidechains are polar:* Asparagine and glutamine contain amide groups; serine, threonine, and tyrosine, hydroxyl groups. Polar sidechains, like mainchain peptides, can participate in hydrogen bonding.

- *Other sidechains are charged:* Aspartate and glutamate (= ionized aspartic acid and glutamic acid) are negatively charged; lysine and arginine (and, usually, histidine) are positively charged. The charged atoms occur at or near the ends of the sidechains, which, except for histidine, are relatively long and flexible. The atoms nearest to the backbone are non-polar. Two sidechains with positive and negative charge can approach each other in space to form a 'salt bridge'.

BOX 2.1   **Naming the atoms**

Atoms in the mainchain of each residue are denoted N, Cα, C and O. The sidechain is attached to the Cα. Sidechain atoms are identified by their chemical symbol, and by successive letters from the Greek alphabet, proceeding out from the Cα. Thus, the sidechain of methionine has atoms Cβ, Cγ, Sδ, Cε:

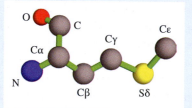

- It is worth learning by heart the structures of the amino acids. Organizing them by physicochemical categories will help.

BOX 2.2   **The chemical structures of the amino acid residues in proteins**

A free amino acid contains an amino and a carboxyl group. Upon formation of peptide bonds, the residue is reduced to –N–Cα–C(=O)– plus a sidechain attached to the Cα. This figure shows complete glycine and alanine *residues* (except for hydrogen atoms). For the other amino acids, only the sidechain appears, except for proline, which is special be-

cause the sidechain forms a pyrrolidine ring that includes the backbone N.

Only the structure of glycine is shown completely. The other amino acids all have one hydrogen atom and $^+NH_3$ and COO^- groups (not shown) bound to the Cα, except proline, which is special.

Glycine	$+NH_3$ | $H-C\alpha-H$ | COO^-	Tyrosine	$-C\alpha-CH_2-\langle\bigcirc\rangle-OH$
Alanine	$-C\alpha-CH_3$	Aspartic acid	$-C\alpha-CH_2-COO^-$
Serine	$-C\alpha-CH_2-OH$	Glutamic acid	$-C\alpha-CH_2-CH_2-COO^-$
Cysteine	$-C\alpha-CH_2-SH$	Histidine	$-C\alpha-CH_2C$ (imidazole ring $N^+{=}C$, $C{-}NH$)
Threonine	$-C\alpha-CH(OH)-CH_3$	Asparagine	$-C\alpha-CH_2-CONH_2$
		Glutamine	$-C\alpha-CH_2-CH_2-CONH_2$
Proline	$N-C\alpha$ (pyrrolidine ring)	Lysine	$-C\alpha-CH_2-CH_2-CH_2-CH_2-NH_3^+$
Valine	$-C\alpha-CH-(CH_3)_2$		
Leucine	$-C\alpha-CH_2-CH-(CH_3)_2$	Arginine	$-C\alpha-CH_2-CH_2-CH_2-NH-C\big\langle{}^{NH_2}_{NH_2^+}$
Isoleucine	$-C\alpha-CH(CH_3)-CH_2-CH_3$		
Methionine	$-C\alpha-CH_2-CH_2-S-CH_3$	Tryptophan	$-C\alpha-CH_2$ (indole ring, N–H)
Phenylalanine	$-C\alpha-CH_2-\langle\bigcirc\rangle$		

Some of the sidechains contain ionizable groups: the –COOH of aspartic acid and glutamic acid, which at ordinary pH are deprotonated to –COO⁻, forming asparate and glutamate; the –$^+NH_3$ group of lysine, the guanidinium group of arginine and the imidazole group of histidine. The state of ionization depends on the pH of the solution, and on interactions that may stabilize either the charged or uncharged form, in effect perturbing the pK. Table 2.1 gives the pK values of the amino acids in isolation.

Table 2.1 Acid/base properties of proteins

Amino acid	pK of sidechain
Aspartic acid	3.90
Glutamic acid	4.07
Histidine	6.04
Cysteine	8.37
Tyrosine	10.46
Lysine	10.54
Arginine	12.48
Terminal amino group	~8.0
Terminal carboxyl group	~2.4

pK values of ionizable groups can vary by about 0.5-1.0 unit (occasionally substantially more), depending on their environments within the protein.

The pK of an ionizable group is the equilibrium constant for deprotonation. The higher the pK, the more reluctant the group is to release a proton. For aspartic acid and glutamic acid, the carboxyl of the sidechain releases the proton, and, as the pK's are below 7, the ionized forms—aspartate and glutamate—are prevalent at neutral pH. For lysine, the positively charged distal ammonium group of the sidechain could release a proton, to form an uncharged amino group, but at ordinary pH lysine sidechains are in the charged, ammonium form.

The resting pH of blood is about 7.4. This is between the pK values of the sidechains, His and Cys. Even so, variation by more than ~0.2 pH units is dangerous. Strenuous exercise tends to acidify blood, and can overwhelm the buffer systems, which include bicarbonate/carbonate, and blood proteins such as haemoglobin. The immediate danger is titration of histidine sidechains, threatening denaturation of proteins.

- The Henderson–Hasselbalch equation: For an acid that dissociates according to the equilibrium $HA = H^+ + A^-$, the relationship between the extent of ionization and the pH of the solution is:

$$pH = pK + \log_{10}\frac{[A^-]}{[HA]}$$

where HA is the protonated form and A^- the deprotonated form. For sidechains containing titratable amino groups, such as lysine, or which have multiple ionizable protons, such as arginine, the equation must be modified, as it is not true that the protonated form is uncharged and that the deprotonated form has a single negative charge. $[A^-]$ = the concentration of A^-.

Protein conformation

By the *conformation* of a polypeptide chain we mean the space curve that the backbone traces out. The mainchain is flexible, and the number of possible spatial patterns is very large. Indeed, in the denatured state, the chain takes advantage of this freedom. The conformations of the molecules in the sample are all different, and the chain can be spread out over a large volume. The native states of proteins, in contrast, are special. Under physiological conditions of solvent and temperature, all molecules with the same amino acid sequence adopt the *same* conformation, the *native state*. Suppression of conformational freedom in the native state is thermodynamically costly. For this reason, random polypeptides do not adopt unique native states. Proteins have evolved coherent sets of interactions to stabilize their native states (see Box 2.3).

The native state conformation of a protein consists of a compact assembly of residues. Features contributing to the stability of the native state include satisfaction of hydrogen-bonding potential of polar groups, burying of non-polar atoms, and dense packing of residues in the interior. The optimal solution of this thermodynamic jigsaw puzzle is a special conformation in which the backbone describes a curve traversing the space occupied by the molecule. The shape of this curve, called the protein's **folding pattern,** usually includes standard elements of secondary structure—α-helices and β-sheets (see Fig. 1.9)—and makes use of a common repertoire of ways in which helices and sheets pack against one another.

BOX 2.3   **Characteristics of the native states of proteins**

- They have a definite three-dimensional structure, dictated by the amino acid sequence.

- Many similar amino acid sequences produce similar structures. That is, protein structures are robust to mutation—not to *all* possible mutations, but to many. This provides pathways for evolution to explore.

- The stabilization of native states is only marginal, about 20–60 kJ mol⁻¹—the equivalent of about two hydrogen bonds.

- The conformations of most of the individual residues are unstrained.

- Native states are compact, well-packed structures. Although they often have surface clefts—for instance, active sites—only rarely are there holes or channels.

- Burial of hydrophobic, or oil-like, sidechains removes them from unfavourable contacts with water (see 'hydrophobic effect' in Stabilization of the native state section).

- Burial of mainchain polar groups requires satisfying their hydrogen-bonding capacity. This is achieved primarily by forming *secondary structures:* α-helices and β-sheets.

• Properties of the native state contributing to stability include: satisfaction of hydrogen-bonding potential of polar groups, burying of non-polar atoms, and dense packing of residues in the interior. The greater conformational freedom of the denatured state has a destabilizing effect on the native state.

Protein architecture is the study of how protein folding patterns may be compared and classified, the analysis of the relationships between the local interactions and the overall folding pattern of the chain, and the elucidation of how the entire structure is determined by the amino acid sequence. Protein architecture also illuminates function, by showing how form follows function and *vice versa*.

Conformational angles and the Sasisekharan–Ramakrishnan–Ramachandran plot

The conformation of a polypeptide chain can be described quantitatively in terms of angles of internal rotation around the bonds in the mainchain (see Fig. 2.2). The bonds between the N and $C\alpha$, and between the $C\alpha$ and C, are single bonds. Internal rotation around these bonds is not restricted by the electronic structure of the bond, but only by possible steric collisions in the conformations produced (see Box 2.4). The peptide bond has partial double-bond character. It therefore has two possible conformations: *trans* (by far the more common) and *cis* (rare).

The most familiar objects that illustrate **dihedral angles** are an ordinary swinging door, or a book. Figure 2.3 shows two planes: a doorframe (blue) and a door (red). The plane of the door rotates with respect to the plane containing the doorframe, around an axis containing the hinges. The dihedral angle is the angle of opening of the door. With the door closed the dihedral angle between the door and the doorframe is 0°. The dihedral angle is 90° when the door is opened far enough to be perpendicular to the doorframe.

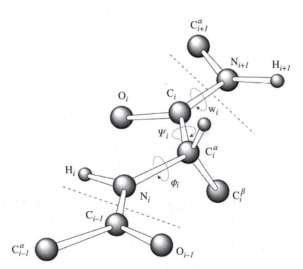

Figure 2.2 Conformational angles describing the folding of the polypeptide chain. Broken lines delimit the residue. The mainchain of each residue (except the C-terminal residue) contains three chemical bonds: N–$C\alpha$, $C\alpha$–C, and the peptide bond C–N linking the residue to its successor. The conformation of the mainchain is described by the angles of rotation around these three bonds:

Rotation around:	N–$C\alpha$ bond	$C\alpha$–C bond	peptide bond (C–N)
Name of angle:	ϕ	ψ	ω

Geometrically, the mainchain of a protein is a succession of points in space: N–$C\alpha$–C–N–$C\alpha$–C (the carbonyl oxygens are not in the mainchain itself). To a good approximation, the bond lengths and angles—the distances between every two successive points, and the angles between every two successive bonds—are constant. The degrees of freedom of the chain then involve *four* successive atoms, and consist of rotations in which the first three atoms are held fixed, and the fourth atom is rotated around the bond linking the second and third. (An ordinary paper clip unwinds easily to give a chain with four vertices, and is a useful object with which to practice.) By convention, the *cis* conformation is 0°, and the positive sense of each angle corresponds to looking down the bond that serves as the axis of rotation, and turning the distant atom in a clockwise direction.

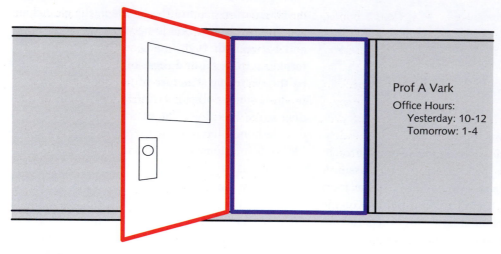

Figure 2.3 An open door illustrates a dihedral angle between the plane of the door and the plane of the doorframe. The door rotates around an axis through the hinges, in the line of intersection between the two planes.

Prof A Vark

Office Hours:
    Yesterday: 10-12
    Tomorrow: 1-4

 BOX 2.4 ## The Sasisekharan–Ramakrishnan–Ramachandran diagram

The mainchain conformation of each residue is determined primarily by the two angles ϕ and ψ. The angle of rotation around the peptide bond, ω, usually has the value $\omega = 180°$ *(trans)* and occasionally (most often before a proline residue) $\omega = 0°$ *(cis)*.

For some combinations of ϕ and ψ, with the *trans* conformation of the peptide bond, atoms would collide, a physical impossibility. V. Sasisekharan, C. Ramakrishnan, and G.N. Ramachandran first plotted the *sterically allowed regions* (see Fig. 2.4). There are two main allowed regions of residue conformation. These regions, α_R and β, correspond to the two major types of secondary structure: α-helix and β-sheet (see Fig. 1.9).

A succession of residues in the α_R conformation generates an α-helix. The α-helix is right-handed, like the threads of an ordinary bolt. A succession of residues in the β-conformation generates a strand that is nearly fully extended. This conformation appears in strands of β-sheets. A succession of residues alternating between α_R and β conformation is sterically impossible.

The α_R region is centred around $\phi = -57°$, $\psi = -47°$. The β region is centred around $\phi = -125°$, $\psi = +125°$. There is a 'neck' between them. The mirror image of the α_R conformation, α_L, is allowed for glycine residues only. (Because glycine is **achiral**—that is, identical with its mirror image—a Ramachandran plot specialized to glycine must be right-left symmetric. For non-glycine residues, collisions of the $C\beta$ atom forbid the α_L conformation.)

A graph showing the ϕ, ψ angles for the residues of a protein, against the background of the allowed regions,

is a **Sasisekharan–Ramakrishnan–Ramachandran plot**, the name often shortened to Ramachandran plot (see Fig. 2.4).

It is no coincidence that the same conformations that correspond to low-energy states of individual residues also permit the formation of structures with extensive mainchain hydrogen bonding. The two effects thereby cooperate to stabilize the native state.

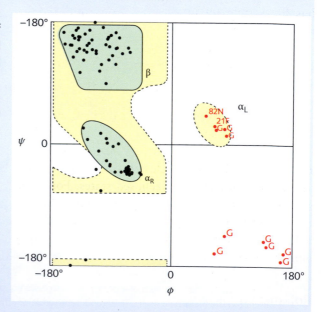

Figure 2.4 Sasisekharan–Ramakrishnan–Ramachandran plot of bovine acylphosphatase [2ACY]. Sterically most favourable regions in green, sterically allowed regions in yellow. Residues with $\phi > 0$, mostly glycines, are identified in red.

Rotations about the N–Cα and Cα–C bonds, and the isomers of the peptide group, describe to a good approximation the mainchain conformational freedom of 19 of the 20 amino acids. Proline is an exception: the sidechain of proline is linked back to the N of the mainchain to form a pyrrolidine ring. This restricts the mainchain conformation of proline residues. It disqualifies the N atom of proline as a hydrogen-bond donor, for instance in helices or sheets. Also, the energy difference between *cis* and *trans* conformations is less for proline residues than for others. Therefore, most *cis* peptides in proteins appear before prolines.

The entire conformation of a protein can be described by these angles of internal rotation. Each set of four successive atoms in the mainchain defines a dihedral angle. In each residue i (except for the residues at the N-and C-termini) the angle ϕ_i is the angle defined by atoms C(of residue i – 1)–N–Cα–C, and the angle ψ_i is the angle defined by atoms N–Cα–C–N(of residue i + 1). ω_i is the angle around the peptide bond itself, defined by the atoms Cα–C–N(of residue i + 1) – Cα(of residue i + 1) (see Fig. 2.2). Each ω_i is restricted to be close to 180° (*trans*) or 0° (*cis*).

- The conformational angles describing the mainchain conformation of residue i of a polypeptide chain are the dihedral angles, ϕ, ψ, and ω, which are defined by the following terms.
  ϕ: C(of residue i – 1)–N–Cα–C
  ψ: N–Cα–C–N(of residue i + 1)
  ω: Cα–C–N(of residue i + 1) – Cα(of residue i + 1)

Sidechain conformation

Angles of internal rotation also describe sidechain conformations. Different sidechains have different numbers of degrees of freedom. An Arg sidechain has four angles of internal rotation, and Gly has none at all. The sidechain conformational angles are denoted χ_1, χ_2, ... For example, for a lysine sidechain χ_1 is the angle determined by the atoms N, Cα, Cβ, Cγ; and χ_2 is the angle determined by the atoms Cα, Cβ, Cγ, Cδ.

Different conformations of any sidechain are called **rotamers**. Are there preferred conformations of sidechains? For many sidechains, the rotation around the Cα – Cβ bond tends to cluster around $\chi_1 = 180°$, +60°, and −60°, avoiding eclipsed states. For some sidechains, the distributions of observed conformational angles may be very highly skewed indeed. For valine, the conformation where $\chi_1 \sim 180°$ is populated almost exclusively.

Rotamer libraries

Most sidechains have relatively small individual repertoires of preferred conformations. **Rotamer libraries** are collections of preferred sidechain conformations. The local backbone structure influences sidechain conformation, because placement of the backbone and Cβ atoms of successive residues creates specific loci for potential steric collision. Secondary structure can thereby bias the sidechain rotamer distribution. For example, the $\chi_1 = 60°$ conformation around the Cα – Cβ bond is hardly ever observed for any residue in a helix, except for serine, which can form a hydrogen bond to the C=O of the preceding residue. *Backbone-dependent rotamer libraries* specify the sidechain conformational states preferred for different backbone conformations.

Rotamer libraries are useful for modelling, because one needs to consider only a relatively small, discrete set of possible conformations. Also useful for modelling is the observation that as proteins evolve, sidechain conformation tends to be conserved. That is, corresponding residues in related proteins tend to have similar sidechain conformations, even when they are mutated. The reason is that sidechains pack against their neighbours. Each sidechain fits into a 'cage' created by neighbouring sidechains. When a residue is changed by mutation, the new sidechain, in order to fit into the old cage, tends to adopt a conformation similar to that of the sidechain it replaces.

From the positions of the backbone atoms in a native protein structure, computer programs can do a good job of reconstructing the sidechain angles. For a test set of 180 X-ray crystal structures of proteins, the program SCWRL4 correctly predicted the value of sidechain conformational angle χ_1 to within 40° of experimental values for 89.3% of 90432 non-glycine, non-alanine residues in 379 proteins (http://dunbrack.fccc.edu/scwrl4).

> • Many sidechains have preferred conformations, tabulated in rotamer libraries. The preferred conformation of a residue may depend on secondary structure context. These preferences are useful in model building.

Stabilization of the native state

The general principle is that native states of proteins are equilibrium states.

Chemical equilibrium at constant temperature and pressure is a compromise:

- Attractive interactions between residues are maximized if the molecules adopt the single common native conformation with the most favourable set of interactions. These include interactions among residues, and between protein and solvent.

- In contrast, the tendency of the protein chains to take as full advantage as possible of available conformational freedom is maximized in the denatured state. Greater conformational freedom = higher entropy.

- The effect of the protein on the solvent is an essential part of the picture. Hydrocarbons are insoluble in water because dissolved non-polar solutes restrict the conformational freedom of the water. Sequestering non-polar sidechains inside the protein, out of contact with water, mitigates this *hydrophobic effect*, and provides a driving force for folding.

Proteins have evolved so that, at equilibrium, the attractive interactions among the residues in the folded state, and the burial of non-polar atoms in protein interiors, overcome the unfavourable restriction of conformational freedom of the polypeptide chain. The protein 'goes native'.

Analyses of protein structures have revealed many details about how Nature stabilizes native states of proteins:

- *Covalent and coordinate chemical bonds:* Many proteins contain covalent chemical bonds in addition to those of the polypeptide backbone and the sidechains.

Disulphide bridges, between cysteine residues, are quite common. The small protein crambin contains three disulphide bridges (see Fig. 2.5). Disulphide bridges can also link different polypeptide chains, as in insulin (see Fig. 2.6) and immunoglobulins.

- *Metal ions* are integral parts of the structures of many proteins. The 4-zinc form of the pig insulin hexamer illustrates both disulphide bridges and metal-ion binding directly to sidechains (see Fig. 2.6). In other cases, the metal is not bound directly to the protein, but is part of a larger ligand. Cytochrome *c* includes an iron-containing haem group (see Fig. 2.7).

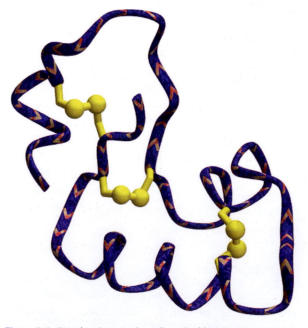

Figure 2.5 Crambin [1CRN]. The yellow double-lollipops are disulphide bridges.

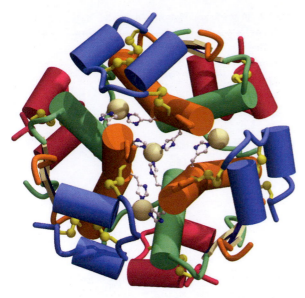

Figure 2.6 Pig insulin hexamer, 4-zinc form [4ZNI]. Insulin contains A and B chains, linked by disulphide bridges. In this figure the A chains are blue and purple, and the helices in the B chains are green and orange. Zn^{2+} ions are light yellow balls. (Which amino acid is ligating the Zn^{2+} ions?) The colours help in recognizing the symmetry of the structure. You are looking down the threefold axis.

- *Hydrogen bonding:* The polar groups of proteins make hydrogen bonds to water in the unfolded state. In the folded state, the hydrogen-bonding potential of groups buried in the interior of the protein must somehow be satisfied. The main chain,

made of polar peptide groups, passes through the interior. Some polar sidechains are also buried. The buried polar atoms lose their interactions with water. To recover the energy, buried polar atoms must form protein–protein hydrogen bonds.

The standard secondary structures—α-helices and β-sheets—achieve hydrogen-bond formation by the mainchain atoms. Therefore *any* amino acid sequence can potentially form helices and sheets—except that the mainchain N atom of proline cannot form a hydrogen bond. This implies that prolines in regions of secondary structure are likely to appear at the N-terminal position of the helix or strand. In practice, secondary structures in proteins can occasionally accommodate internal prolines.

Hydrogen bonds also contribute to the binding of cofactors and substrates (see Fig. 2.8).

- *The hydrophobic effect:* The *hydrophobicity* of an amino acid is a measure of the interaction between the sidechain and water. Hydrocarbon sidechains such as those of leucine and phenylalanine interact unfavourably with water. Just as oil–water

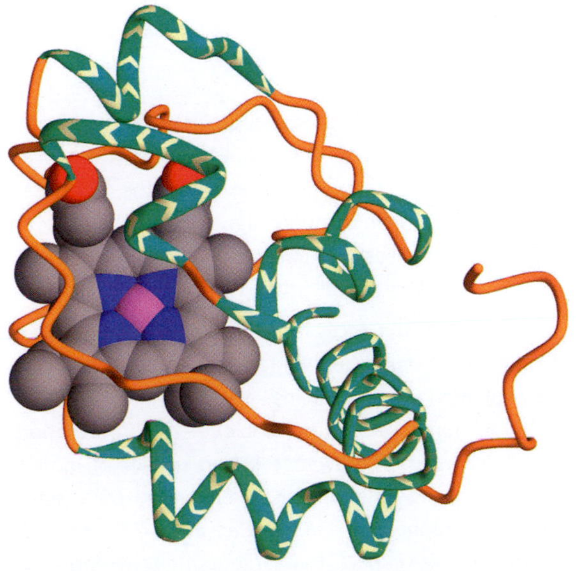

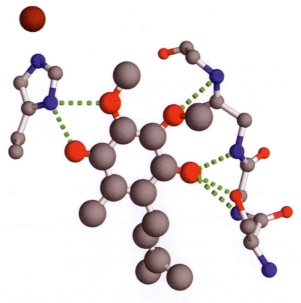

Figure 2.8 Hydrogen bonding between the protein and ubiquinone-1 ligand in the photosynthetic reaction centre of *Rhodopseudomonas viridis* [1PRC]. Small spheres = protein atoms; large spheres = ubiquinone atoms. Maroon sphere at upper left is iron. Note that most crystal structures do not report positions of hydrogen atoms—see Exercise 2.12.

Figure 2.7 Rice cytochrome *c* [1CCR]. The haem group is in a shaded sphere representation, and the purple sphere is the iron.

mixtures separate spontaneously into two phases, there is a tendency for the hydrophobic sidechains to sequester themselves in the interiors of proteins, away from contact with water. This effect provides an important driving force for protein folding. Not all hydrophobic sidechains are buried, but the ones that are contribute to the stability of the native state.

The source of the hydrophobic effect is the ordering of water around non-polar solutes. Pure water contains an internal ordered structure derived from the geometric requirements of water–water hydrogen bonds. However, around non-polar solutes such as methane, water forms clathrate (cage) complexes that are even more ordered than pure water. States of increased order are thermodynamically unfavourable, because *increased order* corresponds to *lower entropy*.

Different residues buried inside proteins contribute to protein stability according to a **hydrophobicity scale** (see Table 2.2). The **accessible surface area** of the protein, calculatable from a set of atomic coordinates, measures the thermodynamic interaction between the protein and water (see Box 2.5). Surface-area calculations identify *which* residues are buried, contributing to the hydrophobic stabilization of a protein structure.

Table 2.2 An amino acid hydrophobicity scale*

2.25	Trp
1.80	Ile
1.79	Phe
1.70	Leu
1.54	Cys
1.23	Met
1.22	Val
0.96	Tyr
0.72	Pro
0.31	Ala
0.26	Thr
0.13	His
0.00	Gly
−0.04	Ser
−0.22	Gln
−0.60	Asn
−0.64	Glu
−0.77	Asp
−0.99	Lys
−1.01	Arg

*Fauchère, J. and Pliška, V. (1983). Hydrophobic parameters π of amino acid side chains from the partitioning of N-acetyl-amino acid amides. *Eur. J. Med. Chem.*, **18**, 369–375. (Fauchère and Pliška based these parameters on the octanol-water partition coefficients of N-acetyl amino acid amides. Many other hydrophobicity scales have been proposed. Some reflect the physicochemical properties of the sidechains; others the statistical distribution of exposed and buried residues in known protein structures.)

BOX 2.5   **Accessible and buried surface area**

The accessible surface area of a protein or protein complex is the area swept out by a water molecule (modelled as a sphere of 1.4 Å in radius) rolling around the outside of the protein. This is proportional to the number of water molecules in contact with exposed atoms. The accessible surface will include nooks and crannies in the protein surface that are larger than 2.8 Å wide, but will smooth over finer wrinkles.

Accessible surface-area calculations rationalize the hydrophobic contribution to the thermodynamics of protein folding and interactions. Observed regularities include:

1. A basic calibration: each $Å^2$ of **buried surface area** contributes ~6 J (25 cal) of free energy of stabilization.

2. The accessible surface area (A.S.A.) of monomeric proteins of up to about 300 residues varies as the 2/3 power of the relative molecular mass M: A.S.A. $= 11.1M^{2/3}$.

3. The formation of oligomeric proteins from monomers buries an additional 1000 $Å^2$–5000 $Å^2$ of surface. Lower values characterize proteins for which the monomer structure is stable in isolation; higher values characterize proteins in which association must stabilize the structure of the monomers as well as the complex.

4. The nature of the accessible and buried area in native protein structures: the average solvent-accessible surface of monomeric proteins—the protein *exterior*—is ~ 58% non-polar (hydrophobic), ~29% polar, and ~13% charged. The average buried surface of monomeric proteins—the protein *interior*—is ~60% non-polar (hydrophobic), ~33% polar, and ~7% charged. Many people expect the large *buried* hydrophobic surface but are surprised at how large the *exterior*

hydrophobic area is. This is partly because there are non-polar residues on the surface, and partly because even polar and charged surface residues contain carbon atoms.

In fact, the main 'take-home message' about the difference between the surface and the interior is that proteins almost never bury charged groups (see Box 2.6).

BOX 2.6   Forces between ions and water are very strong

Ions in aqueous solution bind water tightly, as a result of the strong electrical attraction between a charged ion and the polar water molecules. Have you ever wondered why it is so difficult to break down the crystal lattice of ordinary table salt by heating—the melting point of NaCl is 801°C— even though it is so easy to take the ions apart simply by

adding water? The crystal lattice forces are very strong, *but so are the ion–water forces in the solution.*

For this reason it is energetically very unfavourable to bury an isolated charged sidechain in a hydrophobic environment inside a protein. Conversely, buried salt bridges are very strong.

In general, mutations in buried residues threaten to destabilize protein structures more than mutations in surface residues. Conversion of a buried hydrophobic residue to a charged residue could well prevent a protein from folding.

- *Van der Waals forces and dense packing of protein interiors*: The packing of atoms in the protein interior contributes in two ways to the stability of the structure. One is the exclusion of non-polar atoms from contact with water—the hydrophobic effect. The other is the enhancement of the force of attraction between the protein atoms.

The observed cohesion of ordinary substances shows that there are *attractive* forces between atoms and molecules. Conversely, the fact that substances do not collapse completely—indeed that there are limits to how far they can be compressed—shows that at short range these forces must be *repulsive*. The most general type of interatomic force, the **Van der Waals force**, reflects this principle: the nearer the atoms, the stronger the force, until the atoms are actually 'in contact', at which point the forces become repulsive and strong (see Fig. 2.9).

To maximize the total cohesive force, therefore, as many atoms as possible must be brought as close together as possible. It is the requirement for dense packing that imposes a necessity for *structure* on the protein interior. It produces a jigsaw-puzzle-like fit of interfaces between elements

of secondary structure packed together in protein interiors (see Fig. 2.10). Such a complementarity of opposing surfaces is also responsible for the specificity with which many enzymes bind their substrates, as in Fischer's 'lock-and-key' analogy.

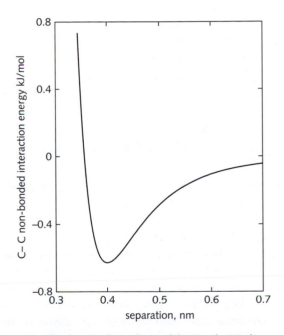

Figure 2.9 The distance dependence of the Van der Waals interaction between two non-bonded carbon atoms. This curve has a single minimum, rises very steeply at distances below the minimum, and rises more slowly above the minimum. If two carbon atoms were separated by the distance of minimum energy, attempts to push them closer together would be resisted by an increasingly strong force. Attempts to pull them apart would be resisted by an increasingly weak force.

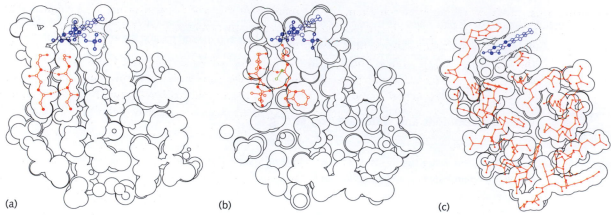

(a)   (b)   (c)

Figure 2.10 Serial sections through the β-sheet of flavodoxin [5NLL] show the dense packing of protein interiors. Flavodoxin contains a five-stranded β-sheet with two helices packed against each side, and the prosthetic group flavin mononucleotide (FMN). Each drawing shows three serial sections through Van der Waals envelopes of the atoms. The sections are cut 1 Å apart. The FMN is shown in blue, at the top of the picture, with broken contours. (a) Sections passing through the mean plane of the sheet. The mainchains of three residues from each of two adjacent strands of sheet are also shown in red. (b) Sections passing through the sidechains above the sheet (in this orientation). The mainchains *and sidechains* of three residues from each of two adjacent strands of sheet are also shown in red. The distal atoms of a methionine, from a helix packing against the sheet (shown in green), insert between sidechains of residues on the two adjacent strands. (c) Sections passing through two helices packed against the sheet (below the sheet in this orientation). In this drawing only the flavin ring of the FMN is shown. Atoms of the polypeptide are shown in red.

• An effect that *opposes* the formation of the native state is that it is thermodynamically unfavourable for all the molecules in a sample with a flexible chain to adopt the same conformation.

> • Because the formation of native states is the result of opposing forces, the stabilization of the native states of protein structures is relatively small, typically 20–60 kJ mol^{-1}. This is the equivalent of no more than about 2–3 hydrogen bonds.

The close packing shown by flavodoxin characterizes many but not all proteins. There are two classes of exceptions: (1) Many proteins have disordered regions, introduced in Chapter 1. (2) The conformations of some proteins are not sharply divisible into some tightly-packed and other disordered regions. The contrast and distinction may be fuzzier. In the 'molten globule' state of a protein there may be a general loosening of the packing, with maintenance of the overall topology of the folding pattern. Depending on the degree of loosening, water may or may not penetrate the protein interior.

Oleg Ptitsyn first hypothesized (and, inspiredly, named) the molten globule state as an intermediate in folding. However, there is now evidence that molten globule states can, under certain circumstances, be isolated in forms stable enough to study. Indeed some show ligand affinities, and even enzymatic activity comparable to the native, close-packed states of the same proteins. Cases in which the ligated forms become close-packed are examples of 'induced fit'. The enzyme–substrate complex is consistent with the 'lock-and-key' model. But in other cases, the molten globule persists in the complex, and the affinity constants are nevertheless comparable in magnitude with those of the native state.

A question posed by the rigid-looking packing in proteins such as flavodoxin is: How is such dense packing compatible with the observation that many proteins undergo conformational changes? For proteins with disordered regions, or at least flexible ones, or which are in the molten globule state, the problem is not so severe—they have 'room to manoeuvre'. But for highly structured proteins, it appears that the amino acid sequence is compatible with multiple conformations, often switching between them as a function of the state of ligation. The classic example is human haemoglobin, which has different native structures in the oxy- and deoxy- states. Such proteins are showing not general disorder, but two—or sometimes more—*discrete* close-packed states.

Structure determinations of particular proteins in two or more conformational states have allowed analysis of the types of conformational changes that proteins can achieve. With few exceptions, experiments give us the structure of the endpoints of a transition; for instance, ligated and unligated forms. In some cases, molecular dynamics can illuminate the trajectory—or trajectories—between them, by simulating the conversion between the initial and final conformations.

Conformational change

Many proteins undergo conformational changes as part of their activities. In many but not all cases binding of ligands triggers the change. For example, the enzyme citrate synthase binds substrate and cofactor in a cleft between two domains. In the unligated form, the two domains swing open to receive the ligands. In the ligated form, the domains close over the ligands, excluding water from the active site. Allosteric proteins show another class of conformational changes induced by ligand binding.

Structurally, various types of conformational changes have been observed.

- *Localized conformational changes in loop regions.* Loops are generally the most flexible regions in proteins. In enzymes, such as triose phosphate isomerase and p21 Ras, a change in state of ligation affects the conformation of local mobile regions.

- *Hinge motions reorientating two domains.* Lactoferrin shows a hinge motion in response to binding iron. Two domains remain individually rigid but change their relative orientation, by means of structural changes in only a few residues in the regions linking the domains (see Fig. 2.11). Hinge motion in myosin is responsible for the impulse in muscle contraction (see Chapter 5).

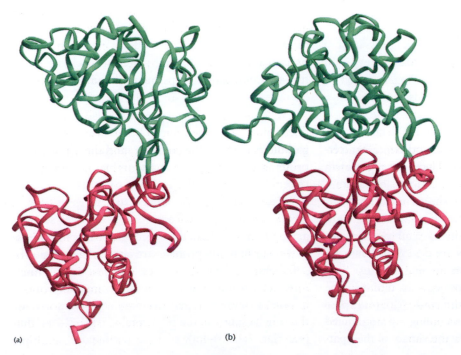

(a)   (b)

Figure 2.11 Lactoferrin is an iron-binding protein found in secretions such as milk and tears. It contains a single polypeptide chain of approximately 700 residues, folded into two lobes of related sequence and similar structure. Each lobe contains two domains. The binding of iron causes the N-terminal lobe to change from (a) an 'open' to (b) a 'closed' conformation. This figure shows the two conformations of the N-terminal lobe. The individual domains are rigid during the conformational change. The only residues that alter their conformation are those at the hinge regions between the two domains. [1LFH, 1 LFG].

- *Cumulative shifts of secondary structures packed at interfaces.* As Fig. 2.11 shows, hinge motion is possible only if the domains are free to move as rigid bodies. In some proteins, a well-packed and extensive interface constrains the relative motion of two domains that are to move with respect to each other. A rule of thumb states that two packed helices can move by no more than about 2 Å with respect to each other, by small readjustments of sidechain conformations. Think of this 2 Å threshold as the limit of plastic deformation of a helix interface. In order to achieve a large structural change, relative motions of 2 Å in individual pairs of helices can be coupled to give large cumulative effects. The ligand-induced conformational change in citrate synthase is an example of a large motion that is the resultant of several small motions at individual helix–helix interfaces.

- *Changes in the relative geometry of subunits.* Many changes, including allosteric changes, involve changes in quaternary structure. Haemoglobin shows a relatively small change—a rotation of about 15° of one $\alpha\beta$ dimer with respect to the other. ATP synthase shows a very large change—it is a rotatory motor in which the γ subunit turns with respect to the $\alpha_3\beta_3$ hexamer, successively inducing conformational changes in the subunits of the hexamer (see Chapter 5). The GroEL–GroES complex is a large reciprocating motor that depends for its activity on a crucial conformational change in the GroEL protein. GroEL–GroES is a 'chaperone' complex that rescues misfolded proteins (see Chapter 8).

- *A few proteins appear to have more than one possible folding pattern, with different topologies.* These include the prion protein and the serpins (see Serpins, Chapter 5).

- *Amyloid aggregates.* Many proteins can form fibrillar aggregates with structures different from their free native state. In at least some cases the structure adopted has a 'crossed-β' form (see Fig. 6.4).

> - Many proteins undergo conformational changes as part of the mechanisms of their normal activities. In most cases the different conformations retain the same folding pattern. Serpins and prion proteins are exceptions, in which the topology of the tertiary structure changes. Proteins can also change conformation to adopt a common β-sheet structure in amyloid fibril formation.

Protein folding patterns

Analysis of protein structures, like connoisseurship in art, requires both a sensitive and trained eye, and technical scientific analysis.

The polypeptide chains of proteins in their native states describe graceful curves in space. These are best appreciated by temporarily ignoring the detailed interatomic interactions and focusing on the calligraphy of the patterns. The major differences among protein structures come not at the level of local interactions, such as formation of helices and sheets; in most proteins these are quite similar. The differences appear rather at the level of the three-dimensional assembly of regions, in which similar substructures are deployed differently in space to give different protein folding patterns.

Study of protein folding patterns in terms of the curves the mainchain traces out in space can (1) expose recurring structural patterns, including but not limited to helices and sheets, and show the variety of their patterns of combination; (2) provide an approach to comparison of different folding patterns, to clarify structure–function relationships and evolutionary pathways.

Supersecondary structures

In addition to helices and sheets themselves, many proteins show recurrent patterns of interaction between helices and sheets close together in the sequence. These supersecondary structures appear in the hierarchy of protein architecture between secondary and tertiary structures. Small structural patterns that recur in many proteins are also known as motifs.

Supersecondary structures are *local* structures, formed by residues in a contiguous segment of the sequence. Common supersecondary structures include the α-helix hairpin, the β-hairpin, and the β–α–β unit (see Figs. 2.12(a)–(c)).

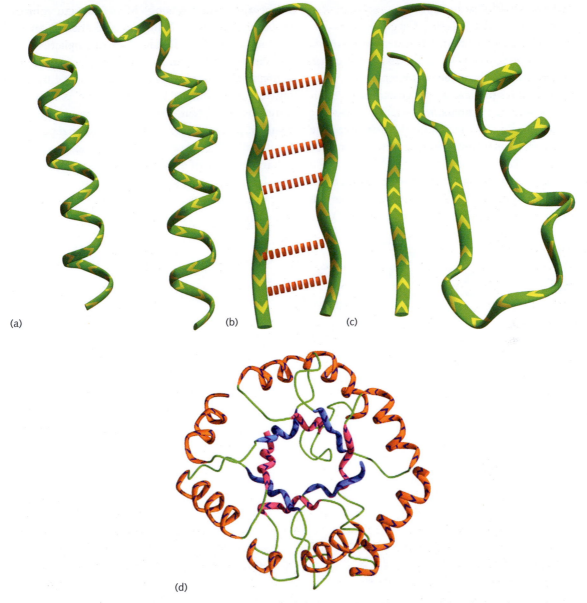

Figure 2.12 Common supersecondary structures: (a) a helix hairpin, (b) a β-hairpin, (c) a β–α–β unit. (d) Spinach glycolate oxidase is an example of a very common β-barrel structure (a β–αbarrel) formed from a succession of β–α–β units. The eight-stranded sheet is closed into a cylinder. Alternate strands in the β-barrel are shown in blue and pink, with dark blue chevrons. Helices are shown in orange, with dark blue chevrons [1GOX].

A succession of β-hairpins can generate an extended antiparallel sheet. By a succession of β–α–β units, a protein can form a very common structure in which eight strands of a parallel sheet form a closed cylindrical array, strand 8 hydrogen bonded to strand 1 (as well as to strand 7) (see Fig. 2.12(d)). The structure is called a β–α barrel. The helices are arranged around the outside of the barrel. (An example of a β-barrel formed from an antiparallel sheet will appear in Fig. 2.28.)

• Supersecondary structures are interacting combinations, of secondary structures that appear successively in the chain. (Interactions may include hydrogen bonding, or Van der Waals packing.) Common examples include: α-helix hairpins, β-hairpins, β–α–β units, and β–α barrels containing several β–α–β units.

An album of small structures

The basic approach to parsing a protein structure is learning to recognize helices and sheets, supersecondary structures, and their patterns of interaction. Many molecular-graphics programs draw cartoons—representing α-helices as cylinders and strands of β-sheet as large arrows (as in Fig. 1.10(c)). This takes the work—and the fun!—out of analysing pictures of protein structures.

Figure 2.13 contains a selection of small protein structures or domains. These pictures illustrate the structural themes that we have discussed. Quite deliberately, the 'parsing' of the structure by representing helices as cylinders and strands of sheet as thick arrows is not done in these pictures. Appreciation of the form, and assembly of the components, of these structures that will gained by studying and analysing the pictures 'by eye' is well worth the effort. The reader is urged to trace the chains visually, picking out the helices and sheets, and disulphide bridges (are the residues they link close or distant in the sequence?) Can you see supersecondary structures? Consider these pictures as exercises in training your eye to recognize the important spatial patterns.

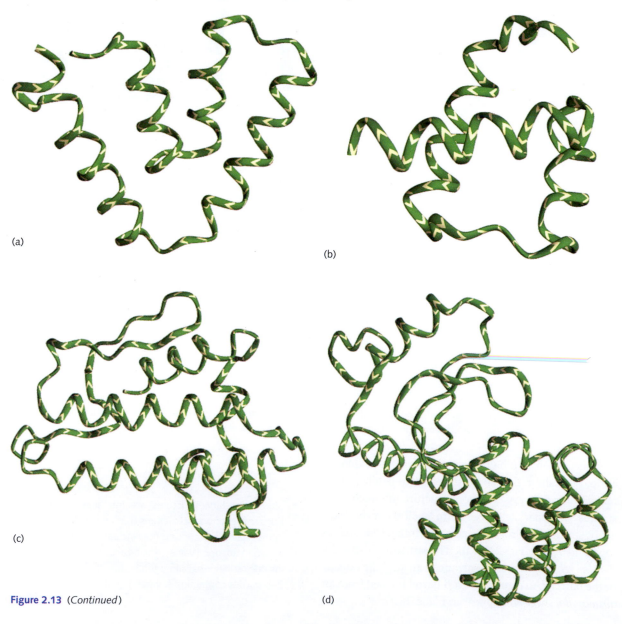

(a)

(b)

(c)

(d)

Figure 2.13 (*Continued*)

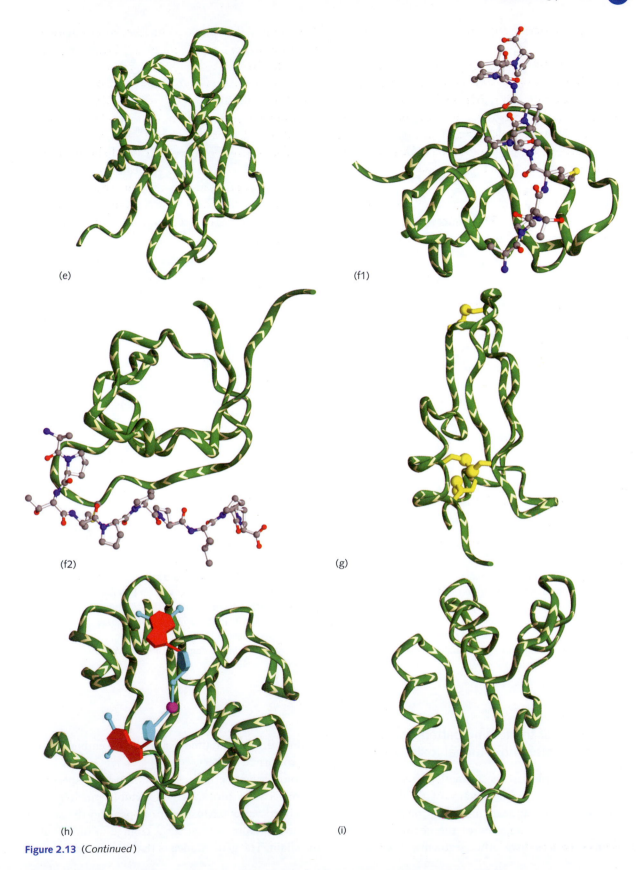

(e)

(f1)

(f2)

(g)

(h)

(i)

Figure 2.13 (*Continued*)

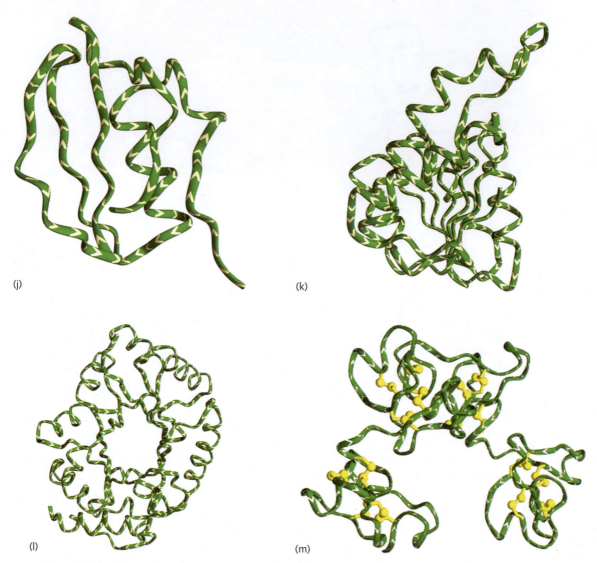

(j)

(k)

(l)

(m)

Figure 2.13 An album of small protein domains, which will repay study. 'To see is itself a creative operation, requiring an effort.'—H. Matisse. (a) Uteroglobin [1UTG]. (b) Engrailed homeodomain [1ENH]. (c) Phospholipase A₂ [1BP2]. (d) T4 Lysozyme [3LZM]. (e) Timothy grass pollen allergen [1WHO]. (f) Abl tyrosine kinase/peptide complex (two views) [1ABO]. (g) Bovine pancreatic trypsin inhibitor [5PTI]. (h) Ribonuclease T1 [2RNT]. (i) Ribosomal protein L7/L12 [1CTF]. (j) Barley chymotrypsin inhibitor [2CI2]. (k) Malate dehydrogenase, NAD-binding domain [1MLD]. (l) Alanine racemase, N-terminal domain [1SFT]. (m) Wheat-germ agglutinin [9WGA].

Comparison of the folding patterns of acylphosphatase and a fungal toxin

Cow acylphosphatase is a small protein—only 98 residues—but it contains many of the basic structural themes that appear in larger proteins also (see Fig. 2.14(a)). Two regions of the chain (blue) curl up into α-helices. Five other segments, in which the chain is drawn out almost straight, assemble side-by-side to form a β-sheet.

Connecting the helices and strands of sheets are regions called **loops**, which exhibit much greater structural variety. Most loops appear on the surface of the structure, and effect a change in direction of the chain. The general idea is that helices and strands

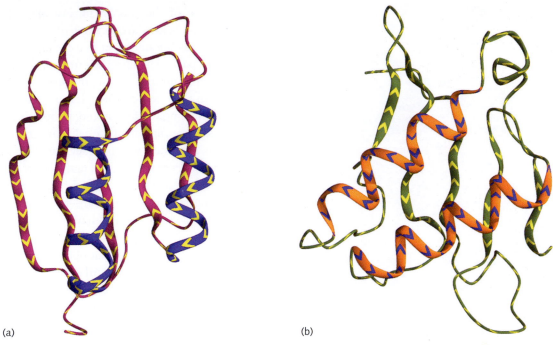

(a)   (b)

Figure 2.14 Comparison of the folding patterns of two small proteins. (a) Cow acylphosphatase [2ACY]. (b) Viral toxin from corn smut fungus *(Ustilago maydis)* [1KPT]. Although there are many superficial similarities between the folding patterns of these two proteins, they have different topologies (unlike the two related proteins in Figure 1.4).

of a sheet pass through the protein from one side to the other, then the chain turns around through a loop and begins another helix, or strand of sheet, to pass through the structure again. Many small proteins follow this pattern; larger proteins are often more complex.

Figure 2.14(b) shows another small protein, KP4, a toxin from corn smut fungus *Ustilago maydis*, encoded by a symbiotic virus. This 105-residue protein also contains two helices in front of a sheet. As in acylphosphatase, the helices are close enough to the sheet that sidechains from the helices and strands of sheet are in contact; we say that the helices are packed against the sheet.

In both proteins the helices are about four turns long (~15 residues) and the strands about 10 residues long. (These lengths are typical, although helices and sheets show great variation in length even in globular proteins—a helix in myosin is over 50 residues long. Helices in fibrous proteins can be much longer.) Indeed, much of the *local* mainchain structure is very similar in these and many other proteins. In fact, in these two structures even the *order* along the se-

quence of the elements of secondary structure is virtually the same:

$$\beta_1 - (\alpha) - \alpha_1 - \beta_2 - \beta_3 - \alpha_2 - \beta_4 - \beta_5,$$

in which the helices and strands of sheet are numbered according to their positions in the *sequence*, not the structure. The (α) in parentheses indicates a short turn of helix within the first loop of the toxin structure. (Can you see it?) This is followed, within that loop, by the full-blown helix α_1.

In acylphosphatase the axes of the helices are roughly parallel to the strands of the sheet. In the toxin, the helix axes are both tipped by about 45° to the strands of the sheet. The reason why the helices appear approximately parallel to each other, in both cases, is that this arrangement provides a large interface for the packing together of the helices.

In these pictures, the structures look empty. This is because only the tracing of the mainchain is shown. A picture containing all the atoms makes it clear that the structures are fairly compact globules of atoms (see Fig. 2.15; see also Fig. 2.10).

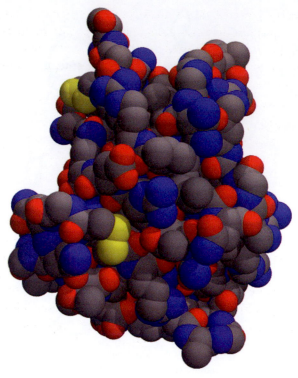

Figure 2.15 All-atom representation of corn smut fungal toxin [1KPT]. In contrast to the representation of this protein in Figure 2.14b, this picture shows the compactness of the packing of the structure, and the topography of the surface; but it would be difficult to trace the chain in this picture. Can you see an α-helix? (Not easy, but it is there.) Most people find the representation in Figure 2.14 more attractive. It is certainly more perspicuous. But as Figure 2.15 shows, it is in some essential respects inaccurate. One cannot help being reminded of Yevtushenko's (admittedly politically incorrect) remark: 'Translation is like a woman. If it is beautiful, it is not faithful. If it is faithful, it is most certainly not beautiful.'

Let us trace the chains through the two structures. So far, we have emphasized the similarities of the structures. As we examine the correlation between the order of secondary structural elements in the sequence and their relative disposition in space, we shall see that the helices and strands of sheet are 'wired up' differently.

> Before reading the following paragraphs the reader must do 'warm-up' exercises by visually tracing the chains in Fig. 2.14.

In acylphosphatase, strand 1—the first strand in order of appearance in the *sequence*—is second from the left (in the orientation depicted. The reader must be careful to distinguish statements that refer only to the orientation illustrated, from those characterizing the structure of the protein intrinsically. The statement that the helices are in front of the sheet is true of this picture. The statement that the protein contains two helices packed against the same surface of a sheet would be true no matter how the structure were orientated.) In the toxin the first strand is the rightmost strand. Following the first strand, in both proteins the chain passes through the first helix to the strand second in the sequence, which in acylphosphatase is the fourth strand from the left, and in the toxin the first on the left. Note that the intrastrand region in the toxin has the longest distance to cover—from the right edge of the molecule to the left. Perhaps this is a partial explanation of why the helices are roughly parallel to the strands in acylphosphatase but oblique in the toxin.

In acylphosphatase the third strand is adjacent to the second strand, and occupies the position third from the left in the sheet. In the toxin the third strand is also adjacent to the second but on its *right*, second from the left in the sheet. This combination of two adjacent antiparallel strands, connected by a loop, occurs frequently in β-sheets. It is called a *β-hairpin* (see Fig. 2.12(b)).

After the third strand, in both structures the chain passes through the second helix to the fourth strand. This is the leftmost strand in acylphosphatase, and fourth from the left in the toxin. Then, in acylphosphatase the chain takes a long excursion to the fifth and last strand, the rightmost. In the toxin, the fourth strand is connected by a second hairpin to the fifth strand, the third from the left.

In the toxin every pair of adjacent strands points in opposite directions—for instance, strand 3 points up and strand 4 points down. This is an *antiparallel β-sheet* (see Fig. 1.9(c)). One way to construct an antiparallel β-sheet would be to assemble a succession of hairpins, in which the order of the strands in the sequence and the structure would match. (The sheet in the toxin, although antiparallel, is not constructed in this manner.) In other proteins, sheets occur with all strands pointing in the same direction, called *parallel β-sheets* (see Figs. 1.9(b), and 2.12(c, d)). The sheet in acylphosphatase is neither parallel nor antiparallel. It is a *mixed β-sheet*. β-sheets in proteins are free to mix parallel and antiparallel strands.

These relationships can be summarized in diagrammatic representations of the topologies of (a) acylphosphatase and (b) the toxin.

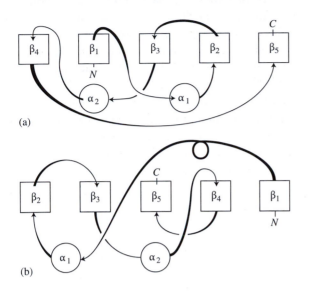

(a)

(b)

in the superposition of the structures of *E. coli* and *S. faecalis* histidine-containing phosphocarrier proteins in Fig. 1.4. This implies that it would not be easy for evolution to interconvert these two structures, and therefore that they are not related.

- It is difficult to decide whether or not two proteins that show significant structural similarity are related. A crucial question: are the differences such that there is no simple evolutionary pathway between them?

We conclude that acylphosphatase and the viral toxin are two *non-homologous* proteins that, nevertheless, have very nearly the same secondary structure but different tertiary structures. An illustrative example. The reader is urged to study the figures, and verify, visually, the verbal description and comparison of these structures presented in this section.

- Compare this example, of two proteins with similarities in secondary and tertiary structure, but which could not be globally superimposed, with the example of Fig. 1.4, in which the two proteins not only have similar secondary structure, but also could be globally superimposed.

There is an interesting substructural symmetry in acylphosphatase. The first strand (in the sequence) is followed by a helix, followed by a second strand that is positioned two strands to the right of the first strand. The third strand (in the sequence) begins the same pattern, but upside down in this picture. A variant of this pattern appears in the toxin: a strand followed by a helix followed by a strand positioned four strands to the *left*.

It would not be possible to overlay the structures of acylphosphatase and the toxin by rigid motions as

Classification of protein structures

When Linnaeus created his classification of living things, he made a catalogue of similarities among the objects in the corpus of material he was studying. Only later did it emerge that the hierarchy of relationships that Linnaeus observed was indeed induced by evolutionary processes and reflected biological kinship.

With protein structures, we can observe similarities in topology, or even in structural details. But only in some cases can we infer genuine biological relationships in the sense of descent from a common ancestor. The alternative is 'convergent evolution'—the independent generation of favourable structural features. Given two protein structures with apparently

similar conformations, but dissimilar function and no clear amino acid sequence similarity, it is usually very difficult to distinguish homology from convergence.

But not always. There are many well-characterized protein 'families'. For example, we know the structures of globins from several hundred species, and the sequences of over 5000. Analysis of the sequences and structures of the globins and other protein families shows that one can construct evolutionary trees on the basis of molecular data from related molecules in different species. These are in most cases equivalent in their branching structure to the evolutionary trees constructed by the classical taxonomic methods of comparative anatomy, embryology, and the fossil record.

It may be illuminating to pursue the analogy with biological classification a little further. Comparing human haemoglobin with dog haemoglobin would be like comparing the skeleton of a human hand with the skeleton of a dog's forepaw; it is well known that the skeletons contain a bone-for-bone correspondence, with a topologically similar spatial relationship, but that quantitatively the sizes, shapes, and exact arrangement of the corresponding bones differ. But comparing human haemoglobin with human (or dog) chymotrypsin is like comparing a lung with a stomach. Only with an understanding of the evolutionary history can one assess the nature of a relationship. Superficial similarities in structure and function may not reflect a true kinship—for example, the gross anatomical features of the eye of a human and the eye of an insect arose independently. Conversely, divergent evolution may conceal a homology—for example, bones of the human ear are derived from bones in the jaws of fishes. Similar problems lurk among the similarities of protein structures.

Databanks of protein structure classifications

There are now so many known protein structures that we need directories and catalogues to keep track of them. 'White pages,' in which the structures are listed by name, are useful provided we know what we are looking for. More generally helpful are 'Yellow pages,' which organize the entries in some reasonable classification.

In 1976, M. Levitt and C. Chothia proposed the first general classification of protein folding patterns. This was based on the 31 protein structures known then! (See Table 2.3. The examples mentioned in the third column are illustrated in this chapter.) Because the individual domains in multidomain or modular proteins often have different structures, each domain must be classified individually.

SCOP (Structural Classification of Proteins), CATH (Class, Architecture, Topology, Homologous superfamily), the DALI Database, and ECOD (Evolutionary Classification of protein Domains) offer classifications of protein structures. These websites have many useful features such as information-retrieval engines, including search by keyword or sequence, presentation of structure pictures, and links to other related sites including bibliographical databases.

SCOP

SCOP (http://scop.mrc-lmb.cam.ac.uk/scop/) organizes protein structures in a hierarchy according to evolutionary origin and structural similarity. It is based on protein domains rather than full protein structures. At the lowest level of the SCOP hierarchy, then, are the individual *domains*, extracted from the Protein Data Bank entries. Sets of domains are grouped into *families* of homologues. (homologues = proteins related by evolutionary descent from a

Table 2.3 Classification of protein folding patterns according to secondary and tertiary structure

Class	Characteristic	Examples
α-helical	secondary structure exclusively or almost exclusively α-helical	myoglobin, cytochrome c, uteroglobin
β-sheet	secondary structure exclusively or almost exclusively β-sheet	timothy grass pollen allergen
α + β	α-helices and β-sheets separated in different parts of molecule. Absence of β–α–β supersecondary structure	T4 lysozyme, staphylococcal nuclease
α/β	helices and sheet assembled into β–α–β units	
α/β-linear	line through centres of strands of sheet roughly linear	flavodoxin; NAD-binding domain of malate dehydrogenase
α/β-barrels	line through centres of strands of sheet roughly circular	glycolate oxidase, alanine racemase
Little or no secondary structure		wheat-germ agglutinin

common ancestor.) These comprise domains for which the similarities in structure, function and sequence imply a common evolutionary origin. Families that share a common structure, or even a common structure and a common function, but lack adequate sequence similarity—so that the evidence for evolutionary relationship is suggestive but not compelling—are grouped into *superfamilies*. Sometimes the subsequent discovery of an intermediate form—a 'missing link' between two members of a superfamily—strengthens the evidence for true homology. This permits merging the superfamily structures into a family.

Superfamilies that share a common folding topology, for at least a large central portion of the structure, are grouped as *folds*. Finally, each fold group falls into one of the general *classes*. The major classes in SCOP are α, β, $\alpha + \beta$, α/β, and 'small proteins', which often have little secondary structure and are held together by disulphide bridges or ligands (for instance, wheat-germ agglutinin). The first four classes follow the original scheme of Levitt and Chothia.

Figure 2.16 shows the SCOP classification of flavodoxin from *Clostridium beijerinckii*.

To give some idea of the nature of the similarities expressed by the different levels of the hierarchy, Fig. 2.17 compares a flavodoxin with other protein domains that are classified, in SCOP, into the same superfamily, fold and class. Flavodoxin from *Clostridium beijerinckii* [5NLL] and NADPH-cytochrome P450 reductase [1AMO] are in the same superfamily, but different families. Flavodoxin and the signal-transduction protein CHEY [1CHN] are in the same fold category, but different superfamilies. Flavodoxin and spinach ferredoxin reductase [1FNB] are in the same class—(α/β)—but have different folds.

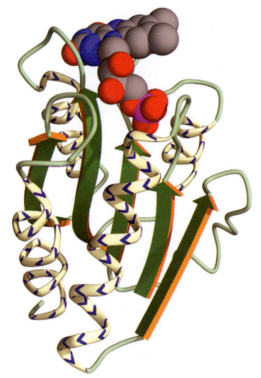

Class	α and β (α/β)
Fold	Flavodoxin-like
core contains a-helices packed onto both sides of a sheet to form a three-layered α–β–α structure; β-sheet contains five parallel strands, with (from left to right) order of appearance in the sequence: 21345	
Superfamily	Flavoproteins
Family	Flavodoxin
Protein	Flavodoxin
Species	*Clostridium beijerinckii*
Domain	PDB entry 5nll

Figure 2.16 *Clostridium beijerinckii* flavodoxin [5NLL] and its classification in SCOP.

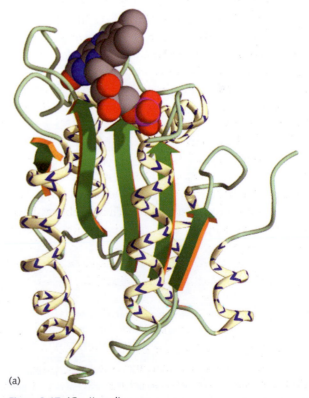

(a)

Figure 2.17 (*Continued*)

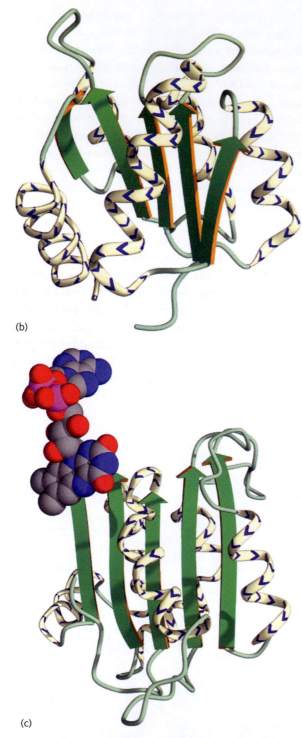

(b)

(c)

Figure 2.17 (a) NADPH-cytochrome P450 reductase [1AMO]. This is in the same SCOP superfamily as flavodoxin, but they are in different SCOP families. (b) Signal-transduction protein CHEY [1CHN]. This is in the same SCOP fold category as flavodoxin, but they are in different SCOP superfamilies. (c) Spinach ferredoxin reductase [1FNB]. This is in the same SCOP class as flavodoxin, but they are assigned to different SCOP folds.

The February 2009 release of SCOP classified 110 800 domains from 38 221 PDB entries (see Table 2.4).

Table 2.4 The SCOP hierarchy (February 2009)

Level	Number of cases
Class	7
Fold	1195
Superfamily	1962
Family	3902
Domain	110 800

SCOP2

There is consensus that SCOP has made a major contribution to our understanding of protein structure and evolution. Testimony includes the plaudits of its user community and the fidelity of its imitators. The products of this very success, plus the growth in quantity and variety of structural data, have guided the way forward.

SCOP2 (scop2.mrc-lmb.cam.ac.uk) is a redesigned successor to SCOP. In presenting SCOP2, Andreeva et al.*write:

> '... it has become clear that relationships between proteins are more complex than anticipated and that protein evolutionary pathways do not always conform to the same rules. The vast amount of structural information also provided new insights into the mechanisms underlying molecular recognition and evolution of protein structure.
>
> ...
>
> '... Therefore, we endeavor to develop a more advanced framework for presentation of protein relationships, a new classification scheme that can be adapted to any particular case and evolutionary scenario.'*

SCOP2 is a Directed Acyclic Graph, in which the nodes represent relationships between selected regions of proteins. This form of graph is similar to those of The Gene Ontology Consortium classifications (see Chapter 7), but represents a substantial difference from SCOP, in which the graph was a tree. In SCOP2, nodes can have multiple 'parents', allowing for a richer network of relationships. For instance, in SCOP2, relation-

*Andreeva, A., Howorth, D., Chothia, C., Kulesha, E. and Murzin, A. (2014). SCOP2 prototype: a new approach to protein structure mining. *Nucl. Acid Res.*, **42**, D310–D314.

ships indicating structural similarity are separate from evolutionary events.

The lowest-level type of node in the SCOP2 graph is the *Species*. A *Species* corresponds in almost all cases to the full amino acid sequence of an individual gene product from a particular species, linked to UniProt. One *Species* may correspond to many wwPDB entries—for instance, the wwPDB contains 239 sperm whale myoglobin structures.

In SCOP2, different substructures extracted from any *Species* can participate in different evolutionary or structural relationships. This is an important generalization from SCOP.

In the SCOP2 graph, some edges connect *Species* nodes with other relationship nodes. The edges select regions of the *Species* that participate in, for example, relationships linking regions from different *Species* showing a common *Fold*. Thus the protein corresponding to one *Species* may contain several structural domains assigned to different *Folds*. A *Family* region can comprise several of these structural domains, provided that they occur in the same arrangement in all members of the *Family*. In addition, each of these structural domains may participate individually, or in combinations with others, in different *Superfamily* relationships. The substructure corresponding to a *Family* is typically longer than the substructures that correspond to *Superfamilies*.

Other edges in the graph link general categories of relationships with the more specific ones that they contain. For instance, *Evolutionary relationships* contains: *Hyperfamily, Superfamily, Family, Protein,* and *Species*. (A hyperfamily in SCOP2 is a region—perhaps fragmentary—shared by different superfamilies.)

Two presentations of SCOP2 available on the website are *Browser mode* and *Graph mode* (scop2.mrc-lmb.cam.ac.uk). Both are navigable by selection using the mouse. SCOP2 links to a molecular graphics viewer and to UniProt (that is, to the subset of UniProt containing proteins for which structural data are available). The description of SCOP2 appearing in current protocols contains a number of scripts for sessions, with the authors holding your hand as you go through them.* Readers should work through these examples.

*Andreeva, A., Howorth, D., Chothia, C., Kulesha, E. and Murzin, A.G. (2015). Investigating protein structure and evolution with SCOP2. *Curr. Protoc. Bioinform.* 49, 1.26.1–1.26.21

CATH

CATH (http://www.cathdb.info/) presents a classification scheme similar to that of SCOP. The letters in its name stand for the levels of its hierarchy: Class, Architecture, Topology, Homologous superfamily. In the CATH classification, proteins with very similar structures, sequences and functions are grouped into *sequence families*. An *homologous superfamily* contains proteins for which there is evidence of common ancestry, based on similarity of sequence and structure. A *topology* or *fold family* comprises sets of homologous superfamilies that share the spatial arrangement *and* connectivity of helices and strands of sheet. In CATH, *architectures* are groups of proteins with similar arrangements of helices and sheets, but with different spatial layout. For instance, different four-α-helix bundles with different layout would share the same architecture but not the same topology in CATH (see Fig. 2.18). Finally, the general *classes* of architectures in CATH are: α, β, $\alpha - \beta$ (subsuming the $\alpha + \beta$ and α/β classes of SCOP), and domains of low secondary structure content.

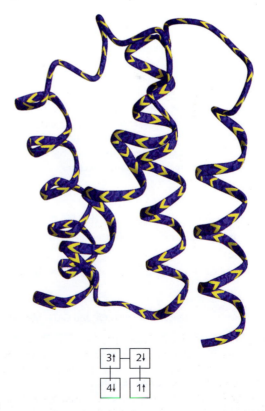

(a)

Figure 2.18 (*Continued*)

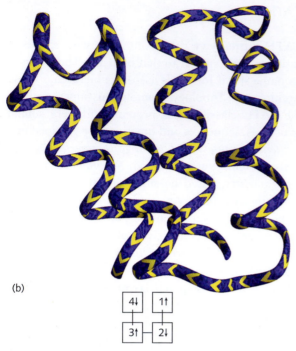

(b)

Figure 2.18 Two four-α-helix bundles with similar arrangements of helices but different spatial layout. (a) Domain from signal sequence recognition protein FFH [1FFH]. (b) Region from T4 lysozyme [1LYD]. These structures could not be globally superimposed. In the structural pictures the axes of the helices are approximately *parallel* to the plane of the page. The squares show the layout of the four helices, as projected onto a plane *perpendicular* to the page, and looking *down* onto the page. The arrows show the direction of the chain in the helix, as in the structure pictures. Lines connecting squares enter the squares if they are at the top of the domains and are occluded by the squares if they are at the bottom of the domains. Note, for example, that in Fig. 2.17(a) the first and fourth helices point in the same direction (N → C going up the page) whereas in Fig. 2.17(b) the first and fourth helices point in opposite directions.

The correspondence between the levels of the hierarchy in SCOP and CATH is rough but not exact (see Table 2.5).

CATH is developing a focus on protein function.

Table 2.5 Correspondence between SCOP and CATH hierarchies

SCOP	CATH
Class	Class
Fold	Architecture
Superfamily	Topology/Homologous superfamily
Family	Sequence family
Domain	Domain

The DALI Database

Both SCOP and CATH emphasize the *presentation* of the results of the classification. The methods for comparing proteins that underly the classification remain internal to the project. A third classification of protein structures, the DALI Database (http://ekhidna.biocenter.helsinki.fi/dali/start), is built around a method for comparing proteins. Application of the method to the entire PDB induces a complete classification of known protein structures.

The program DALI, by L. Holm and C. Sander, provides a general and sensitive method for comparing protein structures. Given two sets of coordinates, it determines the maximal common substructure and provides an alignment of the common residues. DALI is among the best of many programs that address this problem, because of its ability to recognize distant relationships, and its speed of execution. DALI is fast enough to scan the entire PDB routinely for proteins similar to a probe structure.

Crystallographers and NMR spectroscopists who solve a new structure routinely submit the coordinates to the DALI server, to detect similarities to known structures. If the function of the new protein is unknown, a successful hit will often but not always permit a good guess, or at least suggest experiments.

Application of DALI to the entire Protein Data Bank produces a classification of protein structures, the DALI Database.

Do the different classification schemes agree? Recognize that to classify protein structures (or any other set of objects) you need to be able to measure the similarities among them. The measure of similarity induces a tree-like representation of the relationships. CATH, SCOP, DALI and the others agree, for the most part, on what is similar, and the tree structures of their classifications are therefore also similar. However, what an objective measure of similarity does *not* specify is how to define the different levels of the hierarchy. These are interpretative decisions, and any apparent differences in the names and distinctions between the levels only disguise the underlying general agreement about what is similar and what is different.

- Several websites offer classifications of protein structures, including SCOP, CATH and the DALI Database. The construction of SCOP2 to supersede SCOP is in progress.

A survey of protein structures and functions

General classes of protein function, and the types of structures that provide them are described in the following sections.

Fibrous proteins

Traditionally, textbook presentations of **fibrous proteins** begin with hair and silk. These are familiar from daily life, and have the longest history of scientific study, stimulated by the textile industry. Hair and silk give the impression that fibrous proteins have purely structural roles. However, *intracellular* fibrous proteins show that the biological roles of fibrous proteins should be thought of as mechanical, rather than structural. Fibrous proteins are in many cases dynamic rather than passive.

Eukaryotic cells contain three types of internal fibrous structures:

1. *Actin-containing microfilaments* are thread-like fibres, 3–6 nm in diameter, composed pre-dominantly of polymers of actin. Microfilaments associated with myosin are responsible for muscle contraction.

2. *Microtubules* are cylindrical tubes ~25 nm in diameter, polymers of α- and β-tubulin, involved in mitosis, motility, and transport. Cilia and flagella are assemblies of microtubules. Microtubules provide 'tracks' for the movements of motor proteins such as dynein and kinesin.

3. *Intermediate filaments* create an intracellular three-dimensional network between the nucleus and outer cell membrane, determining the shape, form, and mechanical properties of the cell. They are a dynamic assembly, breaking up and reforming during the cell cycle.

FtsZ is a filamentous protein involved in cell division in prokaryotes. It is the major component of the septum that appears between the nascent progeny cells and pinches off to separate them. FtsZ occurs in all bacteria and archaea, and in chloroplasts but not mitochondria. It is related to the eukaryotic protein tubulin.

The FtsZ protofilament contains a linear array of monomers (see Fig. 2.19). Protofilaments combine to form filaments.

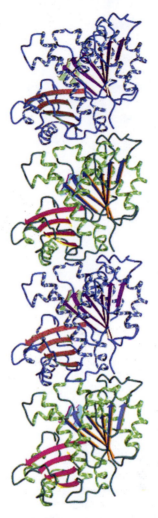

Figure 2.19 Model of the FtsZ protofilament, based on the crystal structure of the *Methanococcus jannaschii* monomer [1FSZ], four subunits assembled by homology with tubulin. I thank J. Löwe and L. Amos for the coordinates.

Traditional fibrous proteins

When you look at other people, what you see is mostly fibrous protein. Their hair contains α-keratin, a rope assembled from a **coiled-coil** of α-helices, a type of structure seen also in transcription activators (see Fig. 2.20). (Helices are a common structural theme in biology: see Box 2.7.) The helical section of a single subunit of α-keratin typically contains ~300 residues in each of the two strands, and is 48 nm long. There are also small capping domains at head and tail. These units assemble into

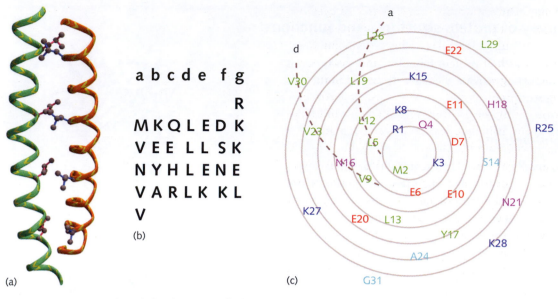

(a)

a b c d e f g
            R
M K Q L E D K
V E E L L S K
N Y H L E N E
V A R L K K L
V

(b)

(c)

Figure 2.20 (a) The coiled-coil structure of α-keratin also appears in the eukaryotic transcriptional activator GCN4 [2ZTA]. This structure contains two helices coiled around each other. It is known as the **leucine zipper** because the leucine repeats every seven residues (shown in ball-and-stick representation). The pitch is 14.7 nm. (b) The sequences of most proteins containing coiled-coils show *heptad repeats*—seven-residue patterns, the positions labelled *abcdefg*, in which the first and fourth positions (a and d) are usually hydrophobic. (c) If the sequence of GCN4 is plotted on a *helical wheel* corresponding to the geometry of a straight α-helix, the hydrophobic positions form a stripe up one face of the helix. (Colour coding: green, medium-sized and large hydrophobic; cyan, small; magenta, polar; blue, positively charged; red, negatively charged.) The a and d positions are indicated in this figure. The supercoiling of the helices around each other brings the a and d positions into even better register than suggested by this figure (see Problem 2.1).

BOX 2.7

Describing the geometry of a helix

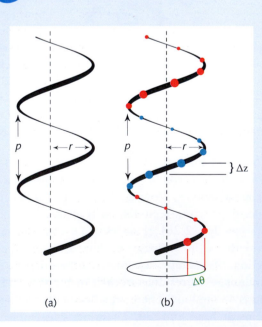

(a)   (b)

Left: A helix formed by the coiling of a continuous line. The broken line indicates the helix axis. The structure is described completely by two numbers: the pitch p = the distance along the axis between successive turns, and the radius r; and the **enantiomorph** or 'hand': right- or left-handed. Right: A helix formed by assembly of discrete units, such as the bases in one of the strands of the DNA double helix, or the amino acids in a polypeptide helix. The structure is defined by the pitch, p, the radius, r, and the rise per residue, Δz. The number of residues per turn is equal to $p/\Delta z$. The angular difference between successive residues is $\Delta\theta = 360° \times \Delta z/p$. In this diagram there appear to be seven residues per turn (count the light blue dots). However, the number of residues per turn need not be a whole number.

More complicated types of helix geometry include two or more helices wound around each other, as in DNA; or supercoiling, as in leucine zippers (see Fig. 2.20) or collagen, (see Fig. 2.21), or circular DNA (see Chapter 5.)

large aggregates, embedded in a matrix of cysteine-rich proteins to form a network, extensively cross-linked by disulphide bridges. The 'permanent wave' treatment of hair first breaks the cross-links using a reducing agent; then after setting the hair into the desired conformation (that is, *macroscopic* conformation), the links are reformed by a mild oxidizing agent, usually dilute hydrogen peroxide.

The horny outer layer of the skin, and fingernails, are other forms of keratin, differing in amino acid composition and sequence. Fingernails are less flexible than skin or hair because of a greater abundance of cysteine residues forming cross-links. The layer of the skin just beneath the surface contains collagen, a glycoprotein (see Fig. 2.21 and Box 2.8). The cornea of the eye is another form of collagen.

Many traditional clothing materials—before the development of synthetic fibres—are also protein. Woollen clothes are α-keratin—made from the hair of sheep or other animals. (Cotton and linen, in contrast, are plant fibres, formed primarily of cellulose.) Animal hair is of course chemically similar to human

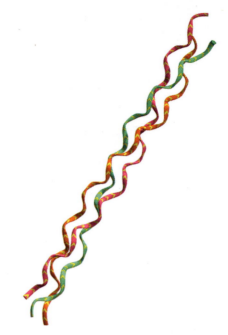

Figure 2.21 The structure of collagen, a three-stranded supercoil formed by plaiting together three polypeptide chains. In each strand, the rise per residue is approximately 0.3 nm. Each polypeptide chain itself forms a helix, with approximately 3.3 residues per turn. The repeat distance of the supercoil is approximately 1 nm [1BKV].

BOX 2.8   **Proteins in health and disease**

Collagen

The many different types of collagen have a common basic structure, a triple-stranded helix, with three polypeptide chains coiled around one another in a braid or plait (see Fig. 2.21). Individual chains are each ~1000 amino acids long. They have a glycine every *third* residue—the sequence scheme is (Gly-X-Y)*n*. X and Y are mostly alanine or special modified amino acids. X is often a 3- or 4-hydroxyproline and Y a hydroxyproline or hydroxylysine. In hydroxylysine the hydroxyl is on the Cδ atom, making it chiral. The prolines and hydroxyprolines effectively prevent the chains from forming α-helices.

Individual molecules assemble into fibrils in different ways suitable for the differing mechanical requirements of the tissues in which they appear. Twelve types of human collagen have been distinguished, encoded by 28 genes on 12 different chromosomes. The differences in amino acid composition and sequence govern their modes of processing and as-

sembly. Hydroxylysines provide sites of linkage to sugars or short polysaccharides. Covalent cross-links are formed after lysine or hydroxylysine sidechains have been oxidized enzymatically to aldehydes in the extracellular space. These aldehydes react either with other aldehydes or amino groups of lysine or hydroxylysines on neighbouring collagen chains. The cross-links contribute to the mechanical strength of the fibre. Differing amounts of hydroxylysine account for the different carbohydrate content of different collagen types. Indeed, not all types of collagen are fibrillar—for instance, type-IV collagen, found in basement membrane, is not. It contains smaller stretches of repeated Gly-X-Y triplets.

Several genetic abnormalities affect the structure of connective tissue. In one form of **Ehlers–Danlos syndrome**, mutations in the gene for the enzyme lysine hydroxylase cause defective post-translational modification of lysine to hydroxylysine. The consequent reduction in cross-linking lowers the mechanical strength. The symptoms of Ehlers–Danlos syndrome include spidery fingers and unusually high flexibility of the joints. The nineteenth-century violinist Niccolò Paganini exhibited these anatomical characters—and very good use he put them to! It has been speculated that he had either Ehlers–Danlos syndrome or **Marfan's syndrome**, a condition affecting another connective tissue protein, fibrillin.

The symptoms of scurvy, the deficiency disease caused by lack of vitamin C, result from defective hydroxylation of prolines and lysines, weakening collagen. For example, loss of integrity of gum tissue and of dentin, inside teeth (dentin forms by mineral deposition on a collagen matrix), leads to loosening and ultimately falling out of teeth.

Sailors on diets restricted by what could be carried on long voyages traditionally suffered from scurvy. An eighteenth-century Scots surgeon, James Lind, discovered that citrus fruit could prevent the disease. This is the origin of the term 'limey', referring originally to British sailors, and, by extension, to the entire nation.

Aside from most primates, many animals can synthesize vitamin C and do not need to obtain it from their diets. Humans lack a single enzymatic activity in the synthetic pathway from glucose to vitamin C: L-gulonolactone oxidase, the enzyme that catalyses the last step in the pathway.

hair, but varies in physical properties. (Paintbrushes have been made from the fur of many animals, including sable, ox, squirrel, pony, goat, hog, camel, and mongoose. Artists are exquisitely sensitive to the variations in stiffness, and to the differences in retention and delivery of paint.)

Lanital was a synthetic fibre based on natural protein: cross-linked casein. It was used widely in Italy during the Second World War.

Silk is β-fibroin, with the repetitive sequence … Gly–(Ala or Ser)–Gly–(Ala or Ser)…, forming an extended β-sheet. The cocoons of moths contain fibres of β-fibroin glued together by a second protein, sericin; adult moths secrete the proteolytic enzyme cocoonase to dissolve the sericin and let them out. It is a paradox that moths can digest the keratin of woollen sweaters, and the fabric of their own cocoons, but not silk scarves! The explanation is that most silk cloth contains the β-fibroin but not the sericin.

Fibrous proteins are ubiquitous beneath the skin also. Connective tissue, such as tendons, also contains collagen—indeed, collagens make up about a third of the protein content of the human body (see Box 2.8).

Many proteins function by interacting with other proteins or with small metabolites. These include the following classes.

Enzymes—proteins that catalyse chemical reactions

Essential features of enzymes are:

- *Catalysis:* enzymes speed up reaction rates.
- *Substrate specificity:* enzymes select specific substrates for catalysis.
- *Reaction specificity:* enzymes selectively produce only one possible product, of many that could be produced from one starting substrate.

The rate of a reaction is typically limited by the highest-energy point on a trajectory from substrate to product, called the **transition state**. One way in which enzymes speed up reactions is by altering the energy barrier between the substrate and the product, by binding transition states more tightly than they bind substrates. Enzymes achieve substrate specificity by spatial complementarity to reactants and cofactors. Figure 2.22 gives an example of enzyme–substrate complementarity.

In Chapter 5, we shall explore the structures and mechanisms of enzymes in detail.

The functions of many enzymes are well understood, in terms of the relationship between the three-dimensional conformation and the physical-organic chemistry of their mechanism. (Biochemists have had an advantage over cell biologists in that many enzymes can be purified and studied in isolation from the cellular context.) We shall discuss the mechanism of enzymatic catalysis in Chapter 5.

- Essential features of enzymes are: catalysis, substrate specificity, and reaction specificity.

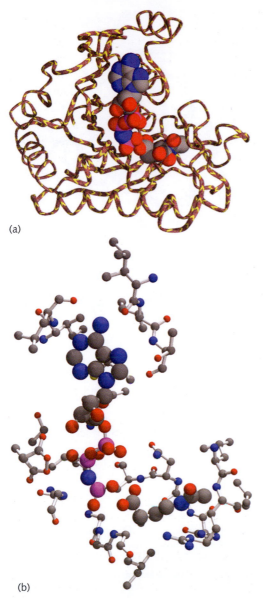

(a)

(b)

Figure 2.22 An enzyme–substrate complex: *E. coli* N-acetyl-L-glutamate kinase binding the substrate N-acetylglutamate and the inhibitory cofactor analogue AMPPNP (instead of the natural cofactor ATP) [1GS5]. The substrate and inhibitor nestle snugly into the enzyme, which holds them in proper proximity for phosphate transfer. (a) The substrate and cofactor analogue occupy a crevice in the molecule. (b) The mainchain and sidechains that surround the ligands. N-acetyl-L-glutamate kinase catalyses a step in the biosynthesis of arginine:

$$\begin{array}{l} \text{COO}^- \\ | \\ \text{HC-NH-C-CH}_3 \\ |\quad\ \| \\ \text{CH}_2\ \ \text{O} \quad + \text{ATP} \longrightarrow \\ | \\ \text{CH}_2 \\ | \\ \text{COO}^- \end{array} \qquad \begin{array}{l} \text{COO}^- \\ | \\ \text{HC-NH-C-CH}_3 \\ |\quad\ \| \\ \text{CH}_2\ \ \text{O} \quad + \text{ADP} \\ | \\ \text{CH}_2 \\ | \\ \text{COOPO}_3\text{H}^{2-} \end{array}$$

Antibodies

Antibodies are family of proteins from vertebrates that function by binding molecules foreign to the organism, notably the surface proteins of pathogens, and triggering cellular mechanisms for killing the invaders.

Enzymes bind substrates, and juxtapose them with catalytic residues. Most antibodies show binding alone. For this reason, some attempts to design artificial enzymes have started with antibodies raised against transition-state analogues (see Chapter 6). The antibody provides the binding; the chemist can insert catalytic residues if necessary. (Catalytic antibodies are called **abzymes**.) In fact, some natural antibodies have proteolytic catalytic activity based on a mechanism similar to that of chymotrypsin.

Inhibitors

Inhibitors are common weapons in the biochemical wars between species. Many drugs are based on adaptations of natural inhibitors; others are the products of chemists' ingenuity and the comprehensive screening procedures of the pharmaceutical industry.

- Leeches use the thrombin inhibitor hirudin to prevent their victims' blood from clotting (see Fig. 2.23).
- Many drugs for the treatment of AIDS are inhibitors of the HIV-1 proteinase (see Fig. 2.24). One general approach to inhibitor design uses **peptidomimetics**, compounds that are sufficiently similar to peptides to compete for a binding site, but cannot undergo reaction.
- Inhibitors downregulate enzymatic activity. Many natural inhibitors exist. Many drugs are natural or designed inhibitors.

Carrier proteins

Retinol-binding protein circulates in the blood plasma in complex with transthyretin, to transport retinol (see Fig. 2.25). The best known transport protein is haemoglobin. Haemoglobin transports O_2 from

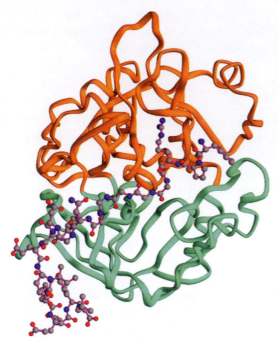

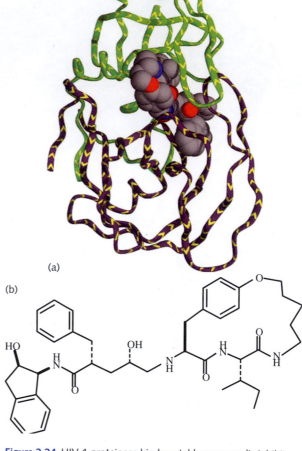

(a)

(b)

Figure 2.23 Thrombin, a key player in the control of blood coagulation, is a member of the chymotrypsin family of serine proteinases. The active site lies in a cleft between two domains. The two domains are homologues that arose by gene duplication and divergence. Human thrombin binds the synthetic inhibitor hirulog-3, a 20-residue peptide related to the natural inhibitor hirudin from the leech [1ABI]. Hirulog interacts with both the catalytic site common to the protease family and an anion-binding exosite specific to thrombin.

the lungs where it is absorbed, to the tissues; and CO_2 from the tissues back to the lungs, where it is excreted.

Figure 2.24 HIV-1 proteinase binds a stable macrocyclic inhibitor that mimics a tripeptide moiety of the natural substrate [1D4K]. (a) The proteinase is a homodimer, with a binding site shared between the monomers. (In order to show the ligand, the orientation is oblique to the axis of symmetry.) (b) Chemical structure of inhibitor.

Membrane proteins

Membranes are the wrapping of cells and of subcellular compartments and organelles (see Fig. 2.26). They contain **phospholipid bilayers**. The acyl tails of the phospholipids mimic a non-polar organic medium, about 3 nm thick. Between this layer and the surrounding aqueous medium, on either side, is a transitional layer approximately 1.5 nm thick.

It is estimated that in the human genome, approximately 30% of genes encode membrane proteins.

As an environment for proteins, the membrane differs from conventional aqueous solution not only in containing an organic medium, but in being **anisotropic**—that is, having a favoured direction. Proteins

interacting with membranes have definite orientations not only parallel to and perpendicular to the membrane, but with respect to the inside and outside of a cell or compartment.

Many proteins are designed to sit within membranes. Their adaptations include regions of continuous non-polar residues that interact with the organic layer of the membrane. These regions, which thread their way through the membrane, are connected through regions in the transitional layers that interact with the phospholipid head groups, and other regions entirely outside the membrane that have physicochemical properties similar to those of soluble proteins.

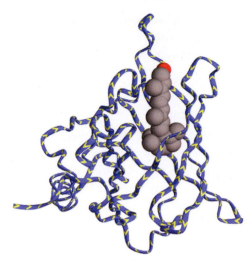

Figure 2.25 Retinol-binding protein [1RBP] is an example of a β-barrel, in which the strands of a β-sheet are wrapped around into a cylinder, with continuous lateral hydrogen bonding around the circumference of the barrel (see also Fig. 2.12d). By forming barrels of different sizes, from different numbers of strands, proteins can create interior cavities of different sizes to bind different ligands.

• Membrane proteins are adapted for a specialized environment. The environment within a membrane is hydrophobic rather than aqueous.

Structurally, membrane proteins fall into two major classes: proteins containing transmembrane helices, such as **bacteriorhodopsin** (see Fig. 2.27), and

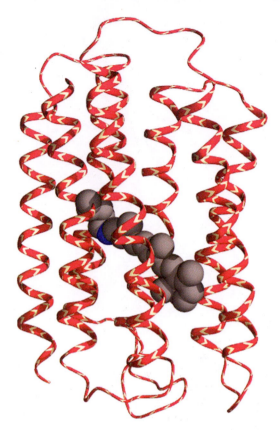

Figure 2.27 The structure of bacteriorhodopsin [2BRD] from *Halobacterium salinarum*, illustrating the common theme of a 7-transmembrane-helix structure. Bacteriorhodopsin is a light-driven pump, converting light energy absorbed by the chromophore, **retinal**, to a proton gradient across the membrane.

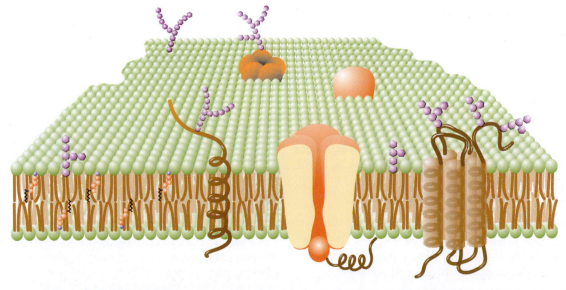

Figure 2.26 Schematic diagram of the structure of a membrane. This figure, showing the phospholipid bilayer and embedded proteins, gives an unwarranted impression that the membrane is a perfectly ordered and relatively rigid structure. Branched chains of small spheres represent polysaccharides.

Membrane proteins mediate exchange of matter, energy, and information between cell or organelle interiors and surroundings:

- **Channels, pores, pumps,** and **carriers** selectively control the import and export of molecules through membranes.

- **Energy transduction** is the function of a number of membrane-resident proteins and protein complexes, including opsins, **photosynthetic reaction centres**, and ATP synthases. A common feature of their mechanism is the generation or release of concentration gradients across membranes. The electrochemical potential energy of a concentration disequilibrium across a membrane is an intermediate in a number of energy transformations; notably the synthesis of ATP, coupled to proteins moving down the concentration gradient.

Receptors

Another important class of membrane proteins detects and reports the arrival of signals at the cell surface.

How can a cell detect a signal molecule in the external medium, and report its arrival to the cell interior, without the signal molecule itself ever needing to enter the cell? Many receptors use an ingenious dimerization mechanism (see Fig. 2.29). The receptor has external, transmembrane, and internal segments. An external ligand binds to *two* molecules of receptor. The juxtaposition of the external portions,

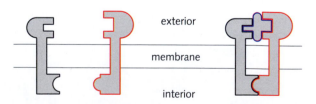

Figure 2.29 Schematic diagram of dimerization mechanism for transmission of a signal across a cell membrane. Receptor molecules contain exterior, transmembrane, and interior segments. In the absence of a ligand they are monomeric (left). Binding of a ligand brings together the exterior *and* the interior domains. Dimerization of the interior domains causes conformational changes, and/or reactions such as phosphorylation, that activate processes inside the cell, as a consequence of binding of the signal molecule outside the cell.

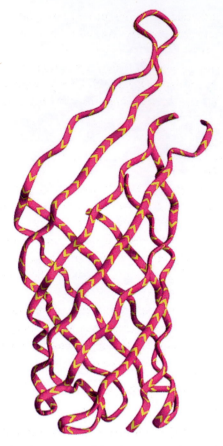

Figure 2.28 The structure of *E. coli* outer membrane protein A (OmpA) [1QJP], a β-barrel protein traversing the cell membrane. OmpA appears in gram-negative bacteria as a structural membrane protein interacting with lipoproteins, and also serves as a docking site for the bacteriocidal protein colicin, and some phages. It is also involved in conjugation.

proteins containing β-barrels, such as the transmembrane domain of *E. coli* outer membrane protein A (OmpA) (see Fig. 2.28; compare this β-barrel with that of retinol-binding protein in Fig. 2.25).

Structural characteristics of membrane proteins include:

- A typical helical membrane protein contains α-helices approximately 20 residues long, connected by loops typically 10–15 residues long (although there are many exceptions.)

- The **positive-inside rule**: the loops between helices live either entirely inside or entirely outside the cell or organelle. Those that are inside contain a preponderance of positively charged residues.

- Regions that interact with head groups, in the transitional region, tend to be enriched in the large polar aromatic residues tryptophan and tyrosine.

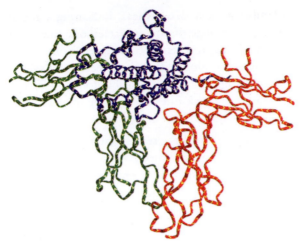

Figure 2.30 Human growth hormone (blue) in complex with two molecules illustrating the dimerized exterior domain of its receptor (green, orange) [3HHR].

The activities of many proteins respond to differences in conditions by **allosteric changes** in conformation.

In addition to controls over activities of proteins, a very heavy industry within cells governs regulation of gene transcription. We shall discuss gene regulation in Chapter 9.

Motor proteins

Motor proteins convert chemical energy to mechanical energy. There are two requirements: (a) coupling ATP hydrolysis (or some other energy source) to conformational change, to generate force, and (b) organizing a cycle of attachment and detachment to a mechanical substrate—a fulcrum—to allow the force to generate movement.

Some motor proteins propel themselves—and their cargo—by exerting force against a stationary object, such as a cytoskeletal filament. Others remain stationary, and propel movable objects.

- *Myosins* interact with actin during muscle contraction (see Chapter 5).

- *Kinesins and dyneins* interact with microtubules, mediating organelle transport, chromosome separation in mitosis, and movements of cilia and flagella.

Myosins, kinesins, and dyneins are primarily *linear* motors. In contrast:

- ATP synthase is a *rotatory* motor. The mechanical step is part of the mechanism for converting the osmotic energy of the proton gradient across a membrane to the high-energy phosphate bond of ATP (see Chapter 5).

> • This section has sketched out broad categories of protein structure and function. One respect in which it is an oversimplification is that the categories overlap!

coordinated by ligation, brings the internal portions together also, because they are tethered to the external regions by the transmembrane segments. Interaction between the interior segments triggers a process such as phosphorylation of a protein. This may initiate a **signal-transduction cascade** that can amplify the original stimulus. The network through which the signal propagates is usually branched.

Figure 2.30 shows the binding of one molecule of human growth hormone (blue) to two molecules of the external segment of the human growth hormone receptor (orange and green).

Regulatory proteins

Most proteins are regulatory proteins, in that they participate in the web of control mechanisms that pervade living processes. Examples of regulatory proteins appear throughout this book. The human growth hormone receptor appeared in Fig. 2.30. Bacteriorhodopsin (see Fig. 2.27), itself a transducer of light energy, has the seven-transmembrane-helix structure typical of a family of proteins called G-protein-coupled receptors.

Control of protein activity

Regulation of enzyme activity is essential to control the flow of matter and energy through metabolic pathways. Many mechanisms are available, including:

- *Feedback inhibition:* Given a sequence of reactions:

$$A \rightarrow B \rightarrow C \rightarrow D$$
$$\uparrow$$
$$D$$

If D inhibits the enzyme catalysing the conversion of A to B, then flow through this pathway will be downregulated when product D is in adequate supply. It is possible for D to be a competitive inhibitor of the step A → B *only* if there is sufficient structural similarity between D and A for them to bind to the same active site. If not, D might interact with the enzyme that converts A → B as an **allosteric effector**.

- *Control of concentrations of inhibitors* is a mechanism of regulating the activity of their protein targets. For instance, the diurnal rhythm of the activity of ribulose-1,5-bisphosphate carboxylase (the enzyme that fixes CO_2 in photosynthesis in many higher plants) is under the control of levels of an endogenous inhibitor 2-carboxy-D-arabinitol-1-phosphate.

- *Proteolytic cleavage* is a mechanism for activation of digestive enzymes. Trypsin is synthesized in the pancreas in an inactive form, trypsinogen. In the gut, cleavage of trypsin between residues 15 and 16 produces a charged amino terminus at residue 16, which can form a salt bridge to Asp194. This induces a conformational change to the active form. Similar mechanisms participate in the very complex control network regulating the enzymes involved in blood clotting (see Chapter 5).

- *Allosteric regulation.* Allosteric proteins give proteins an alternative to the simple non-cooperative relationship between ligand concentration and enzyme activity or ligand binding, as seen in enzymes that follow Michaelis-Menten kinetics (see Chapter 5). In haemoglobin, **positive cooperativity** in ligand binding allows for efficient uptake of oxygen in the lungs, where it is plentiful, and efficient transfer to the tissues, where it is not. Oxygen requirements during periods of intense physical activity are critical, indicating clear selective advantage of effective release to the tissues. Normally the oxygen is transferred from haemoglobin to myoglobin. It is the release of oxygen by haemoglobin that is responsible for the efficiency of transfer: knock-out mice lacking myoglobin appear quite normal (even, surprisingly, in the gym).

Allosteric proteins are also subject to feedback inhibition or stimulation. **Aspartate carbamoyltransferase** is an allosteric enzyme at the first step of the pathway of pyrimidine biosynthesis. CTP, a product of the pathway, is an allosteric effector that downregulates the activity of aspartate carbamoyltransferase. ATP upregulates it.

These examples illustrate the effects of endogenous molecules directly affecting the activity of proteins to which they themselves bind. Protein activity can also be regulated by signals external to the metabolic pathway in which the protein participates, and also even external to the cell.

- *Control of activity through reversible dimerization.* Some enzymes are active only in dimeric form. These can provide a mechanism for transmission of a signal from the cell surface, as illustrated in Figs. 2.29 and 2.30. Conversely, others are active only when *dissociated*; for instance, the **heterodimeric** G proteins (see Box 2.9) and protein kinase A.

- *Covalent modification* can affect activity. Phosphorylation at a specific Ser, Thr, or Tyr residue is a common mechanism for reversible inactivation of proteins in signal-transduction cascades.

These mechanisms of regulation operate within seconds or minutes. Control over gene expression also controls enzyme activity, on longer timescales.

BOX 2.9   *G proteins and G-protein-coupled receptors*

G proteins and G-protein-coupled receptors mediate the initial steps of many signal-transduction cascades. In the specific details of their structure and interaction they illustrate general principles common to many regulatory and signal-transduction processes.

GTP-binding proteins (or G proteins) are an important class of signal transducers. One of them, p21 Ras (see Fig. 2.31), is a molecular switch in pathways controlling cell growth and differentiation. Ras has two conformational states. The resting, inactive, state binds GDP. Membrane-bound G-protein-coupled receptors—proteins related to bacteriorhodopsin—trigger a GDP–GTP exchange transition, associated with a conformational change (see Fig. 2.32). Ras, thus activated, binds Raf-1, a serine/threonine kinase. The Ras–Raf-1 complex then initiates the MAP (mitogen-activated protein) **kinase** phosphorylation cascade. This pathway involves several proteins that activate each other sequentially by phosphorylation. Ultimately, the signal enters the nucleus, where it activates transcription factors regulating gene expression.

Ras has a GTPase activity to reset it to the inactive state. Mutations that inactivate the GTPase activity are oncogenic. Such mutants are trapped in the active state, continuously triggering proliferation. Mutations in Ras appear in ~30% of human tumours.

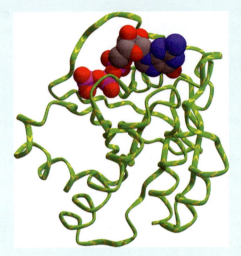

Figure 2.31 p21 Ras binding GTP. Although an active GTPase, the complex was stabilized for crystal-structure analysis by cooling to 100 K [1QRA].

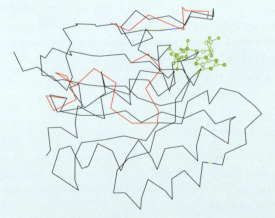

Figure 2.32 The conformational change in p21 Ras from the inactive GDP-binding conformation to the active GTP-binding conformation primarily involves two regions (shown here in red) that form a patch on the molecular surface [1QRA, 1q21].

G-protein-coupled receptors (GPCRs) are a large family of transmembrane proteins involved in reception of messages at cell surfaces and internalization of the signals. They share a substructure containing seven transmembrane helices, arranged in a common topology (similar to that of bacteriorhodopsin, see Fig. 2.27). The transmembrane part is generally flanked by N- and C-terminal domains, external to the nucleus. The N-terminal domain is always outside the cell, and the C-terminal domain always inside. Some GPCRs are involved in sensory reception, including vision, smell, and taste. Others respond to hormones and neurotransmitters. Some, like opsin and bacteriorhodopsin, bind chromophores. (Bacteriorhodopsin is not a signalling molecule but a light-driven proton pump.)

GPCRs constitute the largest known family of receptors. The family is as old as the eukaryotes, and is large and diverse. Mammalian genomes contain ~1500–2000 GPCRs, accounting for about 3–5% of the genome: a similar fraction of the *C. elegans* genome codes for GPCRs. GPCRs are the targets for many drugs, including some used in the treatment of high blood pressure, asthma, and allergies.

The downstream partners of GPCRs in signal-transduction pathways are heterotrimeric G proteins. These consist of three subunits: G_α, G_β, and G_γ. G_α is homologous to monomeric G proteins such as p21 Ras. G_α and G_γ are anchored to the membrane. In the resting, inactive state, G_α binds GDP. Reception of a signal by a GPCR induces a conformational change, activating the GPCR. The activated GPCR binds to a specific G protein, and catalyses GTP–GDP exchange in the G_α subunit. This destabilizes the trimer, dissociating G_α:

$$G_\alpha(GDP)G_\beta G_\gamma \rightarrow G_a(GTP) + G_\beta G_\gamma$$

The two components G_α and $G_\beta G_\gamma$ activate downstream targets, such as adenylate cyclase.

Different GPCRs have different mechanisms for restoring the resting state. The heterotrimeric G proteins are reset via the GTPase activity of G_α, converting $G_\alpha(GTP) \rightarrow G_\alpha(GDP)$. $G_\alpha(GDP)$ does not bind to its receptors—shutting down that pathway of signal transmission. Instead, $G_\alpha(GDP)$ rebinds the $G_\beta G_\gamma$ subunits. This resets the system to await the next stimulus. The GTPase activity of G_α is stimulated by a class of proteins called regulators of G-protein signalling (RGSs).

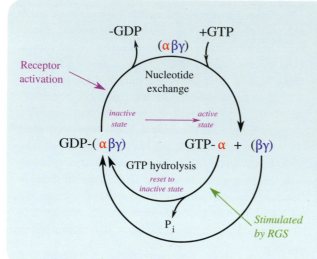

An activated GPCR can interact successively with over 100 G-protein molecules, amplifying the signal. It is essential to turn them off—mutations that render a GPCR constitutively active cause a number of diseases, the symptoms emerging from a war between the rogue receptor and the feedback mechanisms that are unequal to the task of restraining its effects.

Each GPCR interacts with specific G-protein targets. Mammalian genomes contain several homologues of G_α, G_β, and G_γ subunits, giving the potential for a large number of combinations. Although it is not known how many actually exist, it is likely to be fewer than the number of GPCRs. Therefore, many GPCRs must target a single G protein. For instance, all odorant receptors—several hundred in the human—target the same G_α.

• Regulation of protein activity is essential for the health of cells and organisms. There are many mechanisms of regulation. Some operate on proteins directly; others alter levels of expression.

Regulation of tyrosine hydroxylase illustrates several control mechanisms common to many proteins

Tyrosine hydroxylase catalyses the conversion of L-tyrosine to L-3,4-dihydroxyphenylalanine (L-dopa) in neurons, a step in the synthesis of neurotransmitters dopamine and adrenaline. Production of dopamine is reduced in **Parkinson's disease**, and L-dopa is administered to treat it.

Tyrosine hydroxylase is the focus of many diverse forms of regulation, including control over transcription and RNA processing and turnover. One regulatory pathway is triggered by arrival of a neurotransmitter at the external cell surface:

• External binding to a receptor activates adenylate cyclase inside the cell.

• The cyclic AMP produced activates protein kinase A by a mechanism involving subunit dissociation: The resting, inactive form of protein kinase A is a tetramer of two catalytic subunits C and two regulatory subunits R. In the resting state the regulatory subunits inhibit the activity of the catalytic subunits. Binding of cyclic AMP to protein kinase

A dissociates the tetramer, releasing the catalytic subunits in active form.

• Active protein kinase A phosphorylates tyrosine hydroxylase at Ser40, upregulating its activity.

• The mechanism balancing this stimulation is the specific dephosphorylation of Ser40 by phosphatase 2A.

Control cascades

There may be many successive steps between the detection of a signal and the action that a cell takes in response. A *signal-transduction cascade* is a pathway for the flow and transformation of information, just as a metabolic sequence such as glycolysis is a pathway for flow and transformation of matter.

Reasons for a multistep route from stimulus to target may include:

• *Physical separation:* a stimulus may arrive at the outer surface of a cell, and the response—perhaps a change in transcriptional activity—takes place in the nucleus.

• Separating the control pathway into several successive steps permits *amplification* of the signal. Each molecule activated at some step in the cascade can activate many individual downstream target molecules. Examples include successive phosphorylations, as in the MAP kinases, involved in cellular responses to growth factors; or successive proteolytic cleavages, as in blood clotting (see Chapter 5). The amplification factors at each step are multiplied together.

• The pathway may branch at any step, to produce multiple responses to a stimulus.

• Conversely, a pathway may be reticulated—neither strictly linear nor branched into only a tree-like structure. This permits integration of many different stimuli, to fine tune the response. The cell can weigh many considerations in deciding what action to take.

● RECOMMENDED READING

Baldwin, R.L. and Rose, G.D. (2012). Molten globules, entropy-driven conformational change and protein folding. *Curr. Opin. Struc. Biol.* **23**, 1–7. Superb overview.

Chothia, C. (1984). Principles that determine the structure of proteins. *Annu. Rev. Biochem.*, **53**, 537–72. A classic discussion of the essential features that underlie native structures of proteins. Insights are still valid and central to our understanding of protein architecture.

Andreeva, A. , Hovorth, D. , Chothia, C. , Kulesha, E. and Murzin, A.G. (2014). SCOP2 prototype: a new approach to protein structure mining. *Nucl. Acids Res.* **36**, D419–25. Description of the development of the new SCOP system.

Lesk, A.M. (2001). *Introduction to protein architecture: The structural biology of proteins.* Oxford University Press, Oxford. A somewhat more detailed treatment of the topics in this chapter. Particularly useful are the pictures. In 2001 it was possible to try to publish an album showing almost all the protein folding patterns then known. Not today!

Orengo, C., Bateman, A. & Uversky, V. (2014). *Protein Families: Relating Protein Sequence, Structure, and Function.* Wiley, Chichester. A review volume treating in detail many groups of proteins.

Richards, F.M. (1977). Areas, volumes, packing and protein structure. *Annu. Rev. Biophys. Bioeng.* **6**, 151–76, and

Richards, F.M. (1997). Whatever happened to the fun? An autobiographical investigation. *Annu. Rev. Biophys. Biomol. Struct.*, **26**, 1–25. Two articles—one more formal, the other more personal—from the late Prof. Richards. The Scientific American article cited in Chapter 1 is at a more introductory level than the Annual Review article.

Sussman, J.L. and Silman, I. (ed.) (2008). *Structural proteomics and its impact on the life sciences.* World Scientific Publishing, Singapore. A collection of chapters offering a comprehensive description of the state-of-the-art of protein science and its context in molecular and cell biology.

Petsko, G.A. and Ringe, D. (2004). *Protein structure and function.* New Science Press, London. Good general treatment of the relationships between sequence, structure and function.

Tsai, C.-J. , del Sol, A. and Nussinov, R. (2009). Protein allostery, signal transmission and dynamics: a classification scheme of allosteric mechanisms. *Molec. BioSyst.*, **5**, 207–16. A review of different allosteric mechanisms.

● EXERCISES AND PROBLEMS

Exercise 2.1 Write the chemical structural formula for selenocysteine.

Exercise 2.2 On a photocopy of Box 2.2, circle those sidechains capable of forming hydrogen bonds.

Exercise 2.3 Which amino acids have chiral C atoms in their sidechains?

Exercise 2.4 Which hydrophobic amino acids have sidechain volumes larger than that of alanine and smaller than that of phenylalanine?

Exercise 2.5 Write the Henderson–Hasselbalch equation for aspartic acid and for lysine, inserting the correct chemical species for HA and A$^-$ in the generic form of the equation.

Exercise 2.6 On a photocopy of the picture of methionine in Box 2.1, indicate the bond that is the axis of rotation for the conformational angles (a) ϕ, (b) χ_1, (c) χ_3.

Exercise 2.7 Suppose the door and the doorframe in Fig. 2.3 correspond to the angle ϕ of an amino acid residue. (a) On a photocopy of Fig. 2.3 add the positions of the appropriate atoms that would correspond to this dihedral angle. (b) Estimate the value of the dihedral angle shown in the figure.

Exercise 2.8 On a photocopy of the picture of methionine in Box 2.1, sketch in the additional atoms and bonds required to extend it to a picture of the dipeptide Met-Ile, in the extended conformation, with a *trans* peptide bond.

Exercise 2.9 Draw the chemical structures of tripeptides (a) Ala-Leu-Phe, (b) Ser-Pro-Asn (assuming that the peptide preceding the Pro is in the *cis* conformation).

Exercise 2.10 Consider a sample of a protein that contains a histidine sidechain on the surface that makes no specific interactions with other residues. What fraction of the histidine sidechains at that position, in the sample, is in the charged form at pH 7.04?

Exercise 2.11 The insulin hexamer shown in Fig. 2.6 has threefold symmetry. It contains four Zn^{2+} ions. Why must at least one of the Zn^{2+} ions line on the threefold axis?

Exercise 2.12 On a photocopy of Fig. 2.8, showing the hydrogen bonding between the protein and the ligand ubiquinone-1 in the reaction centre of *Rhodopseudomonas viridis*, sketch in estimated positions of the hydrogen atoms in the hydrogen bonds.

Exercise 2.13 On photocopies of Fig. 2.14a and the corresponding topology diagram, indicate the regions of substructural symmetry described in 'Comparison of the folding patterns of acylphosphatase and a fungal toxin'.

Exercise 2.14 What is the amino acid sequence of a short peptide that the inhibitor of Fig. 2.24 resembles most closely? On a photocopy of the structural formula in Fig. 2.24(b), highlight the atoms that the inhibitor has in common with the peptide you proposed.

Exercise 2.15 What is the sequence of the peptide shown in this figure, in one-letter code? Colour code: mainchain, magenta; sidechain—carbon, black; nitrogen, blue; oxygen, red; sulphur, green. Remember that the amino acid sequence is always stated in the direction from the N-terminus to the C-terminus.

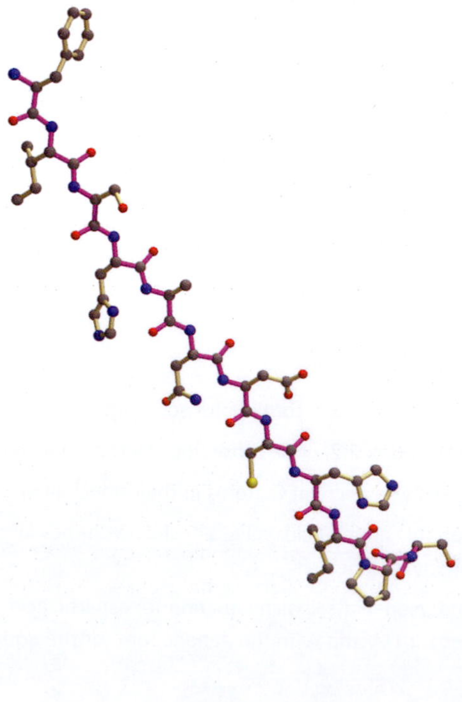

Exercise 2.16 Which tripeptide would you expect to be more water soluble: (a) Ala–Thr–Ser or (b) Phe–Ile–Trp?

Exercise 2.17 What is the net charge on the tetrapeptide Gly–Asp–His–Lys: (a) at pH 1.0, (b) at pH 3.0, (c) at pH 5.5, (d) at pH 8.0, (e) at pH 11.0?

Exercise 2.18 Write the chemical structure of the peptide Trp–Val–Glu–Tyr–His–Arg, indicating the expected state of ionization at (a) pH 3, (b) pH 5, (c) pH 7, (d) pH 9, (e) pH 11. Ignore changes in pK arising from interactions between charged groups.

Exercise 2.19 How many angles of internal rotation are there in a lysine sidechain?

Exercise 2.20 A Sasisekharan–Ramakrishnan–Ramachandran plot specialized to particular side-chains can be constructed as follows: For any amino acid, e.g., Val, create in a computer a model of the tripeptide Ala–Val–Ala. For each combination of ϕ—ψ values of the central residue, find the conformation with lowest energy, and plot that energy as a function of ϕ and ψ. Regions of steric collision will appear as high energy, and can be marked as disallowed regions. Suppose such plots are constructed for Ala and Val. Which would contain the larger disallowed area?

Exercise 2.21 On two separate photocopies of the drawing of the helix on the left in Box 2.7, indicate residue positions, as in the drawing on the right, for a helix containing (a) 2 residues/turn, (b) 1.5 residues per turn.

Exercise 2.22 Suppose that a protein adopts a stable native structure at room temperature. Suppose that this structure contains three buried sidechain–sidechain hydrogen bonds between tyrosine and threonine residues. Suppose that all three threonine residues are mutated to valines. (Valine and threonine have very approximately the same size and shape.) Would you expect the mutant to form the same stable native structure as the original protein at room temperature?

Exercise 2.23 A modified form of chorismate mutase forms a molten globule (mgCM) that binds a transition-state analogue (TSA) with affinity comparable to that of the native state (nCM). Assume that in both cases the complex has a native-like structure. Which ligation:

$$mgCM + TSA = complex$$

$$nCM + TSA = complex$$

would you expect to have the higher entropy change?

Exercise 2.24 On a photocopy of Fig. 2.13, indicate and identify the supersecondary structures that appear in these structures.

Exercise 2.25 Figure 2.33 shows one of the structures shown in Fig. 2.13, but in a different orientation. Which one?

Problem 2.1 In a multiple sequence alignment of a family of proteins, one position contains alanine in 50% of the sequences and phenylalanine in 50% of the sequences. What other amino acids would you expect to be able to appear at that position, and retain the structure and function of the protein?

Problem 2.2 The sequences of the leucine zipper domains of jun and fos are:

fos: RRELTDTLQAETDQLEDEKSALQTEIANLLKEKEK

jun: RIARLEEKVKTLKAQNSELASTANMLREQVAQL

(a) Draw helical wheels for each, and identify the zipper positions. Use an angular difference of 103°, corresponding to the value appropriate for a supercoil, rather than the 100° appropriate for a straight α-helix. This will improve the register (compare Figure 2.20(c)). (b) Photocopy and cut out the two helical wheels, and assemble them in the relative orientation expected for formation of the jun–fos dimer.

Figure 2.33 One of the structures illustrated in Fig. 2.13 shown in a different orientation. Which one?

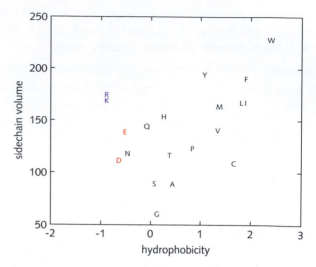

Figure 2.34 Relationship between hydrophobicity and sidechain volume.

Problem 2.3 One of the determinants of hydrophobicity is the size of the sidechain. Fig. 2.34 shows a graph of hydrophobicity and sidechain volume for the 20 natural amino acids. (a) For which subset of sidechains does there appear to be a fairly accurate linear relationship between hydrophobicity and sidechain volume? (b) Which subset of amino acids deviates most from this relationship? (c) For the subset identified in part (a), determine the slope and intercept of a linear fit to the data, and give a formula for the dependence of hydrophobicity on sidechain volume.

Problem 2.4. Describe how you would expect the CD spectra of solutions of uteroglobin (see Fig. 2.13(a)) and timothy grass pollen allergen (see Fig. 2.13(e)), to differ (see Fig. 1.16)?

Problem 2.5 The protease ervatamin C from the evergreen shrub crepe jasmine (*Abernaemontana divaricata*) is a dimer containing two chains of 208 residues each. The amino acid sequence, and the residues on the surface (s) and buried in the interior (b) are:

Chain A:

```
          10        20        30        40        50        60
LPEQIDWRKKGAVTPVKNQGSCGSCWAFSTVSTVESINQIRTGNLISLSEQELVDCDKKN
sssssssssssbbbbsssssbssbsbsbbbbbbbbbbbbbbsssssssssbsbbbbbbbbssbssb

          70        80        90       100       110       120
HGCLGGAFVFAYQYIINNGGIDTQANYPYKAVQGPCQAASKVVSIDGYNGVPFCNEAALK
sbsssssssssbbssbsssssbbbbbsssbssssssssssssssssbsbsbsssbsssssssssbs

         130       140       150       160       170       180
QAVAVQPSTVAIDASSAQFQQYSSGIFSGPCGTKLNHGVTIVGYQANYWIVRNSWGRYWG
sbbssbbbbbbbbbbbssssbssssssssssssssssssssbbbbbbbbbbbsssssbbbbbbbbsbssss

         190       200
EKGYIRMLRVGGCGLCGIARLPYYPTKA
ssbsbsbsbssssssssbsbbbbbbssbssbsss
```

Chain B:

```
        10          20          30          40          50          60
   LPEQIDWRKKGAVTPVKNQGSCGSCWAFSTVSTVESINQIRTGNLISLSEQELVDCDKKN
   sssssssssssbbbbbssssbssbsbsbsbbbbbbbbbbbbbbbbssssssssbsbbbbbbbbbbbbbssb

        70          80          90          100         110         120
   HGCLGGAFVFAYQYIINNGGIDTQANYPYKAVQGPCQAASKVVSIDGYNGVPFCNEAALK
   sbsssssssssbbssbsssssbbbbbsssbsssssssssssssssssbsbsbsssbsssssssssbs

        130         140         150         160         170         180
   QAVAVQPSTVAIDASSAQFQQYSSGIFSGPCGTKLNHGVTIVGYQANYWIVRNSWGRYWG
   sbbssbbbbbbbbbbbssssbsssssssssssssssssssssbbbbbbbbssssbbbbbbbsbssss

        190         200
   EKGYIRMLRVGGCGLCGIARLPYYPTKA
   Ssbsbsbssssssssbsbbbbbssbssbsss
```

Note that the results for the two chains are almost but not exactly identical.

(a) What percentage of the residues is buried? (b) For each of the following physicochemical classes (1) non-polar (GAVILPFM), (2) polar (SCTNQYW), and (3) charged (DEKRH), what percentage of the amino acids in each of these classes is buried and what percentage is on the surface?

Problem 2.6 A protein binds a single ligand in a cleft in its surface. There are no buried polar atoms in the interface. The complex buries 400 Å² of hydrophobic surface area. Estimate the dissociation constant at 300 K.

Problem 2.7 Draw a configuration of four serine sidechains that can form four sidechain–sidechain hydrogen bonds. Indicate the positions of the hydrogens and the lone-pair electrons.

Problem 2.8 (a) Make a table giving the number of sidechain degrees of freedom of each of the 20 sidechains. (b) Table 2.6 gives the observed statistical propensity for amino acids to appear in α-helices. Is there a correlation with the number of degrees of freedom? (c) Is any single amino acid a spectacular exception? If so, can you suggest why? In Table 2.6 of helix propensities of amino acids, the *lower* the value the *higher* the helix-forming propensity. (From: Pace, C.N. and Scholtz, J.M. (1998) A helix propensity scale based on experimental studies of peptides and proteins. *Biophys. J.*, **75**, 422–427.)

Problem 2.9 The accessible surface area (A.S.A.) of small monomeric proteins varies with molecular weight M according to the relationship: A.S.A. $= 11.1M^{2/3}$. If the amino acid sequence of these proteins is placed on a polypeptide in the extended chain conformation, the A.S.A. is of course higher, and is given by: A.S.A.$_{\text{extended chain}} = 1.45M$. (a) Explain why the accessible surface area of a native protein varies as the 2/3 power of M, and that of the extended chain varies linearly with M. (b) What is the formula for the *buried* surface area—relative to the extended chain—of proteins as a function of M? (c) What is the expected buried surface area per residue for a monomeric protein of 100 residues? (Assume that the average M of a residue is 110.) (d) For this example, what is the approximate contribution to the free energy of stabilization of the native state from the hydrophobic effect?

Problem 2.10 On a photocopy of Fig. 2.14(b), label the secondary structures:
$\beta_1 - (\alpha) - \alpha_1 - \beta_2 - \beta_3 - \alpha_2 - \beta_4 - \beta_5$

Problem 2.11 Human head hair grows about 15 cm/year. (a) If human hair were a single non-supercoiled α-helix, how many residues per second would have to be added to each growing hair? (b) In the structure of a single hair, the α-helices are supercoiled, making an angle of ~28° with the fibre axis. If an average diameter of a human hair is ~7.5 μm, and there are ~10^5 hairs on the head

Table 2.6 Helix propensity

Ala	0.00
Leu	0.21
Met	0.24
Arg	0.21
Lys	0.26
Gln	0.39
Glu	0.40
Ile	0.41
Ser	0.50
Trp	0.49
Tyr	0.53
Phe	0.54
Val	0.61
Thr	0.66
His⁰	0.56
His⁺	0.66
Cys	0.68
Asn	0.65
Asp	0.69
Gly	1.00
Pro	3.10

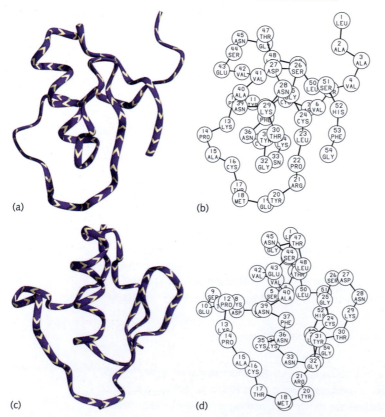

(a) (b) (c) (d)

Figure 2.35 Identify the helices and strands of sheet in this protein. The viewpoints in parts (a) and (b) differ from those of parts (c) and (d) by a rotation of 90°. Use parts (a) and (c) to identify the secondary structure, parts (b) and (d) to determine the residue numbers.

of an average person (in the absence of hair loss from pattern baldness or other causes), estimate the number of residues added to the α-helices on a human head per second.

Problem 2.12 Figure 2.13(e) shows the structure of the timothy grass pollen allergen Phl p II [1WHO], a double β-sheet sandwich structure. Draw a diagram similar in style to the topology diagrams in 'Comparison of the folding patterns of acylphosphatase and a fungal toxin' for the tertiary structure.

Problem 2.13 Figure 2.13(j) shows the structure of barley chymotrypsin inhibitor [2CI2]. Draw a diagram similar in style to the topology diagrams in 'Comparison of the folding patterns of acylphosphatase and a fungal toxin' for its tertiary structure.

Problem 2.14 (a) Identify the helices and strands of sheet in the protein shown in Fig. 2.35. Give the residue numbers at which they begin and end. (b) State how the strands are arranged in the sheet. Draw a simple diagram showing the positions of the strands and their relative directions.

Problem 2.15 (a) Identify the helices and strands of sheet in the protein shown in Fig. 2.36. Give the residue numbers at which they begin and end. (b) State how the strands are arranged in the sheet. Draw a simple diagram showing the positions of the strands and their relative directions. This example is somewhat more difficult than the preceding one.

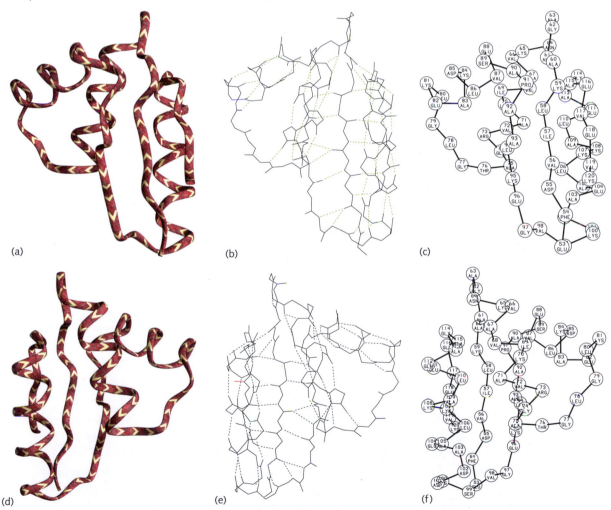

(a)   (b)   (c)
(d)   (e)   (f)

Figure 2.36 Identify the helices and strands of sheet in this protein. The viewpoint of parts (a)–(c) differs from that of parts (d)–(f) by a rotation of 180° around an axis vertical in the page. Use parts (a), (b), (d) and (e) to identify the secondary structure, parts (c) and (f) to determine the residue numbers. This is a somewhat more complicated example.

CHAPTER 3

Protein structure determination

LEARNING GOALS

- *Familiarity with experimental methods of protein structure determination:* X-ray crystallography, nuclear magnetic resonance (NMR) spectroscopy, cryo-electron microscopy (cryo-EM); and with methods for computational prediction of protein structures and knowing their domains of applicability.

- *Understanding the basic experimental data of X-ray crystallography,* and their relation to the chemical structure of the molecules in the crystal.

- *Knowing the basic steps of a protein structure determination by X-ray crystallography:* purification, crystallization, data collection, methods of solution of the phase problem, map interpretation, and refinement.

- *Understanding the background of X-ray diffraction* and its relation to the distribution of atoms in the crystal.

- *Knowing the basic experimental data of NMR spectroscopy,* and their relation to the chemical structure of the molecules studied; including the nuclear Overhauser effect, which provides information about neighbouring residues that may be far apart in the amino acid sequence. Appreciating the special features of solid-state NMR.

- *Following the basic steps of a protein structure determination by NMR spectroscopy:* purification, data collection, inference of secondary structure, assignment of spectra, and calculation of structure.

- *Appreciating the differences between X-ray crystallographic and NMR structure determinations of proteins,* including the lack of requirement for crystals in NMR spectroscopy, and, generally, the multiplicity of models arising from NMR structure determination.

- *Recognizing recent advances in cryo-electron microscopy,* which have taken it to the verge of producing atomic-resolution structure determinations.

- *Understanding the role of computational methods both in predicting protein structure from amino acid sequence,* the classic problem of theoretical molecular biology, and the contributions of computational methods to experimental structure determinations.

- *Appreciating the difficulties and the achievements of methods for structure prediction.* Knowing about the programmes for blind tests of prediction methods, such as the Critical Assessment of Structure Prediction (CASP) programme.

- *Knowing which prediction method or methods to use in practical circumstances:* homology modelling if a near relative of a known structure is available, fold-recognition or *a priori* methods if not. Knowing specialized methods that apply to particular types of structures.

Introduction

Protein structure determinations involve experimental and computational methods, and combined applications of both.

The first three-dimensional protein structures, of myoglobin and haemoglobin, were determined in the late 1950s by J.C. Kendrew, M.F. Perutz, and coworkers, using **X-ray crystallography**. For many years, X-ray crystallography and **fibre diffraction** were the only methods for determining the positions of individual atoms in macromolecular structures. A companion appeared in the 1980s, when K. Wüthrich, R.R. Ernst, and their coworkers developed methods for solving protein structures by **nuclear magnetic resonance (NMR) spectroscopy**. With a third technique, **cryo-electron microscopy (cryo-EM),** it has been possible to determine structures of large aggregates, including viral capsids, large protein complexes, and intact ribosomes. Cryo-EM of large multiprotein complexes plus X-ray crystal structures of individual components has proved a powerful combination. With improvements in the technique, cryo-EM is within shouting distance of determining protein structures to atomic resolution.

Taken together, these methods have given us over 100 000 experimentally-determined protein structures, archived and distributed by the world-wide Protein Data Bank (wwPDB).

In principle, computational methods should be able to deduce protein structures from amino acid sequences. After all, Nature knows such an algorithm, and we expect to discover it, and to write computer programs to implement it. At the moment, however, structure determination is firmly rooted in experiments.

Indeed, some readers will react in surprise or even indignation at the idea of the combined treatment, in this chapter, of experimental and theoretical approaches to protein structure determination. Let there then be no ambiguity: experimental methods of structure determination give results on which we can rely, within limits of error that we understand quite well. Given a protein, experiment will provide an answer. True, some proteins are harder to crystallize than others, but few proteins have failed to yield

eventually. NMR and cryo-EM structure determinations do not even require crystals.

> A crystallographer once boasted in a public lecture that, in the near future, when a new strain of influenza virus emerged, we could crystallize a surface protein, solve the structure, computationally design a drug to bind it, and nip any epidemic in the bud. A member of the audience suggested that in that event the flu virus would evolve so that its surface proteins would resist crystallization.

On the other hand, depending on the protein, computations may get the answer right, or may provide no answer at all, or may provide an incorrect answer without any way to recognize that it is wrong. Let us emphasize: computers will not put experimental methods out of business (in my lifetime at least). If experimentalists take this as an admission, so be it. If computational structural biologists take this as a recantation, so be it.

Eppur si muove: theoretical methods of protein structure prediction have made great strides. However, the point is not to see theoretical and experimental methods as rivals; rather to recognize that they have a more intimate partnership than is generally presented. Few experimental structure determinations are 'untainted' by contributions from theory.

- Almost all sets of X-ray crystallographic measurements, and all sets of NMR measurements, underdetermine the structures and require conformational energy calculations to supplement them. The conformational energy terms enforce proper stereochemistry, and speed the convergence to a correct model that is consistent with the experimental data. The only exceptions are very high-resolution crystal structures—a fascinating and important speciality that is responsible, however, for only a small minority of solved structures.

- Also, ~70% of crystal structures are currently solved using molecular replacement, a method that requires a model of the structure, which in many cases computational model building can provide.

- Protein structure prediction is becoming more precise and reliable. Granted, this is in large part—although by no means exclusively—because of the growing corpus of experimental structures on which many prediction methods are based. Experimental methods are sowing the seeds, not of their own destruction, but of their enhancement.

In what follows, we discuss first the experimental methods of protein structure determination, and then the purely computational approaches.

The most important methods for determination of protein structure include:

- Experimental approaches:
  - X-ray crystallography
  - nuclear magnetic resonance (NMR) spectroscopy
  - cryo-electron microscopy (cryo-EM)
- Computational methods
  - homology modelling
  - *a priori* prediction
  - methods based on inferences of residue contact patterns from correlated mutations

---

X-ray crystallography

One evening in late April 1934, J.D. Bernal developed a photograph he had taken of the X-ray diffraction pattern of a crystal of pepsin, a proteolytic enzyme. Amazed, he recognized from this picture that someday a complete atomic structural model of a protein would be revealed:

> … The wet crystals gave individual X-ray reflections, which were rather blurred owing to the large size of the crystal unit cell, but which extended all over the films to spacings of about 2 Å. That night, Bernal, full of excitement, wandered around the streets of Cambridge, thinking of the future and of how much it might be possible to know about the structure of proteins if the photographs he had just taken could be interpreted in every detail. (Hodgkin and Riley (1968), p.15).

What is it about crystals and X-ray diffraction that caused Bernal such excitement?

Crystals had been known since antiquity. Gems have adorned many men, women, and even animals in life, and have been included in their burial sites. The attractive optical properties of many jewels depend crucially on their crystallinity. Abbé René Haüy, an eighteenth-century mineralogist, proposed in 1784 that the regular external forms of crystals are the manifestation of regular internal arrangements of identical building blocks (see Fig. 3.1). Proteins can form crystals too: F.L. Hünefeld first described haemoglobin crystals in 1840.

It was the combination of Haüy's insights into crystal architecture with early ideas of the structural chemistry of molecules that proved to be such a powerful combination.

Pasteur was one of the first scientists to think in three-dimensional terms about the structure of molecules in general, and biological molecules in particular. Readers will know of his classic experiment in separating racemic tartaric acid, by manual selection of crystals of different mirror image forms. Pasteur recognized that the different crystal forms reflect a difference in underlying chemical constitution. Later, in studying the fermentation of tartaric acid, he observed that microorganisms also were selecting only one form, discriminating between the two kinds of tartaric acid molecules on the basis of their three-dimensional structure, at the molecular level. This marked the beginning of the field that is now called structural biology.

The fact that proteins could crystallize showed that they had—or at least could adopt—a unique structure compatible with regular packing into a crystal, as in Haüy's models. What was unknown was the level of detail at which the different unit cells resembled one another. Of course, in 1850 no one had any idea of the nature of protein structures (or much of any other chemicals for that matter). E. Fischer pessimistically stated in 1913 that '…the existence of crystals does not in itself guarantee chemical individuality, since isomorphous mixtures may be involved, as is frequently evident in mineralogy for the silicates.' Even Bernal's photograph in 1934 preceded F. Sanger's demonstration that insulin had a unique amino acid sequence. However, as early as 1913, S. Nishikawa

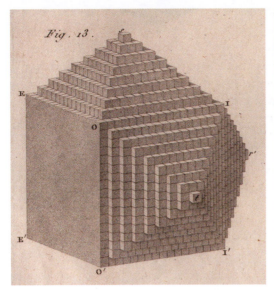 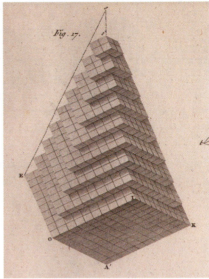

Figure 3.1 Models of crystal structure by Abbé R.J. Haüy. He proposed in 1784 that the regular external forms of crystals are the manifestation of regular internal arrangements of identical building blocks. Haüy tried to explain the interfacial angles of crystals in terms of stacking patterns of the microscopic units.

[From R.J. Haüy (1801). *Traité de Minéralogie.* Paris: Chez Louis, Vol. 5, Plate II, Figs. 13 and 17. Reproduced with the permission of Rare Books and Manuscripts, Special Collections Library, the Pennsylvania State University Libraries.]

and S. Ono had taken X-ray diffraction photographs of silk—the first diffraction pattern of a protein—and concluded, albeit qualitatively, that the material must contain some ordered structure at the molecular level.

X-ray structure determination

Application of X-ray crystallography to structure determination began slightly more than a century ago. Following M. von Laue's suggestion that, because the interatomic spacings in crystals were commensurate with the wavelengths of X-rays, crystals should diffract X-rays, in 1912 W. Friedrich and P. Knipping photographed diffraction from a crystal of copper sulphate pentahydrate.*

A year later, W.L. Bragg and W.H. Bragg (son and father) showed how to use the diffraction pattern to determine the spatial arrangements of atoms in crystals. They first solved the structures of crystalline NaCl and KCl. World War I interrupted scientific

work—W.L. Bragg went to work on sound-ranging methods for locating enemy guns. Serving in the British army in France, he was unable to attend the ceremony at which he was awarded the 1915 Nobel Prize for Physics. After the armistice he turned his attention to the structures of silicate minerals.

After World War II, W.L. Bragg assumed the Cavendish Chair of Experimental Physics at Cambridge. As head of the Cavendish Laboratory, he headed the department in which Max Perutz had initiated work on the crystal structure of haemoglobin. Unquestionably it was Bragg's interest, and faith, in the methods he had invented 40 years earlier that elicited his unflagging encouragement and support of Perutz's work, for many years during which the difficulties appeared insuperable. Bernal's pepsin picture of April 1934 had revealed the richness of the information about protein structure that X-ray crystallography could provide. If this information could only be harvested, it would be possible to specify individual coordinates of all the atoms in a protein, as the Braggs had done for salts and silicates.

It took 25 years.

J. Kendrew, M.F. Perutz, and colleagues solved the structure of myoglobin in 1959, and haemoglobin shortly thereafter—followed by first a trickle and then a flood of crystal structures of other proteins.

*Photographs of the apparatus with which Friedrich, Knipping, and von Laue discovered the diffraction of X-rays by crystals in 1912 appear in: Hoddeson L., Braun, E., Teichmann, J. and Weart, S. (eds.) (1992). Out of the crystal maze: Chapters from the history of solid state physics. Oxford University Press, Oxford. p. 49. (Compare this with Fig. 3.5.)

X-ray crystallography of proteins

The X-ray structure determination of a protein begins with its isolation, purification, and crystallization. If a suitable crystal is placed in an X-ray beam, diffraction will be observed, arising from the regular microscopic arrangement of the molecules in the crystal (see Fig. 3.2). The pattern can be recorded on film (see Fig. 3.3), or recorded in digital form by detectors.

The higher the angular deflection of the diffracted ray from the direction of the main beam, the more detailed the information about the structure it contains (assuming that the unit cell edges are not too far from equal). The angular range out to which data can be collected is called the **resolution** of the diffraction pattern, measured in Å. If two atoms are 2 Å apart in the structure, they will appear as two separate peaks—that is, they will be resolved—in a structure derived from a 2-Å data set. In contrast, the peaks would overlap—be unresolved—in a data set of lower resolution, for instance a 3-Å data set. Note that the numerical measure of the resolution of a data set, in Å, is *inversely* related to the detail of the structure derivable from it.

What then did Bernal see that was so exciting? The X-ray diffraction pattern appeared as a set of spots

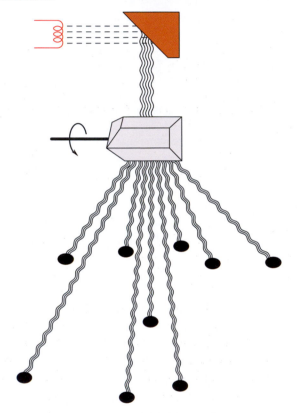

Figure 3.2 X-ray diffraction by a single crystal. X-rays generated by bombardment of a metal target with a beam of electrons (see Box 3.1) pass through a crystal to a detector. Diffraction scatters small components of the radiation in discrete directions. The angles of deflection depend on the structure of the crystal and its orientation relative to the X-ray beam, and on the wavelength of the X-rays. Rotation of the crystal would permit observation of additional diffracted radiation. One simplification of this picture is that only the diffracted beams are shown. In real experiments, the intensity of the transmitted incident beam is very much higher than the diffraction, and makes a giant splash in the centre of the pattern, despite the attempt to block it with a beam stop (see Fig. 3.3).

[After G.A. Jeffrey, X-ray crystallography, in *The science of ionizing radiation*, ed. L.E. Etter, 1965, pp. 573–597. Figure 5, p. 580. Courtesy of Charles C. Thomas, Publisher, Springfield, Illinois.]

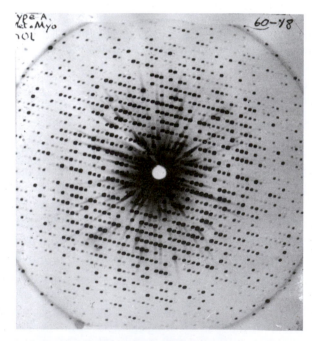

Figure 3.3 Photograph of X-ray diffraction of myoglobin, from work by J.C. Kendrew and his collaborators. Notice, as Bernal did in his 1934 picture, how the spots extend 'all over the film.' The distribution of the reflections in a diffraction pattern gives clues to the size and symmetry of the unit cell, and of the relative orientations of the subunits that it contains (see Problem 3.4). The white splodge in the centre is the 'shadow' of the beam stop, surrounded by splashes from the main beam.

[Reproduced courtesy of the MRC Laboratory of Molecular Biology.]

on a film. A photograph containing only a few spots, petering out in intensity at angles not far from the incident beam, would signify that the molecules in different unit cells are similar only to *low resolution*—perhaps they are 'blobs' of similar general shape, but different in detail. When Bernal saw spots 'which extended all over the films to spacings of about 2 Å', he realized that the proteins in different unit cells were identical in details not much larger than individual atomic spacings at least.

Protein structure determination by X-ray crystallography is now a mature technique. The equipment and technique of data collection, and software for data reduction and structure solution, are now integrated into an effective high-throughput technology. The rate-limiting steps are now in the preparation—expression and purification, and getting good crystals.

Steps in a protein structure determination by X-ray crystallography are:

1. *Purify the protein.* Classically, biochemists isolated proteins from natural samples. Typical sources include beef heart or liver, or suspensions of large quantities of microorganisms. Blood samples, or placentas, or post-mortem material, provided access to human proteins. The wide variety of separation methods suitable for preparative scales include ammonium sulphate fractionation, gel-exclusion chromatography and affinity chromatography.

   An alternative to isolating proteins from their natural sources is to insert a gene into a microorganism and overexpress the proteins in a host. However, mammalian proteins expressed in *E. coli*, for example, will lack post-translational modifications such as glycosylation. Expression of recombinant proteins by yeast, or baculovirus reproducing in insect cells, can circumvent this problem.

   How much pure protein do you need for a structure determination? You need enough to explore different crystallization conditions. How much protein this requires depends in part on whether you have some idea what conditions might work, from crystallization of related molecules, and whether you have access to specialized instruments that explore a range of conditions using only very small amounts of material. Pragmatically speaking, if you approach a crystallographer to propose a project, you are likely to be advised to bring along 1 mg of purified protein.

2. *Crystallize the protein.* The goal is a crystal showing complete internal order, and of adequate size. For data collection on a synchrotron, a crystal 0.1 mm in diameter is acceptable. For other methods of data collection 0.3–0.5 mm is better. By pushing things, microcrystallography can be carried out with ~0.04 mm crystals.

> Purity of a sample is important for crystal formation. Whereas in classical organic chemistry one crystallizes a molecule in order to purify it; in working with proteins one purifies a molecule in order to be able to crystallize it.

The basic idea in crystallizing a protein is to start with the molecule in solution, and to increase the concentration of protein (or of precipitant) until the solubility is exceeded and the protein comes out of solution. Precipitation occurs by formation of nuclei followed by their growth. The size and quality of the crystals depends on the relative rates of nucleation

---

 BOX 3.1   ## X-ray wavelengths

An atom ionized by bombardment with high-energy electrons will emit X-rays when an outer-shell electron drops to the inner-shell orbital from which the electron was ejected. The wavelength of the X-rays is characteristic of the atom. (In 1913, H.G.J. Moseley took advantage of this relationship to determine atomic numbers.) Most early crystallography was done with X-rays emitted from copper. The characteristic Kα line from Cu has a wavelength of 1.54 Å, almost identical to the carbon–carbon single bond length.

X-rays of other wavelengths are available by two means. It is possible to replace copper by another metal—often, molybdenum, with Kα wavelength 0.7071 Å; occasionally other metals. The use of synchrotrons as tunable X-ray sources permits a wider choice of wavelengths. Collection of data at multiple wavelengths offers a very powerful approach to solving crystal structures (see Box 3.4).

and growth. The result is exquisitely sensitive to the exact details of the solvent and procedure.

An experimental survey of crystallization conditions may reveal a system that produces small crystals. One would then (1) explore neighbouring conditions to try to optimize the size and quality of the crystals, and (2) use the tiny crystals to 'seed' the medium.

In some cases, experience with similar molecules suggests conditions. Scientists at the U.S. National Institutes of Standards and Technology maintain the Biological Macromolecule Crystallization Database (http://xpdb.nist.gov: 8060/BMCD4/index.faces).

Crystallization used to be as much an art as a science, demanding both insight and patience. More than many other laboratory skills, there was great variation among scientists in their ability to crystallize proteins.

Recently, crystallization 'robots' have greatly simplified the search for optimal conditions. These instruments prepare an array of varying conditions and monitor crystal formation in parallel. In addition to taking over some of the repetitive steps in the operation, robots have the advantage of working with smaller amounts of material.

It cannot be emphasized too strongly that the size of the crystal is far less important than the quality, meaning the degree of internal order. It is not necessarily true that the larger the crystal the better the diffraction.

3. **Data collection.** If a collimated X-ray beam impinges on a crystal, most of the beam will pass straight through. A small amount will be deflected, through an angle depending on the relative orientation of the crystal and the beam. A detector, such as a photographic film placed behind the crystal, will show discrete spots (see Figs. 3.2 and 3.3).

Recording of data

The three essential components of the experiment are: (1) a source of X-rays, (2) a crystal, and (3) a detector to measure the diffracted X-rays as a function of crystal orientation.

X-ray sources

In the early days, crystallographers produced X-rays by bombarding pieces of metal with high-energy electrons. To increase the brightness of the X-ray beam one can increase the energy of the electron beam. The danger is that the anode may overheat and melt. Cooling and rotating the metal anode allowed higher brightness by reducing local heating. Now, **synchrotrons** provide the preferred X-ray sources for structural biology.

Synchrotrons have two advantages as X-ray sources. First they are *much* brighter than laboratory generators. Laboratory devices can typically produce beams of 5 kW intensity. Synchrotrons can be 10^{10} times brighter! This gives the second advantage, that data can be collected faster using smaller, and fewer, crystals. Data collection can now take only minutes. An advantage of fast data collection is that many crystals degrade upon exposure to X-rays. The radiation creates reactive oxygen species that damage the protein. Protein crystals vary in their radiation resistance but in principle the faster the data are collected the better. If the lifetime of a single crystal is too short to collect an entire data set, it is necessary to collect partial data sets from several crystals—assuming that you have enough of them. To merge the partial data sets requires inter-crystal calibration.

Freezing extends the lifetimes of crystals; not slow freezing as this produces normal ice I, and the volume change would crack the crystal. However, protein crystals can survive flash freezing in liquid nitrogen, with the solvent forming vitreous ice. This is now standard practice.

Of course, another advantage of fast data collection is the saving of human labour. Data collection and reduction that might have taken weeks or months using film may require only hours on a synchrotron.

There are about 130 synchrotrons around the world. Therefore, for most crystallographers, use of a synchrotron requires a trip. Whether this is an advantage or disadvantage, in human terms, depends on your point of view—see Fig. 3.4. Attractive as the environments of some synchrotrons may be, trips are expensive and exhausting. There can be problems in transporting materials across international borders, especially pathogenic samples. An alternative is to post your crystals to synchrotron staff; you stay at home. In any case, currently about 90% of macromolecular

Figure 3.4 Aerial view of the European Synchrotron Radiation Facility, ESRF, in Grenoble, France. ESRF is supported and shared by 19 European countries.

[Photo by D. Morel.]

structures solved by X-ray crystallography make use of data collected at synchrotrons.

Even more powerful than synchrotron radiation are X-ray free-electron lasers (XFELS). In **serial femtosecond crystallography** (SFX) a stream of droplets of protein microcrystals, suspended in aqueous or lipid solvent, passes through the beam. A single laser pulse of a few femtoseconds duration illuminates each crystal, which emits a diffraction pattern as a dying declaration just before the intense radiation destroys it. It is possible to produce observable diffraction from a protein crystal of volume less that 1 μm³.

Detectors

Classically, X-ray diffraction patterns were recorded on film. The relative intensities of the spots were estimated by eye in comparison with a fixed scale of spots of increasing darkness. Later, scanners that measured the darkening of the film at a dense lattice of points allowed computer programs to determine the integrated intensity of the spot. Diffractometers automated data collection by coupling the motion of the crystal with that of a detector, to record the intensity of scattered rays one reflection at a time. Image plates and charged-coupled devices have proved superior, because they record many reflections at once, and have superior dynamic range (see Fig. 3.5).

4. *Solving the structure.* The ultimate goal is a list of the positions of the atoms in the protein, and in favourable cases even their vibrational amplitudes. The first step towards this goal is to generate from the data an *electron-density map* within the crystal. Because it is the electrons in atoms that diffract X-rays, the distribution of electrons in the crystal contains a superposition of peaks (in the three-dimensional distribution), each peak surrounding an atom. In favourable cases, the peaks will be *resolved*—that is, the peaks corresponding to adjacent atoms will overlap each other minimally. In this case, the positions of the atoms can be 'read off' the electron-density map.

Unfortunately, the conversion of the experimental data to an electron-density map encounters a severe obstacle: the **phase problem**. The basic idea is that the data represent complex numbers (quantities of the form $x + iy$ where $i = \sqrt{-1}$, or, equivalently $x + iy = re^{i\phi}$ where r is the *magnitude* and ϕ is called the *phase*.). The problem is that the experiment determines only the magnitudes of the data elements but

(a)   (b)

Figure 3.5 Diffractometer at Pennsylvania State University. The crystal is mounted on a goniometer, in a loop of fibre. The tube pointing in from the left is the collimator of the X-ray beam. The tube pointing diagonally down from the upper left supplies cold nitrogen to keep the crystal frozen. A microscope is pointed at the crystal from just in front of the collimator; a small black stub is visible. After passing through the crystal, a CCD detector (manufactured by Rigaku) images the diffraction pattern. A beam stop to intercept the undiffracted beam appears between the crystal and the detector. (a) View of components. (b) Close-up of crystal.

not their phases. The next section: *X-ray crystallography: the theoretical background* develops this topic. For now, let us assume that phases have been determined, at least approximately, and that we have an electron-density map.

Interpretation of the electron density: model building and improvement

Given the experimental structure factor magnitudes, and even a reasonable estimate of phases, it is possible to calculate a map of the electron density in a unit cell of the crystal. With sufficient resolution, and sufficiently accurate phases, the map contains peaks that correspond—at least approximately—to positions of atoms. It is necessary to interpret the electron density in terms of a molecular structure.

For small molecules, the classic method of interpreting the pattern of peaks in the electron-density map used a 'needle and thread' approach: crystallographers would lay out x and y coordinates of the peaks on a sheet of graph paper pasted onto a styrofoam base. They would insert knitting needles perpendicularly into the styrofoam: for each peak at coordinates (x, y, z), they would push a needle into the styrofoam, at position (x, y), to a depth of $L - z$, where L is the length of the needle. Tying woollen threads between peaks nearby in space would

create a three-dimensional model of the molecular structure.

Interpretation of the first resolved electron-density maps of proteins involved an extension of this approach. To interpret electron-density maps of myoglobin, J.C. Kendrew designed brass wire molecular models of amino acids, and assembled them on a set of vertical rods, at a scale of 5 cm/Å (see Fig. 3.6). To display the electron-density map, contour maps of serial sections—like topographic maps—were drawn on transparent plexiglass plates, and stacked. To make it possible to superpose the model and electron density visually, F.M. Richards introduced the use of a half-silvered mirror placed at 45° between the model and the electron-density map.

Computer graphics superseded this. Programs make use of interactive computer display devices that show selected portions of the electron density, together with a model of part or all of the protein structure. For many years the program FRODO, by T. Alwyn Jones, achieved a near monopoly in the protein crystallographic community. ('If FRODO did not exist, it would have to be invented.') Later, Jones's successor program 'O' superseded FRODO (see Fig. 3.7). Devices such as dials allow manipulation of the model to adjust the fit to the experiment. The latest software for these tasks is COOT (Crystallographic Object-Oriented Toolkit) by P. Emsley and other contributors.

Figure 3.7 T. Alwyn Jones, author of FRODO and O, at a display showing part of an electron-density map, part of a protein model, and menus for manipulating them. A set of four dials stands at the left of the screen. Dials are suitable for interactive control of overall orientation of a molecule, or for internal rotation around selected bonds.

[Photo courtesy Mark Harris, frozentime.se]

Figure 3.6 J.C. Kendrew (left) and M.F. Perutz examining a model of sperm-whale myoglobin. (Note that models of several proteins require separate allocations of materials, and of space to house them; another advantage of computer graphics.)

[Reproduced courtesy of the MRC Laboratory of Molecular Biology.]

From an improved model of the structure, it is possible to recompute phases. If these phases are more accurate than the previous set, then by combining the new model phases with experimental structure factor magnitudes one can compute a better electron-density map. Multiple cycles of rebuilding at the computer graphics device and map recalculation can progressively improve the quality of the model (see Fig. 3.8). This used to require many difficult, uncomfortable, labour-intensive and time-consuming sessions. Crystallographers would emerge from the dark and cold computer graphics rooms rubbing their eyes and breathing heavily. (One could almost hear the strains of the Prisoner's Chorus from Fidelio, 'O welche Lust'.)

The application of molecular dynamics has changed all this. Experimental structure factor magnitudes are in themselves insufficient to determine

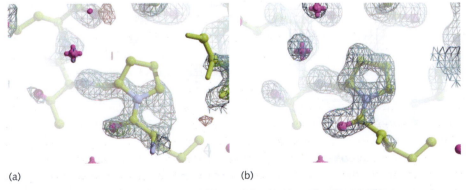

(a)   (b)

Figure 3.8 Improvement of molecular model. (a) Before rebuilding. (b) After rebuilding.

[From: Dodson, E. (2008). The befores and afters of molecular replacement. *Acta Cryst.*, **D64**, 17–24.]

the protein structure, because of the phase problem. The alternative, purely theoretical, approach, to calculate a protein structure by minimizing a conformational energy function or simulating the motion of the system by molecular dynamics, is also usually inadequate to determine the structure. But the combination is extremely powerful. What a crystallographer at a graphics device is doing is exploring conformational space, under the guidance of his or her insight and experience. What molecular dynamics can do is to explore conformational space computationally, under the guidance of *both* the experimental data and the computed interatomic forces. The task of progressive model rebuilding is shared by crystallographer and computer, and the time required reduced from weeks or months to days or even hours.

For data sets of adequate resolution, the program suites ARP/wARP by V.S. Lamzin and A. Perrakis, or Phenix (**P**ython-based **h**ierarchical **en**vironment for **i**ntegrated **x**tallography) by P.D. Adams and colleagues, are capable of fitting a fairly high-quality model automatically. The combination of structure prediction by homology modelling or by ROBETTA (see ROSETTA section) → molecular replacement → ARP/wARP is a very potent approach. A computational-experimental symbiosis.

The endgame—refinement

Once the atoms are positioned at close to their proper positions, it is possible to optimize the fit of experimental structure factors to those computed from the model by refinement. One is comparing the experimental observed structure factor magnitudes $|F_o|$ with the *magnitudes* of the structure factors computed from the model $|F_c|$. Ignore the fact that one can compute model phases as well, because there are usually no experimental data with which to compare them. The usual function to be minimized includes the **R-factor** (*R* for residual or reliability):

$$R = \frac{\sum \| F_o | - | F_c \|}{\sum | F_o |},$$

and also contains terms expressing the conformational energy.

> As a rule of thumb, one expects the final R-factor of the structure to be about 1/10 the resolution of the data. That is, if you collect data on a protein to 2 Å the expected R-factor should be ~20%.

Vary atomic coordinates (x_i, y_i, z_i) to minimize:

$$R + E,$$

where R is a function of experimental $|F_o|$ and computed $|F_c(x_i, y_i, z_i)|$, and E, the conformational energy, is a function of (x_i, y_i, z_i)

A certain leeway is allowed in the relative weighting of these two terms. We shall discuss the exact form of E later in this chapter.

A sensitive test of the progress of refinement is the **free R-factor**. Randomly select a small set (5%) of the reflections to *exclude* from the refinement calculation. Then, during refinement with respect to the non-excluded set of *F*s, monitor the value of:

$$R_{\text{free}} = \frac{\sum \| F_o | - | F_c \|}{\sum | F_o |},$$

with the sums taken over the excluded set. Minimizing the full R-factor—that is, omitting no reflections—in cases where there is a large ratio of adjustable parameters (atomic coordinates) to experimental observations (structure factor magnitudes) runs the risk of overfitting the data. Values of R_{free} tend to be slightly larger than corresponding values of R; a large discrepancy signals trouble.

Advances in computational methods have improved the power of refinement methods, to the point of blurring the distinction between the model building and refinement steps of structure determination.

How accurate are the structures?

Most protein scientists are consumers rather than providers of crystal structures. It is essential to appreciate what details of the coordinates can be trusted.

Crystal-structure determinations are at the mercy of the degree of order in different parts of the molecule. (Order is the extent to which different unit cells of the crystal are *exact* copies of one another.)

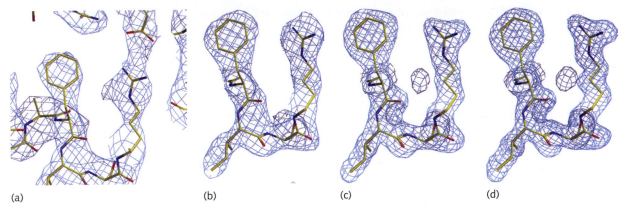

(a) (b) (c) (d)

Figure 3.9 Electron-density maps and model fitting at different resolutions. (a) 4.0 Å. It is possible to trace the general course of the bonding but not to position individual atoms reliably. The 'lollipop' of the ring of the phenylalanine sidechain is identifiable. The arginine and the carboxyl group with which it is interacting are blurred together. The guanidinium group of the arginine is particularly ill-represented. (b) 2.8 Å. It is possible to place the mainchain and sidechains from their overall sizes and shapes. With a resolution of 2.8 Å or better, it is generally possible to build a model into the electron density. (Traditionally, growing crystals that would diffract to 2.8 Å or better was the protein crystallographer's goal.) (c) 2.3 Å. One can see better definition of some features than in the 2.8 Å map. The position of the guanidinium group of the arginine is clearer. The orientations of the carbonyl oxygens are better defined. (d) 1.8 Å. Model building into this map is a piece of cake! The zig-zag of the arginine sidechain is clear, the carbonyl oxygens are quite well defined. Even the lower electron density at the centre of the phenylalanine ring is beginning to appear. Notice that phenylalanine and arginine sidechains have very distinctive sizes and shapes, and can be recognized even at relatively low resolution. To decide whether a region of density is valine, isoleucine or leucine would require quite high resolution.

[Courtesy of Prof. J. Wedekind, University of Rochester.]

The degree of order governs the available resolution of the experimental data. Resolution is an index of potential quality of an X-ray structure determination. Resolution, expressed in Å, measures the fineness of the details that can be distinguished (see Fig. 3.9). (Remember: the lower the number, the higher the resolution.) The higher the resolution, the greater the ratio of the number of observations to the number of atomic coordinates to be determined, and in principle the more precise the results. In structure determinations of small organic molecules or of minerals, the ratio of the number of parameters to be determined to the number of observations is usually generous: ~10. But for a typical protein crystal:

	Low resolution	...	High
Resolution in Å	4.0 3.5 3.0	2.5	2.0 1.5
Ratio of observations to parameters	0.3 0.4 0.6	1.1	2.2 3.8

It was the recognition that the pepsin crystal diffracted to high resolution that so excited Bernal in 1934.

Accuracy and precision of X-ray structure determinations of proteins depend primarily on (1) quality of crystals (= degree of order), (2) care in analysing data, and (3) resolution of data collection.

An animation showing how varying resolution changes the appearance of electron-density maps appears at http://ucxray.berkeley.edu/~jamesh/movies/. The effect is not dissimilar to changing the focus of a camera on an image. The site also contains other movies on related topics.

Experience has shown that the confident determination of different structural features is dependent on different thresholds of resolution (see Table 3.1).

Crystallographers now deposit their experimental data along with the solved structures. This permits detailed checks on the results. But in many cases the experimental data have not been available. How can one then assess the quality of a structure?

There are a number of fairly reliable indicators of the accuracy of the atomic coordinates in published protein structures. Some are derived during

Table 3.1 Confidence in structural features of proteins determined by X-ray crystallography at different resolutions

Structural feature	Resolution				
	5 Å	3 Å	2.5 Å	2.0 Å	1.5 Å
Chain tracing	–	Fair	Good	Good	Good
Secondary structure	Helices Fair	Fair	Good	Good	Good
Sidechain conformations	–	–	Fair	Good	Good
Orientation of peptide planes	–	–	Fair	Good	Good
Protein hydrogen atoms visible	–	–	–	–	–

These are *rough* estimates, and depend strongly on the quality of the data and care of analysis. Experimental determination of hydrogen positions by X-ray crystallography requires sub-1-Å resolution. Hydrogen position can be determined at lower resolution by neutron diffraction, for which the paucity of electrons in hydrogen is not a handicap.

the process of the structure determination and reflect experimental observations; others are derived from the atomic coordinates themselves. The free R-factor is an example of a parameter that assesses the agreement of the structure with the experimental data.

X-ray crystal structure analysis produces estimates of the positions and effective sizes of the atoms in a molecule, known as *B-factors*. B-factors provide important clues; high B-factors in an entire region suggest that the region has not been well determined. This usually reflects imperfect order in the crystal.

In addition to disorder, errors in crystal structures reflect both errors in data and ambiguities in solving the structure. A comparison of four independently solved structures of interleukin-1β, by four crystallographers, starting from the same set of measured structure factor magnitudes, showed an average variation in atomic position of 0.84 Å, higher than the expected experimental error.

Now that one has seen enough well-determined protein structures to know what they should look like, it is possible to subject atomic coordinate sets to a scrutiny independent of the experimental data. Good protein structures (1) are compact, as measured by their surface area and packing density, (2) have hydrogen bonds of reasonable geometry, with

few hydrogen bonds 'missing' in places where they would be expected, e.g. secondary structures, and (3) have residues with a distribution of backbone conformation angles confined almost entirely to the allowed regions of the Sasisekharan–Ramakrishnan–Ramachandran plot (see Fig. 2.4). Programs can flag stereochemical outliers—exceptions to regularities common to well-determined protein structures. Programs are available to carry out this analysis, and the results are available on the web. The entries corresponding to the PDB entries in http://swift.cmbi.ru.nl/gv/pdbreport/ describe diagnostic analysis and identification of problems and outliers.

Outliers are relatively easy to *detect*. However, it is difficult to decide whether they are correct but unusual features of the structure, or the result of errors in building the model, or the inevitable result of crystal disorder. Proper assessment requires access to the experimental data, and identifying and fixing real errors may well require the attention of an experienced crystallographer. The conclusion seems inescapable that structure factors should be archived and available. Starting 1 February, 2008, the wwPDB made deposition of experimental data mandatory. The data are the structure factor amplitudes/intensities (for crystal structures) and the restraints (for NMR structures).

X-ray crystallography—the theoretical background*

What are the data and what is their origin?

The measurements in an X-ray crystal structure determination are the pattern of intensities in the diffraction pattern (see Figs. 3.2 and 3.3).

When the X-ray beam impinges on a crystal, each atom in the crystal is transiently excited, and reradiates

energy in all directions. However, in most directions, the scattered X-rays from different atoms cancel one another out. Only in certain specific directions do the

---

*This section may be skipped on a first reading

X-rays scattered from different atoms reinforce one another. This gives rise to the discrete pattern of spots.

What creates the diffraction pattern?

The intensity of the scattered beam in any direction depends on the superposition of the waves scattered in that direction from all the individual atoms. The intensity of a superposition of waves depends on their relative *phase* (see Box 3.2 and Fig. 3.10). Waves will reinforce each other only if they are in phase—that is, if the peaks and troughs of the waves from different unit cells add up rather than cancel each other out (see Fig. 3.11).

Why are large areas of the diffraction pattern blank? Not because they are not illuminated by X-rays reradiated from the atoms of the crystal, but because in most directions the phases of these reradiated X-rays produce a net zero intensity, like the simple case of two waves that are out of phase in Fig. 3.11b.

Bragg's law interprets the diffraction from a crystal in terms of mirror-like scattering from parallel planes of atoms (see Fig. 3.12). Bragg derived the condition for constructive interference: $n\lambda = 2d \sin\theta$. This requires a relationship between the wavelength of the X-rays (λ), the distance (d) between the selected planes in the crystal, and the orientation of the crystal relative to the X-ray beam. $n\lambda$ is the difference in pathlength travelled by waves scattered from neighbouring rows of atoms. That n be integral is the condition for constructive interference (see Fig. 3.11a). That is, the difference in path length for waves scattered from different atoms must be an integral number of wavelengths (for Cu X-rays, that is a multiple of 1.54 Å).

For any specific orientation of the crystal, only in certain specific directions will the scattered X-rays interfere constructively. To observe the entire diffraction pattern using monochromatic X-rays, it

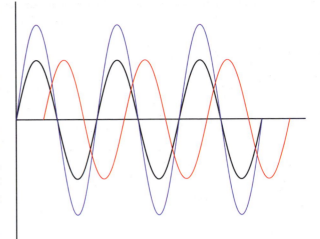

Figure 3.10 The amplitude and phase characterize a wave. What is the phase of a wave? Relative to the black curve as a reference, the blue one differs in amplitude–the peaks are higher and the troughs are lower. But the blue curve matches the reference curve in the *positions* of its crests and troughs. The black and blue curves have the same phase. In contrast, the peaks and troughs of the red curve are displaced from those of the other two. The red curve has the same amplitude as the black reference curve, but a different phase. X-ray detectors record only the integrated intensity of a wave impingent on them. The difference in amplitude of the black and blue curves would be detectable. The phase difference between the black and red curves cannot be measured, but it is essential information that must somehow be reconstructed to solve the structure.

is necessary to rotate the crystal to bring different planes into the right orientation to produce diffraction. Data are collected cumulatively as the crystal rotates.

From Bragg's law, crystallographers refer to individual diffraction spots as 'reflections'.

BOX 3.2 **The phase of a wave, and the combination of waves 'in phase' and 'out of phase'**

Fig. 3.10 shows what is meant by the phase of a wave. The net effect of two (or more) waves is the sum of their amplitudes, which depends crucially on their relative phases. Fig. 3.11 compares the resultant of two waves that are 'in phase' (that is, having the same phase), producing the *sum* of the intensities of the components, and of two waves that

are 'out of phase' (that is, where the peaks of one wave match the troughs of the other), so that the two waves cancel each other out, leaving zero intensity. Waves that are neither perfectly in phase nor perfectly out of phase show an intermediate result: for example, the red and black curves of Figure 3.10 (see Problem 3.1)

(a) Constructive Interference
(waves in phase)

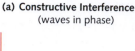
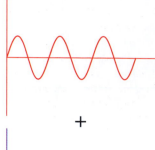

(b) Destructive Interference
(waves out of phase)

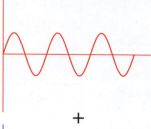
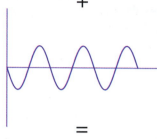

Figure 3.11 (a) Waves that are 'in phase' reinforce each other. (b) Waves that are 'out of phase' can cancel each other out. This figure shows the resultant intensity if the waves are either perfectly in phase or perfectly out of phase. If you were to do the corresponding calculations with only two waves that differed in phase by less than 180° you would find considerable remaining intensity. In diffraction from a crystal, the number of waves is equal to the number of atoms excited. The larger the array of atoms, the more critical the dependence of scattering intensity on angle. For macroscopic crystals, containing very large numbers of atoms, the angular constraint for constructive interference is very precise, leading to a pattern of tiny discrete spots.

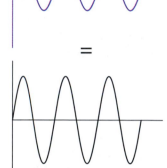
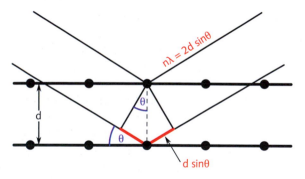

Figure 3.12 Bragg's law, $n\lambda = 2d\sin\theta$, expresses the condition that the difference in path length between scattered X-rays from adjacent parallel planes must be equal to an integral number of wavelengths, for there to be constructive interference. d is the distance between the planes, and θ is the angle between the incident (and reflected) rays and the scattering planes. The broken line is also equal in length to d. The difference in distance travelled along the two paths is equal to the sum of the lengths of the red segments, each of which is $d\sin\theta$.

Each spot contains information from an entire plane of atoms. The information contained in any spot is a composite indication of the positions of many individual atoms. Conversely, the information about the position of any particular atom is 'delocalized' in the diffraction pattern. To solve the structure, one must apply the data from the entire diffraction pattern.

What is the relationship between the diffraction pattern and the structure of the molecule in the crystal?

The X-ray reflections can be thought of as distributed on a **reciprocal lattice**, not to be confused with the crystal lattice itself, but allowing the reflections to be identified by three indices, h, k, and l. To each reciprocal lattice point corresponds a **structure factor, $F(h, k, l)$**. (h, k, and l indicate the orientation of the planes in the Bragg construction that are responsible for the reflection (see Fig. 3.12)).

The electron density in the molecule is related to the diffraction pattern by a pair of relationships called the **Fourier transform:** (see Box 3.3)

- The structure factors $F(h, k, l)$ are the Fourier transform of $\rho(x, y, z)$, the electron density in a unit cell of the crystal.

• The electron density in a unit cell of the crystal $\rho(x, y, z)$ is the *inverse* Fourier transform of the structure factors $F(h, k, l)$.

Knowing $\rho(x, y, z)$ we can compute the $F(h, k, l)$; knowing the $F(h, k, l)$ we can compute $\rho(x, y, z)$.

Thus in principle, if we can measure the $F(h, k, l)$, we could use the Fourier transform relationship to compute the electron density in the crystal. Peaks in the electron density correspond to positions of atoms. Picking the peaks in the electron density gives us the coordinates of the atoms. We must figure out which atoms go with which peaks, but knowing the primary structure of the protein this should not be too difficult.

Now for the bad news.

The experimental data are incomplete—we can measure only the *intensities* of the diffracted rays, but not their *phases*. The difficulty of solving the **phase problem** for proteins accounts for the 25-year interval between Bernal's picture and the atomic structure of sperm-whale myoglobin.

> The structure factors $F(h, k, l)$ are in general *complex numbers*—of the form $a + bi$, where $i = \sqrt{-1}$ or equivalently $F = |F|^{\exp(i\phi)}$, where $|F|$ is the absolute value of the structure factor and ϕ is its phase angle. The experiment determines $|F|$ but not ϕ. This is the notorious *phase problem* of X-ray crystallography.

The phase problem is simpler for small-molecule crystallography, and indeed the structures of simple salts and minerals were solved very early on. It is much harder for proteins. However, there are now quite a few approaches that have proved successful.

 BOX 3.3　　**Fourier and Fourier series**

Jean-Baptiste Joseph Fourier (1768–1830) lived through exciting times: the revolution and the era of Napoléon. He made many notable contributions to mathematics and science, as a result of which his name is one of those to appear on the Tour Eiffel http://en.wikipedia.org/wiki/List_of_the_72_names_on_the_Eiffel_Tower.* His most important achievement was the discovery of the expansion in trigonometric functions—sines and cosines—such that (in one dimension) a function $y(x)$ can be approximated by a series:

$$y(x) = F_0 + F_1 \cos x + F_1' \sin x + F_2 \cos 2x$$
$$+ F_2' \sin 2x + F_3 \cos 3x + F_3' \sin 3x + \ldots$$

The numbers $F_0, F_1, F_1' \ldots$ are constants that characterize different functions $y(x)$. The function $y(x)$ is expressed as a sum of discrete terms $\cos nx$ and $\sin nx$; the constant term is just $F_0 \cos 0x = F_0 \cos 0 = F_0$. Figure 3.13 contains an example.

In X-ray crystallography, in three dimensions, Fourier series relate the structure factors $F(h, k, l)$ to the electron density $\rho(x, y, z)$. Again the structure factors are discrete and the electron density is a continuous function. The electron density is real and positive; the structure factors are in general complex numbers, each with a magnitude and a phase.

The rules of the game are:

1. If you know $\rho(x, y, z)$ you can compute the $F(h, k, l)$, both the magnitudes and the phases.

2. If you know both the magnitudes and the phases of $F(h, k, l)$, you can compute $\rho(x, y, z)$.

3. But, if you know only the magnitudes *but not the phases* of the $F(h, k, l)$, you *cannot* compute $\rho(x, y, z)$. The fact that only the magnitudes of $F(h, k, l)$ but not the phases are determinable from the intensity pattern in a diffraction pattern creates the phase problem. (The magnitude of an observed F is simply proportional to the square root of the corresponding diffraction intensity.)

4. From (1), if you have a candidate structure $\rho(x, y, z)$—in the form of a list of atomic positions—you can compute the $F(h, k, l)$, and compare the magnitudes of the computed F's with the experimentally measured intensities. This provides a way of evaluating the candidate structure. Computational methods can explore ways of changing the candidate structure to improve the agreement of calculated and experimental structure factor magnitudes.

---

*Fourier was a scientific advisor to Napoléon's expedition to Egypt, launched in 1798. Among the achievements of the scientific/scholarly component of that expedition was the discovery of the Rosetta Stone, containing parallel inscriptions in Greek, hieroglyphics, and demotic (colloquial) Egyptian. It was the key to the decipherment of hieroglyphics. The French navy was less successful; the fleet was almost entirely lost to the British, under Nelson, in the Battle of the Nile, in Aboukir Bay. In return for safe passage back to France for the surviving members of the expedition, the British took over the archaeological artefacts collected, including the Rosetta Stone, which is now on display in the British Museum. However, Fourier brought back an inked impression of the inscriptions. He later showed this to Champollion, then an 11-year-old student. Champollion ultimately achieved its decipherment.

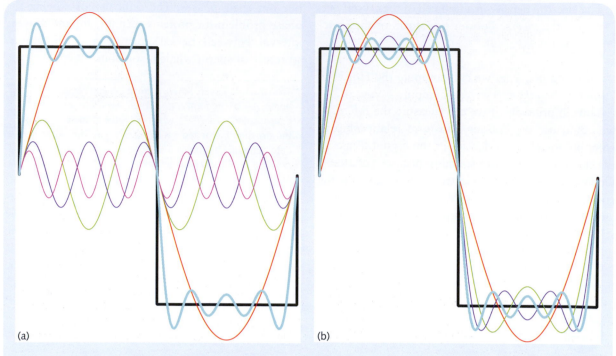

(a)

(b)

Figure 3.13 Fourier series approximation to a 'step function', or 'square wave': $y(x) = 1$ for $0 \leq x \leq 180°$ and $y(x) = -1$ for $180° \leq x \leq 360°$. (a) Solid black line is $y(x)$, the function to be approximated. The values of the first four terms of the Fourier series are shown in red, green, blue and magenta, respectively. The cyan curve (line somewhat thicker) is the sum of the first four terms. (b) Again solid black line is $y(x)$, the function to be approximated. The *sums* of the values of the first n terms of the Fourier series, $n = 1, 2, 3$ and 4, are shown in red, green, blue and cyan, respectively. The cyan curve (line somewhat thicker), the sum of the first four terms, is the same as in (a).

Solving the phase problem in protein crystallography

• The earliest successes in solving the phase problem for proteins used the method of **isomorphous replacement**. Phases were determined by combining diffraction data from a native crystal with data from other crystals containing the same protein, packed in the same way, but modified by the addition of a heavy atom. By comparing the diffraction patterns of the different crystals, it is possible first to determine the positions of the heavy atoms, and from them to compute phases of other structure factors.

• Many new proteins are similar to proteins of known structure. Calculation of the complete diffraction pattern—intensities and phases—expected from the related structure provides approximate phases with which to solve the new one. This method is called **molecular replacement** (see Box 3.4).

BOX 3.4   **The Patterson function and molecular replacement**

Given only the absolute values of the intensities of the X-ray structure factors, or reflections, it is not possible to compute the electron density in the unit cell of the crystal, because the phases are unknown. What *is* possible to compute, without knowing the phases, is the Patterson function, a map giving the distribution of the interatomic distances in a structure. Figure 3.14 shows the relationship between the real structure (in practice, unknown) and the **Patterson function** (directly computable from the data). It is possible to solve the structures of small molecules from Patterson functions alone. But, because the number of peaks in the Patterson function goes up as the square of the number of atoms, Patterson functions of macromolecules are very complex.

Thus, for any model, we can compute its Patterson function. Given the measured structure factors for an unknown structure, we can compute *its* Patterson function. Figure 3.15 shows (a) the model (known), (b) the true structure *(unknown)*, (c) the Patterson function of the model, and (d) the Patterson function of the true structure. Even if the model is correct, to a good approximation, we have no idea of the orientation of the unknown structure in the unit cell. Therefore the two Patterson functions do not appear identical. However, if we rotate the Patterson function of the model, we can get it to match the Patterson function of the unknown.

The results of such a 'Patterson search' calculation are to position and orientate the model in the unit cell. Knowing this, we can compute the Fourier transform of the model, *including the phases,* and transfer the phases to the unknown to compute an electron density:

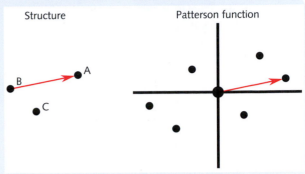

Figure 3.14 The Patterson function of a three-atom structure. The Patterson function has a peak corresponding to the vector between every pair of atoms. The red arrow from atom B to atom A in the structure corresponds to the rightmost peak in the Patterson function. Because there is a vector of zero length between every atom and itself, the Patterson function has a strong peak at the origin. Note that the Patterson function *contains* the true structure. But the number of peaks in the Patterson increases as the square of the number of atoms. For this 'toy' three-atom structure, there are only six peaks other than the origin peak. But obviously for macromolecules the Patterson is quite complicated. (In some cases, the vectors between symmetry-related molecules give a real—albeit scaled—view of the molecule in projection.)

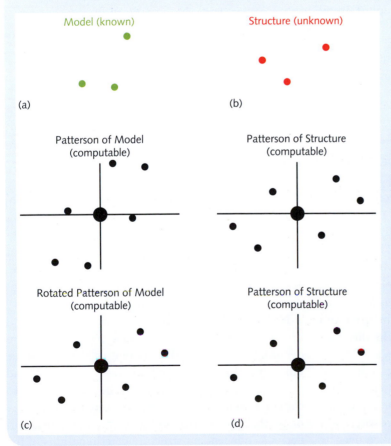

Figure 3.15 Solving a structure by molecular replacement. Patterson search to orientate model in a cell. The idea is that we know the atomic coordinates of a model (upper left, green) of a structure for which we have X-ray diffraction data but we do *not* have the atomic coordinates of the structure (upper right, red = unknown). In particular we have no idea of the position and orientation of the structure in its unit cell. We can compute Patterson functions of both the known and unknown (centre left and right). Because our model is exact, the Patterson functions are identical except for a difference in orientation. By rotating the Patterson of the model to find a match (bottom row), we can determine the orientation. Applying the same rotation to the model (green) would convert it to the orientation of the structure (red). We would also have to determine the *position* of the model in the unit cell by related methods.

Observed structure factor magnitudes: $|F_o(h, k, l)|$

Computed structure factors of model, magnitudes *and* phases:

$$F_c(h,k,l) = |F_c(h,k,l)|e^{i\phi_c(h,k,l)}$$

$$\phi_c(h,k,l) = \text{phase of } F_c(h, k, l)$$

Transfer computed phases to observed magnitudes: $|F_o(h,k,l)|e^{i\phi_c(h,k,l)}$

Compute electron density:

$\rho(x,y,z) = $ Fourier transform of $|F_o(h,k,l)|e^{\phi_c(h,k,l)}$

Sources of molecular replacement models include homology models, or theoretical predictions using programs such as **ROSETTA**.

- Certain atoms absorb as well as scatter X-rays, with effects on the diffraction pattern that contain phase information. This is called **anomalous dispersion**. Solution of crystal structures from measurements of the variation of the intensity distribution in the diffraction pattern over a range of wavelengths is called the method of **multiwavelength anomalous dispersion**, or **MAD**. It requires a tunable X-ray source, available at a synchrotron. Some structures have been solved using anomalous dispersion from sulphur atoms in proteins.

Most newly determined structures are solved using molecular replacement or anomalous dispersion methods.

- Knowledge of general features of electron-density distributions in crystals—for instance, that they must always be positive, and have certain statistical properties—permits calculation of phases directly from the experimental data. **Direct methods** have routinely solved the structures of small molecules for many years, and are now applicable in protein crystallography.

The phase problem of X-ray crystallography arises because the experimental data are incomplete. Methods for solving the phase problem in protein structure determination include:

- isomorphous replacement
- molecular replacement
- multiwavelength anomalous dispersion
- direct methods

Nuclear magnetic resonance spectroscopy in structural biology

Nuclear magnetic resonance (NMR) spectra measure the transitions between energy levels of the magnetic nuclei in atoms placed in a strong homogeneous external magnetic field (see Box 3.5). These energy levels are sensitive to the chemical environment of the atom. They are therefore very rich in information about chemical structure and dynamics. Moreover, nuclei of certain atoms close together in space can detectably exchange excitation energy. In this way, NMR can identify neighbouring residues in proteins, even if they are separated widely in the amino acid sequence. This information contains essential clues for determining the folding pattern of the chain.

Like X-ray diffraction, NMR spectroscopy is a technique invented by physicists, but taken over by chemists. NMR was first observed in 1945–1946, independently by E.M. Purcell, R.V. Pound and H.C. Torrey; and by F. Bloch, W. Hansen and M. Packard. It was one of a very large number of important discoveries made by scientists returning to basic research after participating in projects related to support of the military during World War II.

NMR was first applied to studies of peptides and proteins in the 1970s. The first protein structure determination by NMR was that of the 57-residue bull-seminal protease inhibitor, by K. Wüthrich and collaborators, in 1984. To prove the method so as to convince even the most hardened sceptics, the groups of Wüthrich in Zürich and R. Huber in Munich undertook independent structure determinations of the α-amylase inhibitor, tendamistat, by NMR spectroscopy and X-ray crystallography, respectively. The structures were virtually identical. Subsequently, NMR spectroscopy has taken its place in structural biology alongside X-ray crystallography. There are currently about 10000 NMR entries in the wwPDB, including proteins and nucleic acids.

As methods of macromolecular structure determination, X-ray crystallography and NMR spectroscopy each have advantages and disadvantages. The main advantage of NMR is that it is not necessary to produce crystals; this is occasionally a severe impediment to X-ray structure determination. NMR also gives us a window into protein dynamics, on timescales of about 10^{-9}–10^{-6} s and slower than 10^{-3} s.

Proteins in solution are not subject to the constraints of crystal packing, and are expected to flounce around somewhat. Indeed, typically the result of an NMR structure determination is a set of similar but not identical models (often ~15–20 of them), all of which are comparably consistent with the

combination of experimental data and general stereochemical restraints. This may be a manifestation of dynamics, or incompleteness in the experimental data that leave the structure underdetermined.

NMR has found a large and varied set of applications, not only in physics, chemistry, and molecular biology, but in medicine (magnetic resonance imaging), geology (lowering NMR spectrometers into bore holes to determine the composition and mobility of fluids in underground reservoirs), and materials science (chemical structure determination, imaging of structure, dynamics). NMR has important applications in drug development, for example as a method of screening protein–ligand interactions. The NMR–MOUSE is a portable spectrometer useful for subjects that cannot be brought into the laboratory, for example paintings in art museums, or rare books.

NMR spectra of proteins

The energy levels of a magnetic nucleus depend on the nuclear magnetic moment, characteristic of the atom, and on the *local* magnetic-field strength at the nucleus. This local field is a combination of the external applied field, and its perturbation by electrons in the vicinity of the nucleus. For example, the shielding of a proton in a methyl group will be different from that of a proton bound to a nitrogen.

It follows that each type of chemical group will appear in a different region of an NMR spectrum. In a peptide, or a protein in a denatured state, different

---

Characteristics of protein structure determination by NMR:

- no crystals are required
- very large structures now solvable
- structures in solid state solvable, especially important for membrane proteins
- requires isotopic labelling
- gives multiple models

BOX 3.5   **Nuclear spin, magnetism, and energy levels**

Nuclear magnetism is a consequence of the 'spin' angular momentum of each nucleon—proton or neutron. In nuclei that contain even numbers of protons and even numbers of neutrons, the nucleon spins pair, and cancel; just as in most atoms and molecules with even numbers of electrons the electron spins pair, and cancel to produce a singlet (*i.e.*, net spin = 0) state. Different isotopes of the same element can have different nuclear magnetic moments.

Nuclei with non-zero total angular momentum have a nuclear magnetic moment. Nuclei of use in protein and nucleic acid structure determination include ^1H (the proton),

^2H (the deuteron), ^{13}C, ^{15}N, ^{19}F, and ^{31}P. Because all these except ^1H and ^{31}P are present in natural materials in only trace amounts, NMR spectroscopy often makes use of isotopically labelled samples.

In a magnetic field, the orientations of the nuclear spins, and of the magnetic moments, are quantized: only a few discrete orientations are possible. These states have different energies. In magnetic fields of the strengths typical of NMR experiments, spectroscopy at radio frequencies can induce, and allow the measurement of, transitions between different nuclear spin states.

amino acids will make individual, distinguishable contributions to the NMR spectrum. In the denatured state, all amino acids of the same type (for instance, all threonines) show nearly the same transition energies and, therefore, overlapping peaks in the spectrum. In a protein that adopts a specific tertiary structure, different microenvironments will further perturb the nuclear energy levels. Especially important are the effects of ring currents from aromatic sidechains, and hydrogen-bonding interactions. In the NMR spectrum of a folded protein, different amino acids of the same type will *not* give identical contributions. Spectral dispersion (that is, deviation between the observed spectrum and the spectrum expected for a denatured polypeptide) is indicative of a folded structure.

Figure 3.16(a) contains the NMR spectrum of *E. coli* IscU, a protein that plays a major role in the formation of iron–sulphur clusters. This spectrum, showing as is conventional the absorption intensity as a function of chemical shift—that is, of frequency—is in the context of NMR called a one-dimensional (1D) spectrum. Its richness—the very large numbers of resolved peaks—shows that IscU is in a folded state.

The scale on the abscissa of Fig. 3.16(a) is the **chemical shift**. Chemical shifts are expressed as the difference, in parts per million (ppm), of the frequency of the observed transition, v, relative to the frequency of a reference standard, v_{ref}:

$$\text{Chemical shift (in ppm)} = \delta = \frac{v - v_{ref}}{v_{ref}} \times 10^6.$$

Because the chemical shift is a ratio, it is independent of the strength of the magnetic field at which the spectrum was collected. This allows direct comparison between experiments on different instruments. (For aqueous solutions of biomolecules, the recommended reference standard is the methyl ^1H resonance of 2,2-dimethyl-2-silapentane-5-sulphonic acid.)

One-dimensional spectra such as Fig. 3.16(a) provide very detailed structural information for small molecules, but for proteins the peaks overlap and cannot be resolved. It is necessary to go to two, or even more, dimensions, to increase the resolution of the experiment, as in Figs. 3.16(b) and (c). The basic goal of multidimensional NMR spectroscopy is

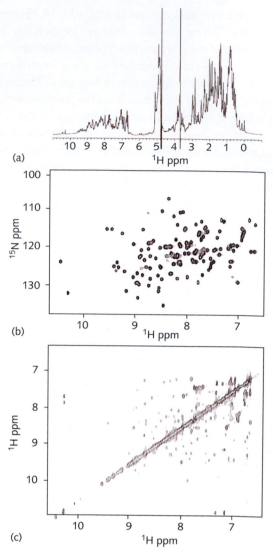

(a)

(b)

(c)

Figure 3.16 NMR spectra of *E. coli* iron–sulphur cluster scaffold homologue (IscU). IscU plays a major role in the assembly of iron–sulphur clusters. (a) One-dimensional ^1H NMR spectrum of the apo form. (b) ^1H–^{15}N heteronuclear single quantum coherence (HSQC) spectrum of uniformly ^{15}N-labelled IscU. The HSQC experiment takes advantage of the scalar (or J) coupling between directly bonded nuclei, in this case ^1H and ^{15}N. (Isotopic labelling is necessary because the abundant ^{14}N nucleus is quadrupolar— spin quantum number I = 1—and has unfavourable spectroscopic properties. Nuclei with spin I = $\frac{1}{2}$ are better behaved in NMR experiments.) The sharpness and dispersion of the peaks show that the protein is well structured without significant disordered regions. (c) Amide region of ^1H homonuclear NOESY spectrum. The off-diagonal peaks in this spectrum correspond to protons close enough together in space for effective excitation energy transfer during the experiment.

[From: Adinolfi, S., Rizzo, F., Masino, L., Nair, M., Martin, S.R., Pastore, A. and Temussi, P.A. (2004). Bacterial IscU is a well folded and functional single domain protein. *Eur. J. Biochem.*, **271**, 2093–2100.]

to resolve the information contained in a spin system according to selected spin–spin interactions and chemical shifts. For example, in two-dimensional experiments, magnetization is transferred from spin to spin via through-bond or through-space interactions, and the resulting spectrum has two frequency axes, corresponding to the chemical shifts of two nuclei. The appearance of intensity at chemical-shift coordinates (x, y) with $x \neq y$ indicates the presence of the selected interaction between spins resonating at x and y.

Measurement of NMR spectra

The basic idea of an NMR experiment is to start with a set of nuclei at equilibrium in the presence of a constant external magnetic field, to perturb this equilibrium in a controlled fashion, and to detect the resulting signals. A large number of experimental protocols are available to select interactions among spins and reveal information from which inferences about structure can be drawn.

In a sample placed in a constant external magnetic field, nuclear magnetic moments are distributed with a slight preference for the direction of the external magnetic field. As a result, the sample exhibits a net magnetization parallel to the external field. Radio-frequency pulses with magnetic-field component perpendicular to the static field are used to rotate this net magnetization into the transverse plane (the plane perpendicular to the static field). The transverse component of the magnetization precesses around the static field at a frequency that depends on the environment of the nucleus of interest. The precession induces a voltage in the detection coil (see Fig. 3.17).

A **pulse sequence** is a series of pulses and delay times, followed by detection of a signal, designed to extract specific information from a set of interacting spins. For example, COSY (correlation spectroscopy) can measure interactions through chemical bonds, giving information about the covalent network characteristic of each amino acid type. Nuclear Overhauser effect spectroscopy (NOESY) experiments measure interactions between nuclei close together in space. The nuclear Overhauser effect is the result of a through-space dipole–dipole relaxation process and does not require intervening bonds

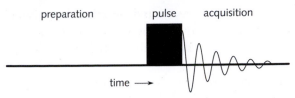

Figure 3.17 In an NMR experiment a sample equilibrated in a constant magnetic field receives a pulse of oscillating magnetic field (radio-frequency pulse), disturbing the equilibrium alignment. The sample's precessing magnetization induces an oscillating voltage in the receiver coil. In addition, the net magnetization component perpendicular to the external field decays as the spins lose synchrony. Note that the Fourier transform of a single-frequency signal is a single peak. Therefore, spectral data are Fourier transformed to produce interpretable results. The rate of decay of the signal, seen in the figure, depends on the transverse relaxation time of the system, and controls the width of the spectral line. Long relaxation times correspond to sharp peaks and short relaxation times correspond to broad ones.

between the interacting nuclei. The one-dimensional version of the experiment involves the selective excitation of a set of spins. The dipole–dipole interaction then transfers the magnetization to neighbouring spins, modifying the populations of the ground and excited states. Because the strength of the NMR signal is proportional to the population difference, intensity changes are observed compared to a reference spectrum for which no excitation was applied. The two-dimensional experiment provides proximity information for all spins at once.

Protein structure determination by NMR

NMR spectra provide three general types of data that make it possible to determine the structures of proteins.

- From effects transmitted between atoms bonded to each other, which affect the precise frequency of the signal from an atom (the chemical shift), NMR can determine the values of conformational angles. In particular, chemical shifts can define secondary structures.

- From interactions through space between non-bonded atoms <5 Å apart (detected by the nuclear Overhauser effect or NOE), NMR can identify pairs of atoms close together in the structure, including those *not* close together in the sequence. The information about atoms distant in the sequence but close together in space is crucial to being able to assemble individual regions into the correct overall folding pattern.

- Residual dipolar couplings are another source of NMR spectral information useful in solving structures and in studying the dynamic state of protein molecules. Residual dipolar couplings are measured on proteins *partially* aligned with respect to the external magnetic field, in gels, or in liquid-crystal media. This contrasts to the freely tumbling solution state with no preferred alignment of the molecules. Because residual dipolar couplings give information about orientations of structural segments with respect to the external magnetic field, they provide information about the *relative* orientation of different parts of the molecule, even if these parts are far apart in the sequence and the structure. This can resolve some of the ambiguity in structure that arises if the chemical shift and NOE data leave parts of the structure underdetermined.

Analysis of residual dipolar coupling can also resolve the question of whether the multiplicity of models emerging in an NMR structure determination is the result of genuine dynamic disorder or underdetermination of the structure by the data.

Determination of mainchain conformational angles

Chemical shifts of H (attached to Cα), Cα, Cβ, carbonyl C, and mainchain N indicate values of mainchain conformational angles ϕ and ψ (see Fig. 3.18(a)).

Values of ϕ and ψ are not enough information for a determination of the structure from the experimental data alone. An NMR spectroscopist would go on to determine the residues neighbouring in space (see below). However, combining chemical-shift data with the most powerful protein structure prediction methods can produce reasonable structures for proteins of up to about 130 residues. This is an example of a hybrid experimental–theoretical method.

Determination of distance restraints

Information important for structure determination that NMR spectroscopy can provide is the identification of pairs of atoms that are close together in space, whether they are nearby or distant in the amino acid sequence. This provides information about the global folding pattern.

Figure 3.16(c) illustrates a NOESY experiment. The diagonal corresponds to the one-dimensional spectrum. Off-diagonal peaks correspond to through-space interactions.

Assignment of the spectrum

To derive explicit structural information from the data (see Fig. 3.18), it is necessary to *assign the spectrum;* that is, to identify which residues are giving rise to the individual peaks. For small systems, homonuclear data, such as shown in Fig. 3.16(c), suffice for complete assignment of the spectrum.

Different amino acids give characteristic covalent patterns in COSY-like experiments. This associates certain frequencies with certain amino acid types. Then, NOEs, typically involving backbone protons, are used to link amino acids with their neighbours in the sequence. This allows correlating peaks in the spectrum with specific dipeptides, which can be placed within the amino acid sequence. With these as anchors, it is then possible to 'walk along the chain' assigning more and more peaks to particular residues. Eventually all the peaks in the spectrum are accounted for, and cross-peaks linking residues distant in the sequence can be catalogued.

As NMR spectroscopists tackle larger proteins, it is generally necessary to label proteins with ^{13}C and ^{15}N, and use 'triple resonance' experiments, with radio-frequency pulses applied at ^{13}C, ^{15}N, and ^{1}H frequencies during the pulse sequence. Heteronuclear data greatly facilitate assignment of the spectrum, and allow solution of structures of larger proteins.

The intensities of the NOE interactions, as measured by the size of the cross-peaks, imply distance constraints between pairs of atoms. The intensity of a cross-peak varies, to a first approximation, as r^{-6}, where r is the distance between the interacting atoms. The stronger the NOE intensity, the shorter the interatomic distance. It is not possible to determine the internuclear distances exactly. Typical inferences take the form shown in Table 3.2.

(a)

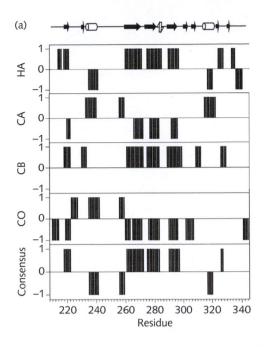

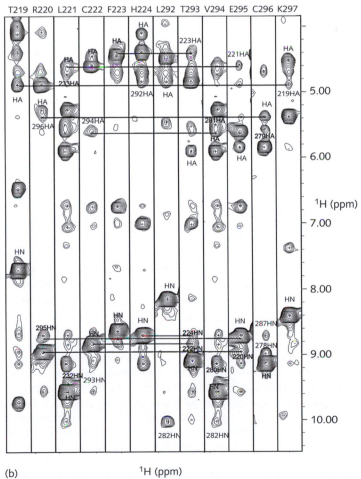

(b) ¹H (ppm)

Figure 3.18 *(Continued)*

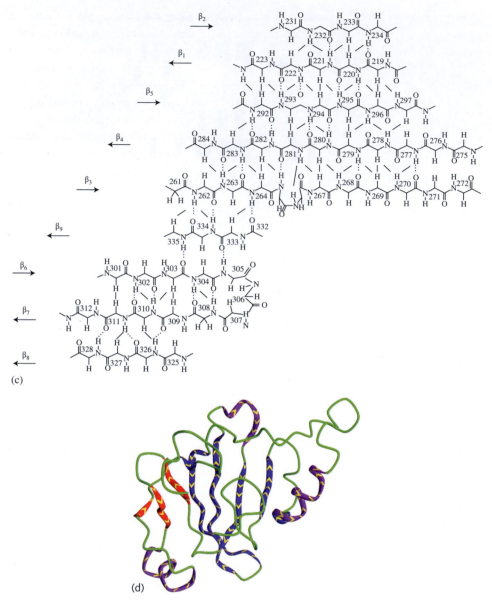

(c)

(d)

Figure 3.18 The AXH domain appears in two proteins, unrelated in overall domain organization and function: the transcription factor HBP1; and ataxin-1. CAG repeats in the gene for ataxin-1 are responsible for spinocerebellar ataxia type-1. Pastore and coworkers have solved the structure of this protein by NMR spectroscopy. (a) The chemical-shift indices for backbone atoms as a function of residue number. The chemical-shift index $\Delta_s = \delta_{native} - \delta_{denatured}$ is the difference between its chemical shift in the native structure and its chemical shift in the denatured state. The sign of the chemical-shift index is sensitive to secondary structure. The inferred secondary structure of AXH appears above the chart. (b) Strips from a 3D ^{1}H–^{15}N NOESY-HSQC. In this experiment, ^{1}H–^{1}H NOE correlations are detected between amide hydrogens and other hydrogens within ~5 Å. The third dimension (not shown) is the ^{15}N chemical shift of the amide nitrogen. The ^{1}H–^{1}H strips are arranged in sequential order with horizontal lines indicating backbone NOEs. (c) The organization of the β-sheet derived from proton–proton contact information from the NOE spectrum. Solid lines indicate contact information. Broken lines indicate hydrogen bonds. (d) The final structure [1v06]. This is a model selected from the largest cluster of deposited structures by OLDERADO. (http://www.ebi.ac.uk/pdbe/olderado/).

[Parts (a), (b) and (c) from: de Chiara, C., Menon, R.P., Adinolfi, S., de Boer, J., Ktistaki, E., Kelly, G., Calder, L., Kioussis. D. and Pastore A. (2005). The AXH domain adopts alternative folds: The solution structure of HBP1 AXH. *Structure*, **13**, 743–753.]

Table 3.2 Distance constraints from NOESY

NOE intensity	Upper bound (Å) to distance
very strong	2.5
strong	3.1
medium	3.4
weak	3.9
very weak	5.0

Transverse relaxation optimized spectroscopy

Formerly, protein structure determination by NMR suffered severe limitations on the size of the molecule treatable, as a result of the slower tumbling rates of larger molecules. In addition, the larger the molecule, the more interacting pairs of nuclei and the more complex the spectrum. Size limits could be pushed by combinations of isotopic labelling of N, C, and H atoms, but of course scientists had unsated appetites for larger and larger structures.

The TROSY (transverse relaxation optimized spectroscopy) technique has greatly extended the size limit for protein structure determination by NMR, to molecules containing more than 1000 residues. TROSY selects the component of the signal that is less relaxed (and therefore sharper). NMR structures solved using TROSY include the GroEL-GroES chaperonin complex, r.m.m. ~900 000 and the complex of the 20S CP proteasome with the 11S regulator protein, r.m.m. 1 100 000.

A particularly important application of TROSY is the determination of membrane protein structures. Membrane proteins can be difficult to crystallize. TROSY makes it possible to study them in detergent-lipid micelles. Structures solved with this technique include the outer membrane proteins OmpX and OmpA (see Fig. 2.28 showed the crystal structure of OmpA.)

TROSY also enhances our ability to reveal inter-molecular interactions, with applications including, but not limited to, screening for potential drugs. It is possible not only to detect binding, but to map out the residues in contact.

From the data to the structure

Computer programs solve for structures compatible with the experimental data: the distance constraints,

and the inferences about mainchain conformational angles from chemical shifts. Also included in the calculation are conformational energy terms. The structures computed must also have low conformational energies, including proper stereochemistry. Even a structure determination using all information from NMR measurements—dihedral angles, distance constraints, and residual dipolar couplings—is therefore a hybrid experimental-computational method. The predominant technique for determining structure from NMR data is restrained molecular dynamics; that is, molecular dynamics with additional pseudo-energy terms that force agreement with the experimental data.

In general, NMR protein structure determinations produce multiple models, all consistent with the experimental data. Figure 3.19 shows a superposition of 19 models reported for the structure of the *Drosophila* antennapedia protein.

NMR spectroscopists have developed criteria for evaluating their structures. As in the case of X-ray structures, there is a combination of checking the stereochemistry of the result, and the agreement of the result with experimental data. NMR and X-ray structures share the evaluation of deviations from regular stereochemistry, such as unusual bond lengths

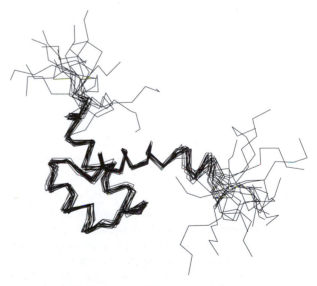

Figure 3.19 Models of *Drosophila* antennapedia protein, determined by NMR spectroscopy [1HOM]. The core of the molecule, comprising the interacting regions of secondary structure, is consistent among the models, but the regions at the N- and C-termini appear variable.

or angles, and the requirement that most residues appear in allowed regions of the Sasisekharan–Ramakrishnan–Ramachandran plot. For NMR structures, structure validation includes checking for agreement with experimental restraints on dihedral angles and interatomic distances and, if residual dipolar couplings are measured, the quality of the fit of RDCs back-calculated from the structure to the experimental data.

Solid-state NMR: magic angle spinning

X-ray structure determinations require the crystalline state, and take advantage of it. The solid state is a handicap to NMR spectroscopy, but special techniques can overcome the difficulties. That membrane proteins are most easily prepared as non-crystalline solids makes application of X-ray diffraction difficult (although not impossible). NMR spectroscopy of samples in the solid state has stepped in to make important contributions to structure determinations of this very important class of proteins.

What is the problem in measuring NMR spectra in the solid state? NMR spectra in solution contain many sharp lines, and are correspondingly rich in information. In the solid state, different molecules are oriented differently with respect to the magnetic field, affecting their shielding and altering their chemical shifts. Instead of sharp lines, NMR spectra of solids contain broad peaks, in part reflecting the dispersion of chemical shifts. In the liquid state, in contrast, sufficiently rapid rotation of the molecules sharpens the lines, as each molecule gives a signal corresponding to the average of its orientations.

Another effect leading to the broadening of the lines in NMR spectra of solids is dipolar coupling. This is the interaction between the nuclear magnetic moments of neighbouring atoms. Dipolar coupling depends on the angle between the internuclear vector and the direction of the external magnetic field. In solids the interatomic vectors point in different directions. Therefore the energy shifts are variable, and contribute to line broadening.

E. R. Andrew and colleagues discovered that spinning the sample sufficiently rapidly around an axis making the so-called 'magic angle' 54.7356° with the external magnetic field will cause broadening effects to average to zero, recovering sharp lines (see

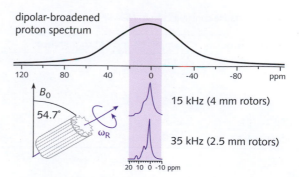

Figure 3.20 Top: the broadened spectrum of protons in a sample at rest in the solid state. Bottom, left: The magic angle, 54.7356°, between the axis of rotation of the sample and the applied magnetic field, B_0. Bottom right: Progressive sharpening of the lines as a function of the angular velocity ω_R of the rotation.

(Picture reproduced courtesy of Dr R. Graf.)

Fig. 3.20). How does this work? If the sample is spinning around an axis that makes an angle θ with the magnetic field, the line broadening arising from dipolar coupling will be proportional to the second Legendre polynomial: $3\cos^2\theta - 1$. For the magic angle, $3\cos^2\theta - 1 = 0$. (In solution, the molecules are spinning around all possible axes. However, for eliminating the line-broadening effects of dipolar coupling, tumbling around axes other than the ones at the magic angle is supererogation.)

Anabaena sensory rhodopsin is a cyanobacterial retinal-binding protein that functions as a photoresponsive regulator of the expression of proteins involved in photosynthesis and circadian rhythms. It is an integral membrane protein, forming a 81 kDa homotrimer. Structure determination of this protein, reconstituted in a synthetic lipid phase, illustrates a successful application of solid-state NMR spectroscopy. Fig. 3.21 shows the structure of the monomer and the pattern of contacts from which it was derived. The folding pattern is a typical 7-transmembrane helix protein, similar to that first observed in the structure bacteriorhodopsin. In the picture of the structure in Fig. 3.21 (right), the retinal is coloured yellow. In the contact diagram (left), the thick blue line represents the retinal, and thin blue lines indicate protein-retinol contacts.

The function of the protein is to respond to changes in illumination by regulating expression of two protein accessory pigments: phycocyanin and phycoerythrin. They are constituents of phycobilisomes,

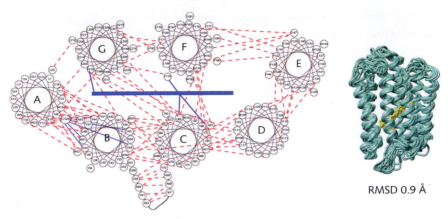

RMSD 0.9 Å

Figure 3.21 Left: network of interresidue contacts derived from solid-state NMR spectra (viewed perpendicular to the membrane, from the periplasmic side). Right: Ensemble of 10 lowest-energy models determined from the data. The average root-mean-square deviation of the Cα atoms in the helices is 0.9 Å. The loops between helices are more variable in conformation: clearly it requires less information to build a rigid object of known structure, such as an α-helix, than a loop of unknown conformation (see also Problem 3.8).

From: Wang, S., Munro, R.A., Shi, L., Kawamura, I., Okitsu, T., Wada, A., Kim, S.Y., Jung, K.H., Brown, L.S. and Ladizhansky V. (2013). Solid-state NMR spectroscopy structure determination of a lipid-embedded heptahelical membrane protein. Nat. Methods 10, 1007–1012.

large protein complexes attached to membranes of thylakoids in chloroplasts. Proteins of the phycobilisome function as 'antenna pigments': their chromophores capture light at wavelengths at which chlorophyll absorbs inefficiently—green, obviously—and transfer the excitation energy to chlorophylls. Cyanobacterial regulation of the expression of these antenna proteins in response to ambient light is called chromatic adaptation.

Anabaena sensory rhodopsin functions as a 'receptor', for light rather than for a chemical signal. It binds a second protein, called the *Anabaena* sensory rhodopsin transducer. Depending on the spectrum and intensity of the ambient light, the rhodopsin changes from the all-*trans* conformation, in the dark, to the 13-*cis* isomer. The ratio of the two isomers depends on the wavelength of the ambient light. The isomerization of the rhodopsin leads to a conformational change in the rhodopsin, and release of the transducer, which down-regulates phycocyanin and phycoerythrin expression.

The regulation of gene expression in response to light cries out for imaginative applications. 'Optogenetics'—as this nascent field is called—has been used to control and monitor activities of individual neurons in living tissues.

Near atomic-resolution low-temperature electron microscopy (cryo-EM)

What are the prospects for extension of structural studies to very large aggregates, which may be difficult to prepare as single crystals suitable for X-ray diffraction, and are beyond the size limit of applicability of NMR?

Electron microscopy of specimens at liquid nitrogen temperatures has revealed structures of assemblies in the range $M_r = 500\ 000$ to 4×10^8, 100–1500 Å in diameter, such as the hepatitis B core shell or the clathrin coat. These results do not achieve atomic resolution—one sees 'blobs' rather than individual atoms. However, at 3–4 Å resolution, attainable in some cases, one can begin to recognize features such as helices and sheets.

An exciting idea is the combination of electron microscopy with separate high-resolution X-ray crystal structures of the component proteins. By determining the positions of the individual proteins within an aggregate, the high-resolution structures can be assembled into a full atomic model of the entire complex.

Characteristics of protein structure determination by cryo-electron microscopy:

• no size limit;

• combination with individual structure determinations of components is very powerful.

In cryo-EM a sample under native conditions is fast-frozen. The solvent is not given time to form the regular hexagonal equilibrium crystal form of ice that we see in snowflakes, but rather remains in a snapshot of the liquid state.

There are two main approaches to deriving a three-dimensional structure.

1. *Diffraction from two-dimensional crystals.* Some proteins form two-dimensional crystals in nature, including bacteriorhodopsin and ribulose-1,5-bisphosphate carboxylase/oxygenase. Others can be persuaded to form artificial two-dimensional crystals, *e.g.* aquaporin. Two-dimensional crystals *diffract* electrons. By measuring the diffraction pattern from tilted samples, collection of an almost complete three-dimensional diffraction pattern is possible. The structure of bacteriorhodopsin (see Fig. 2.27) was solved in this way. This is crystallography without the phase problem, as phases can be measured experimentally.

2. *Image reconstruction.* Many samples do not form crystals. Electron micrographs show images of individual particles. Each image gives a different *projection* of the structure. From multiple images of the same object, in different orientations, a three-dimensional structure can be assembled.

Octameric pyruvate-ferredoxin oxidoreductase from *Desulfovibrio vulgaris* Hildenborough

Pyruvate-ferredoxin oxidoreductase plays a role in the conversion of pyruvate to acetyl-CoA in most anaerobic bacteria, and certain other species. (In aerobic bacteria the pyruvate dehydrogenase complex catalyses this conversion—see Fig. 1.12.)

F. Garczarek, M. Dong, D. Typke, H.E. Witkowska, T.C. Hazen, E. Nogales, M.D. Biggin, and R.D. Glaeser determined the three-dimensional structure of pyruvate-ferredoxin oxidoreductase from *Desulfovibrio vulgaris* Hildenborough. This enzyme is an octamer. Reconstruction by cryo-EM produced a 17-Å resolution structure of the particle.

Although no crystal structure of the enzyme itself or a subunit was available, a related molecule from *Desulfovibrio africanus* had been solved. That enzyme is a dimer. The amino acid sequences of the *D. vulgaris* and *D. africanus* enzymes have 69% identical residues in optimal alignment. Therefore, it was possible to build a homology model of the *D. vulgaris* enzyme, and dock it into the EM structure. This combination of techniques achieved a complete atomic model of the octamer. This result illuminated the determinants of tertiary structure (see Fig. 3.22).

Conformational change in activation of human integrin αVβ3

A combination of X-ray crystallography and electron microscopy has illuminated the conformational change involved in activation of integrins.

Integrins are receptors found in all animals that mediate cell–cell interactions. They transmit information both from outside the cell to inside, to trigger signalling pathways, and from inside the cell to outside, to present the cell status to the environment. An example is their role in platelet aggregation in blood coagulation (see Chapter 5). Integrins on the surface of platelets, normally in an inactive state, become sticky in response to vascular injury, and bridge platelets together in the nascent clot.

Integrins contain multiple copies of two unrelated subunits, α and β. There are many integrins, denoted by their subunit composition. Integrin αVβ3 contains one α and three β subunits. This integrin modulates cell–cell adhesion and cell migration. It is involved in cardiovascular and bone function, and in particular is implicated in cell-growth regulation and angiogenesis (development of blood vessels), both in wound healing and tumour progression.

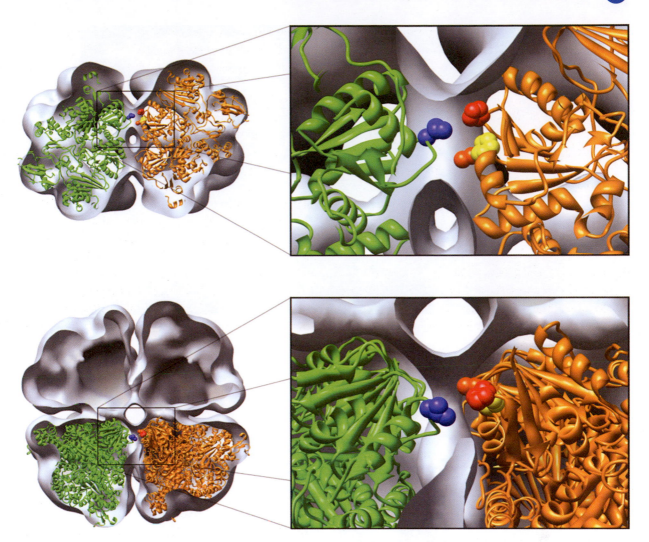

Figure 3.22 Two orthogonal views of pyruvate-ferredoxin oxidoreductase from *Desulfovibrio vulgaris* Hildenborough. This figure shows the interaction between two dimers; that is, it shows a tetramer, or half of the full structure. The detailed intermolecular contact is formed by Val383 (blue) from one dimer packed against Pro318 (yellow) and Ala319 (orange) from another. In the *D. africanus* enzyme, which forms only a dimer, the valine is deleted.

[From: Garczarek, F. *et al.* (2007). Octomeric pyruvate-ferredoxin oxidoreductase from *Desulfovibrio vulgaris. J. Struct. Biol.*, **159**, 9–18.]

The extracellular domains of integrin $\alpha V\beta 3$ contain four domains in the α subunit and eight domains in each β subunit. The first detailed structural information about this protein came from the X-ray crystal structures of the extracellular domains, in one case unliganded and in another binding a pentapeptide. Transition to the state active in ligand binding requires a large-scale conformational change. A three-dimensional structure based on reconstruction from electron-microscope images showed the complex of the extracellular domains of integrin

$\alpha V\beta 3$ with a four-domain fragment of fibronectin (see Fig. 3.23).

Protein structure determinations by cryo-EM are within shouting distance of true atomic resolution. Fig. 3.24 shows high-resolution EM maps and fitted atomic models, from recent work. Whereas Fig. 3.23 showed atomic-resolution structures created by fitting crystal structures into EM maps, the atomic models in the examples appearing in this Fig. 3.24 were constructed *de novo*.

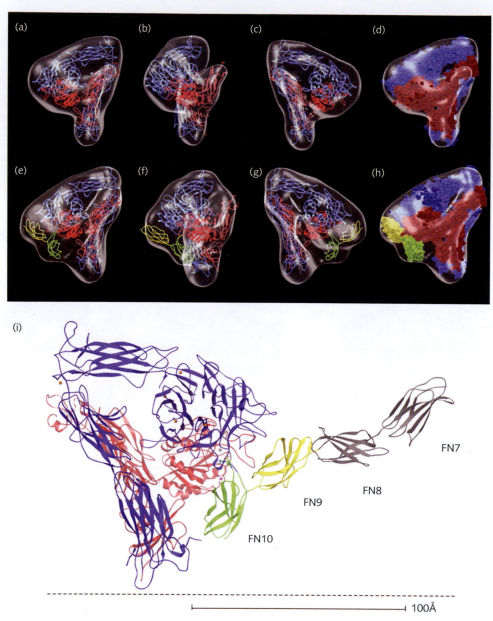

Figure 3.23 Docking of crystal structures into structure reconstructed from electron-microscope images of integrin αVβ3 in complex with fibronectin domains FN9 and FN10. αV chain in red; β3 chain in blue, FN9 domain in yellow, FN10 domain in green. (a), (b), (c): different views of the EM map of the uncomplexed integrin, with crystal structures fitted in. (d) Accessible surface of the crystal structure fitted into the EM map. (e), (f), (g): corresponding views of the EM map of the complexed integrin in complex with two fibronectin domains, with crystal structures fitted in. (h) Accessible surface of the crystal structure fitted into the EM map. (i) Diagram of the model, based on positioning and orientating the domains from the fitting to the EM map as in (g). The broken line below frame (i) is an estimate of the position of the plasma membrane, penetrated by additional transmembrane segments of the integrin not present in the structures determined.

[©**2005** Aidair, B.D., Xiong, J.-P., Maddock, C., Goodman, S.L., Arnaout, M.A. and Yeager, M., Three-dimensional EM structure of the ectodomain of integrin αVβ3 in a complex with fibronectin. **The Journal of Cell Biology, 2005, 168: 1109–1118.** doi:10.1083/jcb.200410068.]

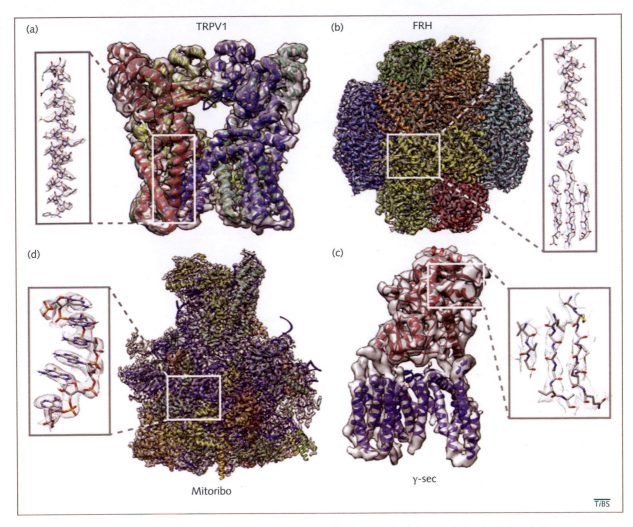

(a) TRPV1 (b) FRH (d) (c)

Mitoribo γ-sec

TiBS

Figure 3.24 (This page) Determination of large macromolecular complexes by cryo-EM. (a) The transient receptor potential cation channel subfamily V member 1 (TRPV1); (b) F420-reducing [NiFe] hydrogenase; (c) large subunit of the yeast mitochondrial ribosome; and (d) γ-secretase. For the first three structures, the resolution approaches 3 Å. In these maps individual sidechains, or RNA bases, are distinguishable. The γ-secretase map, part (d), has a resolution of 4.5 Å. Less detail is interpretable.

From: Bai, X., McMullan, G. and Scheres, S.H.W. (2015). How cryo-EM is revolutionizing structural biology. Trends. Biochem. Sci. 40, 49-57. (A, work of group of Y. Chang at the University of California, San Francisco; B, work of group of J. Vonck of the MPI Frankfurt; C, D, work of the group of S.H.W. Scheres, MRC Laboratory of Molecular Biology, Cambridge).

(Next page) Cryo-EM structure of the 1024-residue *Escherichia coli* β-galactosidase in complex with inhibitor phenylethyl β-D-thiogalactopyranoside (PETG). The enzyme is one of the proteins encoded in the Lac operon, by the *LacZ* gene (see Chapter 9). (a) The entire structure, to give an idea of its size. (b, c and d) Selected regions of the map, showing building of the atomic model into the electron density. '*' represents positions of carbonyl groups, which are difficult to resolve in all but the sharpest maps (see Fig. 3.9). The structure determination elucidated the interactions involved in the binding of the inhibitor. It was even possible to see positions of bound water molecules. The correspondence of the positions of waters in the cryo-EM and crystal structures gives confidence in their placement.

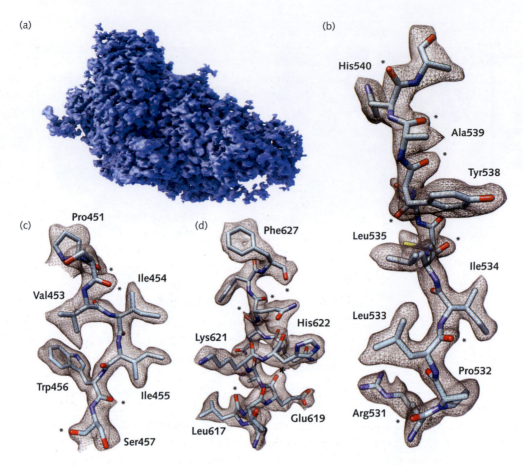

Figure 3.24 (*Continued*)

This figure represents the state of the art—at time of writing! The reader should ask how many of the amino acids could be identified from the electron density alone, were the model not shown, and be impressed by the answer. Note the small hole in the middle of the benzene ring of Phe627 (… not so wide as a church door, but 'twill serve). This does not show up in all but the highest-resolution crystallographic electron-density maps—even Fig. 3.9(d) shows only a 'dimple'. Experience suggests that 1.8-Å resolution is the threshold at which a hole in a Phe ring will sometimes be seen in a map derived from X-ray crystallography. Figure 3.8 (b), showing a genuine hole in a histidine ring, reflects 1.3-Å resolution data. Not all regions of the cryo-EM map are as good as the examples shown here. The authors assign an average resolution of 2.2 Å to their β-galactosidase cryo-EM map.

From: Bartesaghi, A., Merk, A., Banerjee, S., Matthies, Wu, X., Milne, J.L.S. And Subramanian, S. (2015) 2.2 Å resolution cryo-EM structure of β-galactosidase in complex with a cell-permeant inhibitor. Science Express 7 May 2015.

Trajectories of conformational change

The statement that proteins form unique native structures, in which the positions of all atoms are fixed as a consequence solely of the amino acid sequence, embodies a profound and fundamental principle of biology. However, when examined closely it is not perfectly true.

Many proteins undergo conformational changes. Some switch between two or more specific structures. Binding of ligands is in many cases the trigger of the conformational change. The allosteric change in haemoglobin is probably the most famous. However, many enzymes adopt an 'open' conformation in the absence of a ligand, and then close over a substrate after it binds. The closed state may achieve more and/or tighter interactions with the ligand. This is known as **induced fit**. The closed form may exclude water from the binding site, to prevent side reactions. In rhodopsins, light absorption induces conformational change, leading ultimately to a sensory nerve impulse.

- Conformational change is a dynamic characteristic intrinsic to many proteins, in many cases essential for function. Conformational change often needs to be taken into account in docking calculations (prediction of ligand binding). Approaches to predicting conformational changes include molecular dynamics and simplified elastic network models.

Many proteins have disordered regions. It may be that the unligated form is not entirely rigid, but becomes rigidified upon binding. In such cases the non-rigidity is limited to localized regions.

Some conformational changes are more complex than simple two-state switches. ATP synthase contains a rotary motor. Prion proteins, and serpins, undergo more radical structural rearrangements, in which the topology of the tertiary structure changes.

Of course these phenomena threaten to erode our confidence in all methods of structure determination, even the most precise ones. If the object of the structure determination is not in fact a unique structure, what is one actually determining?

When proteins can be prepared in alternative stable states, experimental methods can determine them. Then, if we know the structures in two states, we must (almost always) appeal to theory to adduce the detailed trajectory or trajectories between them. Alternatively, if a protein has a known structure in the unligated state only, then prediction of the ligated state is a '**docking**' challenge. It is necessary to predict *both* the ligated conformation and the mode of binding.

There are two computational approaches to determining the trajectory or trajectories between stable conformational states. One is molecular dynamics. Take a ligated state, delete the ligand, and simulate the motion. Does the system move cleanly to the known unligated state? (It is more difficult to go in the other direction—in that case you have to simulate the capture of the ligand as well as the motion of the protein.) We shall discuss the application of molecular dynamics to conformational change in detail in Chapter 5.

A second computational approach models the dynamics of a protein as an elastic network. Represent the protein by a single 'atom' for each Cα, with springs connecting every pair of Cα atoms. The force constants of the springs depend on the distances between the atoms: the closer the atoms the stronger the spring. The low-energy deformations of the model are particularly easy to compute. They take the form of normal modes of vibration—coupled tiny displacements of the atoms, executed in phase. (The coordinated choreographed movements of the dancers performing a ballet are, in this respect, similar to normal modes.) Although quite simple, this scheme is quite effective at predicting protein dynamics.

Ecotin is an inhibitor of serine proteases. Found naturally in the periplasm of *E. coli*, it has a clinical application as an anticoagulant, binding to Factor Xa (see Chapter 5). Ecotin is unusual in having a rather broad specificity in the proteases it inhibits, attributable to conformational flexibility. This can be thought of as an analogue of 'induced fit', from the point of view of the ligand rather than the enzyme.

S.E. Dobbins, V.I. Lesk, and M.J. Sternberg analysed the conformational change in ecotin upon binding to trypsin in terms of normal modes. They found that a superposition of displacements corresponding to no more than three low-frequency normal modes could account for the conformational change (see Fig. 3.25).

The elastic network model accounts for conformational change in *Mycobacterium tuberculosis* thioredoxin reductase

Thioredoxin reductases catalyse the NADPH-catalysed reduction of thioredoxin, and a variety of other compounds including other proteins, and

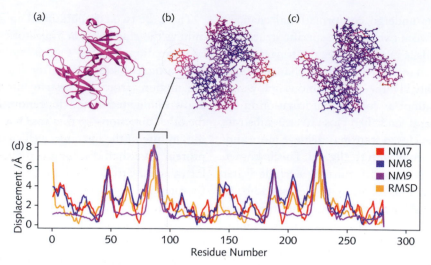

Figure 3.25 (a) Ecotin [1ECZ], regions coloured according to the amplitude of the displacement in the lowest-frequency normal mode, computed in isolation (blue, small amplitude → red, large amplitude). (b) Coloured according to the extent of conformational change upon binding to trypsin (blue, small displacement → red, large displacement). (c) Coloured according to the amplitudes of the three lowest-frequency modes. (d) Linear plot of observed conformational change (orange) and displacements according to three lowest-frequency normal vibrational modes, NM7, NM8 and NM9. ('Normal modes' 1 through 6 account for overall translational and rotational degrees of freedom, with no deformation of the structure.)

[From: Dobbins, S.E., Lesk, V.I. and Sternberg, M.J. (2008). Insights into protein flexibility: The relationship between normal modes and conformational change upon protein–protein docking. *Proc. Natl. Acad. Sci. U.S.A.*, **105**, 10390–10395.]

non-disulphide-containing substrates such as selenite and hydrogen peroxide. In *Mycobacterium tuberculosis,* thioredoxin reductase is involved in defence against oxidative stress.

M. tuberculosis thioredoxin reductase is a dimer of identical subunits. An active site in each subunit appears between two domains, an NADPH-binding domain and a FAD-binding domain. During catalysis, the enzyme switches between two conformations, an oxidized and a reduced form. Independent superpositions of the individual domains show the conformational change to be a hinge motion of the domains with respect to each other, by approximately 11°.

M. Akif, K. Suhre, C. Verma, and S.C. Mande modelled the conformational change using normal mode analysis from the elastic network model. They used the web server *ElNémo*: http://www.sciences.univ-nantes.fr/elnemo/.

The lowest-energy normal modes account almost completely for the conformational change. Figure 3.26 shows the action of one normal mode. Displacements along this mode allow interpolation between the two conformations.

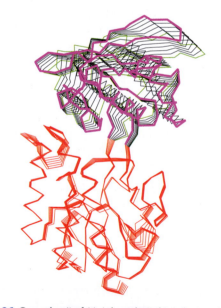

Figure 3.26 One subunit of *M. tuberculosis* thioredoxin reductase, showing motion between oxidized and reduced states. The enzyme has two domains. The two states are superposed on the FAD-binding domain. The NADPH-binding domain is mobile. The interpolants between the end states are computed according to a single low-energy normal mode derived from the elastic network model.

[From: Akif, M., Suhre, K., Verma, C. and Mande, S.C. (2005). Conformational flexibility of M. tuberculosis thioredoxin reductase: crystal structure and normal mode analysis. *Acta Cryst.* **D61**, 1603–1611.]

The relationship between structure determinations of isolated proteins, and protein structure and function *in vivo*

A question that has been asked since the earliest days of protein structure determination is: How do we know that the results on purified samples derived from X-ray crystallography and NMR spectroscopy are relevant to the structure of the molecule in the cell?

Some of the evidence is direct. For a few proteins, it has been possible to diffuse substrate into the crystal and observe enzymatic activity. The implication is that the structure in the crystal is close enough to the native state for the protein to be active. Certainly it justifies interpreting the mechanism of function on the basis of the crystal structure.

In most cases, the interactions within the crystal are only small irritations rather than large structural perturbations. In comparisons of structures of proteins prepared in different crystal forms—with different packing patterns against their neighbours in the crystal—the atomic positions differ, on average, by only a few tenths of an angstrom. There can be local conformational change at the sites of crystal-packing contacts. Moreover, crystal structures and NMR structures of the same protein are reassuringly similar.

On the other hand, most current methods of structure determination are reticent about conformational changes. Many proteins undergo structural changes as part of their function. We can see snapshots of different states, but cannot visualize the pathway between them. In particular, it is difficult to capture short-lived intermediates or transition states. Two attempts to overcome these limitations are (1) slowing down reactions by lowering the temperature, and (2) speeding up the data collection. These specialized methods have been applied to a few systems.

The fact that many proteins undergo conformational changes during their activities does not mean that structure determinations are irrelevant, only that they may be incomplete.

In some cases, molecular dynamics has made it possible to explore the trajectories between the two known end states. The work of M. Karplus and his group on the allosteric transition in haemoglobin, and the internal rotation of ATP synthase, and of M. Levitt and his group on RNA polymerase II translocation, are good examples.

With the recognition that many proteins contain disordered regions, experimental structure determinations can identify disordered regions; but not describe the range of possible conformations of these regions, nor the fraction of molecules in different conformations, and the rates of transitions among different conformational states. For this molecular dynamics is again the appropriate tool.

It may also be pointed out that the concentration of protein inside a cell is much higher than in a classical enzymologist's assay tube. Intracellular concentrations of macromolecules in prokaryotes are estimated at 300–400 mg mL^{-1}; somewhat lower in eukaryotes. Perhaps the question, historically directed at crystallographers, should be turned around—why should we think that measurements on purified proteins in dilute solutions are relevant to the living medium?

Protein structure prediction and modelling

We now turn to purely computational approaches to the prediction of protein structure from amino acid sequence. It must be emphasized that a crucial feature of experimental methods is that measured data serve as an anchor. Agreement with the experimental data of the structure determined provides confirmation of the correctness of the structure, and calibration of the precision. Conversely, theoretical methods, even if they have a consistently successful 'track record', can offer no such proof of correctness in any particular case, in the absence of an experimental structure, or at least experimental data of some relevant kind.

The observation that proteins fold spontaneously into three-dimensional native conformations implies

BOX 3.6 — Overview of modelling methods

Nature computes protein native structure from amino acid sequence. All the information needed to do this computation is contained in the sequence itself: proteins do not need to look things up in databases. We do. Many of the most effective methods for protein structure prediction make use of known structures of homologous proteins. Indeed, the degree of sequence similarity between a protein of unknown structure and its nearest homologue of known structure largely controls what we can achieve in prediction of the unknown structure, and dictates what methods to use.

Generally speaking:

1. If a protein of unknown structure has homologues of known structure with ≥ ~40% identical residues in an optimal alignment, homology-modelling methods are likely to produce a nearly complete structural model. The quality of the model is likely to be good enough to interpret the protein's function. The higher the sequence similarity, the more accurate the model. Moreover, often the active site is well-conserved during evolution. Therefore the model may be most accurate just where you want it to be. Mature software for homology modelling is available, including web servers.

2. If no homologue of known structure has sequence similarity to the unknown with ≥ 40% residue identity, it may still be possible to assign a general folding pattern to the protein of unknown structure. It should be possible to predict its secondary structure with ~70–80% accuracy on a residue by residue basis. It is possible to test the sequence for compatibility with some or all known protein folding patterns, and suggest one or more structures into which the sequence can fold to produce the highest degree of favourable interresidue interactions.

3. If no homologue of known structure is recognizable from the sequences, the last recourse is to use a prediction method general enough to handle novel folds. Such methods include both *a priori* and knowledge-based approaches. At present, the program ROBETTA, by D. Baker and colleagues, is the most effective tool for protein structure prediction whenever homology modelling is not applicable.

4. A multiple sequence alignment may show correlated mutations; that is, pairs of positions that co-vary in that if the residue at one position changes (or stays the same), the residue at the other position changes (or stays the same). It may be possible to infer that such pairs of residues are nearby in space, even if distant in the sequence. Methods developed for solving structures by NMR spectroscopy—which depend on similar information—are applicable to deriving the structure from the set of pairs of neighbouring residues.

5. The ranges of utility of these methods overlap. Many servers will apply several methods to a submitted sequence.

---

that Nature has an algorithm for predicting protein structure from amino acid sequence. Some attempts to understand this algorithm are based solely on general physical principles; others are empirical, based on observations of the known amino acid sequences and protein structures (see Box 3.6). A proof of our understanding would be the ability to reproduce the algorithm in a computer program that could predict protein structure from amino acid sequence. Recently there have been some impressive breakthroughs.

A priori methods of protein structure prediction

Many attempts to predict protein structure from basic physical principles alone try to reproduce the interatomic interactions in proteins, to define a computable energy associated with any conformation. Computationally, the problem of protein structure prediction then becomes a task of finding the global minimum of this conformational energy function. So far this approach has not succeeded consistently, partly because of the inadequacy of the energy function, partly because the minimization algorithms tend to get trapped in local minima, and partly because the calculations require more computer resources than are generally available.

Other *a priori* approaches to structure prediction are based on attempts to simplify the problem, to capture somehow the essentials.

There is a spectrum of approaches, distributed between two extremes:

1. Establish the *most detailed and accurate* model of the interatomic interactions within a protein and between protein and solvent. Apply molecular dynamics to simulate the motion of the system starting with a denatured conformation—perhaps the extended chain—and ending with something in the vicinity of the native state. The idea is that the physics of the problem is fairly well understood, down to the detailed microscopic level. The challenges are computational—how to simulate the system for long enough to attain the native state.

2. Establish the *least detailed* and *least accurate* model that can give the correct answer. If one could identify the essentials, great computational power might not be needed. The idea is that the physics of the problem is *not* well understood, *except* in microscopic detail. Of course everyone accepts the principles of mechanics and thermodynamics, but much of the detail is irrelevant and unilluminating, and this feature is itself a crucial part of the picture. Proteins just aren't that fussy: at the melting temperature half the molecules remain in the native state, despite the very great alteration in the relative strengths of various terms in the free energy relative to normal physiological conditions. This argues for a great robustness in the determinants of structure, which is difficult to capture in detailed calculations, or to explain even if they were captured. It argues for the importance of the distinction between determinants of structure and determinants of stability, also difficult to explain from detailed calculations. Also, many proteins with substantial sequence differences fold to very similar native states. However, in some cases there are large differences in their folding pathways.

The field contains many people widely scattered between these endpoints, linked by a certain creative tension.

It is partly a question of choosing goals. To go beyond a prediction of the native state, to account for trajectories, transition states, intermediates if any, and melting temperatures, a detailed simulation may well be necessary. Conversely, if one wants a perspicuous and satisfying explanation of how amino acid sequence determines protein structure, then even a successful fully detailed calculation may not provide it.

A proponent of molecular dynamics might argue that (1) from a successful fully detailed calculation, one could generate a series of simplified models by making approximations that keep the broad picture intact, and (2) a simplified model that works is—by virtue of its success—interesting, but may be unrealistic, or, even if realistic, incomplete. After all, there may be many simplified models, all of which work, but that do not agree on what is essential. A reason for suspecting that this may be true is the observation that folded proteins solve many problems at once—stereochemistry, packing, hydrogen bonding, conformational entropy compensation. Any spoke may lead to the hub.

Empirical, or 'knowledge-based', methods of protein structure prediction

The alternative to *a priori* methods are approaches based on assembling clues to the structure of a target sequence by finding similarities to known structures. These empirical or 'knowledge-based' methods have become very powerful.

We are coming closer and closer to saturating the set of possible folds with known structures. This is a goal of **structural genomics** projects (see Box 3.7). Once we have a complete set of folds and sequences, and powerful methods for relating them, empirical methods will provide pragmatic solutions of many problems.

What will be the effect of this on attempts to predict protein structure *a priori*? The intellectual appeal of the problem will still be there—we emphasize that Nature folds proteins without searching databases. Moreover, some methods may not merely identify the native conformation, but illuminate function and folding pathways. But it is very likely that the problem will fail to command general interest of the same intensity, nor support of the same largesse, once an alternative pragmatic solution has been found.

- Structural genomics has the goal of complete structure determinations of the proteomes of selected organisms. It aims to provide enough known structures to allow homology modelling of all proteins, of all organisms, from their amino acid sequences. This will require enough protein structures so that the amino acid sequence of any natural protein is sufficiently closely related to some protein of known structure.

BOX 3.7 Structural genomics

In analogy with full-genome sequencing projects, structural genomics has the commitment to deliver the structures of the complete protein repertoire. X-ray crystallographic and NMR experiments will solve a 'dense set' of proteins, such that all proteins are within homology-modelling range of one or more known experimental structures. More so than genomic-sequencing projects, structural genomics projects combine results from different organisms. The human proteome is of course of special interest, as are proteins unique to infectious microorganisms.

The goals of structural genomics have become feasible partly by advances in experimental techniques, which make high-throughput structure determination possible, and partly by advances in our understanding of protein structures, which define reasonable general goals for the experimental work, and suggest specific targets.

The theory and practice of homology modelling suggest that at least 30–40% sequence identity between target and some experimental structure is necessary. This means that experimental structure determinations will be required for an exemplar of every sequence family, including many that share the same basic folding pattern. Experiment will have to deliver the structures of something like 10 000 domains displaying different folds. In the year 2014, 9676 structures were deposited in the wwPDB, so the throughput rate is adequate.

Methods of bioinformatics can help select targets for experimental structure determination that offer the highest payoff in terms of useful information. Goals of target selection include:

- elimination of redundant targets—proteins too similar to known structures;
- identification of sequences with undetectable similarity to proteins of known structure;
- identification of sequences with similarity only to proteins of unknown function;
- proteins of unknown structure with 'interesting' functions; for example, human proteins implicated in disease, or bacterial proteins implicated in antibiotic resistance;
- proteins with properties favourable for structure determination; for instance, likely to be globular and soluble.

The machinery for carrying out the modelling is already up and running. MODBASE http://modbase.compbio.ucsf.edu/ and the SWISS-MODEL Repository http://swissmodel.expasy.org/repository/ collect homology models of proteins of known sequence.

Knowledge-based methods for prediction of protein structure from amino acid sequence are classifiable as:

- Attempts to *predict secondary structure* without attempting to assemble these regions in three dimensions. The results are lists of regions of the sequence predicted to form α-helices and regions predicted to form strands of β-sheet.
- *Homology modelling:* prediction of the three-dimensional structure of a protein from the known structures of one or more related proteins. The results are a complete coordinate set for mainchain and sidechains, intended to be a high-quality model of the structure, comparable to at least a low-resolution experimental structure.
- *Fold recognition:* given a library of known structures—which could be the entire PDB—determine which of them shares a folding pattern with a query protein of known sequence but unknown structure. If the folding pattern of the target protein does not occur in the library, such a method should recognize this. The results are a nomination of a known structure that has the same fold as the query protein, or a statement that no protein in the library has the same fold as the query protein.
- *Prediction of novel folds,* by either *a priori* or knowledge-based methods. The results are a complete coordinate set for at least the mainchain and sometimes the sidechains also. The model is intended to have the correct folding pattern, but would not be expected to be comparable in quality to an experimental structure. D. Jones has likened the distinction between *a priori* modelling and fold recognition to the difference between an essay and a multiple-choice question in an exam.

General categories of prediction methods:

- homology modelling—prediction of an unknown structure from the known structure of one or more related proteins;
- fold recognition—identification of a folding pattern from a library that is predicted to resemble the folding pattern of target protein (or assertion that the library contains no such structure);
- *a priori* methods—prediction of structure from amino acid sequence without explicit reference to any particular known structure.

Secondary structure prediction

It seems obvious that: (1) it should be easier to predict secondary structure than tertiary structure, and (2) to predict tertiary structure a sensible way to proceed would be first to predict the helices and strands of sheet and then to assemble them. Whether or not these propositions are correct, many people have believed in, and acted upon, them. Given the amino acid sequence of a protein of unknown structure, they produce **secondary structure predictions,** the assignment of regions in the sequence as helices or strands of sheet.

The most powerful methods of secondary structure prediction are based on **artificial neural networks** (see Box 3.8). Figure 3.27 shows a specific type of neural

network that has been applied to secondary structure prediction.

A major advance in secondary structure prediction occurred with the application of evolutionary information, the recognition that multiple sequence alignment tables contain much more information than individual sequences. The conservation of secondary structure among related proteins means that the sequence–structure correlations are much more robust when a family as a whole is taken into account. Most neural network-based methods for secondary structure prediction now feed the input layer not simply with the identities of the amino acid at successive positions, but with a profile derived from a multiple sequence alignment.

It has also proved useful to run two neural networks in tandem, to make use of observed correlations among conformations of residues at neighbouring positions. Predictions of the states of several successive residues by one network similar to the one shown in Fig. 3.27 are combined by a second network into a final prediction.

To assess the quality of a secondary structure prediction, classify the residues in the experimental three-dimensional structure into three categories (helix = H, strand = E (extended), and other = −). The percentage of residues predicted correctly is denoted Q3. At the 2000 CASP programme, the PROF server by B. Rost achieved a good prediction of a domain from the *Thermus aquaticus* mismatch repair protein MutS. The value of Q3 for Rost's prediction is 81%.

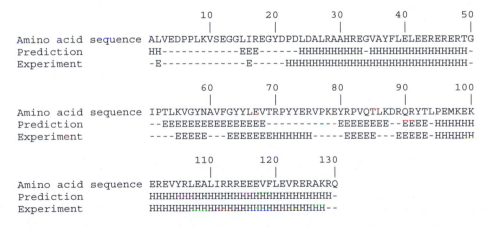

Figure 3.28 shows the experimental structure, with the *predicted* secondary structures distinguished. Except for a short 3$_{10}$ helix, the secondary structural elements are predicted correctly except for some minor

discrepancies in the positions at which they start and end. (Other scoring schemes that check for *segment overlap* are less sensitive to end effects.) The quality of this result is very high but not exceptionally rare.

BOX 3.8 Artificial neural networks

Artificial neural networks have proved to be a very powerful computational method for wide variety of 'machine-learning problems', in molecular biology and many many other fields of science and business. The problems share the feature that it is possible to create a computational structure of sufficient versatility that, upon being shown examples of datasets and told the desired answer (the 'knowns'), a computer program can 'train itself' and subsequently be able to give the right answers for a set of 'unknowns'.

Like natural nervous systems, the power of artificial neural networks lies in their complex connectivity, and the decision logic that each node applies to its inputs.

Each node in a neural network—represented in Fig. 3.27 by a circle—is a mathematical object that can receive one or more input signals, apply some decision process, and either transmit an impulse or remain silent. In this respect they simulate a biological nervous system, with the nodes corresponding to neurons, and the interactions with upstream and downstream nodes corresponding to synapses. Transmission of the signal corresponds to 'firing' of the neuron.

In the neural network illustrated in Fig. 3.27, the amino acid sequence is the external input. The network examines a 15-residue 'window' (and then the sequence slides along and the network examines the next, overlapping, 15-residue window). The input layer is divided into groups of 20 nodes, each corresponding to one position in the sequence. One node in each group of 20 corresponds to an amino acid. Therefore one node in each group of 20 is activated (black circle) according to the amino acid at the corresponding position in the window.

Connections between each of the input nodes to nodes in the intermediate layer are denser than can be shown—in principle there can be complete connectivity, allowing for very complicated decision making. (The intermediate layer is called 'hidden' because it does not interact with the external world either to receive input or to present output.) The nodes of the intermediate layer are connected to an output layer, containing three nodes corresponding to prediction (for the central residue of the window, here a K marked by an arrow) of α-helix, prediction of β-sheet, and prediction of other.

(See: Lesk, *Introduction to Bioinformatics*, 4th edn, Oxford University Press, 2014, Chapter 6, for additional background.)

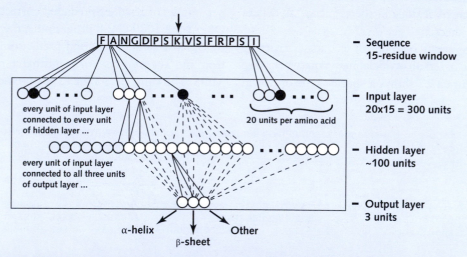

Figure 3.27 A neural network applicable to secondary structure prediction contains three layers:

- The input layer sees a sliding 15-residue window in the sequence. That is, it treats a 15-residue region, predicts the secondary structure of the central residue (marked by an arrow, at the top), and then moves the window one residue along the amino acid sequence and repeats the process. To each of the 15 residues in the current window there correspond 20 nodes in the input layer of the network, one of which will be triggered according to the amino acid in that position.

- A hidden layer of ~100 units connects the input with the output. Each node of the hidden layer is connected to *all* input and output units; not all the connections are shown.

- The output layer consists of only three nodes, which signify prediction that the central residue in the window be in a helix, strand, or other conformation.

This target was classified as being of *medium* difficulty by the CASP4 assessors. At present, PROF is running at an average accuracy of Q3 ~ 77%.

Other secondary structure prediction methods are also doing comparably well. Figure 3.29 shows the results of four different secondary structure prediction methods, on methyl glyoxal synthase from Thermus Sp.Gh5 [2X8W]. All pictures show the same structure, with the observed or predicted α-helices and strands of β-sheet in red and yellow, respectively, mapped onto the structure (a–e) or the sequence (f). Frame (a) shows the secondary structure assignments from the experimentally-determined structure, and assigned by the program DSSP (Dictionary of Secondary Structures of Proteins) by W. Kabsch and C. Sander. The other frames show the results of five different prediction methods. For the most part they are correct and in agreement about the order and approximate lengths of the secondary structure elements. However they differ in precisely where they place the endpoints of these segments.

A test of the maturity of a prediction method is whether it can be made fully automatic (see Critical Assessment of Structure Prediction section). Some computational methods require human intervention and editing of results. Others, including PROF, the system that predicted the secondary structure of MutS, are fully automatic.

Homology modelling

Model building by homology is a useful technique for predicting the structure of a target protein of known sequence, when the target protein is related to at least one other protein of known sequence *and* structure. If the proteins are closely related, the known protein structures—called the parents—can serve as the basis for a model of the target. Although the quality of the model will depend on the degree of similarity of the sequences, it is possible to estimate this quality before experimental testing (see Chapter 7). In consequence, knowing how good a model is necessary for the intended application permits intelligent estimation of the probable success of the exercise.

Although the mechanics of homology modelling are now automated in mature software, the basic steps are:

1. Align the amino acid sequences of the target and the protein or proteins of known structure. It will generally be observed that insertions and deletions lie in the loop regions between helices and sheets.

2. Determine mainchain segments to represent the regions containing insertions or deletions. Stitching these regions into the mainchain of the known protein creates a model for the complete mainchain of the target protein.

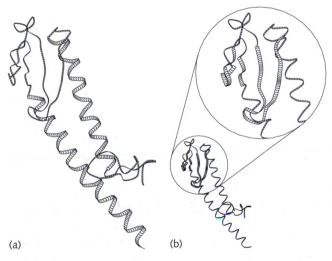

(a)   (b)

Figure 3.28 The structure of the *Thermus aquaticus* mismatch repair protein MutS [1EWQ]. (a) The regions predicted, by the PROF server of Rost, to be helical are shown as wider ribbons. The prediction missed only a short 3₁₀ helix, at the top left of the picture. (b) The regions predicted to be in strands are shown as wider ribbons.

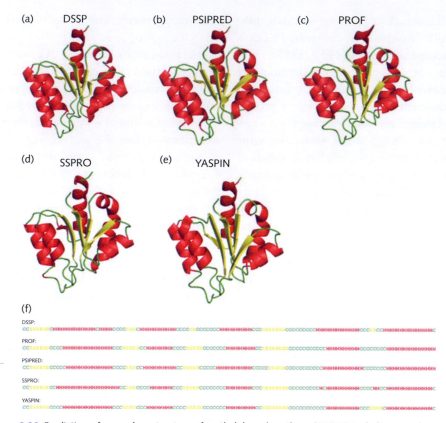

Figure 3.29 Prediction of secondary structure of methylglyoxal synthase [2X8W]. α-helices are shown in red, strands of β-sheet in yellow. (a) The secondary structure assignments derived from the experimental structure, by the program DSSP. (b) Prediction by PSIPRED (D.T. Jones, D. Buchan, T. Nugent, F. Minneci, K. Bryson, A. Lobley, S. Ward, L.J. McGuffin). (c) Prediction by PROF (M. Ouali and R.D. King). (d) Prediction by SSPRO (J. Cheng, A. Randall, M. Sweredoski, P. Baldi). (e) Prediction by YASPIN (K. Lin, V.A. Simossis, W.R. Taylor and J. Heringa). In (a)–(e) the secondary structure elements are mapped onto the experimental structure, which is the same in all cases. (f) The same information mapped onto the sequence.

From: Joseph, A.P. and de Brevern, A.G. (2015). From local structure to a global framework: recognition of protein folds. J.Roy. Soc. Interface 11: 20131147

3. Replace the sidechains of residues that have been mutated. For residues that have not mutated, retain the sidechain conformation. Residues that have mutated tend to keep the same sidechain conformational angles, and could be modelled on this basis. However, computational methods are now available to search over possible combinations of sidechain conformations.

4. Examine the model—both by eye and by programs—to detect any serious collisions between atoms. Relieve these collisions, as far as possible, by manual manipulations.

5. Refine the model by limited energy minimization. The role of this step is to fix up the exact geometrical

relationships at places where regions of mainchain have been joined together, and to allow the sidechains to wriggle around a little to place themselves in comfortable positions. The effect is really only cosmetic—energy refinement will not fix serious errors in such a model.

To a great extent, this procedure produces 'what you get for free' in that it defines the model of the protein of unknown structure by making minimal changes to its known relatives. In some cases, it is possible to make substantial improvements. Indeed, this is one of the challenges in recent CASP programmes. A rule of thumb is that if two or more sequences have at least 40–50% identical amino acids in an optimal alignment of their sequences, the procedure described will produce a model of sufficient accuracy to be useful for many applications.

In most families of proteins the structures contain relatively well-conserved regions and more variable ones. A single-parent structure will permit reasonable modelling of the conserved portion of the target protein, but may fail to produce a satisfactory model of the variable portion. From only one target and one parent sequence, it will not be easy even to predict which are the variable and constant regions. A more favourable situation occurs when several related proteins of known structure provide a basis for modelling a target protein. These reveal the regions of constant and variable structure in the family. The observed distribution of structural variability among the parents dictates an appropriate distribution of constraints to be applied to the model.

SWISS-MODEL hosts a website that will accept the amino acid sequence of a target protein, determine whether a suitable parent or parents for homology modelling exist, and, if so, deliver a set of coordinates for the target. SWISS-MODEL was developed by T. Schwede, M. Peitsch and N. Guex. The program **MODELLER** by A. Šali and coworkers is freely available but must be installed. I-TASSER, developed by Y. Zhang and colleagues, is available both for download and as a web server.

An example of the automatic prediction by SWISS-MODEL is the prediction of the structure of a neurotoxin from red scorpion (*Buthus tamulus*) from the known structure of the neurotoxin from the North African yellow scorpion (*Androctonus australis hector*). These two proteins have 52% identical residues in their sequence alignment. With such a close degree of similarity it is not surprising that the model fits the experimental result very closely, even with respect to the sidechain conformation (see Fig. 3.30).

Another example of a homology model, from the most recent CASP programme, appears in Fig. 3.31.

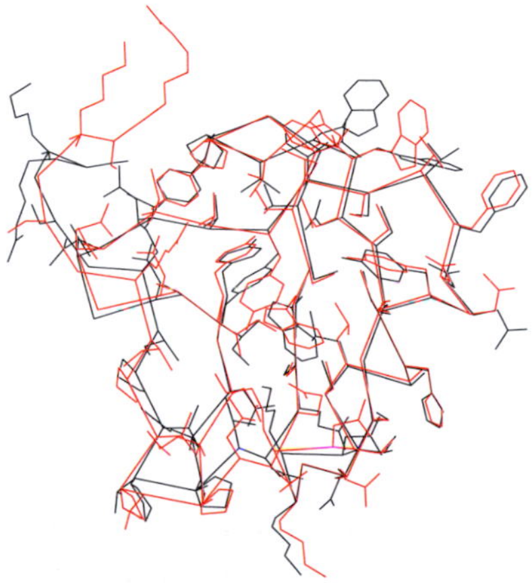

Figure 3.30 SWISS-MODEL predicts the structure of red scorpion (*Buthus tamulus*) neurotoxin, based on the structure of a closely-related protein. Red = prediction based on homologue (toxin II from the scorpion *Androctonus australis* Hector [1PTX]), Black = experimental structure of the target [1DQ7]. The prediction was done *automatically*, and the experimental coordinates of the target structure were shielded from the prediction machinery. Observe that most of the buried sidechains have not mutated, and have very similar conformations. Some sidechains on the surface have different conformations, and the mainchain of the C-terminus is in a different position (upper left). Not shown is a network of disulphide bridges, which constrain the structure. However, a model of this high quality would be expected, for two such closely-related proteins, even without the extra constraints.

Fold recognition

Searching a sequence database for a probe sequence and searching a structure database with a probe structure are problems with known solutions. The mixed problems—probing a sequence database with a structure, or a structure database with a sequence, are less straightforward. They require a method for evaluating the compatibility of a given sequence with a given folding pattern.

The goal is to abstract the essence of a set of sequences or structures. Other proteins that share the pattern are expected to adopt similar structures.

3D profiles

Patterns and profiles derived from multiple sequence alignments are useful in detection of distant homologues. One way to take advantage of available structural information to improve the power of these methods is a type of profile derived from the available sequences *and* structures of a family of proteins.

J.U. Bowie, R. Lüthy, and D. Eisenberg analysed the *environments* of each position in known protein structures and related them to a set of preferences of the 20 amino acids for these structural contexts.

Given a protein structure, classify the environment of each amino acid in three separate categories:

1. its mainchain hydrogen-bonding interactions, that is, its secondary structure,

2. the extent to which it is buried within, or on the surface of, the protein structure,

3. the polar/non-polar nature of its environment.

The secondary structure may be one of three possibilities: *helix, sheet,* and *other.* A sidechain is considered buried if the accessible surface area is less than 40 Å², partially buried if the accessible surface area is between 40 and 114 Å², and exposed if the accessible surface area is greater than 114 Å². The fraction of sidechain area covered by polar groups is measured. The authors define six classes on the basis of accessibility and polarity of the surroundings. Sidechains in each of these six classes may have any of three types of secondary structure assignment: helix, sheet, or other. This gives a total of 18 classes.

Assigning each sidechain to one of 18 categories makes it possible to write a coded description of a protein structure as a message in an alphabet of 18 letters, called a *3D structure profile.* Algorithms developed for sequence searches can thereby be applied to 'sequences' of encoded structures. For example, one could try to align two distantly related proteins by aligning their 3D structure profiles rather than their amino acid sequences. The 3D profile method translates protein structures into one-dimensional probe (or probeable) objects that do not explicitly retain either the sequence or structure of the molecules from which they were derived.

Next, how can one relate the 3D structure profile to the set of known protein folding patterns? It is clear that some amino acids will be unhappy in certain kinds of sites; for example, a charged sidechain would prefer not to be buried in an entirely non-polar environment. Other preferences are not so clear-cut, and it is necessary to derive a preference table from a statistical survey of a library of well-refined protein structures.

Suppose now that we are given a sequence and want to evaluate the likelihood that it takes up, say, the globin fold. From the 3D structure profile of the known sperm-whale myoglobin structure, we know the environment class of each position of the sequence. We must consider all possible alignments of the sequence of the protein of unknown structure with the 3D structure profile of myoglobin. Consider

a particular alignment, and suppose that the residue in the unknown sequence that corresponds to the first residue of myoglobin is phenylalanine. The environment class in the 3D structure profile of the first residue of sperm-whale myoglobin is: exposed, no secondary structure. One can score the probability of finding phenylalanine in this structural environment class from the table of preferences of particular amino acids for this 3D structure profile class. (The fact that the first residue of the sperm-whale myoglobin sequence is actually valine is not used, and in fact that information is not directly accessible to the algorithm. Sperm-whale myoglobin is represented only by the sequence of environment classes of its residues, and the preference table is averaged over proteins with many different folding patterns.) Extension of this calculation to all positions and to all possible alignments (not allowing gaps within regions of secondary structure) gives a score that measures how well the given unknown sequence, upon optimal alignment, fits the sperm-whale myoglobin profile.

A particular advantage of this method is that it can be automated, with a new sequence being scored against every 3D profile in the library of known folds, in essentially the same way as a new sequence is routinely screened against a library of known sequences.

Use of 3D profiles to assess the quality of structures

The 3D profile derived from a structure depends only very indirectly on the amino acid sequence. It is therefore meaningful to ask, not only whether it is possible to identify other amino acid sequences compatible with the given fold, but whether the score of a 3D profile for its own parent sequence gives a high score for the compatibility of the parent sequence with the actual structure. Naturally, if real sequences did not generally appear to be compatible with their own structures, credibility in the method as capturing a valid connection between sequence and structure would be severely impaired. Two interesting results are observed. (1) Protein structures determined correctly do fit their own profiles well, although other, related, proteins, may give *higher* scores. The profile is abstracting properties of the family, not of individual sequences. (2) When a sequence does *not* match a profile computed from an experimental structure of that protein, there is likely to have been an error in the structure determination. The positions in the profile that do not match can identify the regions of error.

Threading

Threading is a method for fold recognition. Given a library of known structures, and a sequence of a query protein of unknown structure, does the query protein share a folding pattern with any of the known structures? The fold library could include some or all of the Protein Data Bank, or even hypothetical folds.

The basic idea of threading is to build many rough models of the query protein, based on each of the known structures and using different possible alignments of the sequences of the known and unknown proteins. This systematic exploration of the many possible alignments gives threading its name: imagine trying out all alignments by pulling the query sequence gently through the three-dimensional framework of any known structure. Gaps must be allowed in the alignments, but if the thread is thought of as being sufficiently elastic the metaphor of threading survives.

Both threading and homology modelling deal with the three-dimensional structure induced by an alignment of the query sequence with known structures of homologues. Homology modelling focuses on one set of alignments and the goal is a very detailed model. Threading explores many alignments and deals with only rough models usually not even constructed explicitly:

Homology modelling	Threading
First, identify homologues	Try many possible folds
Then, determine optimal alignment	Try many possible alignments
Optimize one model	Evaluate many rough models

Successful fold recognition by threading requires:

1. A method to score the models, so that we can select the best one.

2. A method for calibrating the scores, so that we can decide whether the best-scoring model is likely to be correct.

Several approaches to scoring have been tried. One of the most effective is based on empirical patterns of residue neighbours, as derived from known structures. First, we observe the distribution of interresidue distances in known protein structures, for all 210 pairs of residue types. For each pair, derive a probability distribution, as a function of the separation (a) in space, and (b) in the amino acid sequence. For instance for the pair Leu–Ile, consider every Leu and Ile residue in known structures, and, for each Leu–Ile pair, record the distance between their $C\beta$ atoms, and the difference in their positions in the sequence. Collecting these statistics permits estimation of how well the distributions observed in a model agree with the distributions in known structures.

The Boltzmann equation relates probabilities and energies. Usual applications of the Boltzmann equation start from an energy function and predict a probability distribution. (A standard example is the prediction of the density of the atmosphere as a function of altitude, from the gravitational potential-energy function of the air molecules.) For threading, one turns this on its head, and *derives* a pseudo-energy function *from* the probability distribution. This 'energy' function is then used to score threading models.

For each structure in the fold library, the procedure finds the assignment of residues that produces the lowest-energy score.

Fold recognition at CASP

The best methods for fold recognition have been consistently effective. These include, but are not limited to, methods based on threading.

Figure 3.31 shows the prediction by A.G. Murzin of the folding pattern of a protein of unknown function from *H. influenzae*, from the 2000 CASP programme.

Antibody modelling

The antigen-binding region of typical antibodies contains two homologous variable domains from light and heavy chains—V_L and V_H—packed against each other. The binding site itself is formed (in most cases) from six loops, the complementarity-determining regions, or CDRs, three from each of these domains (see Chapter 6).

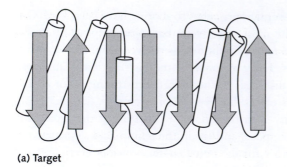

(a) Target

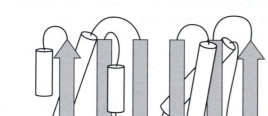

(b) Fold prediction by A.G. Murzin

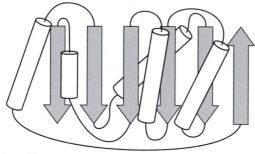

(c) Template of known structure

Figure 3.31 Prediction of structure of *H. influenzae*, hypothetical protein. (a) The folding pattern of the target. (b) Prediction by A.G. Murzin. (c) Folding pattern of the closest homologue of known structure: an N-ethylmaleimide-sensitive fusion protein involved in vesicular transport (PDB entry 1NSF). The topology of Murzin's prediction is closer to the target than that of the closest single parent.

Modelling antibodies presents challenges somewhat different from modelling most other proteins. Corresponding to the core of a family of structures, antibodies contain a framework of relatively constant structure. This is relatively easy to model well. The modeller can expect no kudos for successful framework prediction, *least of all* from immunologists. For many other protein families, a prediction that correctly produced the assembly of core second-

ary structure elements would be considered a success. For antibodies, the more severe challenge is to get the antigen-binding loops correct.

Steps in prediction of antibody structures are:

1. Building the framework, which is usually a standard homology-modelling exercise. One may encounter pitfalls if the antibody is based on an unusual germ-line segment. Predicting the relative orientation of the V_L-V_H domains can also present difficulties.

2. Five of the six CDRs usually have conformations similar to a member of discrete repertoires for each CDR, called **canonical structures**. Signature patterns within the sequence can point to the correct choice.

3. The sixth antigen-binding loop, CDR3 of the heavy chain, presents the hardest problem. The mechanisms that generate antibody diversity impart the greatest variety in length and conformation to this region. In the latest assessment of predictions, approaches that generate this loop with *a priori* approaches, rather than 'knowledge-based' methods, achieved the best results.

Independent of CASP, a specialist group has organized blind tests of antibody modelling. In the current round of this programme, 11 sequences were distributed in a first stage. In a second stage, the experimental structures were distributed, after removing the most difficult region to predict, the third hypervariable loop of the heavy chain. This experiment was designed to determine how useful a guide to predicting V_H CDR3 the exact structural context might be.

The August 2014 issue of the journal 'Proteins: Structure, Function, and Bioinformatics' focused on assessment of the targets and results. One conclusion is that specialized methods for antibody structure prediction outperform even the best general homology-modelling software.

One is awaiting sufficient improvements in these methods that will make it possible to predict, entirely computationally, the sequence of an antibody that will fold properly and bind a prespecified target antigen; for instance, the coat protein of a novel strain of a virus.

Prediction of special categories of structures

Some fold-recognition procedures strive for sufficient generality to identify all known domain structures. Others are specialized to particular types of folds. The best algorithms for prediction of transmembrane helices and coiled-coils make use of hidden Markov models. (For a detailed description of hidden Markov models, see Introduction to Bioinformatics, 4th ed., Chapter 5.)

Prediction of coiled-coils by hidden Markov models

Approaches to prediction of coiled-coiled regions in proteins include:

- profile methods using running windows (PCOILS);
- profile methods, running windows, with residue correlations (PairCoil2);
- hidden Markov models (MARCOIL).

MARCOIL gave the best overall performance in controlled tests.

MARCOIL uses a hidden Markov model trained on a database containing nine classes of proteins:

- Tropomyosins
- Intermediate filaments
- Kinesins
- SNARE proteins
- Myosins
- Dyneins
- Laminins
- Transcription factors
- Others

Submitting to MARCOIL the chicken proto-oncogene protein c-fos, and selecting default parameters:

the program returned the prediction shown in Fig. 3.32. The program is quite confident that the protein contains a coiled-coil domain, between residues ~125 and ~200.

Prediction of transmembrane helices and signal sequences by hidden Markov models

A simple approach to prediction of membrane proteins involves looking for amino acid segments 15–30 residues in length that are rich in hydrophobic residues. However, signal peptides also contain hydrophobic helices: the signal sequence typically comprises a positively charged n-region, followed by a helical hydrophobic h-region, followed by a polar c-region. Methods for recognizing transmembrane helices in amino acid sequences tend to pick up the h-regions of signal peptides as false positives. Methods for recognizing signal peptides in amino acid sequences tend to pick up transmembrane helices as false positives.

L. Käll, A. Krogh, and E. Sonnhammer trained hidden Markov models to test simultaneously for transmembrane helices and signal peptides. The goals are to find both at the same time, to discriminate between them in the results, and to predict not only the positions of the transmembrane helices but the loca-

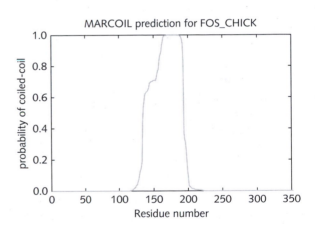

Figure 3.32 Prediction by MARCOIL of a coiled-coil domain in chicken c-fos.

```
>P11939|FOS_CHICK Proto-oncogene protein c-fos - Gallus gallus (Chicken).
MMYQGFAGEYEAPSSRCSSASPAGDSLTYYPSPADSFSSMGSPVNSQDFCTDLAVSSANF
VPTVTAISTSPDLQWLVQPTLISSVAPSQNRGHPYGVPAPAPPAAYSRPAVLKAPGGRGQ
SIGRRGKVEQLSPEEEEKRRIRRERNKMAAAKCRNRRRELTDTLQAETDQLEEEKSALQA
EIANLLKEKEKLEFILAAHRPACKMPEELRFSEELAAATALDLGAPSPAAAEEAFALPLM
TEAPPAVPPKEPSGSGLELKAEPFDELLFSAGPREASRSVPDMDLPGASSFYASDWEPLG
AGSGGELEPLCTPVVTCTPCPSTYTSTFVFTYPEADAFPSCAAAHRKGSSSNEPSSDSLS
SPTLLAL
```

tions—cytoplasmic or interior—of the loops. They called their method *PHOBIUS*.

PHOBIUS is the most successful algorithm currently available for recognizing signal peptides and helical transmembrane proteins, and for predicting the orientation of the transmembrane segments. PHOBIUS is capable of distinguishing h-domains of signal peptides from transmembrane helices: the number of false classifications of signal peptides was 3.9%, and the number of false classifications of transmembrane helices was 7.7%. These results represent a substantial improvement over previous methods. It is interesting that addressing the two problems at once proved to be more successful than treating them separately. Table 3.3 provides a link to PHOBIUS and to other websites about membrane proteins.

Prediction of disordered regions

Many proteins contain disordered regions (see Chapter 1). Some are apparently entirely disordered: for instance, casein has the primary function of nourishing babies, who digest it. (It also inhibits the precipitation of calcium phosphate that is formally supersaturated in milk.) Casein neither needs nor possesses a typical native structure.

A simple definition of disorder is the presence of missing atoms in high-resolution crystal structures. This has the virtue of simplicity, but has the danger that artefacts may arise from crystal-packing interactions.

Prediction of disordered regions is now a challenge in the CASP programme.

Conformational energy calculations and molecular dynamics

A protein is a collection of atoms. The interactions between the atoms create a unique state of maximum stability. Find it, that's all!

The computational difficulties in this approach arise because (a) the model of the interatomic interactions is not complete or exact, and (b) even if the model were exact we face an optimization problem in a large number of variables, involving non-linearities in the objective function and the constraints, creating a very rough energy surface with many local minima. Like a golf course with many bunkers, such problems are very difficult. Simulations by molecular dynamics in principle offer the possibility of exploring conformation space adequately.

The interactions between atoms in a molecule can be divided into:

1. Primary chemical bonds—strong interactions between atoms that must be close together in space. These are regarded as a fixed set of interactions that are not broken or formed when the conformation of a protein changes. However, they are equally consistent with a large number of conformations.

2. Weaker interactions that depend on the conformation of the chain. These can be significant in some conformations and not in others—they involve sets of atoms that are brought into proximity by different folds of the chain.

The conformation of a protein can be specified by giving the list of atoms in the structure, their coordinates, and the set of primary chemical bonds between them (this can be read off, with only slight ambiguity, from the amino acid sequence). Terms used in the evaluation of the energy of a conformation typically include:

- Bond stretching: $\sum_{bonds} K_r (r - r_0)^2$ Here, r_0 is the equilibrium interatomic separation and K_r is the force constant for stretching the bond. r_0 and K_r depend on the type of chemical bond.

- Bond-angle bend: $\sum_{angles} K_\theta (\theta - \theta_0)^2$. For any atom i that is chemically bonded to two (or more) other

Table 3.3 Web resources: Membrane proteins

PHOBIUS (L. Käll, A. Krogh and E. Sonnhammer)	http://phobius.cgb.ki.se/
TOPCONS (A. Elofsson)	http://topcons.cbr.su.se/
PHDhtm (B. Rost)	http://www.predictprotein.org
Membrane Proteins of Known Structure (S. White)	http://www.blanco.biomol.uci.edu/mpstruc/
Membrane Protein EXplorer (S. White)	http://blanco.biomol.uci.edu/mpex/
Membrane Protein Database (P. Raman, V. Cherezov and M. Caffrey)	http://www.mpdb.tcd.ie/

atoms j and k, the angle i–j–k has an equilibrium value, θ_0, and a force constant for bending, K_θ.

- Other terms to enforce proper stereochemistry penalize deviations from planarity of certain groups, or enforce correct chirality (handedness) at certain centres.

- Torsion angle: $\sum_{dihedrals} \frac{1}{2} V_n [1 + \cos n\phi]$. For any four connected atoms: i bonded to j bonded to k bonded to l, the energy barrier to rotation of atom l with respect to atom i around the j–k bond is given by a periodic potential. V_n is the height of the barrier to internal rotation; n barriers are encountered during a full 360° rotation. (For instance, for ethane $n = 3$.) The mainchain conformational angles ϕ, ψ and ω are examples of such torsional rotations (see Fig. 2.2).

- Van der Waals interactions: $A_{ij}R_{ij}^{-12} - B_{ij}R_{ij}^{-6}$. For each pair of non-bonded atoms i and j, the first term accounts for a short-range repulsion and the second term for a long-range attraction between them. The parameters A and B depend on atom type. R_{ij} is the distance between atoms i and j (see Fig. 2.9).

- Hydrogen bonds: $C_{ij}R_{ij}^{-12} - D_{ij}R_{ij}^{-10}$. The hydrogen bond is a weak chemical/electrostatic interaction between two polar atoms. Its strength depends on distance and also on the donor–H–acceptor angle. The approximate hydrogen-bond potential given here does not explicitly reflect the angular dependence of hydrogen-bond strength; other potentials attempt to account for hydrogen-bond geometry more accurately.

- Electrostatics: $Q_iQ_j/(\varepsilon R_{ij})$. For each pair of charged atoms i and j, Q_i and Q_j are the effective charges on the atoms, R_{ij} is the distance between them, and ε is the dielectric 'constant'. This formula applies only approximately to media that are not infinite and isotropic, including proteins.

- Solvent: interactions with the solvent, water, and cosolutes, such as salts and sugars, are crucial for the thermodynamics of protein structures. Attempts to model the solvent as a continuous medium, characterized primarily by a **dielectric constant**, are approximations. With the increase in available computer power, it is now possible to include solvent explicitly, simulating the motion of a protein in a box of water molecules.

There are numerous sets of conformational energy potentials of this or closely-related forms, and a great deal of effort has gone into the tuning of the sets of parameters that specify them. The energy of a conformation is computed by summing these terms over all bonded and non-bonded atoms.

For successful structure prediction, the potential functions must satisfy necessary but not sufficient conditions. One obvious test is to take the right answer—an experimentally-determined protein structure—as a starting conformation, and minimize the energy starting from there. Most high-quality energy functions produce a minimized conformation that is about 1 Å (root-mean-square deviation) away from the starting model. This can be thought of as a measure of the resolution of the force field. Another test has been to take deliberately misfolded proteins and minimize their conformational energies, to see whether the energy value of the local minimum in the vicinity of the correct fold is significantly lower than that of the local minimum in the vicinity of an incorrect fold. Such tests reveal that multiple local minima cannot be reliably distinguished from the correct one on the basis of calculated conformational energies.

Attempts to predict the conformation of a protein by minimization of the conformational energy have so far not provided a general method for predicting protein structure from amino acid sequence. Molecular dynamics offers a way to overcome the problems of getting trapped in local minima, and of the absence of a good static model for protein–solvent interactions. In molecular-dynamics calculations, the protein plus explicit solvent molecules are treated—via the force field—by classical Newtonian mechanics. It is true that this permits exploration of a much larger sector of phase space. However, as an *a priori* method of structure prediction, it has still not succeeded consistently. However, these are calculations that are extremely computationally intensive and here, perhaps more than anywhere else in this field, advances deriving from the increased power of processors will have an effect.

Is lack of computational power the only reason for lack of success in prediction of protein structure by simulation of the folding pathway? There have been several attempts to apply 'brute force', including the IBM Blue Gene supercomputer project and the distributed computing approach of Folding@Home, which makes use of contributions of computer power from over a million participating processors.(A similar approach has been applied to drug design.) An IBM group folded a 20-residue peptide from a fully extended conformation to a state within ~1.5 Å r.m.s.d. of the native state (see Box 3.9).

Scaling of resource requirements for molecular-dynamics calculations

Fully detailed molecular-dynamics calculations perform a series of individual time steps of duration 10^{-15} s (= 1 fs). At each step, the force on each atom is computed, and the motion of the system is predicted from the current set of positions, velocities and accelerations (acceleration = force/mass). The time set chosen is very small to reduce the extent of collision, which would interfere with the extrapolation of the motion of the particles based on the instantaneous values of the velocity and acceleration computed for the beginning of the step.

The computer time required for an individual time step scales approximately as $N \ln N$, where N is the number of residues in the protein. The time required for a protein to fold depends on a number of features, but for the purposes of a 'back-of-the-envelope' calculation it varies with the length N of the protein as $\sim N^{5/3}$.

Therefore, the total computer resources required to fold a protein may be expected to vary approximately as $N^{\frac{2}{3}} \ln N$. This means that if it takes 3 months (of uninterrupted time on a supercomputer running flat out) to fold a protein of length N, it would be expected to require over 1.5 years, on the same system, to fold up a protein of length $3N$ residues.

Successful application of molecular dynamics to protein folding and structure prediction are:

1. Accurate force fields: A concerted effort has gone into calibrating the form and parameters of force fields. They now are very finely tuned.

2. Adequate computing resources to simulate the systems over long time intervals. How long? The discovery of fast-folding proteins has lowered the requirements. For instance, the villin headpiece, a 35-residue fragment, folds to a compact state in ~220 ns at 283 K.

In addition to the growth in general computing power available, D.E. Shaw Research has developed special-purpose hardware for molecular dynamics. Figure 3.33 shows the results of simulations of several small proteins, compared to the experimental results. All are between 10 and 80 residues long. They cover three major classes of fold types: all-α, all-β, and $\alpha + \beta$. The simulation of each protein exhibited multiple folding-unfolding transitions.

In addition to its applications to *a priori* protein structure prediction, molecular dynamics, if supplemented by experimental data, regularly makes extremely important contributions to structure determinations by both X-ray crystallography (usually) and nuclear magnetic resonance (always). How is molecular dynamics integrated into the process of structure determination? For any conformation, one can measure the consistency of the model with the experimental data. In the case of crystallography, the experimental data are the absolute values of the Fourier coefficients (or observed structure factor magnitudes) of the electron density of the molecule. In the case of nuclear magnetic resonance, the experimental data provide constraints on the distances between certain pairs of residues. But in both, typical X-ray crystallography and nuclear magnetic resonance structure determinations, the experimental data underdetermine the protein structure. To solve a structure one must seek a set of coordinates that minimizes a combination of the deviation from the experimental data and the conformational energy. Molecular dynamics is successful at determining such coordinate sets: the dynamics provides adequate coverage of conformation space, and the bias derived from the experimental data channels the calculation quite effectively towards the correct structure.

ROSETTA

ROSETTA is a program by D. Baker and colleagues that predicts protein structure from amino acid sequences by assimilating information from known structures. At recent CASP programmes, ROSETTA has showed consistent success on targets in both the fold recognition and novel fold categories. At present, it leads the field by several lengths.

ROSETTA predicts a protein structure by first generating structures of fragments using known structures, and then combining them. First, for each

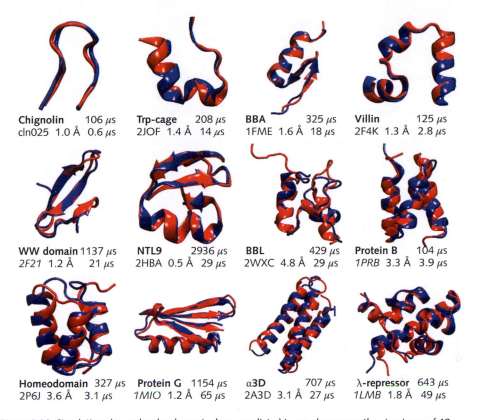

Figure 3.33 Simulations by molecular dynamics have predicted to good accuracy the structures of 12 small proteins. Each frame shows the superposition of the experimental structure (red) and the result of simulation (blue), and reports the protein name, the duration of the simulation, the wwPDB code of the experimental structure (or a close homologue, indicated by an italic wwPDB code), the r.m.s.d. of the superposition, and the folding time (derived from the simulations).

From: Lindorff-Larsen, K., Piana, S., Dror, R.O. and Shaw, D.E. (2011). How fast-folding proteins fold. *Science* **334**, 517–520.

contiguous region of three and nine residues, instances of that sequence and related sequences are identified in proteins of known structure. For fragments this small, there is no assumption of homology to the target protein. The distribution of conformations of the collected fragments serves as a model for the distribution of possible conformations of the corresponding regions of the target structure.

ROSETTA explores the possible combinations of fragments using Monte Carlo calculations (see Box 3.10). The energy function has terms reflecting compactness, paired β-sheets and burial of hydrophobic residues. The procedure carries out 1000 independent simulations, with starting structures chosen from the fragment conformation distribution pattern generated previously. The structures that result from these simulations are clustered, and the centres of the largest clusters presented as predictions of the target structure. The idea is that a structure that emerges many times from independent simulations is likely to have favourable features.

ROBETTA (http://robetta.bakerlab.org) is a web server designed to integrate and implement the best of the protein structure prediction tools. The central pipeline of the software involves first the parsing of a submitted amino acid sequence of a protein of unknown structure into putative domains. Then, homology-modelling techniques are applied to those domains for which suitable parents of known structure exist, and the *de novo* methods developed by Baker and coworkers to other domains. In addition, the user will receive the results of other prediction methods based on software developed outside the ROBETTA group. These include, for example, predictions of secondary structure, coiled-coils, and transmembrane helices.

BOX 3.10 **Monte Carlo algorithms**

Monte Carlo algorithms are used very widely in protein structure calculations to explore conformations efficiently; and in many other optimization problems to search for the minimum of a complicated function. Simple minimization methods based on moving 'downhill' in energy fail, because the calculation can get trapped in a local minimum far from the native state.

In general, Monte Carlo methods make use of random numbers to solve problems for which it is difficult to calculate the answer exactly. The name was invented by J. von Neumann, referring to the applications of random number generators (cards or dice) in the famous gambling casino.

To apply Monte Carlo techniques to find the minimum of a function of many variables—for instance, the minimum energy of a protein as a function of the variables that define its conformation—suppose that the configuration of the system is specified by the variables x, and that for any values of these variables, we can calculate the energy of the conformation, $\varepsilon(x)$. (x stands for a whole set of variables—perhaps the set of atomic coordinates of a protein, or the mainchain and sidechain torsion angles.)

Then, the Metropolis procedure (invented in 1953, allegedly at a dinner party in Los Alamos) prescribes:

1. Generate a random set of values of x, to provide a starting conformation. Calculate the energy of this conformation, $\varepsilon = \varepsilon(x)$. ε will record the lowest value of $\varepsilon(x)$ so far encountered in the calculation.

2. Randomly perturb the variables: $x \rightarrow x'$, to generate a neighbouring conformation.

3. Calculate the energy of the new conformation, $\varepsilon(x')$.

4. Decide whether to *accept* the step; that is, to move $x \rightarrow x'$, or to stay at x and try a different perturbation:

(a) If the energy has decreased; that is, $\varepsilon = \varepsilon(x) > \varepsilon(x')$—that is, the step went *downhill*—always accept it. The perturbed conformation becomes the new current conformation: set $x' \rightarrow x$ and $\varepsilon = \varepsilon(x')$.

(b) If the energy has increased or stayed the same; that is $\varepsilon(x) \leq \varepsilon(x')$—in other words the step goes *uphill*—then, *sometimes* accept the new conformation. If $\Delta = \varepsilon(x') - \varepsilon(x)$, accept the step with a probability $\exp(-\Delta/k_B T)$, where k_B is Boltzmann's constant, and T is an effective temperature.

5. Return to step 2.

It is step 4b that is the ingenious one. It has the potential to get over barriers, out of traps in local minima. The effective temperature, T, controls the chance that an uphill move will be accepted. T is not the physical temperature at which we wish to predict the protein conformation, but simply a numerical parameter that controls the calculation.

For any temperature, the higher the uphill energy difference, the less likely it is that the step will be accepted. For any value of ε, if T is low, then $\varepsilon(x)/(k_B T)$ will be high, and $\exp[-\varepsilon(x)/(k_B T)]$ will be relatively low. If T is high, then $\varepsilon(x)/(k_B T)$ will be low, and $\exp[-\varepsilon(x)/(k_B T)]$ will be relatively high. The higher the temperature, the more probable the acceptance of an uphill move. Two simple simulations, at different temperatures, appear in the companion website of this book.

This relatively simple idea has proved extremely effective, with successful applications including but by no means limited to protein structure calculations.

Simulated annealing is a development of Monte Carlo calculations in which T varies—first it is set high to allow efficient exploration of conformations, then it is reduced to drop the system into a low-energy state.

The Baker group has gone beyond offering a web server. Rosetta@home is a distributed computing project that uses idle computer processing power made available by volunteers. The goals were to use the very large resources thereby applicable, to projects involving protein–protein docking and protein design. Foldit is an online video game about protein folding.

Protein structure prediction from contact maps derived from correlated mutations in multiple sequence alignments

'A single amino acid sequence is silent about its structure; a pair of aligned sequences whispers; a multiple sequence alignment shouts out loud.' But, however soft or loud the utterance, what is it saying?

Multiple sequence alignments of homologous proteins reflect evolutionary pathways. As proteins evolve, mutations perturb structures. As a rule of thumb, a well-packed internal interface in a protein can accommodate an extra methyl group (e.g. Ser→Thr or Val→Leu) but not much more. As evolution proceeds after such a perturbation, the region around the site of mutation may have higher selective constraints, in order that additional mutations do not destroy the structure entirely.

It follows that positions in multiple sequence alignments that show *correlated mutations* can indicate that the positions are nearby in space in the folded structure. Correlated mutations are patterns of amino acid substitutions such that change in the amino acid at some position corresponds to change in the amino acid at one or more other positions. For the sequences in which the amino acid at the first position is constant, there is no mutation at the other positions also (see Fig. 3.34). In other words, each amino acid observed at every position of a set showing correlated mutations, corresponds to a unique amino acid at the other positions.

A reliable theoretical inference of interresidue contact patterns would provide information analogous to that of NMR data. The theoretical results could, ideally, be fed into the protein-NMR pipeline to determine the three-dimensional structure. The potential power

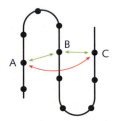

Figure 3.35 Residues shown as black circles, along a stretch of polypeptide chain. Correlated mutations might well correctly identify residues A and B, and residues B and C as neighbours. The danger is that the analysis might also suggest, incorrectly, that A and C are neighbours. A and C might show correlated mutations because they are both correlated with B.

of this approach has been recognized for a long time, but getting it to work proved elusive.

What were the problems? A fundamental difficulty was the false detection of contacts because of mutual-neighbour effects, or, more technically, transitivity. Suppose position A is close in space to position B, but position C is also close to position B. Analysis of correlated mutations may correctly pick up the proximity of A and B and of B and C, but may also—spuriously—suggest proximity of A and C (see Fig. 3.35).

Recognition of this problem led to methods for its solution. Suppose we are given data that depend on many variables—for instance the distribution of amino acids at different residue positions of a multiple

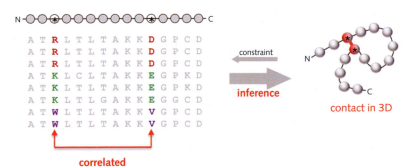

Figure 3.34 Two positions in a multiple sequence alignment may show correlated mutations. That is, a change at one position is echoed by a change at a second position. The inference of a contact between corresponding residues in the three-dimensional structure contributes to a set of distance constraints, similar to those measured by NMR, which can be used to predict the fold. It was formerly hypothesized that correlated mutations would be *compensatory:* That is, if two residues, e.g. Gly-Leu, were in contact in a structure, evolution would be limited to changing them to Leu-Gly, or Ile-Gly, or Ala-Val, in order to hold constant the total sidechain volume of the pair. This could keep the overall backbone conformation unchanged. Compensatory mutations are now known to be the exception rather than the rule. As they evolve, proteins retain their topology but not their precise local conformations.

From: Marks, D.S., Colwell, L.J., Sheridan, R., Hopf, T.A., Pagnani, A., Zecchina, R. and Sander C. (2011). Protein 3D structure computed from evolutionary sequence variation. PLoS One. 6, e28766.

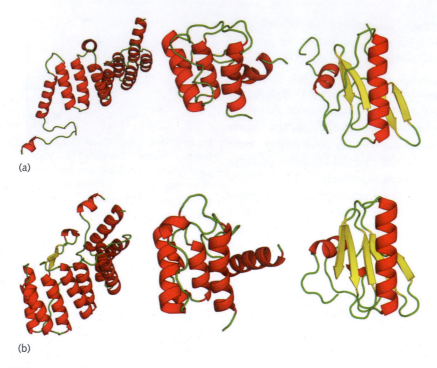

(a)

(b)

Figure 3.36 (a) Predictions and (b) experimental structures of (left to right) N-terminal domain of neutrophil cytosol factor 2 p67phoX A128V (wwPDB entry 1hh8, chain A), calponin homology domain from human β-spectrin (1BKR, chain A), and severin domain (1SVY, chain A).

[From: Kosciolek, T. and Jones, D.T. (2014). *De novo* structure prediction of globular proteins aided by sequence variation-derived contacts. PLoS One, 9, e92197. Reproduced via CC BY 4.0 license]

sequence alignment that show corresponding mutations. Then it is possible to compute *partial correlations*, that is, correlations between each individual pair of variables after subtracting out the dependence on all the other variables. Suppressing these dependencies can eliminate, or at least reduce substantially, the corresponding transitivity effects. (Box 3.11 contains a discussion of the techniques involved.) Use of partial correlations, plus calibrating the number of expected contacts to what is empirically observed in known protein structures of suitable size, improved the accuracy of inference of residues in contact from correlated mutations.

These improvements, bolstered by the general growth in amount of sequence data available, and mature tools for creating accurate multiple sequence alignments, permit impressive predictions of three-dimensional structures from sequences of proteins of a homologous series (see Fig. 3.36).

Determination of residue–residue contacts in proteins is now a challenge of the current CASP programs.

Critical Assessment of Structure Prediction (CASP)

Critical Assessment of Structure Prediction (CASP) is a programme that organizes blind tests of protein structure predictions, in which participating crystallographers and NMR spectroscopists make public the amino acid sequences of the proteins they are investigating, and agree to keep the experimental structure secret until predictors have had a chance to submit their models. CASP runs on a two-year cycle. Every two years the sequences are published in the spring, and predictions are due in the autumn. Assessors assigned to different categories compare the predicted and experimental structures, and judge the predictions.

At the end of the year a gala meeting brings the predictors and the appointed assessors together to discuss the current results and to gauge progress. Speakers at the end-of-year meeting include the organizers, the assessors, and selected predictors, including those who have been particularly successful, or who have an interesting novel method to present.

BOX 3.11   Partial correlation

A study of class performance in a course taught from this book measured, for each student, (a) the grade he or she received, and (b) whether he or she had a hangover the morning of the final exam, and a measure of its severity if non-zero.[*] Thinking geometrically, these data assemble into two vectors, the corresponding entries containing the two values associated with each student. The angle between such vectors indicates their degree of correlation. If the vectors are parallel—perfect correlation—the angle is 0. In contrast, perpendicular vectors correspond to uncorrelated variables.

Figure 3.37(a) shows a reasonable correlation between the vectors, suggesting the advice: 'Just show up at the exam sober, then you'll pass.'

The problem with this conclusion is that there may be other variables involved, which indirectly underlie the correlation observed in Figure 3.37(a). One such variable is whether the student did any studying the night before the exam. Is poor exam performance *really* simply a reflection of drinking, or of drinking *instead of studying* before the exam?

Figure 3.37(b) shows the vectors corresponding to *three* variables. To determine whether performance and hangover are directly correlated, we can subtract out their dependence on study time by projecting the original two vectors (red in Figure 3.37(b)) onto a plane perpendicular to the third vector, that represents number of hours studied (blue in Figure 3.37(b)). Because two perpendicular vectors are uncorrelated, the projection *removes* the components of the grade and hangover variables that are correlated with study time, leaving the components that are *independent of* study time. Note that the projected vectors corresponding to hangover and performance (shown in blue) make a much larger angle than the original ones (52°—red—has widened to 83°—blue). The apparent correlation is thereby weakened, to the point where the advice about how to pass the course should perhaps be modified, or, at least, expanded.

A more general question is: if our goal is to understand the factors underlying exam success, how can we choose from our data the combination of variables that best predicts exam score? Techniques of regression have been developed to address this problem. The simplest example of regression is fitting a straight line to a set of points. But in

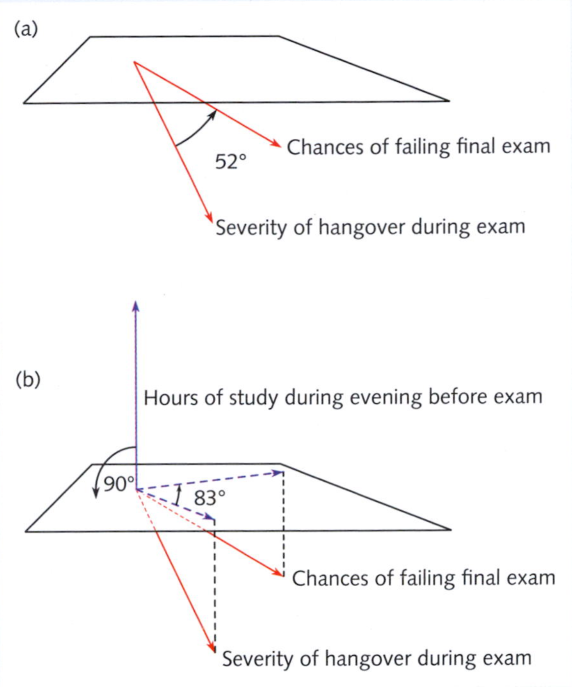

Figure 3.37 (a) Correlation between vectors representing two variables (shown in red), considered independently of other effects. (b) Projection (shown in blue) of the two vectors onto a plane perpendicular to a vector representing a third variable. The apparent correlation between the two original variables, seen in (a), is shown to be the result of a dependence of both on a third variable.

that case, we have no choice: there is only one independent variable. In the multivariate case we want to represent the target data—exam scores—in terms of the other variables, in such a way that (a) we achieve the optimal prediction of the target data from the values of the other variables, and (b) we expose the structure of the system, to identify which variables are really important.

Goal (a) is usually addressed by least-squares techniques. We determine parameters in a function expressing the predicted exam score in terms of other variables, to minimize the sum of the squares of the 'errors', the differences between the predicted and measured values of the target data. In practice, this function often is linear, but need not be so.

Goal (b), if successful, offers the possibility of simplifying the system. It may be possible to identify variables that can be deleted from the model, without degrading too severely

[*]The data are admittedly hypothetical but not unreasonable.

the ability of the remaining variables to account for the data. In the 'LASSO' method (Least Absolute Shrinkage and Selection Operator), we minimize the sum of squares of the errors, *plus* a 'penalty' term proportional to the sum of the absolute values of the parameters that relate the individual variables to the target. In favourable cases, the penalty term forces to zero the parameters relating one or more of the other variables to the target variable. This identifies variables from which the target variable appears to be independent. These variables can be deleted, simplifying the model. Readers will recognize that this is precisely the kind of information we need to exclude certain observed correlated mutations, from candidacy to be true spatial neighbours.

The framework of more general statistical techniques subsumes the calculations we used in analysing exam grades.

Covariance matrices and their inverses

The basic statistical tool for investigating correlations is the covariance matrix. Consider a set of variables $x, y, z, \ldots$ such that a particular set of numbers $(x_i, y_i, z_i, \ldots)$ characterizes each subject i in a sample. For simplicity, suppose that the mean of all individual variables is zero; that is, $\Sigma x_i = \Sigma y_i = \Sigma z_i = 0$, and the variances σ^2 are equal to 1 (The definition of σ^2 of the x_i is $\sigma^2 = \Sigma(x_i - \mu_x)^2/n$; in this definition μ_x is the mean of the values of x—here known to be 0—and n is the number of points.) Then the correlation between variables x and y, over the points i of the sample, is $\Sigma x_i y_i$. Assemble these sums in a matrix in which the rows and columns are labelled, or indexed, by the variables $x, y, z \ldots$.

The terms of the form $\Sigma x_i y_i$ which are off-diagonal terms in this covariance matrix will be large in magnitude if there is a relationship between the corresponding variables. A positive value of $\Sigma x_i y_i$ indicates a *direct correlation;* that is, when x increases, y increases. A negative value of $\Sigma x_i y_i$ indicates an *inverse correlation;* that is, when x increases, y decreases. If $\Sigma x_i y_i = 0$, then there is apparently no relation between x and y that could be useful in predicting y knowing a value of x. For instance, if there are seven points, some of the possibilities are:

Degree of correlation	Formulas for x_i and y_i	Element of covariance matrix
High positive correlation	$x_i = -3/2, -1, -1/2, 0, 1/2, 1, 3/2$ $y_i = -3/2, -1, -1/2, 0, 1/2, 1, 3/2$	$\Sigma x_i y_i = (-3/2) \times (-3/2) + (-1) \times (-1) + (-1/2) \times (-1/2)$ $+ 0 \times 0 + (1/2) \times (1/2) + 1 \times 1 + (3/2) \times (3/2) = 7$ (a positive number)
High negative correlation	$x_i = -3/2, -1, -1/2, 0, 1/2, 1, 3/2$ $y_i = 3/2, 1, 1/2, 0, -1/2, -1, -3/2$	$\Sigma x_i y_i = (-3/2) \times (+3/2) + (-1) \times (+1) + (-1/2) \times (+1/2)$ $+ 0 \times 0 + (1/2) \times (-1/2) + 1 \times (-1) + (3/2) \times (-3/2) = -7$ (a negative number)
Uncorrelated	$x_i = -3/2, -1, -1/2, 0, 1/2, 1, 3/2$ $y_i = 1.08, -1.08, -1.08, 0, 1.08, 1.08, 1.08$	$\Sigma x_i y_i = 0$

For each set of x and y in this table the mean is 0 and the variance is 1.

From its definition, it is clear that the covariance matrix contains information only about *pairwise* relationships between the variables. Each element of the correlation matrix depends only on the values of two variables, for instance, x and y. The covariance matrix contains no information about the interactions between variables. For instance, it does not directly reflect the multivariable interactions that we analysed in the exam grade example. These indirect interactions among the variables were precisely what we needed to sort out mutual influences. What, in terms of the covariance matrix, corresponds to the projection that we did to produce Figure 3.37(b)?

There are several approaches. One is to look at the *inverse* of the covariance matrix. (The inverse of a matrix M is another matrix, M^{-1}, with the same number of rows and columns, such that the matrix product MM^{-1} = the unit matrix; that is, the matrix with all diagonal elements 1 and all off-diagonal elements 0.)

In general, the elements of the inverse matrix combine information from all of the covariance terms. They therefore contain information about the interactions of the variables, in general more than two at a time. In fact, inverting a covariance matrix is equivalent to determining the dependence of each variable on all the other variables. It follows that a 0 in the inverse covariance matrix corresponding to variables x and y implies that variables x and y are independent. Conversely, non-zero off-diagonal elements in the inverse covariance matrix identify genuine 'neighbours'.

These may be literally neighbours in the geometric sense, or, more generally, neighbours in an influence network. The same techniques can be used for analysing neighbouring sites in proteins, and neighbours in signal transduction cascades in regulatory networks (and many, many other phenomena by no means limited to molecular biology).

An example from a lecture by D. Mackay[*] illustrates the ability of inverse covariance matrices to distinguish true neighbours from indirect influences.

Figure 3.38 shows a set of five masses connected by springs. The positions of the masses define a set of variables x, y, z ... The equilibrium positions correspond to $x = 0$, $y = 0$, $z = 0$, ... Assume that the system is jiggling because of random thermal motions. Collect data on the instantaneous values of the variables at many different time points. Thinking ahead, we expect to see correlations between the lengths of adjacent springs, because displacement of a mass will stretch one adjacent spring and compress the other adjacent spring.

A perhaps more familiar real-life version of this would be the jostling of passengers in a crowded tube car (perhaps on the Metropolitan Line). Each rider is displaced by collisions with adjacent people but the impulse is usually not strongly propagated to other passengers. Compare a rugby scrum in which the opposite is true.

The elements of the covariance matrix are sums such as $\Sigma x_i y_i$. We assume that the mean values of x, y, ... are all 0. Mackay represented these variables as a multivariate Gaussian distribution.

[*]http://videolectures.net/gpip06_mackay_gpb/

The covariance matrix has elements that emphasize nearest-neighbour interactions, but are not limited to them: (That the elements of the covariance matrix and its inverse are rational implies that Mackay's numbers, like those in the exam grade example, were not derived from an explicit realistic calculation, much less an experiment. Readers may share my mild incredulity that the end masses will show the same displacement profile as the interior ones, assuming that the walls are fixed.)

Covariance matrix

$$\frac{1}{6} \begin{bmatrix} 5 & 4 & 3 & 2 & 1 \\ 4 & 8 & 6 & 4 & 2 \\ 3 & 6 & 9 & 6 & 3 \\ 2 & 4 & 6 & 8 & 4 \\ 1 & 2 & 3 & 4 & 5 \end{bmatrix}$$

The inverse matrix is 'cleaned up', revealing the structure of the system in terms of its nearest neighbours:

Inverse covariance matrix

$$\begin{bmatrix} 2 & -1 & 0 & 0 & 0 \\ -1 & 2 & -1 & 0 & 0 \\ 0 & -1 & 2 & -1 & 0 \\ 0 & 0 & -1 & 2 & -1 \\ 0 & 0 & 0 & -1 & 2 \end{bmatrix}$$

It is precisely this clarification that is necessary to derive interresidue contacts reliably from correlated mutations in multiple sequence alignments.

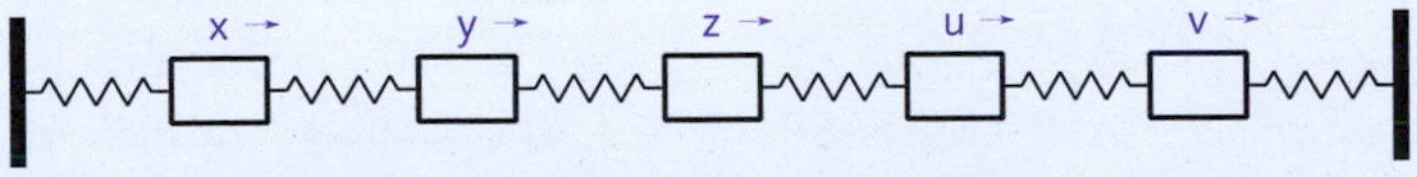

Figure 3.38 Five masses connected by springs, moving in one dimension. The variables are the displacement of the individual masses. Thermal jiggling of the system is described and analysed as a multivariate Gaussian distribution.

The latest CASP programme, CASP11, in 2014, invited predictions in the following categories:

1. *Tertiary structure predictions*. These include *homology modelling* (of targets for which a suitable parent structure or structures can be identified), and *template-free*, or *a priori*, modelling.

2. *Refinement of a prediction*. Can an initial prediction of a target by one method be improved by a different one?

3. Challenges involving residue–residue contacts. One category is *contact-assisted prediction*: Suppose

that a predictor, in addition to the sequence of the target, is given selected information about neighbouring residues from the experimental structure. To what extent can this information improve the prediction? Knowing what contact information is necessary and adequate would set the threshold for success in the related category: *prediction of residue–residue contacts*.

4. *Identifying disordered regions* in target proteins.

5. Methods for *assessing quality of models*, without knowing the experimental structure. (The

correspondence of a 3D structure profile with a target sequence (p. xxx) would be one approach to this category.)

Prediction of secondary structure, and fold recognition, are no longer challenges in the current CASP program.

CASP11 involved 100 targets for the Tertiary Structure prediction challenge. A few were withdrawn, most because of release of the structure from a different group (in other words, the group that submitted the target were committed to delaying publication of the result, and were 'scooped'). There were 37 targets for refinement, and 71 for the 'assisted' categories. A total of 207 'human' groups and 84 servers submitted a total of 58835 models.

Many predictions are prepared by groups of researchers who study the results generated by their computer programs, and select and edit them before submission. In addition, the target sequences are sent to web servers, which return predictions without human intervention. (It has happened that a web server did better than a hand-edited prediction by the author of the server.) **CAFASP** (critical assessment of fully automated structure prediction programme) monitors the quality of these predictions. It is thereby possible to determine to what extent successful procedures can be made fully automatic. There are thus three challenges:

Human against protein	CASP
Computer against protein	CAFASP
Human against computer	CASP v. CAFASP

A separate programme of blind tests of prediction evaluates methods for predicting protein–protein interactions, or 'docking'. This is **CAPRI**: critical assessment of predicted interactions. Both CASP and CAPRI reported assessments in late 2014.

Organized blind tests of prediction methods include:
- critical assessment of structure prediction (CASP)
- critical assessment of fully automatic structure prediction (CAFASP)
- critical assessment of protein interactions (CAPRI).

Figure 3.39 shows a successful automatic prediction in the template-based modelling (homology modelling)

category at CASP11, by I-TASSER server, developed by Y. Zhang and colleagues at the University of Michigan, U.S.A. The target was a 112-residue *de novo* designed protein LFR 1 with a ferredoxin-like fold. The prediction contains 97 residues. (The additional residues, at the C-terminus of the experimental structure, are disordered.) A structural alignment shows 93 Cα atoms fitting to an r.m.s.d. of 1.7 Å. The program constructed the model from multiple templates.

Software from the Zhang group has had fine success in both homology modelling and *a priori* predictions.

Over the last few years, the group of D. Baker at the University of Washington has been outstanding with the quality and consistency of its predictions of protein structures that cannot be built by homology modelling. The programs also excel at difficult homology-modelling tasks.

Figure 3.40 shows a CASP11 prediction from the Baker group of a difficult homology-modelling target, T0782, with sequence similarity between the target structure and the closest template down at 15%. The target is a hypothetical protein (BACUNI_01346) from *Bacteroides uniformis* ATCC 8492. (A hypothetical protein is one which is encoded in a genome but with expression unproved.) Structural alignment shows 84 out of 110 C atoms fitting to an r.m.s.d. of 1.29 Å.

Another fine prediction of this target in CASP11 was from the group of M.J. Skwark of the University of Stockholm (see Fig. 3.41). Structural alignment shows 93 out of 110 C atoms fitting to an r.m.s.d. of 1.48 Å.

A spectacular success of the Baker group at CASP11 was the prediction of the structure of target 806, protein YaaA from *E. coli*. It is an *ab initio* prediction, making use of a number of methods including but not limited to prediction of interresidue contacts. Figure 3.42(a) shows the results of a superposition of 205 C atoms out of 256, or 80%. All residues are shown, but the well-fitting subset is in thicker ribbons. Figure 3.42(b) is extracted from (a), showing only the well-fitting subset. The r.m.s.d. of the 205 C atoms is 2.78 Å. However, some of the deviation is the result of changes in interdomain geometry. Figure 3.42(c) shows the superposition of one of the domains (the one appearing on the left in parts (a) and (b)), comprising residues 1–6, 85–92, and 165–254. The r.m.s.d. is 1.4 Å. This is absolutely bang on (see Weblem 3.4).

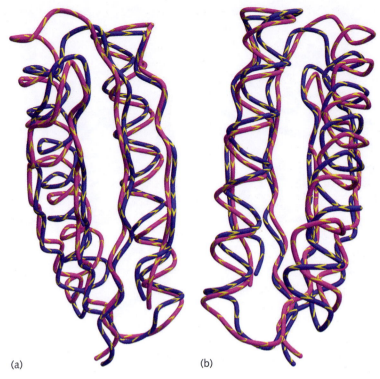

(a)   (b)

Figure 3.39 Prediction of target T0769 (released as wwPDB entry 2MQ8) in the template-based modelling category at CASP11, by I-TASSER, developed by Y. Zhang and coworkers. The model contains 97 residues out of 112. (The additional residues, at the C-terminus of the experimental structure, are disordered.) Superposition of 97 Cα atoms. (a) and (b) show the same pair of molecules, from viewpoints 180° apart ('front' and 'back' views).

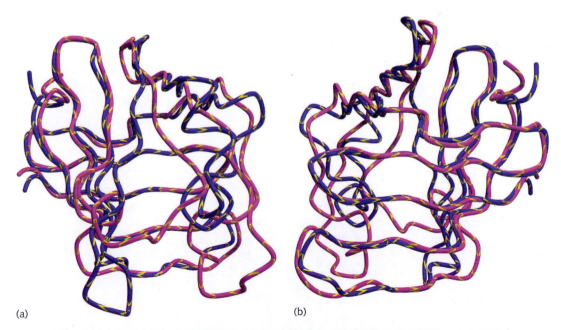

(a)   (b)

Figure 3.40 Prediction of target T0782 (released as wwPDB entry 4QRL) by the ROSETTA server, created by D. Baker and coworkers. The closest template (wwPDB entry 2LA7) shares 15% sequence similarity with the target. (a) and (b) show the same pair of molecules, 'front' and 'back' views.

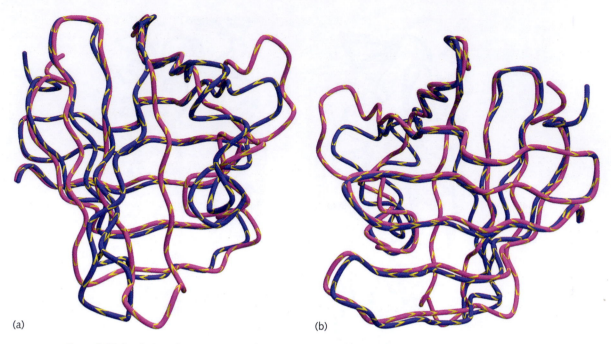

(a)   (b)

Figure 3.41 Prediction of target T0782 (released as wwPDB entry 4QRL) by M. Skwark and coworkers. (a) and (b) show the same pair of molecules, 'front' and 'back' views.

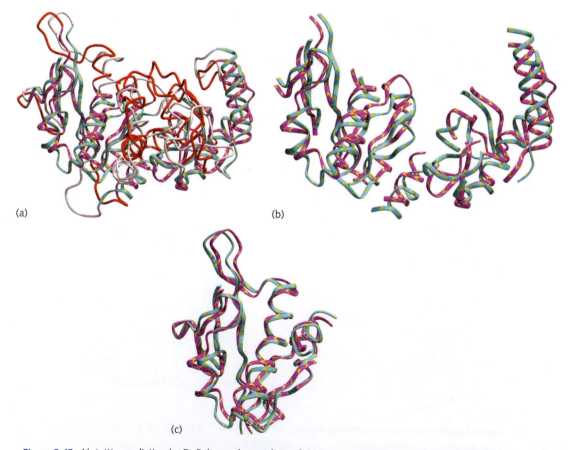

(a)   (b)

(c)

Figure 3.42 *Ab initio* prediction by D. Baker and coworkers of CASP11 target 806, YaaA from *E. coli*. (a) Superposition of selected C atoms from entire structure, showing entire structure. Thicker ribbons show the residues selected for superposition. (b) Only the well-fitting regions, extracted from (a). (c) Superposition of selected atoms from one domain. The fit is extremely impressive.

CAPRI

Results of the latest CAPRI programme show progress in the modelling of complexes. It is now possible to predict the struuctures of protein dimers, based on independent experimental structure determinations of the two partner structures, or even based on homology models.

● RECOMMENDED READING

Al-Lazikani, B. , Hill, E.E. and Morea, V. (2008). Protein structure prediction. *Meth. Molec. Biol.*, **453**, 33–85. A good general overview, now of course somewhat dated.

Baker, E.N. (2004). From penicillin to the ribosome: Revolution in the determination and use of molecular structure in chemistry and biology. *Aust. J. Chem.*, **57**, 829–36. Useful treatment of the development of protein crystallography, and applications of structure determination to understanding protein function.

Berman, H.M. , Kleywegt, G.J. , Nakamura, H. and Markley, J.L. (2013). The future of the protein data bank. *Biopolymers* **99**, 218–22. Where do we go from here?

Bradley, P. , Misura, K.M. and Baker, D. (2005). Toward high-resolution *de novo* structure prediction for small proteins. *Science*, **309**, 1868–71. Description of the most powerful of the protein structure prediction methods not depending on homology modelling from a close relative.

Dodson, G.G. , Lane, D.P. and Verma, C. (2008). Molecular simulations of protein dynamics: new windows on mechanisms in biology. *EMBO Rep.*, **9**, 144–50. Review of molecular dynamics and its applications.

Drenth, J. (1994). *Principles of protein X-ray crystallography.* Springer-Verlag, New York & London. A classic text.

Ekins, S. (2006). *Computer applications in pharmaceutical research and development.* Wiley-Interscience, Hoboken, NJ, U.S.A. Applications of computational methods to drug design.

Evans, P. and McCoy, A. (2008). An introduction to molecular replacement. *Acta Cryst.*, **D64**, 1–10. Description of an important technique for solving X-ray crystal structures of proteins.

Guss, J.M. and King, G.F. (2002). Macromolecular structure determination: Comparison of crystallography and NMR. In: *Encyclopedia of the life sciences*, ed. S. Robertson. Nature Publications Group, London, vol. 11, 290–5. Perspective on two major methods of protein structure determination.

Gutmanas, A. , Oldfield, T.J. , Patwardhan, A. , Sen, S. , Velankar, S. and Kleywegt, G.J. (2013). The role of structural bioinformatics resources in the era of integrative structural biology. *Acta Crystallogr. D Biol. Crystallogr.*, **69**, 710–21. Some interesting history plus a description of the relationships between structure databanks at different levels of resolution.

Hodgkin, D.C. and Riley, D.P. (1968). Some ancient history of protein X-ray analysis. In: *Structural chemistry and molecular biology*. A. Rich and N. Davidson, (eds). W.H. Freeman & Co., San Francisco, pp. 16–28. Early history; the source of the incident involving Bernal and the 1934 pepsin photograph.

de Juan, D. , Pages, F. and Valencia, A. (2013) Emerging methods in protein co-evolution. *Nature Revs. Genet.*, **14**, 249–61. Discussion of methods based on correlated mutations.

Joseph, A.P. and de Brevern, A.G. (2014). From local structure to a global framework: recognition of protein folds. *J. Roy. Soc. Interface*, **11**, 2013.1147. Discussion of methods of predicting local structure and the relationship between local and global structure.

Kleywegt, G.J. and Jones, T.A. (2002). Homo crystallographicus–*Quo vadis? Structure,* **10**, 465–72. Current status and prospects of the field.

Kosciolek, T. and Jones, D.T. (2014). *De novo* structure prediction of globular proteins aided by sequence variation-derived contacts. PLoS One, **9**, e92197. A new method of current high interest: predicting structure from the inference of neighbouring residues from correlated mutations in multiple sequence alignments.

Marion, D. (2013). An introduction to biological NMR spectroscopy. *Mol. Cell. Proteomics*, **12**, 3006–25. An introduction to the field, with historical background, aimed at biologists who wish to use NMR structures of proteins, but not necessarily produce them. There are also a number of specialized textbooks and websites for readers who wish to study the subject in greater depth.

Milne, J.L., Borgnia, M.J., Bartesaghi, A., Tran, E.E., Earl, L.A., Schauder, D.M., Lengyel, J., Pierson, J., Patwardhan, A. and Subramaniam, S. (2013). Cryo-electron microscopy—a primer for the non-microscopist. *FEBS J.*, **280**, 28–45. Introduction to this technique, the results of which are rapidly increasing in quality and, thereby, in importance.

Mourad, R. and Sinoquet, C. (2014). *Probabilistic graphical models for genetics, genomics and postgenomics*. Oxford University Press, Oxford. Mathematical and statistical methods applied to problems of molecular biology.

Perutz, M.F. (1990). How W.L. Bragg invented X-ray analysis. *Acta Cryst.*, **A46**, 633–43. and

Perutz, M.F. (2002). *I wish I'd made you angry earlier: Essays on science, scientists and humanity*. Oxford University Press, Oxford. Contains a description of some of the exciting history, from the leading participant.

Rosenthal, P.B. (2015). From high symmetry to high resolution in biological electron microscopy: a commentary on Crowther (1971) 'Procedures for three-dimensional reconstruction of spherical viruses by Fourier synthesis from electron micrographs'. *Phil. Trans. R. Soc. B.*, **370**, 20140345. History and leading up to the state of the art. R.A. Crowther's 1971 paper introduced the method of three-dimensional reconstruction by combining data from images taken at different orientations.

Šali, A. (1998). 100,000 protein structures for the biologist. *Nature Struc. Biol.*, **5**, 1029–32. Describes the logic underlying structural genomics projects.

Tramontano, A. (2006). *Protein structure prediction. Concepts and applications*. Wiley-VCH, Weinheim. Overview of the subject, by a leading expert, also somewhat dated.

Wlodawer, A., Minor, W., Dauter, Z. and Jaskolski, M. (2013). Protein crystallography for aspiring crystallographers or how to avoid pitfalls and traps in macromolecular structure determination. *FEBS J.*, **280**, 5705–36. A brief outline of technical aspects of crystallography, for aspiring crystallographers, yes, but also for 'consumers' of crystal structures.

Wüthrich, K. (1995). NMR—This other method for protein and nucleic acid structure determination. *Acta Cryst.*, **D51**, 249–70. Description of protein structure determination by NMR, by one of its inventors.

Xiong, Y. (2008). From electron microscopy to X-ray crystallography: molecular replacement case studies. *Acta Cryst.*, **D64**, 76–82. Discussion of the power of the combination of X-ray crystallography and cryo-electron microscopy.

The 11 September 2009 issue of The Journal of Molecular Biology contains a special section commemorating 50 years of protein structure determination including articles by B. Strandberg, R.E. Dickerson, and M.G. Rossman, surviving participants in the original work.

● EXERCISES AND PROBLEMS

Exercise 3.1 Polonium crystallizes in a simple unit cell—one atom per cube of edge length 3.35 Å (335×10^{-6} m). How many polonium atoms are there in a crystal that has the shape of a cube 1 mm on each edge?

Exercise 3.2 The unit cell of a crystal of *E. coli* transaldolase is a rectangular solid with edges 68.9 Å, 91.3 Å, and 130.5 Å that contains eight molecules of a protein of relative molecular mass 35072. What is the concentration of protein in the crystal, in mg mL^{-1}? Compare with the estimates of 300–400 mg mL^{-1} for the intracellular concentration of macromolecules in prokaryotic cells, and a typical protein concentration in an NMR structure determination, 20 mg mL^{-1}.

Exercise 3.3 On a photocopy of Fig. 3.8(a), label the following: (a) position of crystal, (b) collimator of X-rays, (c) beam stop, (d) image plate, (e) source of coolant (N_2).

Exercise 3.4 In the table in Box 3.11, if the order of the numbers in both x and y were changed, keeping the correspondence, would there be any effect on the computed element of the covariance matrix?

Exercise 3.5 Suppose that the chain of five masses (from the same table in Exercise 3.4), instead of being fixed to a wall at both ends, were cyclized into a ring (and perhaps floated on the surface of a pool of still water). Assume that they are constrained to move in a plane and on a circle, so that the problem remains fundamentally one-dimensional. What qualitative differences would you expect this to make in the covariance and inverse covariance matrices in Box 3.11?

Exercise 3.6 In the last line of the table in Box 3.11, the y vector is (1.08, −1.08, −1.08, 0, 1.08, 1.08, −1.08). Why was the value 1.08 chosen, instead of something simpler such as (1, −1, −1, 0, 1, 1, −1)? [Hint: 1.08 = $\sqrt{(7/6)}$]

Problem 3.1 Draw a graph showing the resultant of the red and black curves of Fig. 3.4, and compare with the resultants of waves that are perfectly in phase or out of phase. Assume that the red and black waves of Fig. 3.4 are 70° out of phase, so that the function you wish to plot is cos x + cos $(x − 70)$. (If we take the x-axis in Figs. 3.4 and 3.5 to be an angle, then the waves of Fig. 3.5b are 180° out of phase.)

Problem 3.2 In the structure prediction of the *H. influenzae* hypothetical protein, Fig. 3.26: (a) What are the differences in folding pattern between the target protein and the experimental parent? (b) What are the differences in folding pattern between the prediction by A.G. Murzin and the target? (c) What are the differences in folding pattern between the prediction by A.G. Murzin and the experimental parent? (d) In what respects is Murzin's prediction a better representation of the folding pattern than the experimental parent?

Problem 3.3. In the 2000 Critical Assessment of Structure Prediction (CASP4), one of the targets in the category for which no similar fold was known was the N-terminal domain of the human DNA end-joining protein Xrcc4, residues 1–116. The secondary structure prediction by B. Rost, using the method PROF (profile-based neural network prediction), is as follows (An H under a residue means that that residue is predicted to be in a helix, an E means that that residue is predicted to be in an extended conformation or strand, and—means other):

```
              1              2              3              4              5              6

              0              0              0              0              0              0

Sequence    MERKISRIHLVSEPSITHFLQVSWEKTLESGFVITLTDGHSAWTGTVSESEISQEADDMA
Prediction  ---EEEEEE-----HHHHHH-HHHHHHH--EEEEEE------EE---HHHHHHHHHHHHH

                                                         1              1

              7              8              9              0              1

              0              0              0              0              0

Sequence    MEKGKYVGELRKALLSGAGPADVYTFNFSKESCYFFFEKNLKDVSFRLGSFNLEKV
Prediction  HHH-HHHHHHHHHHHH-----EEEEEE-----EEEEE------EEEE-----HHHH
```

The experimental structure of this domain, released after the predictions were submitted (wwPDB entry [1FU1]) is shown here.

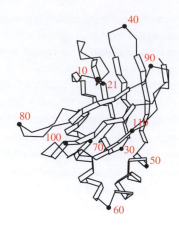

The secondary structure assignments from the wwPDB entry are:

Secondary structure	Residue ranges
Helix	84–88, 95–101, 104–111
Sheet 1	27–29, 49–59, 62–75
Sheet 2	2–8, 18–24, 31–37, 42–48, 114–115

(a) Calculate the value of Q_3, the percentage of residues correctly assigned to helix (H), strand (E) and other (–).

(b) On a photocopy of the picture of Xrcc4, highlight, in separate colours, the regions *predicted* to be in helices and strands. (c) From the result of (b): how many predicted helices overlap with helices in the experimental structure? How many strands overlap with strands in the experimental structure?

Problem 3.4 (a) For the predictions of the secondary structure of methylglyoxyal synthase shown in Fig. 3.29, which of them correctly predicts all assigned secondary structures, in the correct order? (b) What is the value of Q3 for the method that shows the highest value of Q3?

Problem 3.5 C5a is a 74-residue protein fragment from the complement system, a set of proteins that modulates the immune response. C5a is active in response to inflammation. Structures of C5a have been determined both by X-ray crystallography and NMR spectroscopy. The NMR structure shows an N-terminal α-helix in a region disordered in the X-ray structure. (a) Why is it unusual to find *more* secondary structure in an NMR structure than in an X-ray structure? (b) Suggest a possible reason why the N-terminus of the X-ray structure should be disordered if the region is helical in solution. (c) How could you test your hypothesis?

Problem 3.6 Fig. 3.43 contains—not necessarily in this order—the diffraction patterns of periodic arrays of flags of the United States of America, Great Britain, and France. (The calculations were done by creating patterns in which white areas of the flags were assigned the value 0, red areas ½ and blue areas 1, and taking the Fourier transforms of the patterns.) (a) Which is which? How do you know? (b) How could you deduce from the diffraction pattern that the flag of the United States of America has 13 horizontal stripes? (c) From the diffraction pattern of the flag of the United States of America, calculate the ratio of the width of a stripe to the distance between horizontal rows of stars.

Problem 3.7 Describe the successful and any unsuccessful features of the predictions shown in (a) Fig. 3.36, (b) Fig. 3.40.

Problem 3.8 From the list of interresidue contacts in the left frame of Fig. 3.21, how many contacts are between (a) residues in helices with residues in helices, (b) residues in loops with residues in helices, and (c) residues in loops with residues in loops. (d) From the list of interresidue contacts in the left frame of Fig. 3.21, draw a histogram of the separation in the sequences—i.e., how many residues apart—of the residues showing contacts.

Problem 3.9 Kopera et al. have determined the structure of 36-residue *Galleria mellonella* silk proteinase inhibitor 2 by both X-ray crystallography and NMR spectroscopy. The X-ray structure is at unusually high-resolution, 0.98 Å. The wwPDB contains both structures: 4HGU (X-ray) and

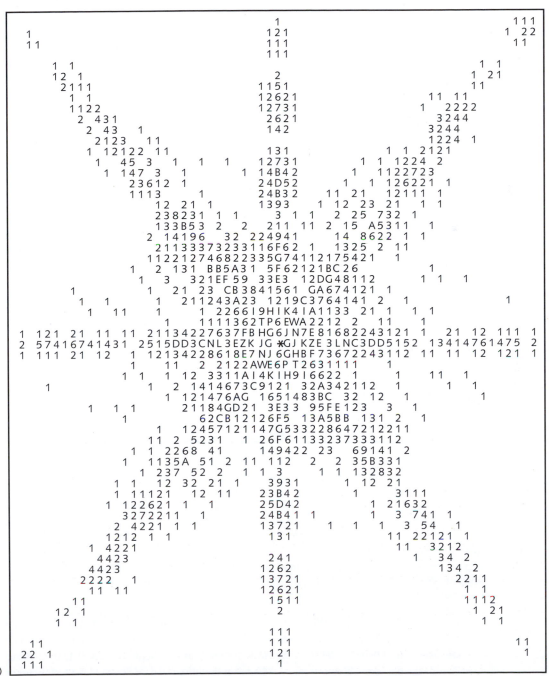

Figure 3.43 (*Continued*)

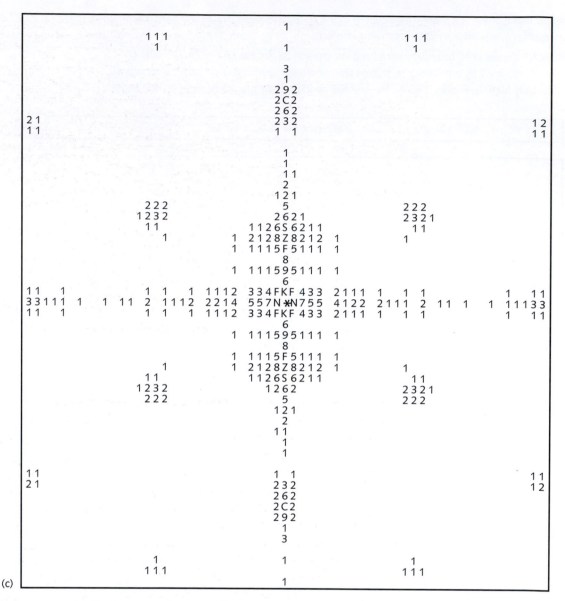

(c)

Figure 3.43 The Fourier transforms, or diffraction patterns, of periodic arrays of the flags of the United States of America, Great Britain, and France (not necessarily in this order). The patterns are oriented as if the flagpole were vertical on the page. The transforms are computed as if the flags were square, rather than longer than they are tall. Each print position corresponds to one 'spot' in the diffraction pattern. The symbols appearing indicate the relative intensity of each reflection: blank is the lowest range, then increasing densities are given digits 1–9, then letters A–Z. Thus, 0 or 1 = absent or weak reflection; Z = strong reflection. '*' indicates the undiffracted beam.

[From: Lesk, A. (1981). *Introduction to Physical Chemistry*. 1st edition, © 1982, pgs. 473–474. Reprinted by permission of Pearson Education, Inc., Upper Saddle River, NJ, U.S.A.]

2M5X (NMR, 17 models). (a) What are the values of the overall r.m.s.d. of the backbone atoms (N, CA, C, O) for the X-ray structure against the different NMR models. (b) Plot a histogram of these values. (c) Taking model 11 of the NMR ensemble as representative, calculate the r.m.s.d. values of the backbone atoms of model 11 against the other 16 models. (d) Plot a histogram of these values—most revealingly, plot this second histogram 'back-to-back' with the other histogram. (e) Can you extract a 'core' of the NMR structures such that if restricted to these selected residues the r.m.s.d. values of the different NMR models is substantially smaller? Try to retain as many residues as is reasonable. (f) What is the distribution of the r.m.s.d. values of this core subset of residues between the NMR models and the X-ray structure?

Kopera, E., Bal, W., Lenarčič Živkovič, M., Dvornyk, A., Kludkiewicz, B., Grzelak, K., Zhukov, I., Zagórski-Ostoja, W., Jaskolski, M. and Krzywda, S. (2014) Atomic resolution structure of a protein prepared by non-enzymatic His-tag removal. Crystallographic and NMR study of GmSPI-2 inhibitor. *PLoS One* **9** e106936

Problem 3.10 In vector terms, the sum $\Sigma x_i y_i$ appearing in the table in Box 3.11 is the dot product of x and y. The correlation coefficient, for distributions with zero mean, is $r = \Sigma x_i y_i \, / \sqrt{(\Sigma x_i^2 \times \Sigma y_i^2)}$. The correlation coefficient must be between 1 (perfect positive correlation) and −1 (perfect negative correlation). (a) Verify that the correlation coefficients corresponding to the three pairs of vectors appearing in the table in Box 3.11 are 1, −1 and 0, respectively. (b) Consider the following sets of y_i (which are the values of the y_i in the first row of the table, perturbed by random noise). Verify that the new y_i have a mean of zero and standard deviation of 1. Compute the correlation coefficient of these y vectors with the vector x that appears in all three rows of the table.

$$y_i = -1.64536, -0.86781, -0.71801, 0.46991, 0.52354, 1.23196, 1.00575$$

$$y_i = -1.65496, -0.40172, -0.68069, 0.95888, -0.14707, 0.31416, 1.61140$$

Problem 3.11 Figure 3.43(a) is set as a single line. Why is this scientifically wrong, and how should Fig. 3.4 (a) have appeared?

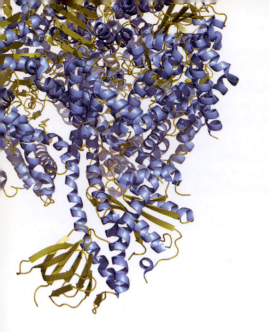

CHAPTER 4

Bioinformatics of protein sequence and structure

LEARNING GOALS

- *Knowing the basic types of data that support contemporary research in protein science,* including nucleic acid sequences of genomes, amino acid sequences of proteins, three-dimensional structures of proteins, protein functions, metabolic pathways and their variations among different organisms, and expression patterns of proteins.

- *Gaining familiarity with standard databases containing information about proteins that are available on the world-wide web,* and how to explore other sources of information and analytic tools.

- *Knowing how to retrieve information from databases* on the basis of keyword, or similarity in sequence or structure.

- *Appreciating the concept of sequence alignment,* the methods for aligning sequences, and facilities for sequence database searching based on them, notably BLAST and PSI-BLAST.

- *Understanding the basis for comparison of structures by optimal superposition,* and its application to structural alignment of sequences of proteins of known structure.

- *Distinguishing the characteristics* of pairwise sequence alignment, multiple sequence alignment, and structural alignment.

- *Knowing the classifications of protein function* by the Enzyme Commission and the Gene Ontology Consortium.

Introduction

Bioinformatics is a new field, a hybrid of biology and computer science. Biology, especially high-throughput data streams such as DNA sequencing and structural genomics projects, provides its input. Computer science permits the effective use of information-processing equipment to support research based on these data. Databases archive and distribute the information. Associated scientists curate the data, to ensure high standards of quality control.

Databases organize knowledge and make it accessible. Algorithms allow analysis of the information, thereby producing additional data streams to be incorporated into the repository. For example, calculation of the accessible surface area of all residues in all known protein structures allows studies of distributions of residues between the surfaces, interiors and interfaces of proteins and protein complexes.

The eminent physical chemist E.A. Molwyn-Hughes wrote long ago that: 'The complete physical chemist blows his own apparatus and solves his own equations.' What would be the modern equivalent?

Perhaps: The complete protein scientist grows his or her own cells, sequences genes from these cells, isolates proteins and solves their structures, and writes his or her own computer programs to analyse how they relate to the totality of biological knowledge. A tall order!

Databases and information retrieval

Databases are central and crucial to contemporary biomedical research. The goals of the information enterprise are to archive, curate, and annotate data; to make the results available; and to impose a structure on the data that facilitates information retrieval. ('A database without effective modes of access is merely a data graveyard')

Information retrieval is easy if you know precisely what you are looking for: for instance, 'What is the amino acid sequence of adenylate kinase from *Micrococcus leuteus*?' The challenge is to allow the greatest generality in framing questions, even if the answers require access to combinations of different types of data, which may appear in different databases. These types of data might include protein sequences, mutations and their correlations with disease, patterns of distribution among tissues of expression levels, structure and function. For example, consider the question: 'For which proteins of known structure involved in diseases of purine biosynthesis in humans, are there related proteins in yeast?' We are setting conditions on: known structure, specified function, detection of relatedness, correlation with disease, specified species.

How can we approach such questions? We require linkages between different categories of data. But several factors, not all scientific, govern the nature and degree of database integration.

Curation and annotation of particular types of data require specific expertise. It is appropriate that crystallographers run protein structure databases. Such considerations push databases towards fragmentation and specialization. Conversely, the requirements of linking collections of different types of data to answer compound queries push databases to consolidate, to allow the imposition of a common structure and retrieval system. Combining databases also allows broad consistency checking, a valuable component of quality control.

There are dangers in database aggregation—analogous, to some extent, to the dangers of protein aggregation. One threat is loss of their individual strengths and the focused expertise of the curators of specialized data archiving projects. Of course it is possible to hire a cabinet of experts on different topics and set them to work down the hall from one another. In any event, the alternative to database consolidation is to devise standards for interoperability, so that a program could parse queries, farm out components to different databases, and collect and organize the results. That approach is an intriguing topic of theoretical research, but not yet a practical tool.

Indeed, at the moment, the trends towards consolidation appear much stronger than the centrifugal forces. Large dedicated institutions are taking charge, and assimilating many smaller databases. ('Resistance is futile.') Indeed, in bioinformatics, only national or commercial rivalries seem to stand in the way of an extrapolation to the formation of a single world-wide comprehensive database. Think Google.

Databases:
- archive and preserve data;
- curate data to ensure accuracy;
- organize the data, imposing a logical structure;
- distribute data;
- provide information-retrieval tools;
- provide links to other databases.

There are other advantages for the user of a comprehensive integrated database, supported by a national or even international organization. Such an institution can marshall the infrastructure to develop users' manuals, tutorial guides, answers to Frequently-Asked Questions (FAQs), and even an online help desk. The National Center for Biotechnological Information (NCBI), a component of the U.S. National Library of Medicine, is a prime example.

Nevertheless, in view of the danger that the result will prove unwieldly, it will be possible to tailor access to the needs of particular projects. Even if the archives are unified, there is no need to limit the ways to access them—colloquially, to design a variety of 'front ends'.

Specialized user communities may extract subsets of the data, or recombine data from different sources, and provide specialized avenues of access. Such information engines will depend on the primary archives as the source of the information, but can redesign the organization and presentation. Indeed, different derived databases can slice and dice the same information in different ways. Another reasonable extrapolation suggests the concept of specialized 'virtual databases' (a concept first suggested in 1981), grounded in the archives but versatile in scope and function, tailored to the needs of individual communities, research groups or even individual scientists. These can be offered as facilities within a comprehensive database—for instance, myNCBI—or can be external and independent.

This chapter first describes the data, their storage and organization. That established, a discussion of methods of data retrieval follows.

Amino acid sequence databases

In 2002, three protein-sequence databases—the **Protein Information Resource (PIR),** at the National Biomedical Research Foundation of the Georgetown University Medical Center in Washington, D.C., U.S.A.; and **SWISS-PROT** and **TrEMBL,** from the Swiss Institute of Bioinformatics in Geneva and the European Bioinformatics Institute in Hinxton, U.K.—coordinated their efforts to form the **UniProt** consortium. The partners in this enterprise share the data but continue to offer separate information-retrieval tools for access.

The PIR grew out of the very first sequence database, developed by Margaret O. Dayhoff, the pioneer of the field of bioinformatics. SWISS-PROT was developed at the Swiss Institute of Bioinformatics. Both these projects were active before the era of high-throughput DNA sequencing. They included data determined from sequencing of proteins themselves. Now, TrEMBL contains the translations of genes identified within DNA sequences in the EMBL Nucleotide Sequence Database, or EMBLBank. TrEMBL entries are regarded as preliminary, and mature into fully-fledged SWISS-PROT entries after curation and extended annotation. Currently (4 Mar 2015) SWISS-PROT contains 547 964 sequences, including a total of 195 174 196 amino acids. TrEMBL is much larger. It contains 92 124 213 sequences, including a total of 29 223 634 881 amino acids.

This disparity is a reflection of the fact that, today, almost all amino acid sequence information arises from translation of gene sequences. However, the amino acid sequences of proteins are *not* entirely inferrable with confidence from a genome sequence. Methods for detection of **open reading frames** are good but not perfect and, in any event, not every open reading frame corresponds to an expressed protein. (An open reading frame is a region of a DNA sequence that contains a start signal, followed by a region long enough putatively to encode a protein, uninterrupted by a Stop codon.) A very serious problem, in eukaryotes, is variation in splicing patterns. In addition, information about ligands, disulphide bridges, subunit associations, post-translational modifications, effects of mRNA editing, etc., is not available from nucleic acid sequences. For instance, from genomic information alone one would not know that human haemoglobin is a tetramer of two α and two β chains, or that insulin is a dimer linked by disulphide bridges. Protein-sequence databanks collect this additional information from the literature and provide suitable annotations.

- Protein-sequence databases contain amino acid sequence data plus annotations. Today, most amino acid sequences are derived by translation of nucleic acid sequences. Annotations include source, literature references, post-translational modifications, assignments of function, and links to other databanks.

The Uniprot Knowledge Base (UniProtKB) entries contain a great richness of annotation, including as major categories:

- names of molecules, including alternative names, and source, including full taxonomy of species of origin;

- protein attributes, such as sequence length and post-translational processing;

- general annotation (comments), including function if known and family relationships;

- function assignment, specified according to the Enzyme Commission and Gene Ontology™ Consortium classifications;

- known ligands;

- subcellular localization;

- sequence annotation, including glycosylation sites; relationship to diseases; and feature tables linked to specific residue ranges, showing for example signal peptides;

- the sequence itself, including tools with which to reformat or analyse the sequence;

- literature references;

- cross-references (links) to other databanks;

- information about provenance and acquisition of entries (database housekeeping);

- references to other relevant documents, including articles published in the scientific literature.

Most items contain links to other databases. Indeed, all the databases in molecular biology form a tight neighbourhood. Instead of the six degrees of separation claimed for humans, the distance between any two databases in molecular biology is never more than one or two clicks.

It is essential that readers familiarize themselves with the appearance and format of entries in UniProtKB and other databases. However, rather than reproducing examples on paper, weblems associated with this chapter in the online resource centre make references to selected sites described in the text.

Protein databases at the U.S. National Center for Biotechnology Information

ENTREZ is the information-retrieval system at the U.S. National Center for Biotechnology Information,

Table 4.1 Catalogue raisonée of the ENTREZ database system of the National Center for Biotechnology Information. From: http://www.ncbi.nlm.nih.gov/gquery/

Literature	
Books	books and reports
MeSH	ontology used for PubMed indexing
NLM Catalog	books, journals and more in the NLM Collections
PubMed	scientific & medical abstracts/citations
PubMed Central	full-text journal articles
Health	
ClinVar	human variations of clinical significance
dbGaP	genotype/phenotype interaction studies
GTR	genetic testing registry
MedGen	medical genetics literature and links
OMIM	online mendelian inheritance in man
PubMed Health	clinical effectiveness, disease and drug reports
Genomes	
Assembly	genomic assembly information
BioProject	biological projects providing data to NCBI
BioSample	descriptions of biological source materials
Clone	genomic and cDNA clones
dbVar	genome structural variation studies
Epigenomics	epigenomic studies and display tools
Genome	genome sequencing projects by organism
GSS	genome survey sequences
Nucleotide	DNA and RNA sequences
Probe	sequence-based probes and primers
SNP	short genetic variations
SRA	high-throughput DNA and RNA sequence read archive
Taxonomy	taxonomic classification and nomenclature catalogue
Genes	
EST	expressed sequence tag sequences
Gene	collected information about gene loci
GEO DataSets	functional genomics studies
GEO Profiles	gene expression and molecular abundance profiles
HomoloGene	homologous gene sets for selected organisms
PopSet	sequence sets from phylogenetic and population studies
UniGene	clusters of expressed transcripts

Table 4.1 (*Continued*)

Proteins	
Conserved Domains	conserved protein domains
Protein	protein sequences
Protein Clusters	sequence similarity-based protein clusters
Structure	experimentally-determined biomolecular structures
Chemicals	
BioSystems	molecular pathways with links to genes, proteins and chemicals
PubChem BioAssay	bioactivity screening studies
PubChem Compound	chemical information with structures, information and links
PubChem Substance	deposited substance and chemical information

part of the U.S. National Library of Medicine. It includes many databases, including but not limited to collections of nucleic acid and genome sequences, protein sequences and structures, and expression patterns. These databases include mutual cross-references. ENTREZ is the prime example of the power of database unification to support multiple-database searches. One of the protein databases, the compilation of reference sequences, overlaps substantially in content and focus with UniProtKB.

ENTREZ offers access via 40 database divisions (see Table 4.1). Also notable are the bibliographic database PUBMED (see 'Web access to scientific literature' in Chapter 1), and Online Mendelian Inheritance in Man™ and Online Mendelian Inheritance in Animals, databases linking genomics and disease. Links between various databases are a strong point of NCBI's system.

Specialized, or 'boutique', databases

Many individuals or groups select, annotate, and recombine data focused on particular subjects, and include links affording streamlined access to information about related topics of interest. For example:

- FlyBase is a database of *Drosophila* genes and genomes. It includes data on many additional aspects of *Drosophila* biology, and is equipped with retrieval tools. It also serves the fly community with news, lists of scientists, meeting announcements, etc. (flybase.org).

- The HIV Structural Database is a resource for information on AIDS-related macromolecules and ligands. This database contains some protein structures not deposited in the wwPDB. It is integrated with Chem-BLAST, a chemical taxonomy-based search engine for browing compounds, and has links to other databases of pharmacological properties of the ligands. (http://xpdb.nist.gov/hivsdb/hivsdb.html)

In the field of immunology:

- IMGT, the International imMunoGeneTics information system®, is a high-quality integrated database specializing in immunoglobulins, T-cell receptors, and major histocompatibility complex molecules of all vertebrate species. The IMGT server provides a common access to all immunogenetics data. At present, it includes two databases: IMGT/LIGM-DB, a comprehensive database of immunoglobulin and T-cell receptor gene sequences from human and other vertebrates, with translation to amino acid sequences for fully annotated entries, and IMGT/HLA-DB, a database of the human MHC referred to as HLA—human leucocyte antigens (http://www.imgt.org).

- IEDB, The Immune Epitope Database and Analysis Resource, curated at the La Jolla Institute for Allergy and Immunology, contains data related to antibody and T-cell epitopes (www.iedb.org).

- MEROPS is a database of proteolytic enzymes and their inhibitors (http://merops.sanger.ac.uk/).

In addition to the databases focusing on specific families of proteins, such as MEROPS, several databases treat enzymes in general:

- Structure–Function Linkage Database (SFLD). Contains sequence alignments, catalytic residues, structures, reaction pathways. Searchable by sequence or partial chemical reaction (http://sfld.rbvi.ucsf.edu/).

- EzCatDB. A compendium of enzyme mechanisms, linked to many related databases, including the Enzyme Commission classification, the world-wide Protein Data Bank, SWISS-PROT, KEGG, PubMed, and a list of compounds that includes metabolites and cofactors; contains classification of enzymes by catalytic mechanism and structure; includes some 3D models of reaction mechanisms (http://mbs.cbrc.jp/EzCatDB/ezcat.html).

- Catalytic Site Atlas (CSA). Contains sequences, structures, catalytic residues, and structural templates for active sites usable as search objects (http://www.ebi.ac.uk/thornton-srv/databases/CSA/).

- Enzyme Classification Browser: Mapping of Enzyme Commission (EC) numbers to structures. (www.rcsb.org/pdb/browse/jbrowse.do?t=3&useMenu=no)

- MACiE. Contains step-by-step reaction mechanisms, indicating catalytic residues; mechanisms shown as animations (http://www-mitchell.ch.cam.ac.uk/maci).

- BRENDA (BRaunschweig ENzyme DAtabase) compiles information on reactions and kinetics, and links with related databases (www.brenda-enzymes.org).

The word boutique describes accurately a diminishing number of these specialized databases. Boutique implies *both* narrow focus and small-size organization. Although the idea of relatively narrow focus does still apply, there are very few independent databases that are the analogues of 'Mom-and-Pop' high-street shops. Most of the surviving specialized databases have large staffs, and—partly as a result of funding patterns—many have been subsumed by large organizations, with varying degrees of preservation of their individuality. Perhaps a better analogy would be a large shopping mall that rents space to specialized establishments.

Many boutique databases are moving towards distributed authorship, more like social media than centrally-compiled offerings.

- Many archival databanks are large national or international institutions. Specialized databases make use of the archives but are free to recombine, repackage and re-present selected information, sometimes together with news and announcements of interest to specialists in the topic of their focus.

Nucleic acid sequence databases

The world-wide nucleic acid sequence archive, the International Nucleotide Sequence Database Collaboration, is a triple partnership of the National Center for Biotechnology Information (U.S.A.); the EMBL Nucleotide Sequence Database, or EMBLBank (European Bioinformatics Institute, U.K.); and the DNA Data Bank of Japan (National Institute of Genetics, Japan). These projects curate, archive and distribute DNA and RNA sequences collected from genome projects, scientific publications, and patent applications. To ensure that these fundamental data are freely available, scientific journals require deposition of new nucleotide sequences, as a condition for publication of an article. Similar conditions apply to protein sequences and structures.

The groups exchange data daily. As a result, the raw data are identical. However, the format in which they are presented, and the nature of the annotation, vary among these databanks.

The nucleic acid sequence databases, as distributed, are collections of entries. Each entry has the form of a text file containing data and annotations for a single contiguous sequence. Many entries are assembled from several published papers reporting overlapping fragments of a complete sequence.

The files are distributed in basic 'vanilla' format, to make it easy to write software to read them. Internal storage formats, supporting curatorial activities within the database, can be more specialized and complex.

Entries have a life history. Because of the desire on the part of the user community for rapid access to data, new entries are made available before completion of annotation and checking. Entries mature through the classes:

Unannotated → Preliminary → Unreviewed → Standard

Rarely, an entry 'dies'—a few have been removed when they are determined to be erroneous.

Databases of macromolecular structure

Structure databases archive, annotate, and distribute sets of atomic coordinates. Started by the late W. Hamilton at Brookhaven National Laboratories, Long Island, New York, U.S.A. in 1971, the major database of biological macromolecular structure is now the world-wide Protein Data Bank (wwPDB). It is a joint effort of the Research Collaboratory for Structural Bioinformatics (RCSB) (a distributed organization based at Rutgers University, in New Jersey; the San Diego Supercomputer Center, in California; and the University of Wisconsin, all in the U.S.A.); the Protein Data Bank Europe (PDBe, at the European Bioinformatics Institute, in the U.K.), and the Protein Data Bank Japan (PDBj, based at Osaka University). The wwPDB contains structures of proteins, nucleic acids, and a few carbohydrates. The parent website is http://www.wwpdb.org.

These archives collect not only the results of structure determinations of biological macromolecules, but also the measurements on which they are based. The wwPDB keeps the data from X-ray structure determinations, and the BioMagResBank those from NMR.

The home pages of the wwPDB partners contain links to the data files themselves, to expository and tutorial material including short news items and a Newsletter, to facilities for deposition of new entries, and to specialized search software for retrieving structures. The general NCBI ENTREZ system also permits retrieval of structural data.

The RCSB, PDBe and PDBj have the same set of structural data at their core. They share a common basic set of search and retrieval facilities, and links. However, they differ in infrastructure; in particular,

in the facilities they support. Here are just a few examples, plucked out of the context of a rich tapestry:

The RCSB offers Gene View, which maps experimental structures onto the human genome.

The PDBe, under the direction of G. Kleywegt, has been especially creative and technically skilful in designing tools for structure retrieval and analysis. These include:

- PDBeMotif combines protein sequence, chemical structure and protein structural data in a single search.
- PDBeFold allows searches for structural similarity and alignment, and for probing the databank with a substructure.
- PDBePISA (Proteins, Interfaces, Structures and Assemblies) allows searching and exploration of macromolecular interfaces.

The PDBj offers SeSAW, allowing identification of putative functional sites in a query via a structural match to the PDB, based either on an experimentally-determined structure or even an homology model (if only a sequence is known).

Some of these are explored in weblems.

The wwPDB overlaps several other databases.

The Electron Microscopy Data Bank (EMDB) collects the results of applying electron microscopy to macromolecular complexes and subcellular structures (www.emdatabank.org).

The Nucleic Acid Structure Databank (NDB) at Rutgers University, New Jersey, U.S.A. complements

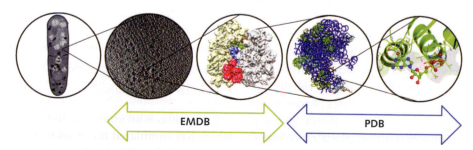

[From: Gutmanas, A., Oldfield, T.J., Patwardhan, A., Sen, S., Velankar, S. and Kleywegt, G.J. (2013). The role of structural bioinformatics resources in the era of integrative structural biology. *Acta Crystallogr D Biol Crystallogr.* **69**, 710–721.]

the wwPDB (http://ndbserver.rutgers.edu). The **Bio-MagResBank**, at the Department of Biochemistry, University of Wisconsin, U.S.A.—now a full partner in the RCSB—archives results of protein structure determinations by nuclear magnetic resonance (http://www.bmrb.wisc.edu/).

The **Cambridge Crystallographic Data Centre** archives the structures of small organic molecules, http://www.ccdc.cam.ac.uk/. It currently contains over 750 000 entries. Now in its fiftieth year, in its early days the CCDC was responsible for pioneering work in structure encoding and searching. The **PubChem** database at the U.S. National Center for Biotechnology Information is another collection of small molecules. It includes the structures and chemical properties of over 50 million compounds and substances, with links to other NCBI databases, including biological activity information, http://pubchem.ncbi.nlm.nih.gov/. Search facilities permit retrieval by compound name or formula, and even an interactive molecular drawing program that allows users to sketch molecular fragments to serve as search objects.

The structural data from these sources are extremely useful in studies of conformations of the component units of biological macromolecules, and for investigations of macromolecule–ligand interactions, including but not limited to applications to drug design.

Organization of wwPDB entries

The wwPDB assigns a four-character identifier to each structure deposited. The first character is a number from 1 to 9. Do not count on mnemonic significance.

Each experimental structure determination of a protein appears as a separate entry in the wwPDB. In many cases several entries correspond to one protein—solved in different states of ligation, or in different crystal forms, or re-solved using better crystals or more accurate data-collection techniques. There have been at least four generations of sperm-whale myoglobin crystal structures.

Each entry includes:

- what protein is the subject of the entry, and what species it came from;
- who solved the structure, and literature references;
- experimental details about the structure determination, including information related to the general

Table 4.3 Websites containing illustrations of protein structures

Summary pages at RCSB	http://www.rcsb.org
jV at PDBj	http://www.pdbj.org/jV/TOP.html
PDBsum	http://www.ebi.ac.uk/pdbsum/
SWISS-3DIMAGE database	http://www.expasy.org/cgi-bin/ sw3d-search-de
Jena Library of Biological Macromolecules	http://www.imb-jena.de/IMAGE. html

quality of the result, such as resolution of an X-ray structure determination and stereochemical statistics;

- the amino acid sequence;
- the atomic coordinates (lines beginning ATOM);
- what additional molecules appear in the structure, other than standard polypeptide chains. These can include cofactors, inhibitors, and water molecules. The keyword HETATM identifies the coordinates of these moieties. Each heterogroup is assigned a three-letter code, to fit into the same field as the standard amino acid names. For example, PSU stands for pseudouridine-5′-monophosphate;
- assignments of secondary structure: helices, sheets;
- disulphide bridges.

Several websites offer albums of pictures of the known protein structures. These include those in Table 4.3. Many databases have links to molecular graphics facilities.

Protein coordinate formats

Databases need several ways to store their information. They need a format for in-house use, which contains internal housekeeping information that relates to database operation but that may be of no interest to the users of the data. The database also needs one or more formats for distribution.

The original Protein Data Bank stored the archived data in a format that was devised in the 1970s. This format was appropriate for its time but began to show its age, increasingly so in this new century. For many years there were also serious problems of consistency and errors. These ranged from simple typographical mistakes (e.g. HO2 for water) to very subtle questions about outliers from typical stereochemistry, some of which were genuine outliers and not errors, but which could not be proven to be errors without detailed appeal to the experimental data. The wwPDB partners have collaborated on 'cleaning up' the files to eliminate at least the simpler problems.

> An outlier is *not* necessarily an error!

A new format, **mmCIF (macromolecule crystallographic information file)** is an adaptation of a standard devised at the Cambridge Crystallographic Data Centre, and promoted by the International Union of Crystallography for small molecules. mmCIF is more compliant with modern principles of database design, notably the conventions of relational databases,

developed by E.F. Codd. This is particularly important for integrating structural databases into general information-retrieval systems in molecular biology. Nevertheless, a large corpus of software keeps the legacy format active.

> • Databases must fully specify a format. The formats must change as database technology evolves, although abrupt changes in format can 'break' software. The format of protein structural data is changing from traditional PDB format to mmCIF format that is more in line with current information-retrieval standards. Nevertheless, a rich literature of software continues to be based on the older format.

Retrieval of sequences and structures

Retrieval means probing one or more databases with a set of search criteria, and receiving the results (see Box 4.1). Search engines return one or more 'hits', sorted in order of quality of match. If the structure

---

 BOX 4.1   **Different categories of retrieval problems in bioinformatics**

It is useful to compare a variety of possible combinations of database searches:

• Given a sequence, or fragment of a sequence, find sequences in the database that are similar to it. This is in fact a string matching problem (possibly with mismatches allowed), and is addressed by sequence similarity search programs such as **BLAST** and **PSI-BLAST** (see section later in this chapter)

• Given a structure, or fragment of a structure, find structures in the database that are similar to it. This is the extension of the similarity search problem to three dimensions. Several programs address this problem, including but not limited to DALI by L. Holm and C. Sander http://ekhidna.biocenter.helsinki.fi/dali_server, SUPER (lcb.infotech.monash.edu.au/super/) or, for short regions, PDBemotif (http://www.ebi.ac.uk/pdbe-site/PDBeMotif/).

• Given a *sequence* of a protein of unknown structure, find other proteins that adopt similar three-dimensional structures. One is tempted to 'cheat'—to look in the sequence data banks for proteins with sequences similar to the probe sequence: for if two proteins have sufficiently

similar sequences, they will have similar structures. However, the converse is not true, and one can hope to create more powerful search techniques that will find proteins of similar structure even among sequences that have diverged beyond the point where they can easily be recognized as similar.

• Given a protein structure, find *sequences* in the data bank that correspond to similar structures. Again, one can cheat by using the structure to probe a structure data bank, but this can give only limited success because there are so many more sequences known than structures. It is therefore desirable to have a method that can pick out the structure from the sequence.

The first two of these tasks can be expressed in purely mathematical terms, and are similar to problems that arise in other fields. For instance, string matching is a common feature of text editors. Mature software, robust and effective, is available to solve them.

The third and fourth problems are much more complicated and are the subjects of current research.

of the database is a set of discrete files, the results of a search will be a list of the relevant files, with links to them. This is familiar from general search engines such as Google.

There are two approaches to identification of sequences: via the annotation and via the sequence itself.

Retrieval of amino acid sequences by keyword

If you know the identifier, for instance Aardvark myoglobin—SWISS-PROT ID 'MYG_ORYAF'—just type that, or 'Aardvark myoglobin', and the search engine will deliver it. (The SWISS-PROT ID combines abbreviations of the molecule name—MYG for myoglobin—and the species name: The Linnaean binomial for aardvark is *Orycteropus afer*.) You could even type 'MYG_ORYAF' or 'Aardvark myoglobin' directly into Google.

More generally, one can probe a database with a set of keywords. For instance, a search in UniProtKB (UniProt Knowledge Base) (http://www.uniprot.org/help/uniprotkb) for human haemoglobin returns 260 results, starting with the α- and β-chains of human haemoglobin, followed by other human haemoglobin chains, such as fetal forms. The database automatically applies both the British and American spelling: haemoglobin and hemoglobin.

Human haemoglobin chains appear at the top of the 260 hits, but the results also include other molecules, such as human phosphoglycerate kinase 1. Human phosphoglycerate kinase 1 appears because its entry contains a list of literature references including a paper published in the journal *Hemoglobin*.

UniProtKB tries to be as comprehensive as possible in collecting information about proteins. This is why a preliminary search can return so much extraneous information. The search engine permits 'follow-up questions', in order to winnow out precisely what the user wants.

Selecting:

'Restrict term "hemoglobin" to protein name'

and

'Restrict term "human" to organism'

produces 53 results, all of which are genuine human haemoglobin chains, including developmental variants and mutants. This narrowed search rejects the

entry containing a reference to a paper in the journal *Hemoglobin*.

UniProt entries contain embedded taxonomy. Therefore, it is possible to search for mammalian haemoglobin without having to specify separately the name of every mammal.

Unless otherwise indicated, the search engine interprets a list of keywords as connected by logical AND. That is 'human haemoglobin' searches for entries that contain both 'human' *and* 'haemoglobin'. It is possible to construct complex searches containing different keywords in different categories, combined with logical AND or OR or NOT, to insist on precisely the combination that you want.

The UniProt Knowledge Base allows broad searches because of the richness of the annotation within *individual* entries.

ENTREZ, the database system of the National Center for Biotechnology Information at the U.S. National Library of Medicine, allows coordinated searching through all or any selected combination of its databases (http://www.ncbi.nlm.nih.gov/sites/gquery). For example, one could identify molecules of known structure implicated in a specific disease. The ENTREZ system is, however, not bilingual: Try a search across databases for haemophilia and see whether any known structures are returned. Then try hemophilia.

The Protein Information Resource (PIR) and associated databases

The Protein Information Resource is one of the partners in UniProt. In addition, the PIR maintains several databases about proteins:

- PIRSF (SF = superfamily): the Protein Family Classification System provides clustering of the sequences in UniProt according to their evolutionary relationships.

- *i*ProClass, an integrated protein knowledge base, is a gateway providing uniform access to over 90 biological databases, with flexible retrieval and navigation facilities.
- *i*ProLINK (*i*ntegrated *p*rotein *l*iterature, *i*nformation and *k*nowledge) is a gateway to the literature.

The Protein Information Resource is an effective combination of a carefully curated database, information-retrieval software, and a workbench for investigations of sequences. The PIR describes itself as an Integrated Protein Informatics Resource for Genomic and Proteomic Research. Think of this as an analysis package sitting on top of a retrieval system. Its functionality includes browsing, searching and similarity analysis, and links to other databases. Users may:

- browse by annotations;

- search selected text fields for different annotations, such as superfamily, family, title, species, taxonomy group, keyword and domain;
- analyse sequences using BLAST or FASTA Searches, Pattern Match, Multiple alignment;
- perform global and domain searches, and annotation-sorted searches;
- view statistics for superfamily, family, title, species, taxonomy group, keywords, domains, features;
- view links to other databases, including wwPDB, COG, KEGG, and BRENDA;
- select specialized sequence groups such as those encoded in the human, mouse, yeast and *E. coli* genomes.

- Two basic ways to identify sequences of interest are (1) search of annotations by keyword and (2) search of sequences for similarity.

Retrieval of structures by keyword

The home pages of the wwPDB partners offer searches by PDB ID or by keyword, and also more advanced search schemes. The advanced search facilities permit searching for different logical combinations of different keywords in different fields, or selected values—for instance, minimum or maximum resolution.

The RCSB presents for each entry a one-page summary of the structure determination experiment and the result. Figure 4.1 shows the summary page for the spinach chloroplast thioredoxin structure, identifier 1FAA. Links from this page take you to:

- the publication in which the entry was described, via the bibliographic database PubMed;
- pictures of the structure (some of these may require that you install a viewing program on your computer);

- access to the file containing the entry itself;
- lists of related structures, according to several different classifications of protein structures;
- stereochemical analysis—the distribution of bond lengths and angles, and conformational angles;
- sources of other information about this entry;
- the sequence and secondary structure assignment;
- details about the crystal form and methods by which the crystals were produced.

In addition to these facilities of the components of the wwPDB, ENTREZ at NCBI also has facilities to search for structures.

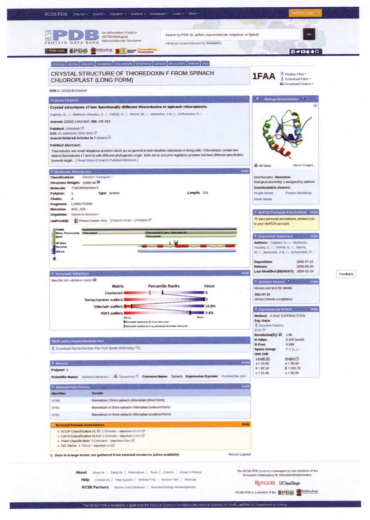

Figure 4.1 The summary page for the wwPDB entry 1FAA, spinach chloroplast thioredoxin. Image from the RCSB PDB (www.rcsb.org) of PDB ID 1FAA (Capitani, G., Markovic-Housley, Z., DelVal, G., Morris, M., Jansonius, J.N., Schurmann, P. (2000) Crystal structures of two functionally different thioredoxins in spinach chloroplasts. J.Mol.Biol. **302**: 135–154).

Probing data banks with sequence information

Although entries in UniProtKB and similar databases contain protein sequences explicitly, they do not respond directly to queries for amino acid sequences or subsequences.

The PIR International site has a facility to search for a specific peptide, of length up to 30 amino acids. In the PIR, one can select PEPTIDE SEARCH in *i*ProClass, and retrieve exact matches for a subsequence. Submitting HKTGPNLNGIFGRKT-GQAPGFTYTDANKN elicits the response that this peptide is unique to the cytochrome *c* of the Eastern grey kangaroo (*Macropus giganteus*).

> Recognizing proteins from sequences of fragments also has applications in proteomics.

In most cases, our interest is not limited to a single molecule, but we wish to know about related molecules. Typical questions would be: how different is the kangaroo sequence from that of related molecules in other marsupials? Are marsupial cytochromes *c* distinguished in systematic ways from placental mammalian cytochromes *c*? What if any residues are conserved in all eukaryotic cytochromes *c*? What is their role in the molecular structure and function?

It is therefore more common to search for sequences that are similar but not necessarily identical to a probe. Most retrieval software allows the user to initiate such a search directly based on the results of probing with a keyword.

The basis of retrieval by sequence similarity is sequence alignment. Fast programs based on sequence alignment include BLAST and PSI-BLAST (see section by the same name). To appreciate these methods we must first discuss the general idea of sequence alignment.

Sequence alignment

An **alignment** of two nucleic acid or amino acid sequences is a correspondence or match between the residues in the full sequences or subsequences.

For example, possible alignments of the short sequences:

g c t g a a c   and   c t a r a a t c

include:

An uninformative alignment:

```
- - - - - - - g c t g a a c g
c t a t a a t c - - - - - - -
```

An alignment without gaps:

```
g c t g a a c g
c t a t a a t c
```

An alignment with gaps:

```
g c t g a - a - - c g
- - c t - a t a a t c
```

And another:

```
g c t g - a a - c g
- c t a t a a t c -
```

Most readers would consider the last of these alignments the best of the four. Is that the best possible alignment? What is the *correct* alignment?

To define what we mean by the correct alignment, we must assume that the sequences are related by evolution. If so, then the correct alignment is the one that correctly reflects the evolutionary pathway relating the two sequences. Rarely can we actually observe this. In most cases, the most we can hope for is to try to find the alignment that corresponds to the most likely evolutionary relationship between the sequences.

How do we know what evolutionary events are the most likely? We must appeal to our knowledge of evolutionary mechanisms. For nucleic acid sequences, transition mutations (purine ↔ purine and pyrimidine ↔ pyrimidine) are more likely events than transversions (purine ↔ pyrimidine). For amino acid sequences, substitutions are likely to be conservative: the replacement of one amino acid by another with similar size and physicochemical properties is more likely than its replacement by another amino acid with less-similar properties. For both types of sequences, the deletion of a set of bases or amino acids in *contiguous* regions in the sequences is more likely than the deletion of the same number of bases or amino acids at *separated* positions. On this basis we can compare two alignments by asking how likely is the minimal set of mutations and/or insertions and deletions that effects the transformation of the sequences indicated by the alignment.

Any alignment corresponds to one or more sequences of mutations and insertions/deletions that transforms one sequence into the other. For instance, the fourth alignment in the example corresponds to insertion of a 'g' at the beginning of the first sequence, deletion of two 't' residues at positions 5 and 7 of the first sequence, and mutation of a 'g' to an 'a' at position 4. An *optimal* alignment corresponds to the most likely set of evolutionary events. It is an assumption that this will be the alignment that reflects the highest possible degree of similarity between the sequences.

But although the sequence of mutations and insertions/deletions in an optimal alignment *may* correspond to an actual evolutionary pathway, it is impossible in general to know that it *does*. (In particular, it is difficult if not impossible to distinguish the *order* of the events.)

> • Alignment is the assignment of residue–residue correspondences. The correct alignment reflects the evolutionary pathway between homologous proteins. In practice, we compute that alignment that optimally reflects the similarities among sequences.

To determine optimal alignments, we need a way to examine all possible alignments systematically. Then, we need to compute a score reflecting the quality of each possible alignment. For amino acid sequences, the **substitution matrix BLOSUM62** gives a statistical picture of the frequency of observed amino acid replacements in regions free of insertions and deletions. It is in common use in constructing scoring schemes for alignments. One must also estimate the relative likelihood of substitutions v. insertions/deletions—known in the trade as **gap weighting**.

Then we can identify the alignment with the optimal score. In many cases, the optimal alignment is not unique: several different alignments may give the same best score. Moreover, even minor variations in the scoring scheme may change the ranking of alignments, causing a different one to emerge as the best.

Methods for determining optimal alignments are most easily understood with reference to a tool called the **dotplot**.

The dotplot

A simple pictorial representation of relationships between two sequences is afforded by the 'dotplot', a matrix giving a quick overview of the similarity. Rows of the matrix correspond to one sequence;

ATPases lamprey/dogfish

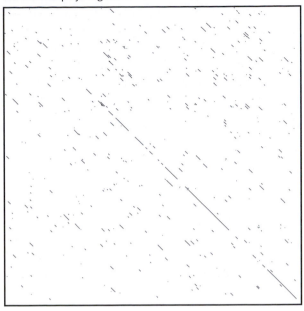

Figure 4.2 Dotplot relating the mitochondrial ATPase-6 genes from a lamprey (*Petromyzon marinus*) and dogfish shark (*Scyliorhinus canicula*). Dotplots give quick pictorial statements of the relationship between two sequences. The long diagonal signals indicate regions of high similarity, standing out from the noise of isolated matches. Note the absence of strong similarity near the beginning of the sequences.

columns to the other. A dot appears in the matrix if the characters at that row and column match, otherwise it is left blank. The dotplot gives a quick pictorial statement of the relationship between two sequences. Obvious features of similarity stand out.

Figure 4.2 shows a dotplot relating the mitochondrial ATPase-6 genes from a lamprey (*Petromyzon marinus*) and dogfish shark (*Scyliorhinus canicula*). Stretches of similar residues show up as diagonals. It is clear that the similarity of these sequences is weakest near the beginning.

> The ATPase-6 gene codes for a subunit of the ATPase complex. In the human, mutations in this gene cause Leigh syndrome, a neurological disorder of infants produced by the effects of impaired oxidative metabolism on the brain during development.

Displacement of diagonals occurs when there is an insertion or deletion between two regions of sequence similarity. The PAX-6 protein of mouse and

Drosophila eyeless

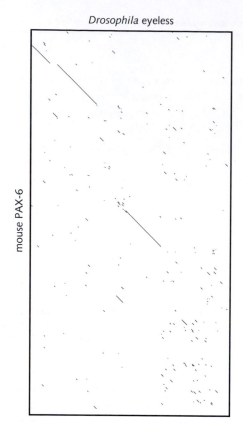

mouse PAX-6

Figure 4.3 Dotplot relating the PAX-6 protein of mouse, and the eyeless protein of *Drosophila melanogaster*. The diagonal signals show three extended regions of similarity. The displacements of the diagonals from collinearity show, in particular, that between the second and third regions of similarity there is a longer intervening region in the mouse protein than in the fly protein.

the eyeless protein of *Drosophila melanogaster* are both involved in eye development. A dotplot relating these sequences (see Fig. 4.3) shows three extended regions of similarity with different lengths of sequence between them, two near the beginning of the sequences and one near the middle. Between the second and third of them, there is a longer intervening region in the mouse than in the *Drosophila* sequence.

If two sequences are sufficiently closely related the alignment can be read directly off the dotplot; in any case, the plot provides a quick overview of the nature of the relationship between the sequences.

A disadvantage of the dotplot is that its 'reach' into the realm of distantly related sequences is poor. In analysing sequences, one should always look at a dotplot to be sure of not missing anything obvious, but be prepared to apply more subtle tools.

An available program, Dotter, by E.L. Sonnhammer, computes and displays dotplots. It allows the user to control the calculation and alter the appearance of the display by adjusting parameters interactively, http://sonnhammer.sbc.su.se/Dotter.html. To use the full set of features of Dotter it is necessary to install it locally.

A website that offers interactive dotplotting is http://athena.bioc.uvic.ca/virology-ca-tools/jdotter/

> • A dotplot gives an overview of the similarities between two sequences. Any path through the dotplot from upper left to lower right corresponds to an alignment of the sequences.

Dotplots and alignments

Dotplots provide the 'playing field' of possible alignments.

Any path through the dotplot from upper left to lower right, moving from each point only in east (horizontal), south (vertical) or southeast (diagonal) directions, corresponds to a possible alignment. If the direction of the path between successive cells is diagonal, two pairs of successive residues appear in the alignment without any gap between them. If the direction is vertical, a gap is introduced in the sequence indexing the rows. If the direction is horizontal, a gap would be introduced in the sequence indexing the columns. Note that no moves can be directed up and/or to the left, as this would correspond to aligning several residues of one sequence with only one residue of the other, or to introduce gaps at the same position in both sequences.

Figure 4.4 shows a 'toy' example, a dotplot of the identities between the short name (DORO-THYHODGKIN) and full name (DOROTHY-CROWFOOTHODGKIN) of a famous protein crystallographer.

The red arrows indicate an obvious path through the dotplot, corresponding to the obvious alignment:

DOROTHY - - - - - - - HODGKIN
DOROTHYCROWFOOTHODGKIN

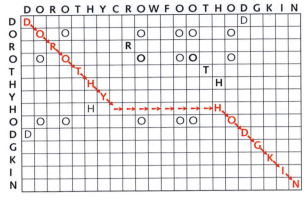

Figure 4.4 Dotplot of sequences DOROTHYHODGKIN and DOROTHYCROWFOOTHODGKIN. Letters corresponding to *isolated* matches are shown in non-bold type. The longest matching regions, shown in red, are the first and last names DOROTHY and HODGKIN. Shorter matching regions, such as the OTH of dorOTHy and crowfoOTHodgkin, or the RO of doROthy and cROwfoot, are noise.

Table 4.4 BLAST searches in sequence databanks

Program	Type of query sequence	Search in database of
BLASTP	amino acid sequence	protein sequences
BLASTX	translated nucleotide sequence	protein sequences
TBLASTN	amino acid sequence	translated nucleotide sequences
TBLASTX	translated nucleotide sequence	translated nucleotide sequences
PSI-BLAST	amino acid sequence	protein sequences

All these programs compare amino acid sequences with amino acid sequences, using by default the BLOSUM62 matrix, a table of the relative likelihoods of different amino acid substitutions. Searches involving nucleotide sequences, either as the query sequence or in the database searched, are carried out by translating nucleotide sequences to amino acid sequences in all six possible reading frames. A related program, BLASTN, compares nucleic acid query sequences with nucleic acid databanks directly.

This is a trivial example. In practice it would be necessary to score every path through the dotplot according to our estimate of the probability of a substitution, or of introducing or extending a gap.

Given a suitable scoring scheme, an algorithm by S.B. Needleman and C.D. Wunsch, based on **dynamic programming**, computes the optimal path or paths through the dotplot. A modification of the Needleman–Wunsch method, by T.F. Smith and M.S. Waterman, allows searching for **local matches**; that is, matching a short subsequence within a long sequence. Programs that implement these algorithms are basic tools of computational molecular biology.

Programs for optimal sequence alignment are powerful, but not fast enough for convenient searching of large databases for sequences similar to a probe sequence. BLAST (Basic Local Alignment Search Tool) was developed as an approximate alignment method, fast enough for database screening.

BLAST and PSI-BLAST

BLAST was designed for rapid sequence searching of large databases. Often the databank contains close matches to the query sequence. Programs such as BLAST, although they are less sensitive than full optimal sequence alignment, are capable of identifying these close matches. If that is what you want, fine. For example, if you want to search for homologues of a

mouse protein in the human genome, the similarity is likely to be high and an approximate method likely to find it. But if you want to search for homologues of a human protein in *C. elegans* or yeast, the relationship may be more tenuous. More sophisticated, slower, methods may be required. (It may come as a surprise, but computer time requirements are still a consideration. For although computing is becoming less expensive, the sizes of the databanks and the number of searches desired, on a world-wide basis, are growing. The net effect is that the pressure on computing resources is increasing.)

BLAST comes in several flavours, depending on the type of query sequence and type of database to search (see Table 4.4).

The method used by BLAST is also based on the dotplot approach, checking for well-matching local regions. For each entry in the database, it checks for short *contiguous* regions that match short *contiguous* regions in the query sequence, using a substitution scoring matrix but allowing no gaps. An approach in which candidate regions of *fixed length* are identified initially can be made very fast by the use of lookup tables. Figure 4.5 shows the process schematically.

Once BLAST identifies a well-fitting region, it tries to extend it. In some versions gaps are allowed. The output of BLAST is the set of local segment matches.

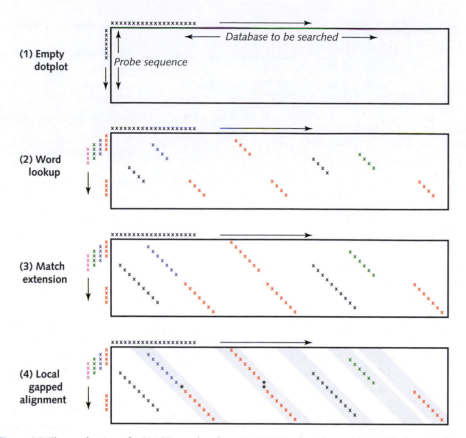

Figure 4.5 The mechanism of a BLAST search, schematic. BLAST solves the problem of finding matches of a probe sequence in a full genome or a full database, much longer than the probe sequence. (1) The algorithm operates within the outline of a dotplot, just as if the problem were going to be solved by application of an exact-alignment method. (2) BLAST first divides the probe sequence into fixed-length words of length k, here $k = 4$. It then identifies all exact occurrences of these words in the full database— no mismatches, no gaps. Note that the same 4-letter word may occur several times in the probe sequence (shown here in red), and of course each 4-letter word may match many times within the database. It is possible to do this step quickly, for many probe sequences, after pre-processing the database to record the positions of appearance of all 4-letter words. (3) Starting with each match, BLAST tries to extend the match in both directions. Still no mismatches, no gaps allowed. (4) Given the extended matches, BLAST tries to put them together by doing alignments *allowing* mismatches and gaps, but only within limited regions containing the preliminary matches (grey areas). The result of this step is to add to the matches the positions shown as *. This produces longer matching regions. It is the restriction of the more complex matching procedure to relatively small regions, rather than applying it to the entire matrix, that gives the method its speed. The price paid is that if a combined match lies outside the grey area, the method will miss it. In the example illustrated, the matching regions coloured red and green, at the right of the matrix, will not be combined, but reported as separate hits.

Even more powerful than BLAST is its development, PSI-BLAST, which looks for common patterns among similar sequences, not just checking individual pairwise comparisons.

PSI-BLAST also searches a databank for sequences similar to a query sequence. BLAST and its variants check each entry in the databank *independently* against a query sequence. PSI-BLAST begins with such a one-at-a-time search. It then derives pattern information from a **multiple sequence alignment** of the initial hits, and reprobes the database using the pattern. Then it repeats the process, fine tuning the pattern in successive cycles (see Fig. 4.6).

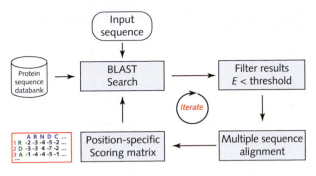

Figure 4.6 Schematic flowchart of a PSI-BLAST calculation, to detect protein sequences in a database similar to a probe sequence. The user submits an input sequence, and chooses a protein-sequence database to probe. First, using the input sequence and a standard substitution matrix, such as BLOSUM62, a BLAST calculation identifies similar sequences in the database, and assigns a statistical measure of significance, E, to each 'hit'. For each sequence retrieved from the database, E is the number of sequences of equal or higher similarity to the probe sequence that would be expected to be found in the database, just by chance. The program will select those sequences for which E is no greater than a specified threshold, often chosen as 0.005, and perform a multiple sequence alignment of them. By counting the relative frequencies of different amino acids in each column of the multiple sequence alignment, the program will derive a position-specific scoring matrix. The red box at the lower left shows a part of a position-specific scoring matrix. The columns are labelled by the 20 natural amino acids, shown in blue. The rows are labelled by the sequence to be scored by the matrix, residue numbers in red and amino acids in green. In this case the N-terminal sequence of the sequence to be scored is RDA. The entries in the column are the log-odds scores, derived from the multiple sequence alignment, of finding any amino acid at any position in the multiple alignment. For instance, the entry under A in row 3 is -1; therefore the probability of finding an A at the third position is proportional to 10^{-1}. To find the score of the sequence, add up the value in the R column of the first row, the D column of the second row, the A column of the third row, etc., to give: $10^{-3} + 10^{-7} + 10^{-1}$. In this example the probabilities are expressed unscaled and as logs to the base 10. Note that the sequence being scored may contain gaps. This matrix can be used as an alternative to the input sequence and substitution matrix in a BLAST search. Each subsequent BLAST search, based on the matrix derived in the previous step, will return a different set of 'hits'. With a sensible choice of input parameters the procedure will usually converge, to produce a more extensive set of similar sequences than would be returned by the simple BLAST search of the input sequence performed in the first step.

Here is a flowchart that PSI-BLAST follows:

1. Probe each sequence in the chosen database independently for local regions of similarity to the query sequence, using a BLAST-type search but allowing gaps.

2. Collect significant hits. Construct a multiple sequence alignment table between the query sequence and the significant local matches.

3. Form a profile from the multiple sequence alignment. A profile contains the distribution of amino acids at each position of the multiple sequence alignment.

4. Reprobe the database with the profile, still looking only for local matches.

5. Decide which hits are statistically significant and retain only these.

6. Go back to step 2, until a cycle produces little or no change. This accounts for the 'Iterated' in the program title.

PSI-BLAST, using an iterated pattern search, is much more powerful than simple pairwise BLAST in picking up distant relationships. PSI-BLAST correctly identifies three times as many homologues as BLAST in the region below 30% sequence identity. It is therefore a very useful method for analysing whole genomes. When the genomes were first sequenced, PSI-BLAST was able to match protein domains of known structure to 39% of the genes in *M. genitalium*, 24% of the genes in yeast, and 21% of the genes in *C. elegans*.

• BLAST and PSI-BLAST are standard and very effective tools for searching databases for sequences similar to a probe sequence.

Significance of alignments

Suppose alignment reveals an intriguing similarity between two sequences. Is the similarity significant or could it have arisen by chance? For some simple phenomena—tossing a coin or rolling dice—it is possible to calculate exactly the expected distribution of results, and the likelihood of any particular result.

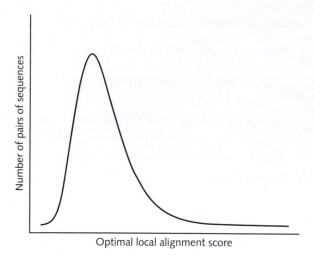

Figure 4.7 Optimal local alignment scores for pairs of random amino acid sequences of the same length follow an extreme-value distribution. For any score x, the probability of observing a score $\geq x$ is:

$$P(\text{Score} \geq x) = 1 - \exp\left(-Ke^{-\lambda x}\right)$$

where K and λ are parameters related to the position of the maximum and the width of the distribution, respectively. Note the long tail at the right. This means that a score several standard deviations above the mean has a higher probability of arising by chance (that is, it is *less* significant) than if the scores followed a normal distribution.

For sequences it is not trivial to define the population from which the alignment is selected. For instance, taking random strings of nucleotides or amino acids as controls ignores the bias arising from non-random composition.

A practical approach to the problem is as follows: If the score of the alignment observed is no better than might be expected from the corresponding alignment of a *random permutation* of the sequence, then it is likely to have arisen by chance. We may randomize one of the sequences, many times, realign each result to the second sequence (held fixed), and collect the distribution of resulting scores. Figure 4.7 shows a typical result.

Clearly if the randomized sequences score as well as the original one, the alignment is unlikely to be significant. We can measure the mean and standard deviation of the scores of the alignments of randomized sequences, and ask whether the score of original sequence is unusually high (see Box 4.2).

Many 'rules of thumb' are expressed in terms of per cent identical residues in the optimal alignment. If two proteins have over 45% identical residues in

their optimal alignment, they will almost certainly have very similar structures, and are very likely to have a common or at least a similar function. If two proteins have over 25% identical residues, they are likely to have a similar general folding pattern, although the mainchain conformation will be expected to show significant distortion. On the other hand, observation of a lower degree of sequence similarity cannot rule out homology.

R.F. Doolittle suggested a general calibration of pairwise sequence similarity, for homology inference. Two full-length protein sequences (≥ 100 residues) that have $\geq 25\%$ identical residues in an optimal alignment are likely to be related. Doolittle defined the range between 18 and 25% identity as the **twilight zone**, where there may be tantalizing suspicion of a genuine evolutionary relationship, but the evidence falls short of proof. In some cases the active site is better conserved than the bulk of the protein. The appearance of a sequence **motif** (a signature, in the sequence, of an active site) can support the case for homology.

Below the twilight zone is a region where pairwise sequence alignments reveal very little. Below ~15% identical residues in an optimal alignment, we are mired in the noise. In this range of similarity, we have no reason to believe that the sequences are related— although they might be.

Although the twilight zone is a treacherous region, we are not entirely helpless. In deciding for or against a genuine relationship, the 'texture' of the alignment is important: are the similar residues isolated and scattered throughout the sequence or are there **icebergs** (another term of Doolittle's); that is, local regions of high similarity that may correspond to a shared active site? We may need to rely on other information, about shared ligands or function. Of course if the structures are known, we could examine them directly. Structures are 'the court of last resort'.

- Whether an alignment that shows intriguing elements of similarity is revealing a genuine evolutionary relationship may depend on a statistical analysis of the results. As a rule of thumb, sequences are likely to be genuine homologues if their similarity after optimal alignment is above R. Doolittle's 'twilight zone'— 18–25% identical residues.

BOX 4.2   How to play with matches but not get burned

Pairwise alignments and database searches often show tenuous but tantalizing sequence similarities. How can we decide whether we are seeing a true relationship? Statistics cannot answer biological questions directly, but can tell us the likelihood that a similarity as good as the one observed would appear, just by chance, among unrelated sequences. To do this we want to compare our result with alignments of the same sequence to a large population. This 'control' population should be similar in general features to our aligned sequences, but should contain few sequences related to them. Only if the observed match stands out from the population can we regard it as significant.

To what population of sequences should we compare our alignment? For pairwise alignments, we can pick one of the two sequences, make many scrambled copies of it using a random-number generator, and align each permuted copy to the second sequence. For probing a database, the entire database provides a comparison population.

Alignments of our sequence to each member of the control population generates a large set of scores. How does the score of our original alignment rate? Several statistical parameters have been used to evaluate the significance of alignments:

- The **Z-score** is a measure of how unusual our original match is, in terms of the mean and standard deviation of the population scores. It reflects the extent to which the original result is an outlier from the population. If the original alignment has score S,

$$Z\text{-score of } S = \frac{S - \text{mean}}{\text{standard deviation}}$$

A Z-score of 0 means that the observed similarity is no better than the average of the control population, and might well have arisen by chance. The higher the Z-score, the greater the probability that the observed alignment

has not arisen simply by chance. Experience suggests that Z-scores ≥ 5 are significant.

- Many programs report P = the probability that the alignment is better than random. The relationship between Z and P depends on the distribution of the scores from the control population, which do *not* follow the normal distribution.

A rough guide to interpreting P values:

$P \leq 10^{-100}$	exact match
P in range 10^{-100}–10^{-50}	sequences very nearly identical, e.g. SNPs or alleles
P in range 10^{-50}–10^{-10}	closely related sequences, homology certain
P in range 10^{-5}–10^{-1}	usually distant relatives
$P > 10^{-1}$	match probably insignificant.

- For database searches, some programs (including PSI-BLAST) report **E-values**. The E-value of an alignment is the expected number of sequences that give the same Z-score or better if the database is probed with a random sequence. E is found by multiplying the value of P by the size of the database probed. Note that E but not P depends on the size of the database. Values of P are between 0 and 1. Values of E are between zero and the number of sequences in the database searched.

A rough guide to interpreting E-values:

$E \leq 0.02$	sequences probably homologous
E between 0.02	homology unproven but can't be ruled out
$E > 1$	you'd have to expect this good a match just by chance.

Statistics are a useful guide, but not a substitute for thinking carefully about the results, and further analysis of ones that look promising!

Multiple sequence alignment

In Nature, even a single sequence contains all the information necessary to dictate the fold of the protein. How does a multiple sequence alignment make that information more intelligible and useful?

As the experience with PSI-BLAST shows, a joint alignment of many related sequences contains far more useful information than an alignment of a single pair of sequences. The basic reason for this is that

Table 4.5 Possible colour scheme for display of multiple sequence alignment

Colour	Residue type	Amino acids
Yellow	Small non-polar	Gly, Ala, Ser, Thr
Green	Hydrophobic	Cys, Val, Ile, Leu, Pro, Phe, Tyr, Met, Trp
Magenta	Polar	Asn, Gln, His
Red	Negatively charged	Asp, Glu
Blue	Positively charged	Lys, Arg, (His)

Program	URL
Clustal-W	http://www.ebi.ac.uk/Tools/msa/clustalw2/
T-Coffee	http://www.tcoffee.org/ or http://www.ebi.ac.uk/Tools/msa/tcoffee/
Muscle	http://www.ebi.ac.uk/Tools/msa/muscle/

it is possible to examine the extent and distribution of variation at individual positions. Alignment tables expose patterns of amino acid conservation, from which distant relationships may be more reliably detected. Patterns of conservation provide important clues to the structural or functional role of different regions of the sequence. Structure prediction tools also give more reliable results when based on a multiple sequence alignments than on single sequences (for example, see 'Secondary structure prediction' in Chapter 3). The method of structure prediction involving identification of neighbouring residues from correlated mutations is based directly on the multiple sequence alignment.

Visual examination of multiple sequence alignment tables is one of the most profitable activities that a molecular biologist can undertake away from the lab bench. Don't even *think* about not displaying them with different colours for amino acids of different physicochemical type. A reasonable colour scheme (not the only possible one) is that shown in Table 4.5.

To be informative a multiple alignment should contain a distribution of closely and distantly related sequences. If all the sequences are very closely related, the information they contain is largely redundant, and few inferences can be drawn. If all the sequences are very distantly related, it will be difficult to construct an accurate alignment (unless all the structures are available), and in such cases the quality of the result, and the inferences it might suggest, are questionable. Ideally, one has a complete range of similarities, including distantly related examples linked through chains of close relationships.

Popular programs for multiple sequence alignment include:

A multiple sequence alignment of thioredoxins shows the importance of conservation patterns

Thioredoxins are enzymes found in all cells. They participate in a broad range of biological processes, including cell proliferation, blood clotting, seed germination, insulin degradation, repair of oxidative damage, and enzyme regulation. The common mechanism of these activities is the reduction of protein disulphide bonds.

The structure of *E. coli* thioredoxin contains a central five-stranded β-sheet flanked on either side by α-helices; these helices and strands are indicated by the symbols α and β in Fig. 4.8(a). Homologues are likely to share most but not all of the secondary structure of the *E. coli* enzyme.

Thioredoxins are members of a superfamily including many more distantly related homologues. These include glutaredoxin (hydrogen donor for ribonucleotide reduction in DNA synthesis), protein disulphide isomerase (which catalyses exchange of mismatched disulphide bridges in protein folding), phosducin (a regulator of G-protein signalling pathways), and glutathione S-transferases (chemical defence proteins). Implicit in the multiple sequence alignment table of the thioredoxins themselves are patterns that should be applicable to identifying these more distant relatives (see Fig. 4.8(a)).

Figure 4.8(b) shows a summary of the alignment as a *sequence logo,* in which letters of different sizes indicate different proportions of amino acids. (T. Schneider and M. Stephens designed sequence logos; this example was produced using the webserver at http://weblogo.berkeley.edu/logo.cgi.)

Structural and functional features of thioredoxins that we might hope to identify from the multiple sequence alignment include (see Fig. 4.8(c)):

- *The most highly conserved regions probably correspond to the active site.* The disulphide bridge between residues 32 and 35 in *E. coli* thioredoxin

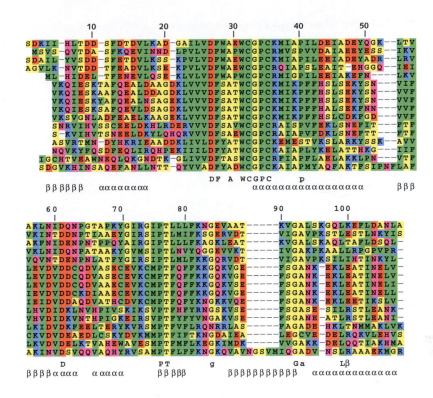

(a)

Thioredoxin

(b)

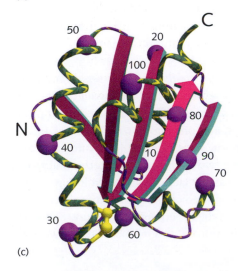

(c)

Figure 4.8 (a) Alignment of amino acid sequences of *E. coli* thioredoxin and homologues. Some of the sequences have been trimmed at their termini. Residue numbers in this table correspond to positions in the *E. coli* sequence (top line). Helix (α) and strand (β) assignments for *E. coli* thioredoxin are from PDB entry 2TRX. (b) Sequence logo derived from this multiple sequence alignment. (c) The structure of *E. coli* thioredoxin [2TRX] contains a central five-stranded β-sheet flanked on either side by α-helices. (Compare this with flavodoxin (see Fig. 2.16)—in what way does the fold differ?) Residue numbers correspond to those in the multiple sequence alignment table. The N- and C-termini are also marked. Spheres indicate positions of the Cα atoms of every tenth residue. The reactive disulphide bridge between Cys32 and Cys35 appears in yellow.

is part of a WCGPC[K or R] motif conserved in the family. This region is active in the redox reaction that the proteins catalyse. Other regions conserved in the sequences, including the PT at residues 75–77 and the GA at residues 92–93, are involved in substrate binding.

- *Regions rich in insertions and deletions probably correspond to surface loops. A position containing a conserved Gly or Pro probably corresponds to a turn.* Turns correlated with insertions and deletions occur at positions 9, 20, 60 and 95. The conserved glycine at position 92 in *E. coli* thioredoxin is indeed part of a turn. It is in an unusual mainchain conformation, one that is easily accessible only to glycine (see Chapter 2). The conserved proline at position 76 in *E. coli* thioredoxin is also associated with a turn. It is in another unusual mainchain conformation, this one easily accessible only to proline: the peptide bond preceding residue 76 is in the *cis* conformation.

- *A conserved pattern of hydrophobicity with spacing 2 (that is, every other residue)—with the intervening residues more variable and including* hydrophilic *residues—suggests a β-strand on the surface.* This pattern is observable in the β-strand between residues 50 and 60.

- *A conserved pattern of hydrophobicity with spacing ~4 suggests a helix.* This pattern is observable in the region of helix between residues 40 and 49.

This example shows how conserved patterns in multiple sequence alignments and structural features illuminate each other.

Analysis of structures

Sequences that have diverged beyond the point where they can reliably be aligned, can, from their failing hands, throw the torch to the structures.

- Protein structure tends to diverge more conservatively than amino acid sequence during evolution. Therefore structure alignment is more powerful than pairwise sequence alignment for detecting homology and aligning the sequences of distantly related proteins.

Superposition of structures

As in the case of sequences, a fundamental question in analysing structures is how to define and compute a measure of similarity. A basic tool is **superposition:** given two or more structures, move them around in space so that as many corresponding atoms as possible are as close to each other as possible. In choosing the corresponding atoms to superpose protein or nucleic acid structures or substructures, it is a great simplification that the residues appear in a linear order. Indeed, choosing corresponding Cα atoms in two proteins is equivalent to aligning their sequences. Such a **structural alignment** is still an alignment; that is, an assignment of residue–residue correspondences.

It may be useful to contrast three related problems in computational structural chemistry.

1. *To measure the similarity of two sets of atoms with known correspondences.* This would arise if one wants to superpose the Cα atoms of two protein structures, and one *knows the alignment* and that it is without gaps. This situation might arise in comparing two conformations of the same protein, as in the case of lactoferrin shown in Fig. 2.11; or evaluating a prediction by comparing a computed model with the experimental structure. Simple and robust solutions of this problem are known.

2. *To measure the similarity of two sets of atoms with unknown correspondences, but for which the molecular structure—specifically the linear order of the residues—restricts the nature of the correspondence.* Just as in sequence alignment, the order of aligned residues must follow the order in the sequences. (If one sequence contains residues in the order AB and the second sequence contains residues in the order CD, it is *not* permissible to align A with D and B with C.)

3. *To measure the similarity between two sets of atoms with unknown correspondence, with no restrictions on the order of the aligned atoms.* This problem arises in the following important case: suppose two (or more) molecules have similar biological effects, such as a common pharmacological activity. It is often the case that the structures share a common constellation of a relatively small subset of their atoms that are responsible for the biological activity, called a **pharmacophore**. To identify pharmacophores it is useful to be able to find the maximal subsets of atoms from the two molecules that have a similar structure, with no restrictions on their order.

A basis for a general approach to all these problems is the operation of superposition of sets of atoms with known alignment, case 1. Case 1 is the simplest.

Calculation of optimal superposition of aligned sets of atoms

Even if two structures are exactly congruent, the atoms may not be at the same positions in space. A measure of similarity of two ordered sets of points is the root-mean-square deviation (r.m.s.d.) after **optimal superposition**. Suppose the original coordinates of the two sets of points are:

$$(X_i, Y_i, Z_i), \ i = 1, \ \ldots N \quad \text{and} \quad (x_i, y_i, z_i), \ i = 1, \ \ldots N$$

The most general motion of a rigid body can be expressed as a rotation plus a translation. Therefore we can find the optimal superposition by applying a rotation and a translation to the coordinates of the

second set of points, (x_i, y_i, z_i), $i = 1, \ \ldots N$, producing the transformed coordinates: (x_i', y_i', z_i'), $i = 1, \ \ldots N$

The **root-mean-square deviation** is:

$$\text{r.m.s.d.} = \sqrt{\frac{\sum_i d_i^2}{N}},$$

where d_i is the distance between (X_i, Y_i, Z_i) and (x_i', y_i', z_i'), *after* applying the rotation and translation.

The optimal r.m.s.d. is the minimal value of the r.m.s.d. over all possible translations and rotations.

Because the root-mean-square deviation contains the *square* of interatomic distances, it is particularly sensitive to outliers.

Finding the optimal superposition of two aligned sets of points is a simple exercise in linear algebra. In the computing literature this is known as the orthogonal Procrustes problem. See, e.g., Golub, G. and van Loan, C. (1989). Matrix computations. 2nd ed., Johns Hopkins Press, Baltimore, MD, U.S.A.

Structural alignment

A simple version of the pairwise structural alignment problem is as follows: given the Cα atoms of two protein structures, select subsets from both proteins, containing the same number of atoms, as many as possible, that can be superposed to a low value of the root-mean-square deviation. The alignment is fixed by taking the atoms in both substructures in order of their appearance in the sequence. (Taking into account atoms other than Cα presents more complicated forms of the problem.)

- A measure of geometric similarity between two sets of points with known correspondence is the root-mean-square deviation after optimal superposition. To perform a structural alignment of two point sets it is in general necessary to extract well-fitting subsets and *deduce* the correspondence.

A general problem in structural alignment is how large a substructure to select. One might be able to choose 100-atom substructures with an r.m.s.d. of 1.5 Å, or 150-atom substructures with an r.m.s.d. of

2.0 Å, or 200-atom substructures with an r.m.s.d. of 3.5 Å. Which is preferred? Sometimes, there is an obvious core of the structure that fits much better than the rest. Such a core was clear in the figure showing antennapedia structures (see Fig. 3.19). In such cases one can easily make intelligent choices.

Some approaches to structural alignment work by converting the problem from three dimensions to one dimension, and applying sequence alignment methods. For example, by calculating conformational angles, the mainchain conformation of a protein may be written as a *one-dimensional* sequence of ϕ, ψ values—two numbers per residue. Alternatively, characterize each residue by the distances from its Cα atom to the Cα atoms in other residues nearby in the sequence—perhaps the eight residues that differ in the sequence by no more than four positions. This produces a sequence of arrays of distances, to which sequence alignment methods can be applied. (In these cases, what is being aligned is not two sequences of characters, but two sequences of sets of numbers.)

Another approach to structure alignment is similar to BLAST—detect short local regions of similar structure and piece them together.

Experience with structure alignment methods has shown that for closely related structures, any reasonable method will work. The challenge is to devise something that can be pushed to be effective on very divergent structures.

The most powerful methods of structure alignment make use of *patterns of interresidue contacts*. As proteins evolve, their structures change. Among the subtle details that evolution has strongly tended to conserve are the patterns of contacts between residues. That is, if two residues are in contact in one protein, the residues corresponding to these two in a related protein are also likely to be in contact. This is true even in very distant homologues, and even if the residues involved change in size. Mutations that change the sizes of packed buried residues produce adjustments in packing of the helices and sheets against one another.

L. Holm and C. Sander applied these observations to the problem of structural alignment of proteins. If the interresidue contact pattern is *preserved* in

Figure 4.9 The regions of common fold, as determined by the program DALI by L. Holm and C. Sander, in the TIM-barrel proteins mouse adenosine deaminase [1FKX] (black) and *Pseudomonas diminuta* phosphotriesterase [1PTA] (red). In the alignment shown in this figure, the sequences have only 13% identical residues—closer to midnight than to the twilight zone.

distantly related proteins, then it should be possible to *identify* distantly related proteins by detecting conserved contact patterns.

Computationally, one constructs matrices of contact patterns in two proteins (this is very easy), and then seeks the maximal matching submatrices (this is hard). Using carefully chosen approximations, Holm and Sander wrote an efficient program called **DALI** (for **distance-matrix alignment**) that is now in common use for identifying proteins with folding patterns similar to that of a query structure. The program runs fast enough to carry out routine screens of the entire Protein Data Bank for structures similar to a newly determined structure, and even to perform a classification of protein domain structures from an all-against-all comparison. Holm and Sander have found several unexpected similarities not detectable at the level of pairwise sequence alignment.

An example of DALI's 'reach' into recognition of very distant structural similarities is its identification of the relation between mouse adenosine deaminase, *Klebsiella aerogenes* urease, and *Pseudomonas diminuta* phosphotriesterase (see Fig. 4.9).

DALI is available over the web. You can submit coordinates to the site http://ekhidna.biocenter.helsinki.fi/dali_server and receive the set of similar structures and their alignments with the query.

Multiple structure alignment

Just as multiple sequence alignment is more powerful than pairwise sequence alignment, so multiple structure alignment is more powerful than pairwise structure alignment.

The basis of multiple structural alignment is multiple structural superposition. Figure 3.19 showed a multiple superposition of different NMR structures of the *Drosophila* protein antennapedia. In this case there was no question of alignment, because the structures were all different conformations of the same protein. Given sets of aligned coordinates, algorithms, to determine exactly their joint optimal superposition, exist that are generalizations of the method for pairwise superposition. Selection of the size of the substructures to align is still a problem, as in the pairwise case.

To superpose the structures of a set of homologues, there is an alignment as well as a superposition problem. There are several approaches to these challenges. **MUSTANG**, written by A.S. Konagurthu, is a development of DALI's distance-matrix approach to *multiple* structural alignment (http://www.csse.monash.edu.au/~karun/Site/mustang.html).

Database searching for structures or fragments

To probe a structural database with either a whole molecule or a fragment is to perform the equivalent of a global or local structural alignment between a probe structure and the database.

PDBemotif and PDBeTemplate

The PDBe group at the European Bioinformatics Institute in Hinxton, U.K., maintains **PDBemotif**, a database of ligands and motifs, both related to binding sites in proteins. Users can retrieve structures by:

- keyword, notably including name of ligand;
- three-dimensional motif—local regions of structures such as β-hairpins. The search is based on Cα atoms only, and therefore is amino acid sequence independent;
- PROSITE motif—a pattern in the amino acid sequences;

- ligand environments—which amino acids are near neighbours and how they interact.

PDBeTemplate allows probing with a set of residues, which need not be consecutive. The user may submit a three-dimensional structure or merely a set of amino acids. For instance, a search in PDBeTemplateBrowse for HIS ASP SER would identify serine proteases.

Common substructures have functional or architectural implications

Screening structural databases for fragments of proteins has illuminated the range of conformations that short regions can display, and has led to a number of useful applications.

Probing databases with short *consecutive* segments from proteins has shown that the main chain conformations of short regions tend to recur in protein structures. Obviously α-helices and strands of β-sheets are common themes, but other, less-regular, conformations also occur. It has been estimated that all eight-residue regions of mainchain in well-determined protein structures are similar in conformation to one of a set of about 60 structures, and all five-residue regions are similar in conformation to one of a set of about 150 structures. The latter result suggests that each residue can have about $150^{1/5} = 2.75$ conformational states per residue; this is reasonable given the Sasisekharan–Ramakrishnan–Ramachandran plot. The fact that eight-residue regions have fewer conformations than would occur if all residue conformations were independent is expected because the chain must avoid collisions.

Probing databases with short regions pick up both homologous regions—that is, regions from related structures—and also examples where regions with similar conformations have entirely different structural contexts and interactions in different proteins. Comparisons among these regions reveals the nature of the determinants of their conformations: for example, a certain type of loop may require certain hydrogen-bonding interactions, but different proteins can provide these interactions in different ways, and the loop can thereby occur in different structural contexts (see Fig. 4.10).

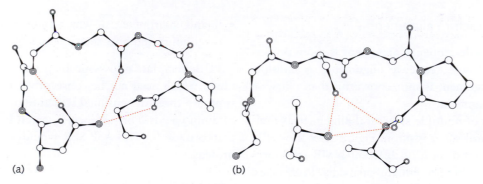

(a)   (b)

Figure 4.10 (a) A loop from the antigen-binding site of an immunoglobulin domain. The conformation of the loop is determined primarily by a *cis*-peptide bond before the Pro at the right, and hydrogen bonds formed by the sidechain of the Asn residue N-terminal to the loop. (b) A loop of very similar conformation occurs in tomato bushy stunt virus. These loops have entirely different structural contexts: in the immunoglobulin it is a hairpin connecting two successive strands of β-sheet; in the virus it connects strands from different sheets. Nevertheless, the structural determinants of the loops in the two proteins are shared. There are *cis*-prolines at equivalent positions. Hydrogen bonds, similar to those made by the Asn sidechain in the immunoglobulin domain, are made in the virus by the carbonyl oxygen of an alanine, distant in the sequence, but occupying the same position in space relative to the loop.

Databases of protein families

Evolutionary relationships are essential for making sense of biological data. Evolution provides the framework for an integrated appreciation of the properties of molecules and processes, and their similarities and differences in various species. Perhaps less obvious is that comparative studies illuminate, in an essential way, even individual molecules. Knowing only a single sequence, or structure, it is difficult to understand the significance of particular features. Patterns of conservation identify features that Nature has found it necessary to retain. The challenge then is to figure out why.

Study of evolutionary patterns must begin with assembling a set of homologues. We again emphasize (1) there is a distinction between *homology* = descent from a common ancestor = a yes-or-no property, and *similarity* = some quantitative measure of the difference between two objects, and (2) that similarity can always be measured but it is rare to be able to observe homology directly; therefore, in most cases homology is an *inference* from similarity. (The methods and results of metagenomics may make it easier to observe homology directly.)

Protein structure changes more conservatively than amino acid sequence. Therefore, inference of

homology from structural similarity can link more distant relatives than sequence similarity can. Cases that lie in the twilight zone where sequence similarity is suggestive but not convincing, are resolvable by examination of structural similarity. In many cases, structural similarity can identify homologues even if no signal whatever—at least no signal detectable by current techniques—remains in the sequences.

It is common to refer to a group of related proteins as a family. Many databases classify proteins into families. These include *sequence-orientated databases* such as InterPro, Pfam, and COG, and *structure-orientated databases* such as SCOP and CATH. The assignment of proteins to families is similar but not identical in different sources.

Most protein families contain many clusters of closer relatives. These form subfamilies. Conversely, two or more families can be grouped into superfamilies. Whereas the distinction between homologous and non-homologous proteins is objective (even if we cannot determine it with confidence in all cases), the clustering of homologues into subfamilies or superfamilies is partially a matter of convention or taste. Definition of subfamilies and superfamilies may legitimately differ among different databases.

> • Many databases cluster proteins into families, on the basis of amino acid sequence, structure, or function—or combinations of these features.

Classifications of protein structures

Several websites offer hierarchical classifications of all proteins of known structure according to their folding patterns (see Chapter 2):

• SCOP: Structural Classification of Proteins, and its successor, SCOP2
• CATH: Class/Architecture/Topology/Homology
• DALI: Based on extraction of similar structures from distance matrices

These sites are useful general entry points to retrieval of protein structural data. For instance, SCOP offers facilities for searching on keywords to identify structures, navigation up and down the hierarchy, generation of pictures, access to the annotation records in the PDB entries, and links to related databases.

Classification and assignment of protein function

The Enzyme Commission

The first detailed classification of protein functions was that of the **Enzyme Commission (EC)**. In 1955, the General Assembly of the International Union of Biochemistry (IUB), in consultation with the International Union of Pure and Applied Chemistry (IUPAC), established an International Commission on Enzymes, to systematize nomenclature. The Enzyme Commission published its classification scheme, first on paper and now on the web: http://www.chem.qmul.ac.uk/iubmb/enzyme/.

EC numbers (looking suspiciously like IP numbers) contain four numeric fields, corresponding to a four-level hierarchy. For example, EC 1.1.1.1 corresponds to the reaction:

$$\text{an alcohol + NAD = the corresponding aldehyde or ketone + NADH}_2.$$

Note that several reactions, involving different alcohols, would share this number (whether or not the same enzyme catalysed them); but that the same dehydrogenation of one of these alcohols by an enzyme using the alternative cofactor NADP would not. It would be assigned EC 1.1.1.2.

The first field in an EC number indicates to which of the six main divisions (classes) the enzyme belongs:

Class 1. Oxidoreductases
Class 2. Transferases
Class 3. Hydrolases
Class 4. Lyases
Class 5. Isomerases
Class 6. Ligases

The significance of the second and third numbers depends on the class. For oxidoreductases the second number describes the substrate and the third number the acceptor. For transferases, the second number describes the class of item transferred, and the third number describes either more specifically what they transfer or in some cases the acceptor. For hydrolases, the second number signifies the kind of bond cleaved (e.g. an ester bond) and the third number the molecular context (e.g. a carboxylic ester or a thiolester). (Proteinases, a type of hydrolase, are treated slightly differently, with the third number including the mechanism: serine proteinases, thiol proteinases and acid proteinases are classified separately.) For lyases the second number signifies the kind of bond formed (e.g. C–C or C–O), and the third number the specific molecular context. For isomerases, the second number indicates the type of reaction and the third number the specific class of reaction. For ligases, the second number indicates the type of bond formed and the third number the type of molecule in which it appears. For example, EC 6.1 for C–O bonds (enzymes acylating tRNA), EC 6.2 for C–S bonds (acyl–CoA derivatives), etc. The fourth number gives the specific enzymatic activity.

The **Enzyme Structures Database** at PDBe links Enzyme Commission numbers to proteins of known structure,http://www.ebi.ac.uk/thornton-srv/databases/enzymes/.

The Gene Ontology™ Consortium protein function classification

In 1999, Michael Ashburner and many coworkers faced the problem of annotating the soon-to-be-completed *Drosophila melanogaster* genome sequence. As a classification of function, the EC classification was unsatisfactory, if only because it was limited to enzymes. Ashburner organized the **Gene Ontology**™ **Consortium** to produce a standardized scheme for describing function.

An ontology is a formal set of well-defined terms with well-defined interrelationships; that is, a dictionary and rules of syntax.

The Gene Ontology™ Consortium (GO; http://www.geneontology.org) has produced a systematic classification of gene function, in the form of a dictionary of terms, and their relationships.

Organizing concepts of the GO project include three categories:

- **Molecular function:** a function associated with what an individual protein or RNA molecule does in itself; either a general description, such as *enzyme*, or a specific one, such as *alcohol dehydrogenase* (specifying a catalytic activity, not a protein). This is function from the biochemists' point of view.

- **Biological process:** a component of the activities of a living system, mediated by a protein or RNA, possibly in concert with other proteins or RNA molecules; either a general term, such as *signal transduction*, or a particular one, such as *cyclic AMP synthesis*. This is function from the cell's point of view.

Because many processes are dependent on location, GO also tracks:

- **Cellular component:** the assignment of site of activity, or partners; this can be a general term such as *nucleus* or a specific one such as *ribosome*.

Figure 4.11 shows examples of the GO classification.

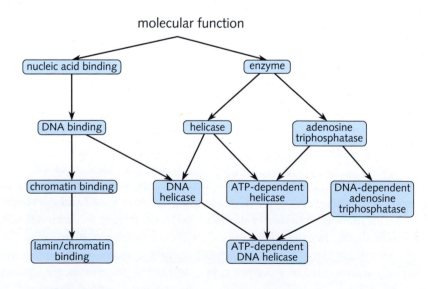

(a)

Figure 4.11 (*Continued*)

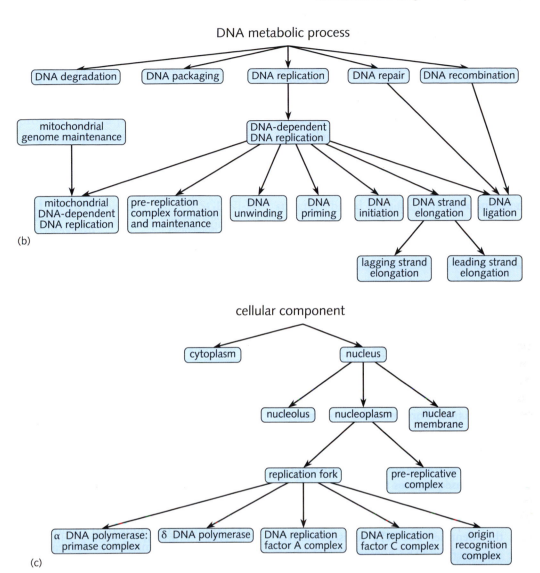

Figure 4.11 Selected portions of the three categories of Gene Ontology (GO), showing classifications of functions of proteins that interact with DNA. (a) *Molecular function:* including general DNA binding by proteins, and enzymatic manipulations of DNA. (b) *Biological process:* DNA metabolism. (c) *Cellular component:* Different places within the cell, or partners. These pictures illustrate the general structure of the GO classification. Each term describing a function is a *node* in a graph. Each node has one or more parents and one or more descendants: arrows indicate direct ancestor–descendant relationships. A path in the graph is a succession of nodes, each node the parent of the next. Nodes can have 'grandparents', and more remote ancestors. Unlike the EC hierarchy, the GO graphs are *not* trees in the technical sense, because there can be more than one path from an ancestor to a descendant. For example, there are two paths in (a) from enzyme to ATP-dependent helicase. Along one path helicase is the intermediate node. Along the other path adenosine triphosphatase is the intermediate node. Although the nodes are shown on discrete levels to clarify the structure of the graph, all the nodes on any given level do not necessarily have a common degree of significance, unlike family, genus and species levels in the Linnaean taxonomic tree, or the ranks in military, industrial, academic, etc. organizations. GO terms could not have such a common degree of significance, given that there can be multiple paths, of different lengths, between different nodes.

$$\frac{1}{2} = e^{-kt_{1/2}} \quad \text{or} \quad t_{1/2} = \ln 2 / k = 0.693 / k.$$

It is a fascinating property of exponentials that $t_{1/2}$ is the same no matter how much A you start with.

For a more complicated reaction, say A+A→B, the rate might be proportional not to [A] but to $[A]^2$. Then $\frac{d[A]}{dt} = -k[A]^2$. In this case the rate of disappearance of A will be a more complicated function of time. In general, if:

$$n_A A + n_B B + n_C C + \ldots \rightarrow n_P P + n_Q Q + n_R R + \ldots$$

we might observe a rate constant of the form:

$$\text{rate} = k \frac{[A]^{m_A} [B]^{m_B} [C]^{m_C} \ldots}{[P]^{m_P} [Q]^{m_Q} [R]^{m_R} \ldots}.$$

Here, the exponents in the rate equation, m_A, m_B, …, define the kinetic order of the reaction. The **molecularity** of the reaction is the number of colliding entities that participate in the reaction. *Order and molecularity are in general not the same.* (And the stoichiometry of the reaction is not a reliable indicator of either! In other words the n_A, etc. in the chemical equation and the m_A, etc. in the rate formula may be the same but need not be.) In a multistep reaction, the order reflects the mechanism of only the slowest step.

It follows that kinetic data alone cannot prove a hypothesized reaction mechanism. However, they can disprove one: if a proposed mechanism is inconsistent with the kinetic data, it must be discarded.

(2) **Substrate specificity:** Exposed to a mixture of potential reactants, enzymes can select a particular substrate or substrates. Substrate specificity may be absolute or partial. For example, urease has no known substrate other than urea itself:

$$(NH_2)_2 CO + H_2O \xrightarrow[\text{urease}]{} CO_2 + 2NH_3.$$

In contrast, hexokinase will catalyse the phosphorylation of several sugars—including mannose and fructose—but not galactose, xylose*, maltose or sucrose.

The basic source of substrate specificity is the spatial complementarity of enzyme and substrate. E. Fischer's classic statement is still worth reading (see Box 5.3). The 'lock-and-key' image suggests that enzymes are rigid. But we now know that many enzymes undergo conformational changes as part of their mechanisms of catalysis. Conformational changes also participate in mechanisms of control of the activity of allosteric enzymes.

R. Feynman once asserted that '… everything that living things do can be understood in terms of the jigglings and wigglings of atoms.' (Note that 'jigglings' are small translations; 'wigglings' are small rotations.)

*In fact xylose *inhibits* yeast hexokinase, which is too bad, because it would be attractive to use inedible byproducts of agriculture as sources of biofuels, instead of starch from maize. This would avoid raising food prices from a 'food or fuel' competition. Xylose is a major component of the hydrolysis of lignocelluloses. Development of xylose-fermenting strains of *Saccharomyces cerevisiae* is another goal of contemporary energy research.

(3) **Reaction specificity**: most compounds can undergo several possible reactions. Different enzymes catalyse different reactions, allowing independent control over the flow patterns through branched metabolic pathways. For instance, pyruvate can enter the Krebs cycle, via the formation of acetyl-CoA by the pyruvate dehydrogenase complex (see Fig. 1.12). Or it may be converted to oxaloacetate in gluconeogenesis. Or it may be fermented to lactic acid, or to ethanol. Or converted to diacetyl (2,3-diketobutane) or acetoin (3-hydroxybutanone).

(4) **It is often said that enzymes, as catalysts, accelerate the approach to equilibrium but do not shift the equilibrium.** It follows that a reaction with a very high equilibrium constant will form products in an effectively irreversible way. Nevertheless, a second enzyme may effect the reverse transformation by coupling it to hydrolysis of ATP. (Of course the new equilibrium corresponds to a different reaction.) An example is the phosphorylation of glucose, the last step in gluconeogenesis and the first step in glycolysis:

gluconeogenesis (synthesis of glucose):

$$\text{glucose-6-phosphate} + H_2O \rightarrow \text{glucose} + P_i$$

catalysed by: *glucose-6-phosphatase*

glycolysis (digestion of glucose):

$$\text{glucose} + ATP \rightarrow \text{glucose-6-phosphate} + ATP$$

catalysed by : *hexokinase*

Both these reactions are spontaneous as written, at physiological substrate concentrations.

BOX 5.3 — Emil Fischer's 'lock-and-key' analogy

'Their restricted action … may therefore be explained on the basis of the assumption that only with a similar geometric structure can the molecules approach each other closely, and thus initiate the chemical reaction. To use a picture, I should say that the enzyme and the glucoside must fit each other like a lock and key, in order to effect a chemical action on each other. … The finding that the activity of enzymes is limited by molecular geometry to so marked a degree should be of some use for physiological research.'
(Fischer, E. 1894, translated by J. Fruton).

(5) Enzymes are subject to control. The activities of thousands of enzymes must be choreographed to produce a smooth and steady flow through metabolic pathways. Control may apply at the level of individual enzyme molecules, via inhibitors or allosteric effectors. Control over transcription can regulate amounts of different enzymes. Breakdown in regulatory mechanisms is a common cause of disease.

In most cases, control does not apply to individual proteins in isolation. Rather, there are complex cascades or networks of control interrelationships, reacting to different kinds of signals and making use of different mechanisms.

Essential features of enzymes:
- catalytic activity—speeding up attainment of equilibrium
- substrate specificity
- reaction specificity
- individual enzymes accelerate the approach to, but do not shift, equilibria
- subject to control of activity

Reaction rates and transition states

Any chemical reaction is the motion of a set of atoms from an initial constellation—the reactants—to another final constellation—the products. If both reactants and products are stable, then before and after the reaction their internal motions are generally only local; that is, small, low-energy, displacements around minimum-energy equilibrium states. A trajectory corresponding to the reaction connects the reactant and product states. Such a trajectory involves large-scale structural changes and high energies.

It has long been conventional to represent such a trajectory by a diagram plotting energy against a **reaction coordinate** (see Fig. 5.2). This graph has several notable features:

1. The 'reaction coordinate' is an abstract entity, generally difficult to define precisely. In the case of a unimolecular *cis–trans* isomerization of a substituted ethene, the angle of rotation around the double bond serves as a reasonable reaction coordinate. But in more complicated reactions a simple physical correlate of the reaction coordinate is harder to define.

2. As drawn in textbooks (including this one) the curve is smooth, with a single maximum. It gives the impression of a unique trajectory from reactants to products, with no subsidiary maxima or minima, which would correspond to stable intermediates, nor even any bumps and dips, which would correspond to unstable ones. This primarily reflects our ignorance. The actual contours of the energy landscape are largely invisible to experimental techniques of structure determination, because the intervening states are transient. In some cases theoretical calculations can trace out true reaction trajectories.

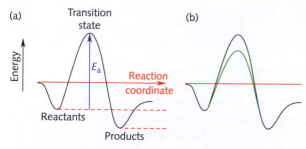

Figure 5.2 (a) A graph of energy (vertical) against 'reaction coordinate' (horizontal). The reaction coordinate is a measure of the progress of the reaction. Both reactants and products are stable. Therefore they appear at local minima in the energy. To convert from reactants to products requires traversing a barrier. The configuration at the top of the barrier is called the transition state. The height of the barrier above the energy level of the reactants—the activation energy E_a—controls the rate of reaction. The higher the barrier, the slower the reaction. (b) Comparison of the uncatalysed reaction (black) with the catalysed reaction (green). In the presence of a catalyst that does not change the energies of reactants and products, but stabilizes the transition state, the barrier is lower and the reaction rate higher. The invariance of the energies of reactants and products implies that the equilibrium constant does not change.

Why would it be difficult to detect a more complex shape of the graph in Fig. 5.2(a), one with multiple maxima and minima, by measurements of reaction rates? Typically, the speed of a multistep reaction depends primarily on the slowest step, the **rate-limiting step**. It is the rate of the slowest step that simple kinetic measurements reveal.

> • Many reactions proceed through multiple steps. The observed reaction rate is the speed of the slowest step, the *rate-limiting* step.

One feature of the curve that is relatively easily accessible to experiment is the height of the barrier—the **activation energy** E_a of the rate-limiting step. It is this quantity that primarily determines the reaction rate.

> • The trajectory of a reaction passes over a barrier called the **transition state**. The height of the barrier determines the reaction rate: the higher the barrier, the slower the reaction rate. Catalysts, including enzymes, can speed up reactions by lowering the height of the barrier.

E_a can be determined from the variation in reaction rate k with temperature T. Many reactions follow the **Arrhenius equation**:

$$\frac{\mathrm{d}\,\ln k}{\mathrm{d}T} = -\frac{E_a}{RT^2}.$$

Do all reactants pass over the same barrier, with the same height E_a? That is, is the trajectory from reactants to products unique? Is the transition state unique? Not precisely. We must envisage a sheaf of trajectories, indistinguishable by experiment, passing through similar but not identical regions of conformation space. The structure and energy of the transition state would also show some variation, depending on the reaction pathway. However, the traffic through different trajectories decreases very steeply as the energy maximum increases. The trajectory corresponding to the lowest activation energy will dominate the traffic. Therefore, the assumption of a unique energy of activation is, in general, a sound one.

How does the energy of activation control the reaction rate?

In a sample of reactants in the gaseous state at thermal equilibrium, the molecules have a distribution of energies, governed by the Maxwell–Boltzmann distribution. Only those few molecules with energy sufficient to surmount the barrier can react. Therefore:

• For a given temperature: *the higher the activation energy barrier, the slower the reaction.*

• For a given reaction: *the higher the temperature, the more molecules with high energies and the faster the reaction.* For many simple uncatalysed organic reactions, the rule of thumb is that the reaction rate doubles with each 10°C rise in temperature. This observation reflects the fact that activation energies of many reactions have roughly the same magnitude.

In some cases, for instance isomerizations or fragmentations, there is only one reactant. For these, a catalyst increases the reaction rate by lowering the activation energy (see Fig. 5.2(b)). J.B.S. Haldane and L. Pauling recognized that enzymes function as catalysts by stabilizing transition states. Enzymes bind transition states more tightly than they bind substrates. (If not, they would *increase*, not decrease, the activation energy, and the reaction rate would slow down.)

In cases involving two or more reactants, additional considerations come into play. For a reaction that requires a collision of reactants to form an activated complex with sufficient energy to get over the barrier, a high-energy collision is a necessary condition for reaction. The probability of a high-energy collision depends on the energy distribution in the reactant molecules; that is, on the temperature. But it is also necessary for the reactants to approach with a proper relative orientation. In addition to stabilizing the transition state, a catalyst may speed up the reaction by bringing reactants together preferentially in the proper orientation. An enzyme may have binding sites for two reactants, that enforce a proper juxtaposition in the bound state (see Fig. 5.3).

This can be thought of as an increase in the **effective concentrations** of reactants. An actual increase in reactant concentrations without changing the temperature would increase the total collision rate but not change the fraction of collisions that lead to reaction. A change in effective concentration, by binding the reactants to an enzyme, leads directly to a very large increase in the fraction of collisions that lead to reaction.

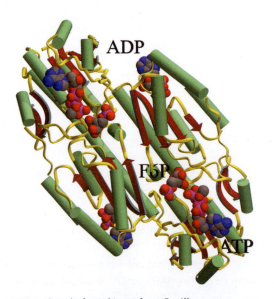

Figure 5.3 Phosphofructokinase from *Bacillus stearothermophilus*, showing substrate fructose-6-phosphate (F6P), cofactor ATP, and allosteric effector ADP. The enzyme is holding the substrate and cofactor in the correct relative geometry for effective phosphate transfer. The ADP acts to regulate the activity of the enzyme; note that it binds in a location distant from the active site itself. The complete molecule is a tetramer, of which only two subunits are shown. Homologous enzymes from mammals are octamers.

The activated complex

In the transition state, the atoms may form a relatively stable molecule called an **activated complex**, although they need not do so. But if such a complex exists and has a lifetime of sufficient duration to attain thermal equilibrium, it is legitimate to talk about its thermodynamic properties, such as its Gibbs free energy of formation. Indeed, the transition-state theory of reaction rates was originally derived on the assumptions that the activated complex is in *thermodynamic* equilibrium with its surroundings and in *chemical* equilibrium with the reactants.

The activated complex is postulated to be in thermal and chemical equilibrium with the reactants, but is also capable of decomposing into products.

$$A + B = AB^{\ddagger} \rightarrow \text{Products}.$$

If the activated complex is in chemical equilibrium with the reactants, we can define an equilibrium constant and a free energy for its formation:

$$A + B = AB^{\ddagger} \quad \Delta G^{\ominus\ddagger} = -RT \ln K^{\ddagger}$$

$$K^{\ddagger} = \frac{[AB^{\ddagger}]/c^{\ominus}}{([A]/c^{\ominus})([B]/c^{\ominus})} = \frac{[AB^{\ddagger}]c^{\ominus}}{[A][B]}.$$

The standard state concentration, $c^{\ominus}$, is conventionally taken as 1 mol L⁻¹ (L = litre). The quantity $[A]/c^{\ominus}$, is then the *ratio of the actual concentration of A to the standard concentration*, a *dimensionless* quantity corresponding to the numerical value of the concentration of [A] expressed in mol L⁻¹. It is necessary to keep $c^{\ominus}$, explicitly in the equilibrium constant to rectify the units.*

> The 'underground' symbol $\ominus$ indicates a property of a standard state. The superscript ‡ indicates a property of the activated complex or transition state.

- **Equilibrium constants must be dimensionless!!** If not, how could you take their logarithms?

---

*If we defined [A] as the *numerical value* of the concentration of A expressed in mol L⁻¹, everything would always be consistent, but it is common to take [A] to mean the concentration of A, often expressed in units of mol L⁻¹. It is common for people who are not physical chemists to write, for example, [Na+] = 1 mol L⁻¹.

We must leave an escape hatch in these assumptions of equilibrium, to allow the formation of products. At least one set of displacements must allow the complex to dissociate. Analysis of this motion, with a low barrier to dissociation, leads to a rate equation for formation of product:

$$-\frac{d[AB^{\ddagger}]}{dt} = \frac{k_B T}{h}[AB^{\ddagger}]$$

where h is Planck's constant and k_B is Boltzmann's constant (not a rate constant).

We can express the concentration of the activated complex in terms of the concentration of reactants and the equilibrium constant for its formation. The rate of formation of products is then:

$$-\frac{d[AB^{\ddagger}]}{dt} = \frac{k_B T}{h}[AB^{\ddagger}]$$
$$= \frac{k_B T}{h} e^{\Delta H^{\ominus\ddagger}/RT} e^{\Delta S^{\ominus\ddagger}/R}[A][B](c^{\ominus})^{-1}.$$

This result relates the rate constant and the energy of activation E_a of the Arrhenius equation to the thermodynamic properties of the activated complex, $\Delta G^{\ominus\ddagger} = \Delta H^{\ominus\ddagger} - T\Delta S^{\ominus\ddagger}$. The reduction in E_a depends on both the enthalpy and entropy of activation, $\Delta H^{\ominus\ddagger}$ and $\Delta S^{\ominus\ddagger}$.

For all enzymes, the enthalpy of activation is lower in the enzyme-catalysed pathway than in the uncatalysed reaction. For many enzymes, including those involving a single substrate, or for which a single substrate reacts with water, the rate enhancement is driven far more by enthalpic effects than by entropic ones.

Many two-substrate reactions show a different profile. In principle, forming a single bound species (ES) from two or more dynamically independent ones (E and S) should exact an entropic cost. However, in some cases dehydration of the ligands causes a compensatory increase in entropy as a result of the hydrophobic effect. For instance, for the phosphorylation of glucose by ATP catalysed by hexokinase, enthalpy and entropy make contributions of comparable magnitudes to the transition-state affinity for the Michaelis complex and the transition state. In contrast, the rate enhancement of the reduction of an imino acid by glutamate dehydrogenase:

$$\Delta^1\text{-pyrroline-2-carboxylic acid} + \text{NADPH} =$$
$$\text{DL-proline} + \text{NADP}^+$$

depends almost entirely on entropy.

- If the activated complex achieves thermal equilibrium (but of course not chemical equilibrium, or the reaction would not proceed to products) then we can interpret its thermodynamic properties. Some rate enhancements are enthalpy driven. Others are entropy driven.

Measurement of reaction rates

There is one inviolable rule: the reaction mixture must not change significantly during the course of acquisition of individual items of data. There are several ways to achieve this. If the reaction is sufficiently slow, there is no problem. For example, fermentation of wine over the course of several months allows easy regular measurements, such as pH or sugar content. The time required for even leisurely collection and analysis of a sample is short compared with the rate of reaction. In many laboratory experiments, it is possible to stop the reaction, perhaps as simply as by placing a test-tube in an ice bath, or precipitating the reaction mixture. Subsequent analysis can proceed at a stately pace.

What is more of a challenge is the measurement of fast transient events that occur on millisecond or microsecond timescales. Following such events often requires access to time intervals comparable to or shorter than typical mixing times. Development of suitable instrumentation has been an important achievement.

To measure fast components of enzymatic reactions, there are two possible approaches: (1) Slow down the reaction. (2) Use methods that capture information in short times. Both approaches have been pursued.

Slow the reaction down

There are several ways to slow down a fast reaction, to the point where it is possible to trap and study an intermediate. An example is a study of the cleavage of indole-3-glycerol by the α-subunit of tryptophan synthase, which combines mutagenesis with low-temperature X-ray crystallography.

The wild-type structure of tryptophan synthase from *Salmonella typhimurium* contains residues, Asp60 and Glu49, implicated in catalysis by the properties of site-directed mutants. It is not possible to solve the crystal structure of the enzyme–substrate complex at room temperature, because the substrate will undergo reaction. Nor did a crystal structure of tryptophan synthase with bound substrate analogue indole-3-propanol phosphate show a direct interaction of the Glu49 sidechain with the ligand.

The Asp60→Gln mutant is active at room temperature, but inactive at 95 K. S. Rhee, E.W. Miles and D.R. Davies froze crystals of the mutant tryptophan synthase Asp60→Gln to 95 K and diffused in the substrate, indole-3-glycerol. They solved the structure to 2.0 Å resolution. The structure showed the interaction of the Glu49 sidechain with the C3′ hydroxyl of the substrate. This supports the mechanism according to which the Glu49 side-chain carboxyl accepts a proton from the C3′ hydroxyl of the substrate, cleaving it to indole and glyceraldehyde-3-phosphate.

Fast methods of data collection

Flow methods permit the study of fast reactions. Figure 5.4 shows, schematically, an apparatus designed to inject reactants into a mixing chamber. Complete mixing can be achieved within about 10^{-3} s. The reacting solution then flows down a tube and arrives at an observation point 'downstream' after a fixed time interval.

A common variation is **stopped-flow** in which the filling of a syringe downstream of the mixing chamber both cuts off the flow and triggers data collection.

Observations of reactions at the microsecond level require still other specialized techniques. It is necessary to eliminate the mixing step. In **relaxation methods,** a sudden change in conditions perturbs a system from an equilibrium state. For example, in the temperature-jump method, the system receives a thermal shock: the discharge of a capacitor through a solution can raise the temperature by 10°C within a time interval of 10^{-6} s. There is no flow or mixing. The adjustment of the properties of the system to the new equilibrium values at the higher temperature can be followed spectrophotometrically.

Flash photolysis is another technique for following fast reactions. A pulse of light perturbs the system, either raising a fraction of the molecules in a mixture to excited states, or dissociating a ligand. Using a laser, the duration of the exciting pulse can be reduced to the picosecond (10^{-12} s) range. Subsequent changes are followed by fluorescence or absorption spectroscopy, or Raman spectroscopy.

Flash photolysis has been applied to study the dynamics of CO in the haem pocket of myoglobin. Starting with the CO bound to the iron, a laser pulse dissociates the ligand into the region of the protein adjacent to the binding site. There is subsequent competition between rebinding the ligand by the iron, and its complete escape into the surroundings. The rebinding kinetics are observable by Raman spectroscopy over a 10 ns to 10 ms timescale.

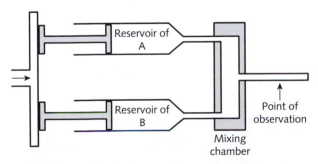

Figure 5.4 A continuous-flow apparatus for studying fast reactions. Enzyme E and substrate S are injected into a mixing chamber. The reaction mixture flows through the exit tube, past one or more downstream points of observation. By varying the distance between the mixing chamber and the point of observation, it is possible to follow the time course of the reaction.

- Techniques for measuring properties of fast reactions include:

Method	Timescale
Flow methods	~10^{-3} s
Relaxation methods	~10^{-6} s
Flash photolysis	~10^{-12} s

Active sites

Many enzymes bind substrates in crevices, often between domains. The pictures of phosphofructokinase (see Fig. 5.3), N–acetyl-L-glutamate kinase (see Fig. 2.22), and thrombin-hirudin (see Fig. 2.23—an inhibitor, not a substrate) are examples. These active sites both bind substrates and juxtapose specific catalytic residues with them.

In most cases the **active site** is a small portion of the protein, perhaps ~10%. Why then is the rest of the protein necessary?

Reasons include:

1. The rest of the protein is required in order to bring the active-site residues into their correct spatial relationship. The active-site residues are generally distant in the sequence, and it is the folding of the chain that brings them into proximity.

2. In many enzyme mechanisms, proteins must undergo conformational changes. The entire structure is needed to provide the levers and fulcra for the mechanical activity.

3. In some proteins active sites are in strained conformations. The rest of the structure must provide the energy to stabilize this. Coupling of relief of this strain to interaction with a substrate can enhance binding affinity and catalytic power. Typically, the enzyme becomes more rigid, thermostable, and protease resistant with substrate bound.

4. Conversely, in some cases the active site is flexible in the absence of ligand. The rest of the structure organizes the structural response to ligation.

- The catalytic activity of an enzyme is often focused on an active site, where substrate is bound, and juxtaposed with catalytic groups.

Cofactors

The natural amino acids have a range of chemical properties, but not enough for many biochemical reactions. Many metal ions and small organic molecules attach to enzymes or enzyme–substrate complexes and participate in catalysis. For example, NAD^+ and $NADP^+$ accept electrons during dehydrogenation reactions. Several metal ions undergo reversible oxidation and reduction, in electron-transport chains of respiration and photosynthesis, and in haemoglobin.

Classes of **cofactors** tend to specialize in different types of reactions, some of which are shown in Table 5.1.

Table 5.1 Typical biological roles of common cofactors

Type of cofactor	Example	Biological role
Redox	NAD^+, $NADP^+$ flavin adenine dinucleotide coenzyme Q	electron or hydrogen transfer
Group transfer	thiamine pyrophosphate coenzyme A pyridoxal phosphate S-adenosyl methionine biotin tetrahydrofolate UDP-glucose	aldehyde transfer acyl transfer amino group transfer methyl group transfer carboxyl group transfer methyl group transfer glucosyl group transfer

The reader will note that many cofactors are vitamins or related to vitamins.

Protein–ligand binding equilibria

Reversible binding of ligands to proteins involves equilibria of the form:

$$\text{Protein} + \text{Ligand} = \text{Protein·Ligand}$$
$$P + L = P·L$$

for a one-to-one complex, or:

$$\text{Protein} + n\,\text{Ligand} = \text{Protein·Ligand}_n$$

for the binding of n identical ligands to a single protein. These do not exhaust the possibilities. Many proteins bind two or more different ligands at the same time: enzymes binding a substrate and a cofactor provide many examples. A common index of the affinity of a complex is the *unitless dissociation constant*, K_D, the equilibrium constant for the *reverse* of the binding reaction:

$$\text{Protein·Ligand} = \text{Protein} + \text{Ligand} \quad K_D = \frac{[P][L]}{[PL]}$$

where [P], [L] and [PL] denote the numerical values of the concentrations of protein, ligand, and protein–ligand complex, respectively, when they are measured in mol L^{-1}. (If the concentration of P is 0.001 mol L^{-1}, [P] = 0.001.) The lower the K_D, the tighter the binding. K_D corresponds to the concentration of free ligand at which half the proteins bind ligand and half are free: [P] = [PL].

The Michaelis constant of an enzyme is the dissociation constant of the enzyme–substrate complex, assumed in the Michaelis–Menten model to be at equilibrium with respect to enzyme + substrate (see next section).

The K_D is related to the Gibbs free energy change of dissociation by the relationship:

$$PL = P + L, \quad \Delta G^{\ominus} = \Delta H^{\ominus} - T\Delta S^{\ominus} = -RT \ln K_D$$

Assuming no structural change on ligation, the entropy term will favour dissociation, because two objects will have greater conformational freedom if they are kinetically independent than if they are tethered. Therefore, to achieve a stable complex, the enthalpy term, and burial of hydrophobic surface, must provide attractive forces adequate to overcome the intrinsic entropic barrier to binding. Raising the temperature, which gives more importance to the entropy term, will promote dissociation.

To get a feel for the numbers, at 300 K, the purely kinetic entropy gain upon dissociation, $T\Delta S^{\ominus}$ is about 20 kJ mol^{-1}. This is equivalent, in terms of attractive interactions, to about a hydrogen bond, or burial of about 200 Å^2 of hydrophobic surface. A value of $\Delta G^{\ominus}$ of 50 kJ mol^{-1} for a dissociation reaction corresponds to a dissociation constant $K_D \sim 2 \times 10^{-9}$ at 300 K.

Dissociation constants of protein–ligand complexes span a very wide range (see Table 5.2).

Several databases collect data on the structures and thermodynamics of interaction of proteins with small ligands. A few examples, of many, are:

Relibase http://www.ccdc.cam.ac.uk/free_services/relibase_free/

http://www.bindingdb.org/bind/index.jsp

http://bioinf-tomcat.charite.de/superligands/

http://zhanglab.ccmb.med.umich.edu/BioLiP/

Protein–protein interaction databases are a separate topic.

The Scatchard plot

Many proteins have multiple binding sites for a ligand. In some cases the n binding sites are identical and non-interacting; in these cases binding is not cooperative. Antibodies are common examples. Let $\bar{v}$ denote the

Table 5.2 Examples of protein–ligand complexes. A *high* $\Delta G^{\ominus}$ of dissociation implies *tight* binding.

Biological context	Ligand	Typical K_D	$\Delta G^{\ominus}$ at 298 K (kJ mol^{-1})
Allosteric activator	Monovalent ion	10^{-4}–10^{-2}	11–23
Coenzyme binding	NAD$^+$, for instance	10^{-7}–10^{-4}	23–40
Antigen–antibody complexes	Various	10^{-4}–10^{-16}	23–91
Thrombin inhibitor	Hirudin	5×10^{-14}	76
Trypsin inhibitor	Bovine pancreatic trypsin inhibitor	10^{-14}	80
Streptavidin	Biotin	10^{-15}	85.6

average number of bound ligands per protein. Then $\bar{v}/n$ is the fraction of all binding sites occupied.

$$\frac{\bar{v}}{n} = \frac{[L]}{K_D + [L]}$$

in which K_D is the dissociation constant for binding of a single ligand to a single site on the protein, provided the dissociation constant is independent of the number of ligands bound:

$$K_D = \frac{[P][L]}{[PL]} = \frac{[PL][L]}{[PL_2]} = \frac{[PL_2][L]}{[PL_3]} = \cdots$$

For such system, a **Scatchard plot**, of $\dfrac{\bar{v}}{[L]}$ against $\bar{v}$, will be linear, with slope K_D and intercept n with the x-axis (see Fig. 5.5). From a Scatchard plot, it is possible to determine both the dissociation constant and the maximum number of binding sites.

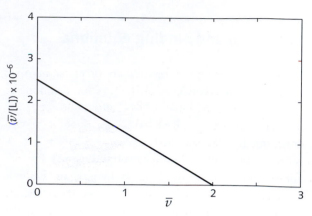

Figure 5.5 A Scatchard plot is a graph of protein–ligand binding data that is linear if the protein has n independent equivalent binding sites. The slope of the line is $-1/K_D$ and the intercept with the x-axis is the number of binding sites. (K_D is the dissociation constant for each individual binding site.)

A non-linear Scatchard plot suggests multiple binding sites with *different* dissociation constants.

Catalysis by enzymes

Enzymes provide examples of protein–ligand complexes. They bind substrates and cofactors *selectively* and in *specific geometric orientations*. In this way, they ensure that a substrate is juxtaposed with catalytic residues of the protein, or that two substrates approach each other in the correct orientation for a favourable reaction. If the same two molecules, free in solution, were to collide in random orientation, the probability of reaction would be much lower.

Some enzyme-catalysed reactions follow the same pathway as the uncatalysed reactions, but with lower activation barriers. Other enzymes substitute different reaction pathways, with intermediates very different from those of the uncatalysed reaction.

To understand rate enhancement by activation-barrier lowering, compare the affinities of the initial enzyme–substrate complex and the enzyme–transition-state complex:

S = substrate, S‡ = transition state, E = enzyme, ES = enzyme–substrate complex, ES‡ = enzyme–transition-state complex, $\Delta G^‡$ = activation energy, $\Delta\Delta G^‡$ = change in activation energy

---

Free energy of activation
  in presence of enzyme $= G(ES^‡) - G(ES)$
Free energy of activation
  in absence of enzyme $= G(S^‡) - G(S)$
Subtracting:
$$
\begin{aligned}
\Delta\Delta G^‡ &= [G(ES^‡) - G(S^‡)] - [G(ES) - G(S)] \\
&= [G(ES^‡) - G(S^‡) - G(E)] \\
&\quad -[G(ES) - G(S) - G(E)] \\
&= \text{binding affinity of transition state S}^‡ \\
&\quad -\text{binding affinity of substrate S}
\end{aligned}
$$

The rate enhancement is directly related to the lowering of the activation energy, $\Delta\Delta G^‡$. The effect of the enzyme on $\Delta G^‡$ is the *difference* between the affinity of the enzyme for the transition state S‡ and for the substrate S. (Here, $\Delta G = G(ES) - G(S) - G(E)$ is the Gibbs free energy change of the *association* reaction E + S = ES.) An efficient enzyme will bind its substrate adequately to get the process started, but bind the transition state more tightly. Some enzymes are rigid, and have better complementarity to the transition state than to the substrate. Others undergo conformational changes upon binding substrate, from a form adapted to bind the substrate to one adapted to bind the transition state. This is known as 'induced fit' (see Chapter 1).

Enzyme kinetics

Kinetics is the measurement of reaction rates and their dependence on conditions, including concentrations of reactants, products, and catalysts. Classically, the measurement of reaction velocity as a function of substrate concentration [S] involved mixing enzyme and substrate, and following the reaction by measuring the disappearance of substrate or the appearance of product. For instance, the fact that NADH but not NAD$^+$ has an absorption maximum at 340 nm made it convenient to follow dehydrogenation reactions of the form:

malate + NAD$^+$ = oxaloacetate + NADH + H$^+$

by running the reaction in a spectrophotometer and taking absorbance readings at 340 nm at regular time intervals.

A simple model accounting for the kinetics of an enzyme-catalysed reaction, by V. Henri, and by L. Michaelis and M. Menten involves enzyme and substrate interacting to form an enzyme–substrate complex. The complex breaks down irreversibly to release product and restore the original free enzyme:

$$E + S \underset{k_{-1}}{\overset{k_1}{\rightleftharpoons}} ES \overset{k_2}{\longrightarrow} P$$

where E = enzyme, S = substrate, ES = enzyme–substrate complex, P = product, and k_1, k_{-1} and k_2 are rate constants for the individual reaction steps. [S] = the concentration of substrate, [E] = the concentration of enzyme, and [ES] = the concentration of enzyme–substrate complex. More precisely, [E] is the concentration of active sites; each enzyme molecule may contain more than one active site. (How would you determine this?)

An important contribution of Michaelis and Menten was to emphasize the importance of determining the *initial* rate, v_0, of the reaction, in the absence of product. In practice, this requires following the early time course of the reaction and extrapolating back to the moment of mixing enzyme and substrate. (There is also a transient stage as the enzyme and substrate mix and interact, before establishment of equilibrium. This stage lasts only milliseconds. It is observable only with special techniques (see 'Measurement of reaction rates'), and does not affect extrapolated inferences of initial rates.) Under these circumstances, there is no back-reaction of product, if only because there is no product there to back-react. This assumption also avoids certain potentially complicating factors, such as product inhibition or enzyme degradation—which were probably not even recognized in 1913.

Michaelis and Menten further assumed that the forward and reverse rates of the first step are faster than the formation of product; that is: $k_1 \gg k_2$, and $k_{-1} \gg k_2$. The picture is that ES is at equilibrium with E + S, with product P 'bleeding off' slowly.

Michaelis and Menten derived from this model the rate equation relating the initial velocity v_0 to the substrate concentration [S]:

$$v_0 = \frac{V_{max}[S]}{K_M + [S]}.$$

Figure 5.6 shows the general shape of this curve. For low values of [S], v_0 and [S] are proportional. In this region, the enzyme is accommodating all substrate molecules equally well. The rate-limiting step is the encounter of substrate and free enzyme. As [S] increases, v_0 as a function of [S] rises less and less steeply. The substrate concentration at which $v_0 = \frac{1}{2}V_{max}$ is equal to K_M. With further increase in [S], v_0 approaches a limiting value V_{max}. This is attributable to saturation of the enzyme. Virtually all the enzyme is in the form of enzyme–substrate complex, ES (apply Le Chatelier's principle to the E + S $\rightleftharpoons$ ES equilibrium). The enzyme is working flat out to produce product. The rate of appearance of product is V_{max}, *independent* of substrate concentration. The observed rate at the plateau corresponds to the rate of some step of the reaction that occurs *after* binding of the substrate.

- L. Michaelis and M. Menten derived an equation relating the initial velocity v_0 of an enzymatic reaction to the substrate concentration [S], in the presence of a constant amount of enzyme:

$$v_0 = \frac{V_{max}[S]}{K_M + [S]}.$$

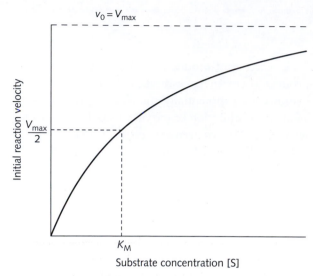

Figure 5.6 Typical dependence of initial reaction velocity of an enzyme-catalysed reaction on substrate concentration in presence of a fixed amount of enzyme. The graph is linear at low substrate concentrations and approaches a maximum value at high substrate concentrations. These curves depend on two parameters, the maximum velocity, V_{max} and the substrate concentration corresponding to half the maximum velocity. In the Michaelis–Menten model the substrate concentration corresponding to the rate $\frac{1}{2}V_{max}$ is the Michaelis constant, K_M, interpreted as the dissociation constant of the enzyme–substrate complex.

The values of V_{max} and K_M characterize the enzyme, the substrate, and the conditions of reaction, such as temperature, pH, and ionic strength.

Curves of this shape are quite common, arising from many phenomena that exhibit saturation or, in general, follow a 'law of diminishing returns'. I. Langmuir derived a version to describe the absorption of molecules on a surface. Such a graph could also describe the grade you will receive in your protein science course, as a function of the number of hours you study (see Box 3.11).

The Michaelis constant, K_M, is also the dissociation constant of the ES complex:

$$ES \rightleftharpoons E + S, \quad K_M = \frac{[E][S]}{[ES]}.$$

After the transient phase, the concentration of ES is approximately constant. G.E. Briggs and J.B.S. Haldane recognized that the 'steady-state' assumption, that [ES] is constant, equivalent to:

$$\frac{d[ES]}{dt} = k_1[S] - k_{-1}[ES] - k_2[ES] = 0$$

permits an alternative derivation of an equation in the same *form* as the **Michaelis–Menten equation**, but with a different interpretation of the constant K_M (see Problem 5.4).

$$\text{Michaelis–Menten} \qquad \text{Briggs–Haldane}$$

$$K_M = \frac{k_{-1}}{k_1} \qquad\qquad K_M = \frac{k_2 + k_{-1}}{k_1}$$

Applying the Michaelis–Menten assumption $k_2 \ll k_{-1}$ reduces the Briggs–Haldane formula for K_M to the Michaelis–Menten formula for K_M.

Derivation of K_M and V_{max} from rate data

Given a set of data recording v_0 as a function of [S] for some enzyme–substrate combination, it is possible to use curve-fitting software to derive K_M and V_{max}. Before such software was available, people transformed the equation to linearize it. Then, plotting the points on a graph and fitting a straight line allowed the two parameters to be derived from the slope and intercept of the line. The best known of these transformations are those of Lineweaver and Burk, and of Eadie and Hofstee. Here is an example:

Determination of K_M and V_{max} for ribulose-1, 5-bisphosphate carboxylase/oxygenase from soybean leaves

The enzyme ribulose 1,5-bisphosphate carboxylase/oxygenase catalyses the initial step in the fixation of CO_2 in the Calvin cycle of green plant photosynthesis.

G. Bowes and W. L. Ogren measured the initial steady-state rate of CO_2 uptake as a function of substrate concentration for an enzyme isolated from soybean leaves (see Table 5.3). Let us determine the Michaelis constant K_M and the saturation velocity V_{max} for soybean ribulose-1,5-bisphosphate carboxylase/oxygenase under the conditions of these experiments.

The Michaelis–Menten equation implies that a graph of $v_0/[S]$ against v_0 will be linear. Such a graph is an Eadie–Hofstee plot. The line has slope $-1/K_M$ and y-intercept V_{max}/K_M.

Figure 5.7 shows an Eadie–Hofstee plot of the kinetics of CO_2 uptake. For this preparation under these conditions, K_M for $CO_2 = 1.3 \times 10^{-4} (\text{mol L}^{-1})$ and $V_{max} = 26.6$ nmol CO_2 incorporated per second per milligram of protein.

Table 5.3 Initial steady-state rates of CO_2 uptake catalysed by soybean ribulose-1,5-bisphosphate carboxylase/oxygenase as a function of CO_2 concentration, at pH 8.0, 298.15 K, in the absence of O_2

CO_2/mmol L^{-1}	Rate of CO_2 uptake/nmol s^{-1}
0.046	6.90
0.093	10.81
0.139	13.66
0.232	17.54
0.463	20.19
0.926	23.39
2.315	25.11

[a] Rates are expressed per milligram of protein.

[b] G. Bowes and W.L. Ogren (1972), Oxygen inhibition and other properties of soybean ribulose-1,5-diphosphate carboxylase, *J. Biol Chem.*, **247**, 2171–2176.

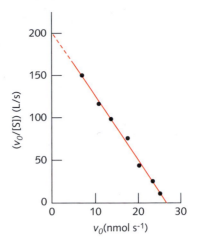

Figure 5.7 Eadie–Hofstee plot of the kinetics of incorporation of CO_2 into phosphoglycerate, catalysed by ribulose-1,5-bisphosphate carboxylase/oxygenase under the reaction conditions specified in Table 5.3.

Measures of effectiveness of enzymes

At high substrate concentrations, the velocities v_0 and V_{max} are proportional to the amount of enzyme present. How can we characterize an enzyme and a set of reaction conditions, independent of the amount of enzyme?

The **turnover number** of an enzyme, k_{cat}, is the ratio $V_{max}/[E]_{total}$, where $[E]_{total}$ is the total concentration of enzyme. The turnover number is a measure of 'throughput' on a molecular basis: It represents the number of substrate molecules converted to product, *per enzyme molecule* under conditions of saturation; that is, at high substrate concentrations.

More usually, enzymes operate at low substrate concentrations. If $[S] << K_M$, $v_0 = \left(\dfrac{k_{cat}}{K_M}\right)[E][S]$. $\left(\dfrac{k_{cat}}{K_M}\right)$ gives a measure of the catalytic efficiency of enzymes at low substrate concentrations. Two factors may contribute to increasing the value of $\left(\dfrac{k_{cat}}{K_M}\right)$:

(1) a low K_M implies a high affinity of substrate and enzyme, and (2) a high k_{cat} implies that the enzyme–substrate complexes formed will turn over rapidly to product.

Different enzymes show a large range of k_{cat}/K_M values. However, no matter how efficient the catalytic mechanism itself, the rate of a reaction is limited by the rate of encounters of enzyme and substrate, which depends on the diffusion rate. If every encounter results in reaction; that is, the catalysis is **diffusion limited**, k_{cat}/K_M would be $\sim 10^8–10^9$ $(mol/L)^{-1}$ s^{-1}, under typical conditions. Some enzymes achieve this. They are called 'perfect' enzymes, because no further evolution of enhanced catalytic efficiency in the enzyme itself could improve turnover rate.

Some enzymes that approach diffusion-limited rates are:

Enzyme	Substrate	$k_{cat}s^{-1}$	$K_M(mol\ L^{-1})$	k_{cat}/K_M $(mol/L)^{-1}$ s^{-1}
acetylcholinesterase	acetylcholine	1.4×10^4	9×10^{-5}	1.6×10^8
catalase	H_2O_2	4×10^7	1	4×10^7
fumarase	fumarate	8×10^2	5×10^{-6}	1.6×10^8

> • The turnover number k_{cat}/K_M is a measure of the efficiency of enzymatic catalysis. Some enzymes are such effective catalysts that the reaction rate is limited by the diffusion rate.

Inhibitors

Inhibitors are molecules that reduce the activity of enzymes. Some inhibitors act by binding reversibly to the active site, but are unable to complete the reaction. Because these bind to the same site as the substrate, so that binding of substrate and inhibitor are mutually exclusive, they are called **competitive inhibitors**. Because competitive inhibitors reduce the concentration of available enzyme, less enzyme–substrate complex is formed for given concentrations of enzyme and substrate than in the uninhibited case. Therefore, the apparent K_M increases. However, a sufficiently high substrate concentration can achieve the same V_{max} as the uninhibited enzyme. Flooding the active site with substrate 'washes out' the inhibitor.

Non-competitive inhibitors bind to other sites on the enzyme, leaving the normal active site available for substrate binding. The inhibitor must interfere with catalytic activity in some way other than competing for substrate binding. Very high substrate concentrations will not recover the uninhibited value of V_{max} as in competitive inhibition. However, if the active site is unaffected, the K_M may be unchanged. Alternatively, if a non-competitive inhibitor binds to an allosteric site on the enzyme, altering the affinity for substrate despite not occupying the active site itself, the K_M may increase or decrease.

Because enzymes bind transition states more tightly than they bind substrates (see Fig. 5.2), many competitive inhibitors resemble transition states. In designing a drug against an enzyme, medicinal chemists will look for something that resembles a transition state. In nature, inhibitors are common weapons in the biochemical battles between species. In our attempts to control pathogens, many drugs are based on adaptations of natural inhibitors of receptors or enzymes; others are the products of chemists' ingenuity. Several drugs used in the treatment of AIDS are inhibitors of the HIV-1 proteinase (see Fig. 2.24).

Aptamers are a relatively new class of molecules that offer great versatility to bind molecules with high affinity and specificity. Aptamers are oligonucleotides—typically 15–40 nucleotides in length. They fold into tertiary structures containing binding sites. It is possible to create an aptamer against a specific target by successive rounds of selective screening. Some aptamers provide effective and selective enzyme inhibitors.

Irreversible inhibitor binding

Many inhibitors bind reversibly to enzymes. The fact that in the presence of a competitive inhibitor V_{max} is recoverable by adding large amounts of substrate implies that the inhibitor can dissociate from the enzyme.

Other inhibitors bind irreversibly to enzymes:

• The mechanism of poisoning by ions of many heavy metals is the irreversible reaction of the metal with sulphydryl groups of proteins. This can inactivate specific catalytic residues. Enzymes in the haem biosynthetic pathway are particularly sensitive to lead. Anaemia is therefore a symptom of even low levels of lead poisoning. Acute exposure to high levels of lead, or chronic lead poisoning, causes many other symptoms.

• Organophosphate insecticides react irreversibly with the catalytic serine of acetylcholinesterase. This serine is present in vertebrate acetylcholinesterases as well as those of insects. Therefore these insecticides are dangerous to humans. Identification of a cysteine residue unique to insect acetylcholinesterases (see Fig. 5.8) permitted design of inhibitors that bind to the active site and form an irreversible link to the sulphydryl group of the cysteine. These compounds are toxic to insects but not to humans.

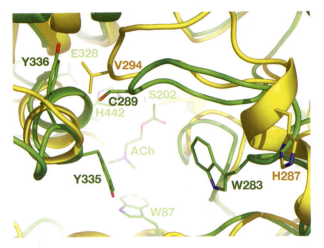

Figure 5.8 Superposition of structures of active sites of modelled acetylcholinesterases from human (yellow) and greenbug *Schizaphis graminum* (green). The substrate appears at the centre of the picture, and the C289 and V294 residues show the target for selective inhibition.

[From: Pang, Y.P., Singh, S.K., Gao, Y., Lassiter, T.L., Mishra, R.K., Zhu, K.Y. and Brimijoin, S. (2009). Selective and irreversible inhibitors of aphid acetylcholinesterases: steps towards human-safe insecticides. *PLoS ONE*, **4**, e4349. Reproduced via CC BY 4.0 License]

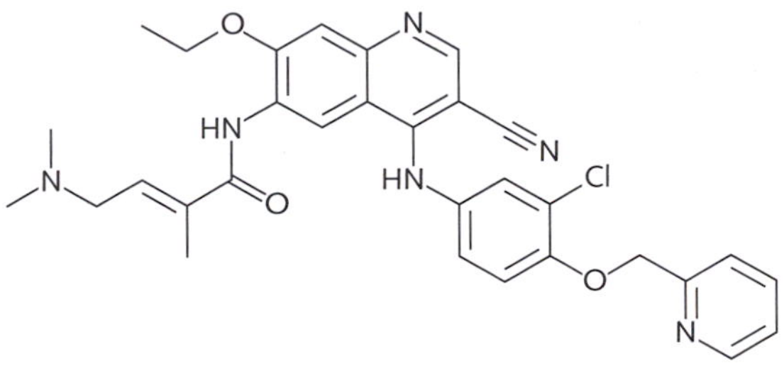

- A drug now in trials for cancer therapy, HKI-272 (see Fig. 5.9), irreversibly inhibits the epidermal growth factor (EGF) receptor by covalent linkage to the receptor binding site. This compound has recently shown positive results in Phase III clinical trials.

Figure 5.9 Structure of anticancer drug, HKI-272 (neratinib), now in clinical trials.

[From: Wissner, A. and Mansour, T.S. (2008). The development of HKI-272 and related compounds for the treatment of cancer. *Archiv der Pharmazie*, **341**, 465–477.]

- The family of serine protease inhibitors called ser-pins show an unusual mechanism of inhibition, discussed later in this chapter.

> An inhibitor is a molecule that slows down an enzymatic reaction. Types include:
> - competitive inhibitors;
> - non-competitive inhibitors, including inhibitors that bind irreversibly.
>
> Inhibitors are useful in the control of the activities in metabolic networks and provide many important drugs.

Multisubstrate reactions

The Michaelis–Menten model relates kinetics and mechanism for the simplest enzymatic reactions—those involving only a single substrate. Most biochemical reactions involve two or more substrates, or a substrate and a cofactor. The reaction catalysed by phosphofructokinase (see Fig. 5.3):

$$\text{fructose-6-phosphate} + \text{ATP} \rightleftharpoons$$

$$\text{fructose-1,6-bisphosphate} + \text{ADP}$$

is a typical *transfer* reaction. Such reactions, involving two substrates and two products, are called **bi-bi** (bi = bimolecular) reactions. (ADP, the **allosteric effector** of phosphofructokinase, does not participate in the reaction itself.)

The mechanistic description of multisubstrate reactions is necessarily more complex than for single-substrate reactions. Several types of scenarios are possible (see Fig. 5.10):

- Does the reaction pass through an intermediate state in which both substrates are simultaneously attached? If so, can the two substrates be bound in either order, or is there a required sequence? Analogous questions apply to product release; is there a preferred sequence or do the products follow Lady Macbeth's command to: 'Stay not upon the order of your going, but go at once.'

- Alternatively, it may be more accurate to describe the process as a sequence of separate reactions, with

The mechanism of action of thymidylate synthase

Investigations of thymidylate synthase provide a good illustration of the complementarity between solution studies to determine the dynamics of the reaction—the sequence of events, the identities of the intermediates, and the rates of individual steps—and crystallography, to provide structural details.

Because thymine nucleotides are essential for cell proliferation, thymidylate synthase is a target for anticancer agents, such as 5-fluorouracil, a substrate analogue; and other compounds, that are related to the cofactor dihydrofolate.

Thymidylate synthetase converts deoxyuridine monophosphate (dUMP) to deoxythymidine monophosphate (dTMP), essential for DNA synthesis and repair (see Fig. 5.11). The cofactor N_5, N_{10}-methylene-5,6,7,8-tetrahydrofolate (CH_2THF) is converted to dihydrofolate (see Fig. 5.12). The cofactor plays roles at two steps of the reaction pathway. First, it donates a one-carbon fragment, to form a 5-methylene derivative of the pyrimidine ring, with conversion of the cofactor to tetrahydrofolate. Subsequently, the tetrahydrofolate reduces the double bond to form the product, deoxythymidine monophosphate, and is itself oxidized to dihydrofolate.

The reaction proceeds in a multistep process. To create the reactive conformation, dUMP binds first, and then CH_2THF. Because *two* reactant molecules are converted to *two* product molecules, and dUMP must bind *before* CH_2THF binds, the reaction is classified as *ordered-sequential bi-bi* (bi = bimolecular). After both ligands arrive, the enzyme undergoes a conformational change to close the active site over the substrate and cofactor (see Fig. 5.13).

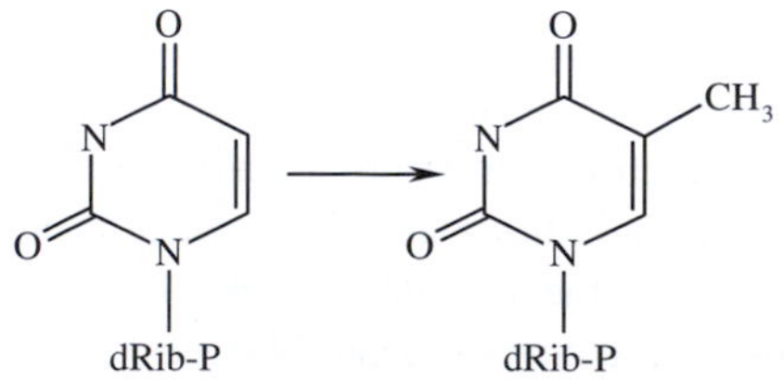

(a)

(b)

(c)

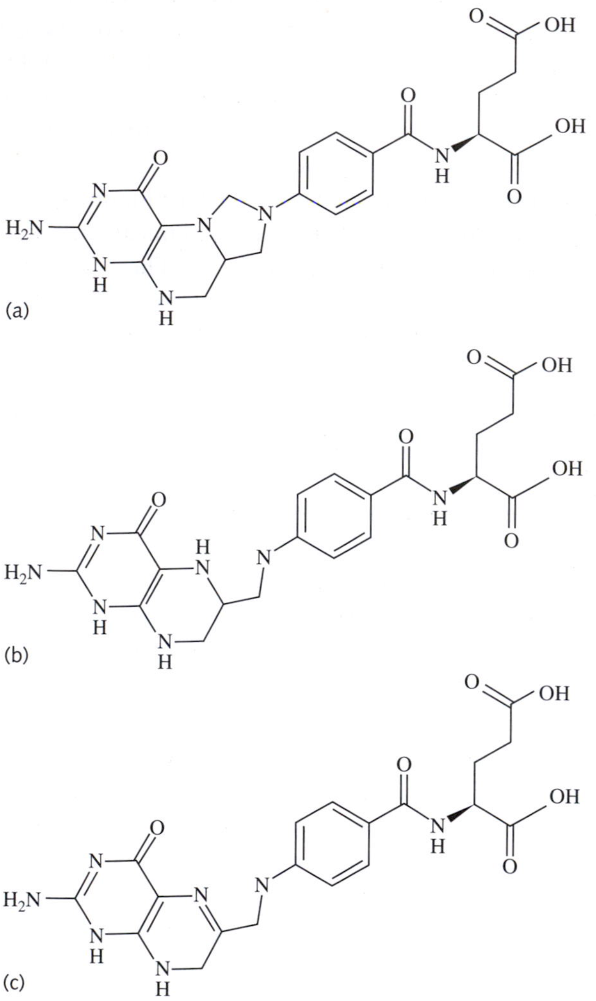

Figure 5.12 Cofactors in steps of the thymidylate synthase reaction. (a) N_5,N_{10}-methylene-5,6,7,8-tetrahydrofolate (CH_2THF). (b) 5,6,7,8-tetrahydrofolate (THF) (intermediate). (c) dihydrofolate (DHF).

Figure 5.11 Thymidylate synthase catalyses the conversion of 2′-deoxyuridine-5′-phosphate to 2′-deoxythymidine-5′-phosphate. dRib-P = 2′-deoxyribose-5′-phosphate.

The following steps ensue: (see Fig. 5.14).

- Nucleophilic attack of the sulphydryl group of Cys198 on C6 of the uracil ring of the dUMP, to form a covalent adduct (see Fig. 5.14(b)).

- Breaking of the methylene-N10 bond of CH_2THF to form a 5-iminium cation, N5-CH$^+$THF, and addition of the iminium ion to C5 of the uracil ring. The substrate now has two covalent attachments: to the enzyme at C6 and to the cofactor at C5. The pyrimidine ring is completely saturated, with two carbonyl substituents (see Fig. 5.14(c)).

- Tautomerization to form an enol (see Fig. 5.14(d)).

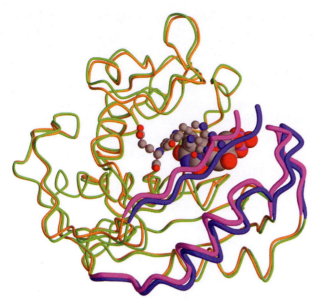

Figure 5.13 Comparison of open and closed forms of thymidylate synthase from *E. coli*. The major changes in mainchain conformation are a shift in an extended region at the C-terminus, in a helix packed against it, and in a loop near the N-terminus. Open form [2VET] in green; closed form [1TSN] in orange, with regions of changed conformation in blue ([2VET]) and magenta ([1TSN]). The cofactor methylenetetrahydrofolate and the inhibitor FdUMP shown in all-atom representation. Small balls represent methylenetetrahydrofolate; large balls represent FdUMP.

- Release of the cofactor in the form of tetrahydro-folate (THF), to leave 5-methylenedeoxyuridylate, containing a *double* bond to a methylene group at the 5-position (see Fig. 5.14(e)).

- Reduction of the double bond by the THF, followed by breakage of the link to Cys198, to form the product deoxythymidine monophosphate (see Fig. 5.14(f)).

This mechanism of action of thymidylate synthase, and the rates of individual steps, were first worked out 'in solution', and subsequently illuminated by crystallography:

- Binding studies showed that thymidylate synthase follows an ordered-sequential **bi-bi** mechanism. In the absence of cofactor CH_2THF, thymidylate synthase binds substrate dUMP (or product dTMP, or substrate analogues such as FdUMP = 5-fluoro-dUMP).

However, in the absence of nucleotides, binding of the cofactor is undetectable. The equilibrium constant for dissociation, K_D, and the rate constants k_{on} and k_{off} for association and dissociation, are measurable from the dependence on substrate concentration of the **quenching** of the fluorescence of

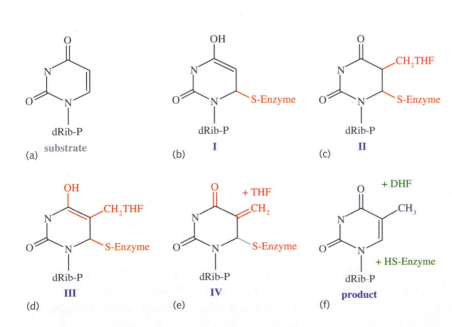

(a) **substrate**   (b) **I**   (c) **II**

(d) **III**   (e) **IV**   (f) **product**

Figure 5.14 Steps of the thymidylate synthase reaction. Red indicates the loci of changes at each step. (a) Substrate, deoxyuridine monophosphate, dUMP. dRib-P = 2′–deoxyribose-5′-phosphate. (b) Binding of enzyme to position 6 of the uracil ring through sidechain of Cys198. (c) Binding of cofactor N_5,N_{10}-methylene-5,6,7,8-tetrahydrofolate (CH_2THF) to position 5. (d) Tautomerization to form an enol. (e) Addition of methylene group from cofactor to position 5 of uracil ring, breaking the covalent attachment of cofactor, now in the form of tetrahydrofolate (THF). (f) Reduction of methylene double bond by THF and release of covalent attachment to Cys198, to form product, deoxythymidine monophosphate, and oxidized cofactor in the form of dihydrofolate (DHF).

one or more tryptophans in the enzyme. The **'on' and 'off' rates** (that is, the kinetics of binding and dissociation) are obtainable from the rate of fluorescence quenching after mixing of enzyme and dUMP, measurable by stopped-flow spectroscopy.

- Equilibrium and rate constants for binding of cofactor N_5, N_{10}-methylene-5,6,7,8-tetrahydrofolate (CH_2THF) to the enzyme-dUMP complex are measurable from changes in the absorbance of the cofactor as a result of the change in its environment. Measurement of the 'on' and 'off' rates requires use of a modified cofactor, 10-propargyl-5,8-dideazafolate.

Once the substrate and cofactor are in place, the reaction is ready to go.

- Evidence that Cys198 (residue numbering in the *Lactobacillus casei* sequence, corresponding to Cys146 in the *E. coli* sequence) is the catalytic nucleophile that attacks dUMP was provided by the studies of the inhibitor FdUMP, which undergoes the first steps of the reaction, but stops before elimination of THF. Proteolytic digestion of the covalent complex between the enzyme, FdUMP, and CH_2THF (analogous to intermediate (c) in Fig. 5.14) yielded a peptide covalently attached to FdUMP and CH_2THF through Cys198. Subsequently, the inactivity of a Cys198→Ala mutant confirmed the implication of the Sγ of Cys198 cysteine in attack on dUMP.

- A transient effect on the absorbance at 340 nm is interpretable as a change in the enzyme conformation, confirmed by studies of mutants (and subsequently by crystal structures).

- The details of the interaction of the substrate and cofactor were revealed by isotope effects on the reaction rate of cofactors selectively deuterated at different positions.

- The rate of proton abstraction from C5 of dUMP is measurable from rates of tritium release from labelled substrate (see Fig. 5.14, steps (c) → (d)).

- Hydride transfer to the C5-methylene intermediate can be measured from monitoring formation of a side product 5-(2-hydroxyethyl)thiomethyl-dUMP. The rate constant of formation of the final product dTMP can be calculated from the partitioning of this step between dTMP and side product (see Fig. 5.14, steps (e) → (f)).

These experiments revealed the chemistry of the reaction, as shown in Fig. 5.14. Crystal structures of the enzyme supplied many structural details. (There are currently 276 thymidylate synthase structures in the wwPDB. These comprise enzymes from 24 species, including mutants, and various complexes including known or potential anticancer drugs.) Figure 5.15 shows schematically the residues around the active site that are responsible for binding and catalysis.

Finer-Moore, Santi, and Stroud identify the five 'most essential' residues for catalysis, primarily on the basis of conservation among natural homologues and from loss of activity of site-directed mutants. Kinetic analyses of site-directed mutants together with crystal structures of mutants in complexes with substrates or substrate analogues demonstrated the roles of these residues in binding and catalysis. Fig. 5.15 shows these residues and their interactions with ligands.

- Cys198 forms a covalent link to C6 of the substrate dUMP.

- Arg218 binds to the 5'–phosphate of the substrate. Other interactions of this residue help maintain the conformations of loops forming part of the active site.

- Tyr146 is essential in removal of the proton from C5 of the substrate dUMP during the step from (c) to (d) in Fig. 5.14.

- The sidechain of Asp221 binds to the pterin ring of the cofactor, required for proper orientation of the cofactor. It also donates a proton during the step from (d) to (e) in Fig 5.14.

- Glu60 forms hydrogen bonds, through bridging water molecules, to O4 of the substrate dUMP. It participates in various proton-shuttling steps.

Computational approaches to enzyme mechanisms

What has now become a standard approach to the determination of a detailed enzyme mechanism is to solve high-resolution structures of the enzyme (1) without the substrate, and (2) in complex with the substrate or an unreactive substrate analogue. These structures reveal which residues are responsible for

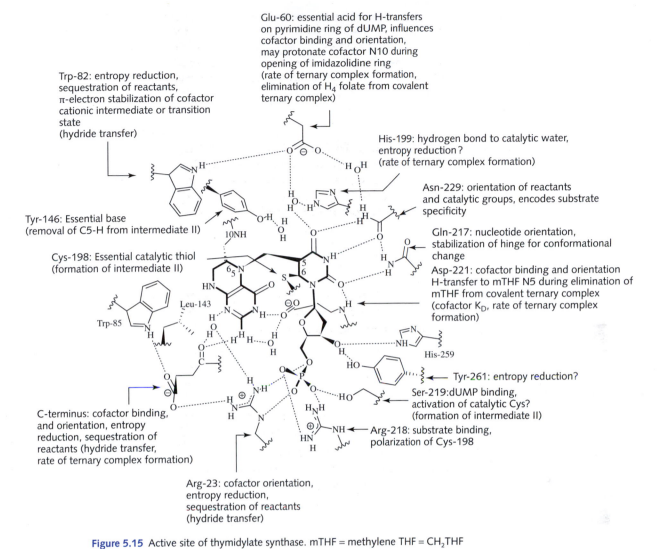

Glu-60: essential acid for H-transfers on pyrimidine ring of dUMP, influences cofactor binding and orientation, may protonate cofactor N10 during opening of imidazolidine ring (rate of ternary complex formation, elimination of H_4 folate from covalent ternary complex)

Trp-82: entropy reduction, sequestration of reactants, π-electron stabilization of cofactor cationic intermediate or transition state (hydride transfer)

His-199: hydrogen bond to catalytic water, entropy reduction? (rate of ternary complex formation)

Asn-229: orientation of reactants and catalytic groups, encodes substrate specificity

Tyr-146: Essential base (removal of C5-H from intermediate II)

Gln-217: nucleotide orientation, stabilization of hinge for conformational change

Cys-198: Essential catalytic thiol (formation of intermediate II)

Asp-221: cofactor binding and orientation H-transfer to mTHF N5 during elimination of mTHF from covalent ternary complex (cofactor K_D, rate of ternary complex formation)

Leu-143

Trp-85

His-259

Tyr-261: entropy reduction?

Ser-219:dUMP binding, activation of catalytic Cys? (formation of intermediate II)

C-terminus: cofactor binding, and orientation, entropy reduction, sequestration of reactants (hydride transfer, rate of ternary complex formation)

Arg-218: substrate binding, polarization of Cys-198

Arg-23: cofactor orientation, entropy reduction, sequestration of reactants (hydride transfer)

Figure 5.15 Active site of thymidylate synthase. mTHF = methylene THF = CH_2THF

[From: Finer-Moore, J.S., Santi, D.V. and Stroud R.M. (2003). Lessons and conclusions from dissecting the mechanism of a bisubstrate enzyme: thymidylate synthase mutagenesis, function, and structure. *Biochemistry* **42**, 248–256.]

substrate binding and catalysis, and the ligand-induced conformational changes in the enzyme, or the 'induced fit.'

Of course, structure determinations can provide only snapshots of several selected points along the reaction pathway. Given the power of molecular dynamics, is it possible to work out a complete mechanistic pathway computationally, given the structures of substrate and enzyme? We're getting there.

- In a detailed study of triosephosphate isomerase (see Fig. 5.16), V. Guallar, M. Jacobson, A. McDermott, and R.A. Friesner started with a very high-resolution low-temperature crystal structure of the enzyme–substrate (dihydroxyacetone phosphate)

complex. To key intermediates along the catalytic pathway, they applied a combination of (1) *quantum mechanics*, applied to the active site itself; (2) *molecular mechanics*, a classical treatment of the rest of the protein; and (3) *protein structure*

$$
\begin{array}{ccc}
CH_2OH & & H\!-\!C\!=\!O \\
| & & | \\
C\!=\!O & \rightleftharpoons & H\!-\!C\!-\!OH \\
| & & | \\
CH_2OPO_3^{2-} & & CH_2OPO_3^{2-} \\
\text{dihydroxyacetone} & & \text{glyceraldehyde} \\
\text{phosphate} & & \text{3-phosphate}
\end{array}
$$

Figure 5.16 The reaction catalysed by triosephosphate isomerase: dihydroxyacetone phosphate $\rightleftharpoons$ D-glyceraldehyde 3-phosphate.

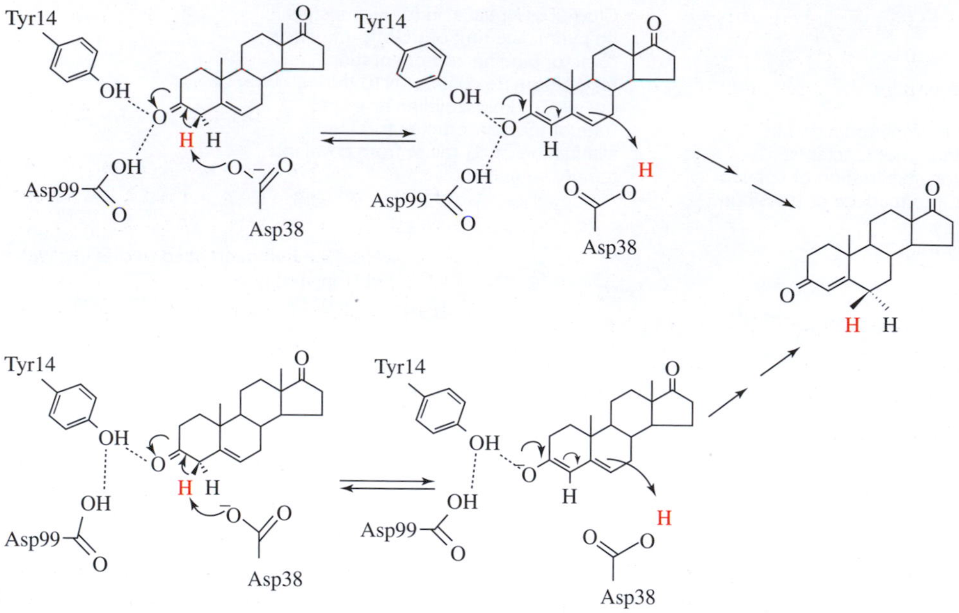

Figure 5.17 Ketosteroid isomerase catalyses the isomerization of selected steroids, with a carbonyl at position 3, in the A ring, and a double bond at positions 5–6 in the B ring. The intermediate is a dienolate stabilized by hydrogen bonding of the O⁻ at position 3 of the steroid to the sidechain hydroxyl of Tyr 14. The carboxyl group of Asp38 is the temporary 'parking place' for a proton. Upper and lower portions of this figure shows alternative hydrogen-bonding interactions between Asp99, Tyr14, and the substrate. Molecular dynamics suggested that the upper scheme is correct.

[From Mazumder, D., Kahn, K. and Bruice, T.C. (2003). *J. Amer. Chem. Soc.*, **126**, 7553–7561.]

prediction methods to analyse conformational changes in the protein. A major conformational change in triosephosphate isomerase is the conversion upon substrate binding from an open to a closed form. This involves the reconformation of a loop to cover the bound substrate.

The results account for a wide range of experimental data. These include the conformational changes in the protein, both the major loop movement and changes in sidechain positions. Values of the energies of the intermediates have estimated accuracies

of within ~8.4–12.6 kJ mol⁻¹. It was possible to discriminate between alternative microscopic pathways for proton transfer.

- In a study of ketosteroid isomerase, D. Mazumder, K. Kahn, and T.C. Bruice carried out a 1.5-ns molecular-dynamics run on an enzyme–substrate complex in water. It was possible to distinguish between alternative mechanisms in favour of a cooperative hydrogen-bonding scheme in which the sidechains of both Tyr14 and Asp99 interact with the O3 of the steroid (see Fig. 5.17).

The mechanism of action of chymotrypsin

Chymotrypsin connects several eras in the study of enzyme mechanisms. It was the subject of many classical studies before its structure was determined in 1968. Chymotrypsin was one of the first protein crystal structures solved. It was one of the first cases that demonstrated the power of protein crystal structures to

enhance the achievable detail in describing the mechanisms of enzymatic catalysis and specificity. Its story is worth recounting in some detail.

Chymotrypsin is an enzyme of the mammalian digestive system that catalyses the hydrolysis of proteins in food. Kinetic and crystallographic studies have established the structural basis of its specificity and the mechanism of its catalytic activity.

Chymotrypsin will cleave peptide and ester bonds adjacent to aromatic sidechains:

$$\underset{peptide}{R_1\overset{O}{\overset{\|}{C}}{+}NHR_2} \qquad \underset{ester}{R_1\overset{O}{\overset{\|}{C}}{+}OR_2}$$

where R_1 must be aromatic. Biochemists have taken advantage of the differences in relative rates of the major steps of the reaction for amides and esters to investigate the intermediates and the mechanism.

The evidence from kinetics

All esters containing the same acyl group have the same steady-state rate of chymotrypsin-catalysed hydrolysis; that is, the rate of production of R_1COOH from R_1COR_2 is independent of R_2 but not independent of R_1. For ester hydrolysis the rate-limiting step must involve *only* the acyl portion of the substrate. B.S. Hartley and E.P. Kilby suggested the formation of an **acyl-enzyme** intermediate:

$$R_1(C{=}O)\text{--Enzyme}$$

containing a covalent bond between the enzyme and a fragment of the substrate, and release at this step of the remainder of the substrate. The rate-limiting step of the ester hydrolysis must occur *after* the formation of this intermediate, to explain why the overall reaction rate depends on the acyl moiety alone.

For the natural peptide substrates, in contrast, deacylation is so much faster than acylation that the acyl enzyme does not accumulate and is difficult to detect. It is possible to prove the existence of the acyl-enzyme intermediate in the amide case by chemical methods, however. These involve experiments using reagents that can compete with H_2O for reaction with an acyl group as it leaves the enzyme, and reactions of amines with an acyl enzyme to synthesize peptides (the reverse reaction of that normally catalysed). The competition experiments give results consistent with a common mechanism for ester and amide hydrolysis. The kinetics of the reverse reaction, together with the equilibrium constant for peptide hydrolysis, permit an estimate of the rate of the forward reaction that is close to the observed value. This proves that

$$E + R_1COX \underset{k_{-1}}{\overset{k_1}{\rightleftharpoons}} E\text{--}R_1COX \overset{k_2}{\longrightarrow}$$

$$E\text{--}OCR_1 + HX \overset{k_3}{\longrightarrow} E + R_1COOH$$

Figure 5.18 The mechanism of hydrolysis of a peptide or ester, catalysed by chymotrypsin, where E is the enzyme, R_1COX is the substrate. The rate constant for the deacylation step, k_3, must be the same for esters and peptides with the same acyl group (R_1) because they form the same acyl-enzyme intermediate. The natural substrates are peptides; $X = NHR_2$. For these $k_2 \le k_3$, and at normal pH, no acyl enzyme accumulates. For synthetic ester substrates, $X = OR_2$, $k_2 > k_3$, and hydrolysis of the acyl enzyme is the rate-limiting step.

the acyl-enzyme mechanism can account for the rate at which chymotrypsin hydrolyses peptides.

Figure 5.18 summarizes the transformation of the substrates in the reaction. Let us next consider the structural basis in the *enzyme* of its catalytic activity and specificity.

The evidence from crystallography

Bovine chymotrypsin is a protein containing 241 amino acids, with a relative molecular mass of about 25 000. D.M. Blow and coworkers determined the structure of chymotrypsin by X-ray crystallography. Like other enzymes, chymotrypsin has a cavity in its surface into which a substrate of the proper structure can fit and bind. In chymotrypsin, this cavity contains a groove that can accommodate the thin, flat, aromatic rings of phenylalanine, tyrosine, and tryptophan. Van der Waals forces between the sidechain of the substrate and the residues of the enzyme that line the binding pocket stabilize the Michaelis complex. This accounts for the specificity, in this case exactly according to Fischer's classic 'lock-and-key' model. There are additional sites of hydrogen bonding between substrate and enzyme, which are complementary to the structure of almost *any* polypeptide chain. These contribute to the binding affinity but not to the specificity.

For ester substrates, there is crystallographic proof of the acyl-enzyme intermediate. R. Henderson determined the structure of a stable acyl-enzyme adduct, indoleacryloyl-chymotrypsin. It contains a covalent bond between the substrate and the residue serine 195 in the enzyme.

The stereochemical specificity of the interactions in the binding pocket account for the catalytic specificity of the enzyme. Bulky sidechains, such as leucine, cannot fit into the pocket; small sidechains, such as glycine and alanine, fail to bind effectively. This illustrates the 'lock-and-key' aspects of the mechanism by which the enzyme limits its activities to substrates of specific structure. In fact, chymotrypsin is one of a set of peptidases, all closely related both structurally and mechanistically. Each has a binding pocket complementary in size, shape, and charge distribution to the sidechain of the amino acid adjacent to the scissile bond in its preferred substrate (see Chapter 7).

- The mechanism of proteolysis by chymotrypsin proceeds through an acyl-enzyme intermediate. The specificity is determined by a pocket in the enzyme adjacent to the site of binding of the scissile bond.

Substrate binding both positions and orientates the target bond correctly with respect to the 'working' portion of the enzyme. The molecule achieves its catalytic activity by interactions between the sensitive portion of the substrate and residues from the protein in the catalytic active site. They prominently involve a **catalytic triad** of residues: Ser195, His57 and Asp102 (see Fig. 5.19).

Serine proteases form a very large and diverse family of enzymes that participate in many different biological processes. In several cases a set of serine proteases coordinate their activities in a cascade of successive activations. The advantages of the cascade organization are that the signal is amplified at each step, and there are many steps subject to control. The cascade controlling blood coagulation is the best known. Other cascades control dorsal-ventral cell fate in *Drosophila melanogaster* embryogenesis, and the complement system in the vertebrate immune response. There is evidence that these systems, so diverse in biological function, are related.

Blood coagulation

Blood clotting is an essential defence against injury. Damage to a blood vessel *must* elicit an effective and immediate response to prevent blood loss. Our bodies plug leaks in blood vessels with clots containing platelets and fibrin. This process must be subject to rigorous control: the light trigger necessary for quick response creates a concomitant danger of formation of unwanted clots.

Following tissue repair, the clot must dissolve. In this respect, coagulation differs from the irreversible aggregation of protein, in prion diseases for instance, that leads to cell death.

Damage to the walls of blood vessels exposes proteins including collagen and **tissue factor**, normally covered by the endothelial cells. Activated platelets bind to the collagen, providing the immediate first step of damage control. Subsequently, a network of fibrin molecules reinforces the platelet plug.

The precursor of the **fibrin clot** is **fibrinogen**, a large soluble protein amounting to ~3% by weight of blood plasma protein. Fibrinogen contains a 'sticky' region that forms the basis for aggregation. An N-terminal peptide masks this region. Upon vascular injury, cleavage of fibrinogen by thrombin exposes the sticky region, leading to aggregation and formation of a clot. Think of taking a box of plasters (= 'band-aids') and removing all the protective strips, after which the plasters will all stick together.

To prevent premature cleavage of fibrinogen, thrombin exists in an inactive state, prothrombin. A different protease, Factor X, activates prothrombin to thrombin by specific cleavage. These are the last steps in an extended cascade of cleavage and activation. The cascade is branched, into **intrinsic** (or contact activation) and **extrinsic** (or tissue factor) pathways. The two pathways converge at the activation of Factor X, which cleaves thrombin (see Fig. 5.20). The extrinsic pathway is the main route for initiation of coagulation. The intrinsic pathway amplifies the cascade, by action of thrombin on Factor XI.

Thrombosis

Thrombosis is the formation of a blood clot within an undamaged blood vessel, obstructing the flow of blood. Blockages of coronary arteries lead to heart attacks; clots that block arteries in the brain lead to strokes. These are major causes of death in developed countries. Clinically, a component of treatment of heart attacks and strokes is administration of 'clot-busting' drugs. These include tissue plasminogen activator (TPA) and streptokinase.

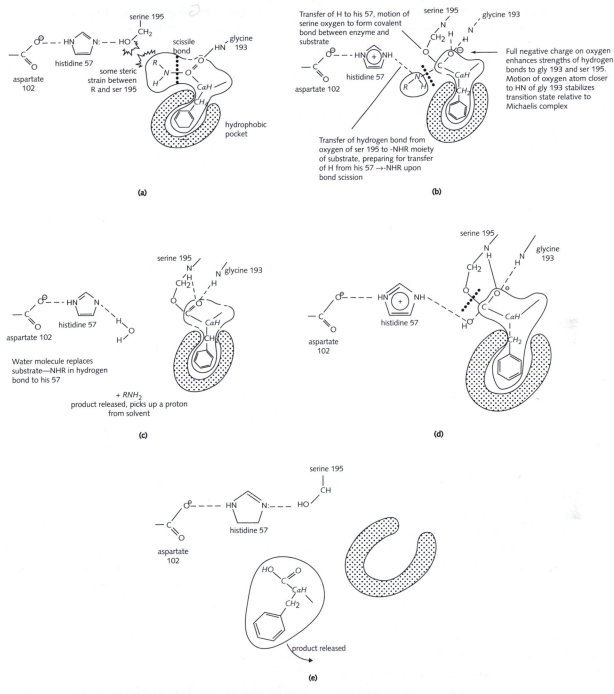

Figure 5.19 The mechanism of action of chymotrypsin. An explanation of the mechanism must specify how the enzyme lowers the activation energy of the reaction and how the active site binds specific substrates, rejecting others. The interactions that account for the initial binding of the substrate in the Michaelis complex must be distinct from the interactions responsible for the catalytic activity that stabilize the transition state. (Forces that stabilize the Michaelis complex threaten to *increase* the activation energy.) Chymotrypsin, like other enzymes, couples the formation of the transition state from the Michaelis complex to *changes* in the interactions between the substrate and the enzyme. Prominent among these changes, in the case of chymotrypsin and related enzymes, is an internal rearrangement of charges in two residues— serine 195 and histidine 57. (a) Formation of the Michaelis complex. (b) Transition state for formation of the acyl enzyme. (c) The acyl-enzyme intermediate. The N-terminal portion of the polypeptide substrate is released. The C-terminal portion is still linked to the enzyme. (d) Transition state for deacylation. (e) Release of product and return to the initial state.

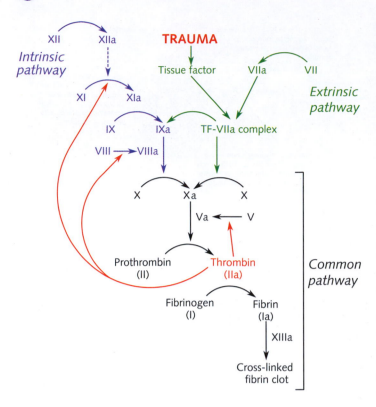

Figure 5.20 Vascular trauma initates a cascade of activations of clotting factors that result in the formation of a clot. The naming of clotting factors by Roman numerals preceded their identification; for instance, we now know that Factor IIa is thrombin. The 'a' indicates the active form. The clotting factors are proteases, and the cascade proceeds by successive cleavages to produce active forms. For example (upper left), Factor XIa cleaves Factor IX to IXa, activated factor IXa cleaves Factor X to Xa, Factor Xa cleaves prothrombin to thrombin. The initial trigger for coagulation is vascular injury. This exposes a protein called *tissue factor*, or Factor III (see Box 5.5). Tissue factor facilitates the activation of Factor VII. A complex between tissue factor and Factor VIIa starts the cascade of protolytic activations.

Two branches of the pathway converge. The extrinsic pathway (green) is the main pathway for initiation of coagulation. The intrinsic pathway acts to amplify the response. (The broken line connecting Factor XIIa with the activation of Factor XI indicates its minor role. The main activator of Factor XI is thrombin.) The common pathway (black) involves the activation of Factor X, which activates thrombin, which cleaves fibrinogen to fibrin, which forms the actual clot. Thrombin has a central role in these processes. Thrombin and its interactions appear in red.

BOX 5.5   **Proteins in health and desease**

Tissue factor: a link between coagulation and cancer

The correlation between cancer and coagulation disorders was first observed in 1865. There are clinical implications in both directions: tumours can activate blood coagulation, leading to thromboembolism in patients. Conversely, coagulation factors can stimulate tumour growth, metastasis, and angiogenesis, and inhibit apoptosis. Underlying this link may be the deep evolutionary relationship between protease cascades in coagulation and development. The link with tissue factor in these processes is now securely established: in addition to its role in blood coagulation, tissue factor can trigger cell-signalling pathways. Exploration of the use of anticoagulants in cancer therapy has become an active field.

Anticoagulant therapy

The earliest anticoagulant drug used clinically was heparin, a natural heterogeneous sulphated oligosaccharide. Heparin does not interact with thrombin directly, but activates an inhibitor of thrombin.

Another anticoagulant in clinical use is warfarin, originally used as a poison for rats and mice (see Fig. 5.21(a)). Warfarin is a vitamin K antagonist. Vitamin K is a cofactor in the carboxylation of thrombin, Factor VII and other proteins of the coagulation cascade. During this process vitamin K is oxidized. Warfarin inhibits vitamin K epoxide reductase, an enzyme that restores oxidized vitamin K to its reduced form.

Clopidogrel (see Fig. 5.21(b)) acts by a different mechanism. It interferes with the platelet ADP receptor, inhibiting the aggregation of platelets.

As a central enzyme in blood coagulation, thrombin itself is a prime target for anticoagulant therapy. At the protein level, thrombin directly catalyses the formation of fibrin from fibrinogen, and also activates clotting Factors V, VIII, XIII, and protein C. Thrombin also has several effects at the cellular level. For instance, thrombin activates platelets by cleaving receptors, an

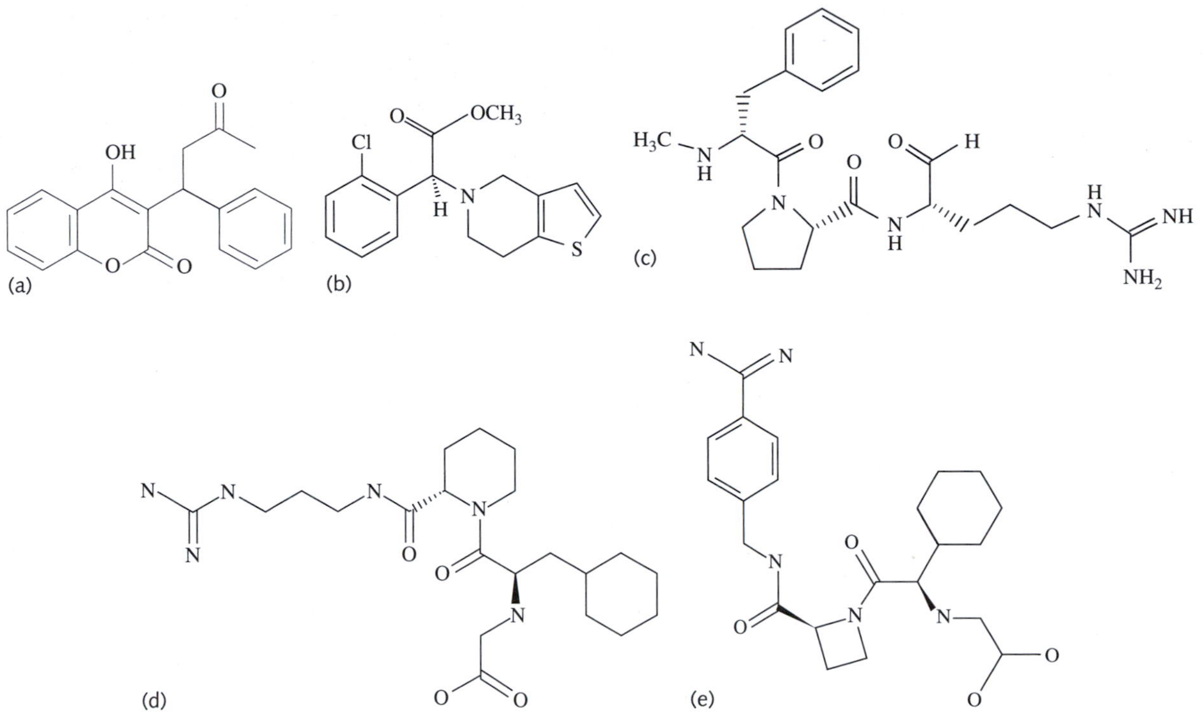

Figure 5.21 The structures of some molecules used clinically as anticoagulants. (a) Warfarin. (b) Clopidogrel. (c) Efegatran. (d) Inogatran. (e) Melagatran (withdrawn because of liver toxicity).

event that initiates intracellular signalling events and cell transformation (see Fig. 5.22). The effect is to enhance recruitment of platelets into a clot at the site of vascular injury.

Some animals that suck blood, such as leeches, produce natural thrombin inhibitors. Injection of inhibitor keeps the blood flowing—the analogue of a 'bottomless cup of coffee'. Moreover, leeches can ingest up to five times their resting body volume of blood. The anticoagulants keep it liquid.

The medicinal leech (*Hirudo medicinalis*) produces specific thrombin inhibitors, called hirudins. Hirudin-1 is a 65-residue polypeptide that binds tightly to thrombin. It forms a very stable non-covalent inhibitory complex with thrombin, $K_D = 20 \times 10^{-15}$. The binding site of thrombin is unusually extensive, containing 'anion-binding exosites'—pockets lined with positively-charged residues. Hirudin binds to both the catalytic site and the anion-binding exosite of thrombin (see Fig. 2.23). Like natural substrates, hirudin forms a short stretch of two-stranded β-sheet with the enzyme in the catalytic site. But whereas for natural substrates the strands are antiparallel, in the hirudin–thrombin complex they are parallel.

Modification of hirudins has produced several oligopeptide thrombin inhibitors that combine (1) the carboxy-terminal residues of hirudin, which bind to the anion-binding exosite of thrombin, and (2) specific active-site inhibitors, such as D-Phe-L-Pro-L-Arg. This tripeptide sequence is also the basis for derivatives including 'PPACK' (= D-Phe-Pro-Arg-CH₂Cl), and others in which the C-terminal carboxyl group is replaced by an aldehyde, abbreviated Arg-H (see Figs. 5.21 (c)–(e)).

D-Phe-Pro-Arg-H	efegatran
HOOCCH₂-D-Cha-Pip-Nag	inogatran
HOOCCH₂-D-Chg-Aze-Pab	melagatran

Other thrombin inhibitors are peptidomimetics—non-peptide compounds that share features of shape and charge distribution with peptide inhibitors. One of these is argatroban:

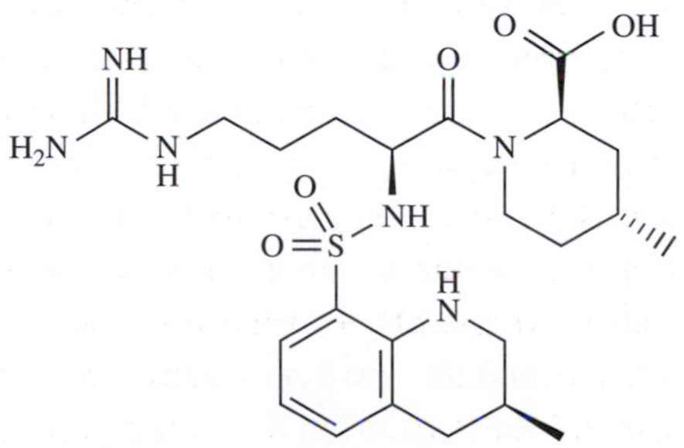

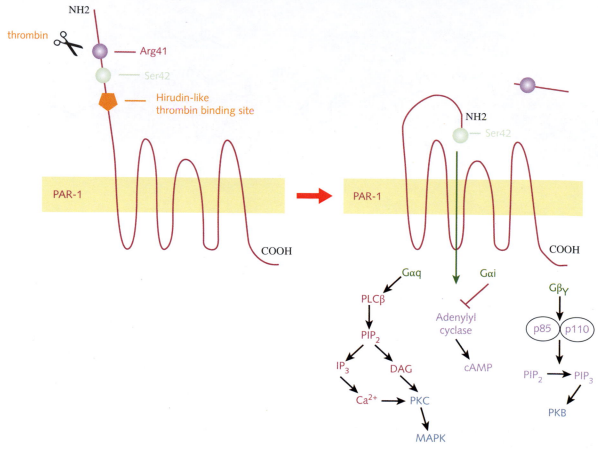

Figure 5.22 The activation of platelet surface receptors upon cleavage by thrombin of a G-protein coupled Protease-Activated Receptor (PAR-1). The extracellular region of the receptor includes a hirudin-like N-terminal sequence that binds to a thrombin exosite. Cleavage by thrombin produces a new N-terminus. This new N-terminal region can bind to another region of PAR-1, causing a conformational change that activates the receptor and triggers a signal cascade. This *tethered ligand* mechanism is also observed in the activation of proteases such as trypsin.

From: Villares, G.J., Zigler, M. and Bar-Eli, M. (2011).The emerging role of the thrombin receptor (PAR-1) in melanoma metastasis—a possible therapeutic target. Oncotarget 2, 8–17.

Serpins: serine proteinase inhibitors—conformational disease

The **serpins** are a family of proteins with a variety of biological roles, not limited to inhibition of proteinases. They are of interest for their medical importance (clinical consequences of serpin dysfunction include blood-clotting disorders, emphysema, cirrhosis, and mental illness), for their very unusual conformational changes between states of different folding pattern, and for their unusual mechanism of action. Unlike inhibitors that bind tightly—but in principle reversibly—to active sites, serpins perform a suicide mission. The serpin is cleaved to form a stable proteinase–inhibitor complex in which the proteinase is partially denatured, not only rendering it inactive but increasing its susceptibility to proteolysis.

That serpins must undergo dramatic conformational changes when cleaved by proteinases was recognized from the first serpin crystal structure, cleaved α_1-antitrypsin (see Fig. 5.23). The residues adjacent to the cleaved peptide bond are 65 Å apart!

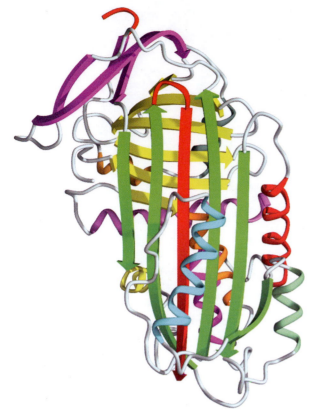

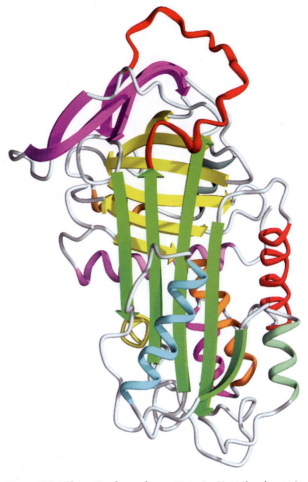

Figure 5.23 Cleaved α_1-antitrypsin [7API]. The secondary structure of the serpin fold includes three β-sheets and nine α-helices. The natural role of α_1-antitrypsin is to inhibit elastase in the lung. In the uncleaved form, a peptide bond links the residues at the C-terminus of the red strand, at the bottom of the picture, with the free end of the red stump at the upper left.

Figure 5.24 The native form of α_1-antitrypsin. Note the change in structural context of the region coloured red, which was a strand in the centre of the sheet in the previous figure [2PSI].

The region N-terminal to the scissile bond, called the reactive centre, formed a strand at the centre of a large six-stranded β-sheet. In the uncleaved structure, these residues, separated by a major element of the secondary structure of the protein in the cleaved form, must somehow come together.

Several conformational states of serpins are known

The activity of serpins involves transitions between different conformational states.

- In the *active* or *'native'* form of serpins, the polypeptide chain is intact. The reactive centre—the segment that interacts with the proteinase—forms an exposed loop at one end of the main β-sheet. In the native form this β-sheet is *five-stranded* (see Fig. 5.24). Comparison with the cleaved form shows that the central strand has been pulled out, and the surrounding strands have closed up together around it. Figure 5.25 shows a superposition of native and cleaved forms of α_1-antitrypsin.

- In the *cleaved* form the conformational change integrates the reactive centre loop into the body of the molecule, as a sixth strand in the central β-sheet (see Fig. 5.23). A free cleaved serpin cannot inhibit proteinase.

- The *latent* state is another uncleaved form, in which the reactive centre loop is inserted into the

Figure 5.25 The native and cleaved forms of α_1–antitrypsin, superposed on the largest common substructure. [2PSI, 7API].

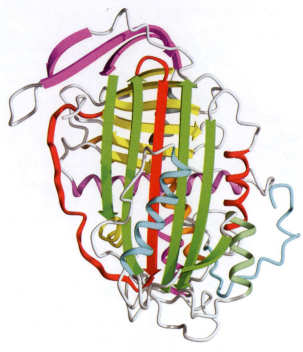

Figure 5.26 The latent state of human antithrombin [1ATH]. Antithrombin plays a role in controlling the proteases involved in blood coagulation.

main β-sheet as in the cleaved form, but the chain is intact (see Fig. 5.26). Under physiological conditions the native states of inhibitory serpins are metastable, converting spontaneously to the latent state at different rates (within about 2 h for plasminogen activator inhibitor–1). The multiple conformational states observed among the serpins are an exception to the rule that homologous proteins have structures containing a core with a unique folding pattern.

• Serpins form *polymers* by intermolecular β-sheet formation. If regions of a serpin molecule do not integrate correctly into the structure, but instead form part of a second molecule, and the process continues, a polymer forms. This is an example of **domain swapping** (see Chapter 7). Polymerization is particularly common in certain mutant serpins, such as Z α_1-antitrypsin (E342K), in which the structure is destabilized. The polymers

form deposits in the liver, leading eventually to cirrhosis. Lack of functional α_1-antitrypsin also causes lung damage as a result of inadequate inhibition of elastase, often leading to emphysema (see Box 1.3).

In relying on structural mobility for their mechanism of action—necessarily involving at least partial unfolding—the serpins have rendered themselves vulnerable to aggregation. From the idea that the lesion involves the dynamics of interconversion of structural states has emerged the concept of **conformational disease**, embracing conditions produced by serpin mutants, prions, and other aggregating proteins.

Mechanism of proteinase inhibition by serpins

Free serpin in the native state presents an exposed reactive centre loop to proteinase attack. The normal mechanism of action of chymotrypsin-like serine proteinases involves cleavage of the substrate, and formation of an intermediate acyl-enzyme complex with the amino-terminal part of the substrate linked to the enzyme. Hydrolysis of the acyl-enzyme complex restores the original state of the enzyme.

Any proteinase that studied this mechanism in the previous section of this book, or elsewhere, is in for a nasty surprise when it encounters a serpin. Cleavage of α_1-antitrypsin by trypsin triggers the serpin conformational change: the reactive centre loop inserts into the main β-sheet, and the (hapless) enzyme, bound in the acyl-enzyme complex, is dragged to the opposite pole of the serpin. Steric clashes break up the proteinase structure. The catalytic triad is forced apart, preventing release of the enzyme by normal deacylation. The salt bridge between the free amino terminus and the sidechain of Asp194, involved in activation of trypsin, is broken, an alternative salt bridge forming between Asp194 of the proteinase and Lys328 of the inhibitor. This reverses the conformational change that activated the proteinase. In fact, almost 40% of the proteinase structure becomes disordered, rendering it susceptible to proteolytic attack. A brutal scenario.

- Serpins are an unusual type of protein that show multiple folding patterns in one family. Their suicide mechanism of inhibition renders them prone to aggregation.

Evolutionary divergence of enzymes

Study of families of enzymes show how proteins can evolve altered or novel functions. Comparisons of related proteins reveal the features of structure and function that change and those that stay the same. In some cases, the catalytic atoms occupy the same positions in molecular space, although the residues that present them are located at different points in the fold. In other cases the positions in space of the catalytic residues are conserved even though the identities and functions of the catalytic residues vary. In these cases, there appears to be a set of **conserved functional positions** within the space of the molecule.

The mechanism of action of malate and lactate dehydrogenases

Malate and lactate dehydrogenases are related enzymes that catalyse similar reactions (see Box 5.6.)

BOX 5.6   Reactions catalysed by lactate dehydrogenase (LDH) and malate dehydrogenase (MDH)

$$\begin{array}{ccccccccc}
\text{COO}^- & & & & & & \text{COO}^- & & \\
| & & & & & & | & & \\
\text{C} = \text{O} & + & \text{NADH} & + & \text{H}^+ & = & \text{HCOH} & + & \text{NAD}^+ \\
| & & & & & \text{LDH} & | & & \\
\text{CH}_3 & & & & & & \text{CH}_3 & & \\
\text{pyruvate} & & & & & & \text{lactate} & &
\end{array}$$

$$\begin{array}{ccccccccc}
\text{COO}^- & & & & & & \text{COO}^- & & \\
| & & & & & & | & & \\
\text{C} = \text{O} & + & \text{NADH} & + & \text{H}^+ & = & \text{HCOH} & + & \text{NAD}^+ \\
| & & & & & \text{MDH} & | & & \\
\text{CH}_2 & & & & & & \text{CH}_2 & & \\
| & & & & & & | & & \\
\text{COO}^- & & & & & & \text{COO}^- & & \\
\text{oxaloacetate} & & & & & & \text{malate} & &
\end{array}$$

They arose by **gene duplication** at an early stage of the history of life, and their sequences have diverged. (In an optimal alignment, human malate and lactate dehydrogenases have ~20% identical residues.) Nevertheless, site-directed mutagenesis showed that a single residue change (Gln→Arg) could change the specificity of *Bacillus stearothermophilus* lactate dehydrogenase to malate (see Fig. 5.27).

> Reports of that work may have been read by a *Trichomonad*, which developed a malate dehydrogenase that, in an evolutionary tree of these enzymes, is much more similar to lactate dehydrogenases than to other malate dehydrogenases.

Arg171 forms a salt bridge with the carboxyl group common to the substrates of the two enzymes. A proton is transferred from the sidechain of His195 to the C=O group common to the substrates, that is the site of reduction. The NADH supplies the other proton.

The mutation that changes the specificity involves 102Arg/Gln at the lower right of Figs. 5.27(a) and (b). The bulkier Arg does not allow room for the larger substrate, oxaloacetate. It is unfortunate that the inhibitor bound in the malate dehydrogenase structure is smaller than the natural substrate, which makes this effect less clear (see Exercise 5.14).

Enolase, mandelate racemase, and muconate lactonizing enzyme catalyse different reactions but have related mechanisms

It is arguable that the relationship between malate and lactate dehydrogenases is really more a change in specificity than a change in the reaction. A family of enzymes more highly diverged in function includes enolase, mandelate racemase, and muconate lactonizing enzyme I. They have a common structure, closely related to the TIM-barrel fold (see Fig. 2.12(d)). However, they catalyse different reactions.

Looking only at sequence and structure runs the risk of overlooking a more subtle similarity. These enzymes share a common feature of their *mechanism*: each acts by abstracting a proton adjacent to a carboxylic acid to form an enolate intermediate (see Fig. 5.28). The stabilization of a negatively-charged transition state is conserved. In contrast, the subsequent reaction pathway, and the nature of the product, vary from enzyme to enzyme. These enzymes have not only a similar overall fold, but each requires a divalent metal ion, bound by structurally equivalent ligands. However, other residues in the active site differ, to produce enzymes that catalyse different reactions.

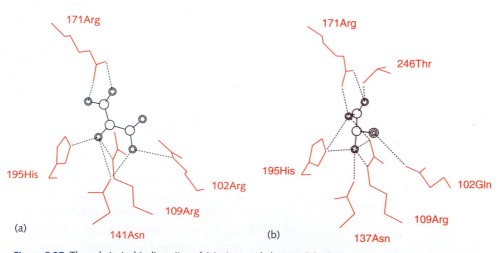

Figure 5.27 The substrate-binding sites of (a) pig muscle lactate dehydrogenase and (b) pig cytoplasmic malate dehydrogenase. The lactate dehydrogenase binds the inhibitor oxamate, CH_3CONH_2, instead of the natural pyruvate product, CH_3COCOO^-. The malate dehydrogenase binds the inhibitor, α-ketomalonate, $^-OOCCOCOO^-$, instead of the natural oxaloacetate product, $^-OOCCH_2COCOO^-$.

As enzymes evolve, they may:

- change kinetic parameters;
- change substrate specificity;
- change reaction catalysed;
- change control mechanism;
- but they usually retain general mechanism of catalysis.

DNA replication and transcription, and RNA splicing and translation are all biological reactions that require efficient and carefully regulated catalysts. Many of these functions are carried out by large, multienzyme complexes. In some cases we know the structures of the complexes, notably the ribosome, and DNA and RNA polymerases, and these make a description of the mechanism of action possible.

These operations are quite complex, and do not lend themselves to short sections in a book of this type. Instead, let us consider a somewhat simpler example, *E. coli* **DNA topoisomerase III**. This illustrates the kind of macroscopic enzyme–substrate interactions involved in DNA processing. It also exemplifies a system in which the 'reaction' catalysed is not a transformation of chemical bonding pattern, but a change in the topology of a DNA double helix.

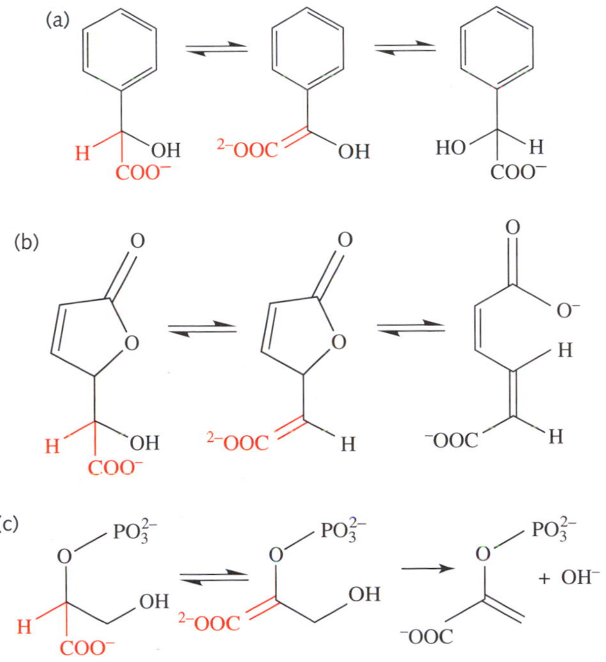

Figure 5.28 Common mechanism in the enolase family of enzymes: (a) mandelate racemase, (b) muconate lactonizing enzyme, (c) enolase.

The structure and mechanism of *E. coli* topoisomerase III

The equilibrium structure of the B form of DNA is the iconic double helix with 10 base pairs per turn. If the two ends of a stretch of linearly extended double-helical DNA were grasped and twisted, the DNA would become either overwound or underwound. If the two ends were then linked to form a circular DNA molecule, the twisting could not be relieved by simple conformational changes.

Supercoiling is a measure of this twisting. Each full turn of *unwinding* counts as one negative unit of supercoiling. Each full turn of *tightening*, or overwinding, counts as one positive unit of supercoiling. The equilibrium form, with zero supercoiling, is *relaxed*.

To relieve the local strain associated with overwinding, a piece of circular DNA may form a superhelix. A similar phenomenon is familiar in the supercoiling of the helical wires attached to telephone receivers.

Supercoiling of DNA occurs during many natural processes, including DNA replication and transcription, segregation of daughter chromosomes, and recombination and repair. Even the packing of DNA into cells presents problems that must be addressed by higher-order structures; the DNA must condense, but yet be available to the replication and transcription apparatus.

DNA topoisomerases are enzymes that manipulate the higher-order levels of DNA structure by changing the supercoiling. In the process, they act as enzymes to catalyse the breakage and reformation of chemical bonds. However, the overall effect of their action is not a change in the primary chemical bonding, but in the supercoiling of the DNA. They operate by temporarily breaking either one or both strands, passing a single strand or both strands through the break, and then rejoining the cleaved strands. This is topologically equivalent to local winding or unwinding. There are several families of topoisomerases, which differ in structure and activity.

E. coli topoisomerase III is type IA. Type IA enzymes act only on underwound—negatively supercoiled—DNA, to reduce the underwinding and bring

positions of the FG corners. In consequence, the deoxy quaternary structure is destabilized because the dimers no longer fit together properly (having changed their shape). Adopting the alternative quaternary structure requires the tertiary structural changes to take place even in subunits not yet liganded.

As a result of the quaternary structural change, these unliganded subunits have been brought to a state of enhanced oxygen affinity. It is important to emphasize that this is a sequence of steps in a logical process and not a description of a temporal pathway of a conformational change.

● RECOMMENDED READING

Benkovic, S. and Hammes-Schiffer, S. (2003). A perspective on enzyme catalysis. *Science*, **301**, 1196–202.

and

Blow, D. (2000). So do we understand how enzymes work? *Structure* **8**, R77–81. These two articles describe modern concepts of the sources of catalytic activity in enzymes.

Cornish-Bowden, A. (ed.) (1997). *New beer in an old bottle: Edward Buchner and the growth of biochemical knowledge*, ed. València, València University. A collection of general articles, from a historical perspective.

Fischer S , Olsen KW , Nam K. and Karplus M. (2011). Unsuspected pathway of the allosteric transition in hemoglobin. *Proc. Nat'l. Acad Sci. USA*. **108**, 5608–13. A simulation of the trajectory of the transition between oxy and deoxy states of haemoglobin.

Lodola, A. and Mulholland, A.J. (2013). Computational enzymology. *Methods Mol. Biol.* **924**, 67–89. Modelling enzymatic reactions, with a view to answering fundamental questions about catalysis, and to applications to drug discovery.

Vendruscolo, M. and Dobson, C.M. (2006). Dynamic visions of enzymatic reactions. *Science*, **313**, 1586–7. The importance of conformational changes in enzyme catalysis.

Walsh, C. (2001). Enabling the chemistry of life. *Nature,* **409**, 226–31. A general article about enzyme structure, mechanism, applications, and directed evolution.

Williams, D.H. , Stephens, E. and Zhou, M. (2003). How can enzymes be so efficient? *Chem. Commun.*, 1973–6. Detailed addressing of questions of details of enzyme structure and thermodynamics of enzyme catalysis.

Wolfenden, R. and Snider, M.J. (2001). The depth of chemical time and the power of enzymes as catalysts. *Acc. Chem. Res.*, **34**, 938–45. A general expository article focusing on the thermodynamics of catalysis.

● EXERCISES AND PROBLEMS

Exercise 5.1 Of those enzymes shown in Fig. 5.1, which have substrates that would be stable for a week in the absence of enzyme? (Take 'stable' to mean retaining 90% or more of initial amount of unreacted substrate.)

Exercise 5.2 How will increasing the viscosity of the medium, perhaps by adding sucrose—keeping temperature, and enzyme and initial substrate concentrations the same—affect the rate of the reaction as measured *in vitro* on purified enzyme and substrate, catalysed by (a) acetylcholinesterase ($k_{cat}/K_M \sim 2 \times 10^8$ mol s^{-1}) and (b) urease ($k_{cat}/K_M \sim 2 \times 10^5$ mol s^{-1}).

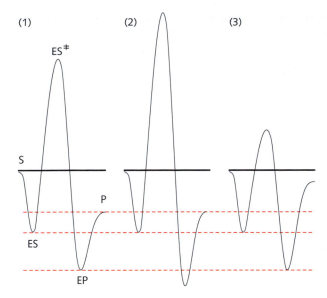

Figure 5.43 These three reaction–coordinate diagrams describe different reactions and different enzymes. However, the same energy scale (*y*-axis) is used in all three. The red broken calibration lines are guides to help compare the graphs. S = substrate, ES = enzyme–substrate complex, ES‡ = transition state, EP = enzyme–product complex, P = product.

Exercise 5.3 Suppose an enzymatically catalysed reaction has a turnover number of 100 s^{-1}. (a) What is the half-life of the substrate? (b) Would observation of formation and decomposition of intermediates on this timescale be possible by continuous-flow or stopped-flow methods or would relaxation methods be required?

Exercise 5.4. Draw a reaction diagram analogous to Fig. 5.2(a) for a reaction in which a stable intermediate appears before the highest-energy maximum (the transition state) and (b) for a reaction in which a stable intermediate appears after the highest-energy maximum.
Exercises 5.5 to 5.8 are based on the reaction coordinate diagrams shown in Fig. 5.43.

Exercise 5.5 What can you say about the relative values of the Michaelis constant for reactions (1), (2) and (3)?

Exercise 5.6 In the *absence* of enzyme, that is, for the uncatalysed reaction, which reaction would have the lowest equilibrium constant?

Exercise 5.7 In the presence of enzyme, which reaction has the lowest forward reaction rate?

Exercise 5.8 In the presence of enzyme, which reaction has the highest reverse reaction rate?

Exercise 5.9 Suppose a reaction S→P has an equilibrium constant <1. A further reaction P→Q has an equilibrium constant >>1, but, in the absence of a catalyst, a very high activation energy. (a) Starting with an equimolar mixture of S and P, and an enzyme that catalyses only the S→P reaction, will the concentration of S decrease? (b) If to this mixture there is added another enzyme that converts P→Q at high rate, will the concentration of S decrease?

Exercise 5.10 Use the Michaelis–Menten equation:

$$v_0 = \frac{V_{max}[S]}{K_M + [S]}$$

to answer the following. (a) In the limit [S] → 0, the equation is effectively linear: v_0 = constant × [S]. What is this constant? (b) In the limit [S] → ∞, v_0 tends to a constant. What is this constant?

Exercise 5.11 From the Scatchard plot in Fig. 5.5, estimate the values of K_D and the number of binding sites n.

Exercise 5.12 O_2 is a competitive inhibitor of ribulose-1,5-bisphosphate carboxylase/oxygenase. Processing of O_2, called *photorespiration*, causes loss of assimilated carbon, which is released as CO_2. It is estimated that photorespiration costs a plant up to 25% loss in CO_2 assimilation. (a) Assuming that you could grow plants under controlled conditions, how would you enhance the efficiency of photosynthesis? (b) Mutant plants deficient in photorespiration enzymes that act downstream of ribulose-1,5-bisphosphate carboxylase/oxygenase do not grow in normal air. How could you rescue them?

Exercise 5.13 From the quenching of enzyme fluorescence as a function of substrate (dUMP) concentration, H.T. Spencer, J.E. Villafranca, and J.R. Appleman determined the dissociation constant of dUMP from *E. coli* thymidylate synthase to be $K_D = 3.9$ μM. From the rate of fluorescence quenching after mixing of enzyme and dUMP, measurable by stopped-flow spectroscopy, they found $k_{on} = 9.8 \times 10^6$ M^{-1} s^{-1} and $k_{off} = 41s^{-1}$. Note that K_D, and k_{on} and k_{off}, were determined in independent experiments. What check on the data would be easy to apply? What is the result of this test?

Exercise 5.14 On a photocopy of the structure of malate dehydrogenase in Fig. 5.27(b), sketch in the natural substrate, oxaloacetate.

Exercise 5.15 How would you go about desiging an **agonist** for the PAR-1 receptor of platelets?

Exercise 5.16 On a photocopy of the Hill plot of haemoglobin and myoglobin (see Fig. 5.40), indicate where fully saturated haemoglobin appears and where fully deoxygenated haemoglobin appears.

Problem 5.1 In the example in Table 5.3 we determined the Michaelis constant and the maximal velocity of ribulose-1,5-bisphosphate carboxylase/oxygenase with respect to the substrate CO_2 in the absence of oxygen. Table 5.4 contains the results of measurements with the same enzyme preparation under an O_2 atmosphere. (a) What are the values of K_M and V_{max} under these conditions? (b) State qualitatively the effect of O_2 on the reaction rate. (c) What can you infer about the mechanism by which O_2 affects the interaction of the enzyme with CO_2?

Readers who do not have local access to an appropriate curve-fitting package should search the web for the following combination of keywords: 'Michaelis Menten gui'.

Table 5.4 Initial steady-state rates of CO_2 uptake catalysed by soybean ribulose-1,5-bisphosphate carboxylase/oxygenase at pH 8.0 and 298.15 K under an O_2 atmosphere. (Rates expressed per mg of protein. Data of G. Bowes and W.L. Ogren; compare Table 5.3.)

$[CO_2]$/mmol L^{-1}	Rate of CO_2 uptake/nmol s^{-1}
0.046	2.74
0.093	5.40
0.139	7.69
0.232	11.11
0.463	15.05
0.926	19.34
2.315	25.72

Problem 5.2 Suppose that the rates of a reaction, in the absence and in the presence of an enzyme, are given by the expressions:

$$\text{without enzyme:} \qquad \text{rate} = ke^{-\Delta G^{\ddagger}/(RT)}$$

$$\text{with enzyme:} \qquad \text{rate} = ke^{-\Delta G_E^{\ddagger}/(RT)}$$

Here, $\Delta G^{\ddagger}$ and $\Delta G_E^{\ddagger}$ are the free energies of activation of the uncatalysed reaction and the enzymatically catalysed reaction. Assume that the constant of proportionality k is the same in both cases. If $\Delta G_E^{\ddagger}$ is 2 kJ mol^{-1} lower than $\Delta G^{\ddagger}$ by what factor will the rate be increased at 310 K (= 37°C)? ($R = 8.314$ J K^{-1} mol^{-1}).

Problem 5.3 On a photocopy of Fig. 5.6, extrapolate the rate of the linear region at low substrate concentration to determine the substrate concentration at which v_0 would equal V_{max} if there were no saturation of the enzyme. Estimate the ratio of the value you determined to K_M.

Problem 5.4 Briggs and Haldane considered the following rate equation for enzymatic catalysis:

$$E + S \underset{k_{-1}}{\overset{k_1}{\rightleftharpoons}} ES \xrightarrow{k_2} P$$

The Briggs–Haldane scheme is more general than the Michaelis–Menten model because it does not assume that $k_{-1} \gg k_2$. Show that this scheme is also consistent with the observed substrate dependence of the rate and reinterpret the parameters derived from fitting the reaction rates, in terms of k_1, k_{-1}, and k_2. Assume that the initial substrate concentration is much greater than the enzyme concentration, and that the concentration of the enzyme–substrate complex, [ES], will be constant during the reaction, until substrate is depleted. Find the initial steady-state rate.

Problem 5.5 For a protein binding a single ligand, PL=P+L, the dissociation constant is $K_D = \frac{[P][L]}{[PL]}$, the total concentration of ligand is $[L]_{tot} = [L]+[PL]$, and the total concentration of protein is $[P]_{tot} = [P]+[PL]$. Suppose $K_D = 10^{-6}$ and $[P]_{tot} = 1$ mM. (a) For what value of $[L]_{tot}$ is the protein half-saturated; that is, $[PL]=0.5 \times ([P]+[PL])$ or $[P]=[PL]$? (b) For what value of $[L]_{tot}$ is the protein three-quarters saturated?

Problem 5.6 Plot the data in Table 5.5 giving rate enhancements and estimated enzyme–transition state dissociation constants. Do the data confirm the idea that enzymes achieve rate enhancement by stabilizing transition states?

Problem 5.7 Triose phosphate isomerase is an enzyme in the glycolytic pathway that catalyses the interconversion of dihydroxyacetone phosphate and glyceraldehyde-3-phosphate (see Fig. 5.16). It is believed to proceed through an enediol intermediate. (a) Draw the structure of the enediol intermediate. Suppose the reaction were to proceed in tritiated water (^3H_2O). Where would you

Table 5.5 Rate enhancements and transition-state dissociation constants

Enzyme	Rate enhancement	Transition-state dissociation constant
β–amylase	1.0×10^{22}	1.0×10^{-22}
Acetylcholinesterase	2.0×10^{17}	5.0×10^{-18}
AMP nucleosidase	5.0×10^{16}	2.0×10^{-17}
Phosphotriesterase	5.3×10^{15}	1.9×10^{-16}
Triosephosphate Isomerase	5.6×10^{13}	1.8×10^{-14}
Chorismate mutase	4.2×10^{10}	2.4×10^{-11}
Carbonic anhydrase	9.2×10^{8}	1.1×10^{-19}

expect labelled hydrogen to appear if (b) the reactior proceeded by a direct hydride transfer, (c) the reaction proceeded by the enediol mechanism, assuming that the intermediate remained bound to the enzyme, (d) the reaction proceeded by the enediol mechanism, assuming that the intermediate were released and rebound?

Problem 5.8 A fundamental principle of enzyme catalysis is that enzymes speed up reactions that would otherwise proceed too slowly at body temperature and 1 atm pressure. Nitrogen fixation takes place at low temperature and pressure in the root nodules of leguminous plants. In contrast, the Haber process to generate ammonia from N_2 and H_2 requires high temperature and pressure. Why is this a bad choice to illustrate this principle?

Problem 5.9 Derive the Michaelis–Menten equation:

$$v_0 = \frac{V_{max}[S]}{K_M + [S]}$$

from the Briggs–Haldane steady-state assumptions.

Show that, with reference to the reaction scheme:

$$E + S \underset{k_{-1}}{\overset{k_1}{\rightleftharpoons}} ES \xrightarrow{k_2} P$$

$$K_M = (k_{-1} + k_2)/k_1.$$

Problem 5.10 The Arrhenius equation:

$$\frac{d\ln k}{dT} = -\frac{E_a}{RT^2}$$

is equivalent to:

$$\frac{d\ln k}{d(1/T)} = -\frac{E_a}{R}$$

where the gas constant $R = 8.31447$ J mol^{-1} K^{-1}.

Suppose a reaction rate doubles upon raising the temperature from 25°C (= 298.15 K) to 35°C (= 308.15 K). What is the value of E_a?

Problem 5.11 How accurately would you have to determine the activation energy of a reaction at 25°C (= 298.15 K) to determine the rate constant correctly to within a factor of 10?

Problem 5.12 Transform the Michaelis–Menten equation into the Eadie–Hofstee equation.

Hint: Solve

$$v_0 = \frac{V_{max}[S]}{K_M + [S]}$$

for [S], to find:

$$[S] = \frac{v_0 K_M}{V_{max} - v_0}.$$

Then invert to find an expression for 1/[S] and multiple both sides of the equation by v_0 to give:

$$\frac{v_0}{[S]} = \frac{V_{max} - v_0}{K_M} = \frac{V_{max}}{K_M} - \frac{v_0}{K_M}.$$

Problem 5.13 You buy a jar of honey, advertised as being made by bees from the nectar of clover (typically *Trifolium repens*). Clover fixes CO_2 by phosphoenolpyruvate (PEP) carboxylase. You suspect that the honey has been adulterated by addition of sucrose from sugar cane. Sugar cane fixes CO_2 by ribulose-1,5-bisphosphate carboxylase/oxygenase. The carbon isotope effect is much higher for ribulose-1,5-bisphosphate carboxylase/oxygenase than for PEP carboxylase. Given a sample of a leaf from clover, a leaf from sugar cane, a sample of honey, and a mass spectrometer how could you identify the source of the sugar in the honey? (An analogous problem stated in terms of pure maple syrup would have the same answer.)

Problem 5.14 Using molecular graphics software, draw one or more pictures that illustrate the interactions of the five most essential residues in catalysis by thymidylate synthase (see Fig. 5.15).

Problem 5.15 From a study of Fig. 5.15, what residues account for the discrimination against dCMP as a substrate, and what is the mechanism by which they achieve this specificity?

Problem 5.16 The dissociation constant for the binding of oxygen to myoglobin is

$$K_{myo} = \frac{[Mb]p\,O_2}{[MbO_2]}.$$ From the Hill plot in Fig. 5.40, and the values for haemoglobin of $K_1 = 35$ mmHg and $K_4 = 0.21$ mmHg, evaluate K_{myo}.

CHAPTER 6

Proteins with partners

LEARNING GOALS

- *Knowing the general features of a typical protein–protein interface,* including size, chemical composition, shape and charge complementarity, and the role of embedded water molecules.

- *Understanding accessible surface area,* how it is measured, and its significance in terms of the thermodynamics of protein folding and interaction.

- *Recognizing the common role of aggregation of proteins in several diseases,* including sickle-cell anaemia, amyloidoses, **Huntington's, Parkinson's,** and **Alzheimer's** diseases, and prion diseases such as 'mad cow disease'.

- *Appreciating the very great variety of immunoglobulin structures required to recognize the entire organic world.* Understanding the genetic mechanisms for generating antibody genes by combining segments.

- *Understanding the domain structure of typical immunoglobulins,* and the distinction between variable and constant domains, the significance of hypervariable regions, and the distribution of framework and complementarity-determining regions (CDRs).

- *Relating structures of antibodies to their function of binding antigens:* how the antigen-binding site is formed from loops from variable domains. Knowing the successes and limitations of the canonical-structure model of the structures of antigen-combining sites.

- *Understanding structure–function relationships in proteins of the major histocompatibility complex, and T-cell receptors,* and how they interact during the immune response.

- *Knowing basic principles of viral capsid assembly,* the purely geometric problems that viruses face in capsid design, and how different viruses solve these problems.

- *Recognizing the very great variety of protein–nucleic acid interactions* and being familiar with some examples in detail.

Introduction

Proteins rarely act alone.

Many proteins form parts of stable aggregates, including oligomeric proteins such as haemoglobin (see Fig. 1.11(c)) or phosphofructokinase (see Fig. 5.3); larger complexes such as GroEL–GroES, which assists in protein folding, or the proteasome, which participates in controlled protein turnover; many fibrous proteins such as keratins or actin (see Chapter 2); and viral coat proteins. Photosynthetic reaction centres are large multiprotein structures involved in the capture of light energy.

In addition to stable long-lived complexes, many metabolic and regulatory processes involve proteins that may associate only transiently. All enzymatic reactions with macromolecular substrates, such as tyrosine kinases are examples. Arrival of a ligand at a cell surface may initiate a signal-transduction cascade by inducing receptor dimerization (see Fig. 2.29).

Central to life are protein–DNA complexes, including the structures of chromosomes and the enzymes that replicate, repair, and transcribe DNA, and protein–RNA complexes, including the spliceosome, which removes introns from pre-messenger RNA, and the ribosome, containing the protein-synthesizing machinery. Regulation of gene transcription depends on interaction between proteins and specific DNA sequences. Protein–protein interactions may modulate transcriptional regulation. For example, interactions with protein cofactors affect the DNA-binding specificity of homeodomain proteins, which are transcription regulators involved in laying down a body plan. In the control of expression of the *lac* operon in *E. coli*, the CAP protein bound to a site within the promoter region interacts with and enhances the binding affinity of RNA polymerase (see Chapter 9).

Stable oligomeric proteins may contain many copies of one protein, or combine different ones. Some prokaryotic proteins containing identical subunits are homologous to eukaryotic proteins containing related but non-identical subunits, arising by gene duplication and divergence. The proteasome is an example. Some viruses achieve diversity *without* duplication, by combining proteins with the same sequence but different conformations.

Interacting proteins span a range of structures and functions:

- Simple dimers in which the two monomers appear to function independently.
- Oligomers with functional 'cross-talk', including ligand-induced dimerization of receptors, and allosteric proteins such as haemoglobin, phosphofructokinase, and asparate carbamoyltransferase.
- Large fibrous proteins such as collagen, actin or keratin.

- Non-fibrous structural aggregates such as viral capsids.
- Large aggregates with dynamic properties such as ATPase, pyruvate kinase, and the GroEL–GroES chaperonin.
- Protein–DNA complexes. In many cases initial binding is followed by recruitment of additional proteins to form large complexes.
- Many proteins, whether monomeric or oligomeric, which function by interacting with other proteins, often transiently. These include all enzymes with protein substrates, and many antibodies, inhibitors, and regulatory proteins.
- Protein interactions are frequently associated with disease, as misfolded or mutant proteins are prone to aggregation. Amyloidoses are diseases characterized by extracellular fibrillar deposits, usually with a common crossed-β-sheet structure. They arise from a variety of causes, including overproduction of a protein, destabilizing mutations, and inadequate clearance in renal failure. Alzheimer's and Huntington's diseases are also associated with protein aggregation.

Complexes of proteins:
- may involve other proteins and/or nucleic acids
- may be long-lived or transient
- vary greatly in size and types of subunits
- may cause disease as a consequence of unwanted aggregation

Structural studies have elucidated several important features of the interactions between soluble proteins:

- *What holds the proteins together?* Burial of hydrophobic surface, hydrogen bonds and salt bridges.
- *Do proteins change conformation upon formation of complexes?* In many cases they do. In these cases the interaction energy has to 'pay for' the conformational change, and the interface tends to be larger.
- *What determines specificity?* Complementarity of the occluding surfaces: in shape, hydrogen-bonding potential, and charge distribution. Prediction of protein complexes from the structures of the partners is the *docking problem* (see Chapter 3).

General properties of protein–protein interfaces

Sets of interacting proteins of known structure include many oligomeric proteins, proteinase–inhibitor complexes, antibody–antigen complexes, and complexes involved in signalling. The interfaces can be characterized by (1) the overall amount of surface buried, (2) the chemical composition of the buried surfaces, (3) the shape and charge complementarity of occluding surfaces, and (4) specific interactions, such as hydrogen bonds.

Burial of protein surface

The accessible surface area (ASA) of a protein is calculated by rolling a probe sphere of radius 1.4 Å over the protein and determining the area of the surface generated (see Chapter 2). The surface buried by formation of a complex is computed by calculating the ASA of the complex and subtracting the ASA of the components separately. The results give the area buried in the interface, and identify the residues or atoms that change accessibility.

If the structure of the complex is known but the structures of the free partners are not, one can calculate a buried surface in the complex by separating the partners artificially in the computer. This would not accurately measure the change in surface area, because it necessarily ignores conformational changes that might take place on association. In a number of cases, a region that is disordered in a component becomes ordered in the complex (and, in the very special case of the serpin–proteinase complex, vice versa!).

The interface is defined by the atoms that lose accessible surface area in forming the complex. Not all interface atoms are completely buried. A picture of a typical interface would be a continuous, relatively flat surface patch, with some atoms at the centre buried completely, and others at the periphery buried only partially. A typical interface might involve 22 residues and 90 atoms, of which 20% would be mainchain atoms, and an occasional water molecule.

A histogram of surface area buried in binary protein complexes shows a peak centred at 1600 Å^2, containing about 70% of the complexes. The minimum buried surface for stability under typical conditions is about 1200 Å^2. Complexes that bury more than 2000 Å^2 tend to involve conformational changes upon complex formation.

The composition of the interface

The chemical character of protein–protein interfaces is intermediate between that of the surfaces and interiors of monomeric globular proteins. Interfaces are enriched in neutral polar atoms at the expense of charged atoms (see Table 6.1).

The amino acid composition of interfaces is enriched in hydrophobic residues—Ile, Val, Leu, Met, Phe, and Tyr—relative to remaining exposed surface. A recent survey of amino acid distribution distinguishes the centre from the periphery of the interface (see Fig. 6.1). The periphery of the interface is enriched in Arg.

Complementarity

Complementarity of interfaces is reponsible for specificity. It involves geometric complementarity, that is, good packing at the occluding surfaces, and also complementarity in hydrogen bond and charge–charge interactions. This is true of many complexes in which portions of the interfaces may be flexible or even disordered in the separated state (see Chapter 1).

Most protein–protein interfaces are as well-packed as protein interiors. This optimizes the Van der Waals interactions that stabilize the complex. In many cases, isolated water molecules occupy sites in the interface, avoiding holes in places where the fit of the proteins is not exact.

Table 6.1 The chemical nature of protein surfaces, interfaces, and interiors

	Atomic composition, %		
	Non-polar	Polar (neutral)	Polar (charged)
Surfaces of globular proteins	57%	24%	19%
Interfaces	56%	29%	15%
Interiors of globular proteins	58%	39%	4%

Mainchain atoms constitute an average of 19% of interfaces.

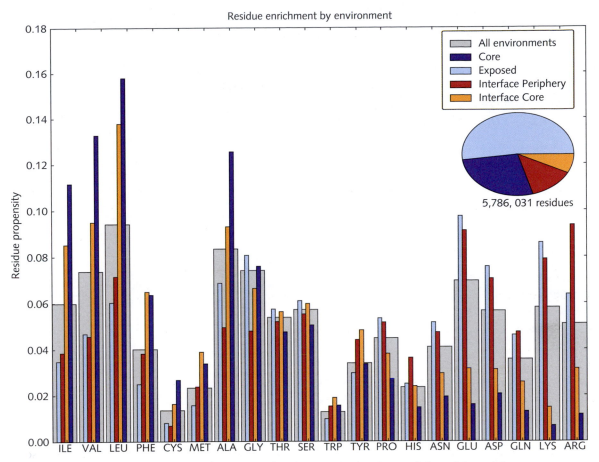

Figure 6.1 Residue propensities for amino acids in four structural environments in protein–protein interfaces: (1) Buried in the core of the domain, not in the interface; (2) solvent-exposed in the domain but not in the interface; (3) at the core of the interface; (4) in the periphery of the interface. Gray bars show the overall observed frequencies of the amino acids; therefore coloured bars standing proud of the corresponding gray bar indicate an enrichment of the amino acid in the corresponding structural environment.From: Bickerton, G.R., Higueruelo, A.P. and Blundell, T.L. (2011). Comprehensive, atomic-level characterization of structurally characterized protein–protein interactions: the PICCOLO database. BMC Bioinformatics 12, 313. Reproduced via CC-BY License

Specific interactions at protein–protein interfaces

Atom–atom contacts include non-polar Van der Waals interactions, hydrogen bonds and salt bridges, and occasionally disulphide bonds. Typically there is one hydrogen bond per 170 Å² of area, one-third of which involve a charged side-chain. There is on average one fixed water molecule per 100 Å² of interface, usually forming hydrogen bonds to both proteins. That is, a typical interface of 1700 Å² would be expected to have about 10 intermolecular hydrogen bonds and about 17 fixed water molecules.

Protein–protein complexes are stabilized by:
- burial of hydrophobic complementary occluding surfaces;
- hydrogen bonding;
- covalent bonds such as disulphide bridges.

Phage M13 gene III protein and *E. coli* TolA

During infection of *E. coli* by phage M13, a complex forms between the N-terminal domain of the minor coat gene 3 protein of the phage and the C-terminal domain of a receptor protein in the bacterial cell

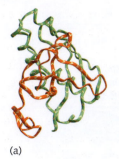
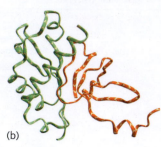
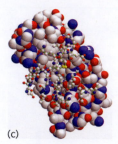

(a)   (b)   (c)   (d)

Figure 6.2 Interface between phage M13 gene III protein (N-terminal domain) and *E. coli* protein TolA (C-terminal domain). (a,b) Folding patterns and relative orientation of domains, viewed approximately (a) perpendicular and (b) parallel to the interface. Note the β-sheet formed from strands contributed by both partners. (c) TolA domain shown as spheres, gene III protein shown in ball-and-stick representation. (d) Slice through interface, TolA black, gene III protein red, water molecule blue. It is possible that another water molecule sits next to the one inside the structure.

membrane, TolA. Figure 6.2 shows different representations of this interface.

The complex is stabilized by burial of 1765 Å² of surface area, by combination of β-sheets from both proteins to form an extended β-sheet (see Fig. 6.2(a)), and by several sidechain–sidechain hydrogen bonds and salt bridges. The area buried in the complex (see Table 6.2) is divided almost evenly between the two partners.

Table 6.2 N-terminal domain of phage M13 gene III protein/C-terminal domain *E. coli* protein TolA

Partner	Accessible surface area		
	Separately	In the complex	Difference
Gene III protein	4270	3382	888
TolA	5311	4434	877
Sum	9581	7816	1765

Multisubunit proteins

Proteins fit together in many ways. It would be useful to browse and revisit the pictures of multisubunit proteins in this book, with the following questions in mind:

- *What is the stoichiometry?* How many different types of subunits appear, and how many of each are present? Most proteins are homodimers or homotetramers. Monomers and heterooligomers are less common. The ribosome is an extreme example of a heterooligomer, containing RNAs as well as proteins. Proteins containing even numbers of subunits are more common than those containing odd numbers of subunits.

- *What is the relationship between the contributions of different subunits to the interface?* Consider a dimer of two identical subunits. In **isologous** binding, the interface is formed from the same sets of residues from both monomers. In **heterologous** binding,

different monomers contribute different sets of residues to the binding site. A handshake is isologous.

- *Is the structure open or closed?* In an open structure, at least one of the sites forming the binding surface is exposed in at least one of the subunits, so that additional subunits could be added on. In a closed structure, all binding surfaces are in contact with partners, and the assembly is saturated. Domain swapping (see Fig. 7.51) often, but not always, produces closed isologous dimers.

An isologous open structure is not possible—why?

- *What is the symmetry of the structure?* Symmetry is the rule, rather than the exception, in structures of oligomeric proteins. The subunits in most dimers are related by an axis of twofold symmetry. Yeast hexokinase is an exception. It forms an asymmetric dimer. The human growth hormone

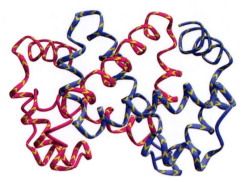

Figure 6.3 Trp repressor, an intimately intertwined dimer [3$_{WRP}$].

receptor, a nearly symmetric dimer, binds an asymmetric ligand (see Fig. 2.30).

- *Do any of the subunits undergo conformational changes on assembly?* Often we don't know. In cases of extensively interlocked interfaces, such as Trp repressor (see Fig. 6.3), we can be confident that the monomers could not adopt the same structure in the absence of their partners. In ATP synthase, the interactions with the γ subunit distort a threefold symmetric complex of αβ subunits (see Fig. 5.37).

- *What features contribute to the stability of oligomeric proteins?* The usual suspects: burial of hydrophobic surface, hydrogen bonds, and salt bridges. Covalent bonds link many oligomeric proteins. Some, such as immunoglobulins, and insulin (see Fig. 2.6) form interchain disulphide bridges. (The chains of insulin also share metal-binding sites between monomers.) Cross-links between lysine and hydroxylysine residues give different collagens their different mechanical properties.

Diseases of protein aggregation

The ability of proteins to form complexes is essential for life. But when this propensity escapes control it can be harmful, even fatal.

The first disease attributable to protein aggregation was sickle-cell anaemia, the first recognized **molecular disease**. Sickle-cell anaemia arises from a single point mutation in the β-chain of haemoglobin, converting the normal form, HbA, to HbS: β6Gln → Val. HbS/HbA heterozygotes have a milder form of the disease than HbS/HbS homozygotes. The mutation creates a sticky hydrophobic patch on the surface, normally interrupted by the polar Gln sidechain. As a result, the deoxy form of haemoglobin forms polymers that precipitate within the red blood cell (see Box 1.3).

Many diseases are now recognized to arise from protein aggregation (see Box 6.1.) Mutations can aggravate the problem. Misfolded proteins are more prone to aggregate, and mutated proteins are more prone to misfold. Overproduction of proteins as a result of breakdown of control mechanisms, as in myelomas that overproduce immunoglobulin light chains, also aggravates the threat of aggregation.

BOX 6.1	Diseases associated with protein aggregates	
Disease	**Aggregating protein**	**Comment**
Sickle-cell anaemia	deoxyhaemoglobin–S	mutation creates hydrophobic patch on surface
Classical amyloidoses	immunoglobulin light chains, transthyretin, many others	extracellular fibrillar deposits
Emphysema associated with Z-antitrypsin	mutant α$_1$-antitrypsin	destabilization of structure facilitates aggregation
Huntington's	altered huntingtin	one of several polyglutamine repeat diseases
Parkinson's	α-synuclein	found in Lewy bodies
Alzheimer's	Aβ, τ	Aβ = 40–43 residue fragment (see Fig. 6.4)
Spongiform encephalopathies	prion proteins	infectious, despite containing no nucleic acid

Amyloidoses

Classical amyloidoses involve extracellular accumulations of insoluble fibrils, 80–100 Å in diameter. Although the name amyloid implies polysaccharide, the aggregates contain protein. (Virchow, in the mid-nineteenth century, named them amyloid because they could be stained with iodine, like starch.) Common features include:

- characteristic microscopic appearance after staining with haematoxylin/eosin;
- regular fibrils seen in electron micrographs;
- β-sheet structure in X-ray fibre diffraction, with the backbone perpendicular to the fibre axis—the **cross-β structure**;
- bright green fluorescence under polarized light after staining with the dye Congo Red, specific for β-sheet structures;
- solubility at low ionic strength.

Many proteins are known to form amyloid fibrils, including immunoglobulin light chains and their fragments, transthyretin, and lysozyme. Although these proteins do not have a common folding pattern in their native state, they can apparently adopt a common structure in the amyloid state. It has been suggested that *all* proteins can form this common cross-β fibrillar structure, under suitable conditions.

Alzheimer's disease

Alzheimer's disease is a neurodegenerative disease common in the elderly. It is associated with two types of deposits:

1. dense insoluble extracellular protein deposits, called **senile plaques**. These contain the Aβ fragment (the N-terminal 40–43 residues) of a cell surface receptor in neurons, the β-protein precursor (β PP), or **amyloid precursor protein (APP)** (see Fig. 6.4).

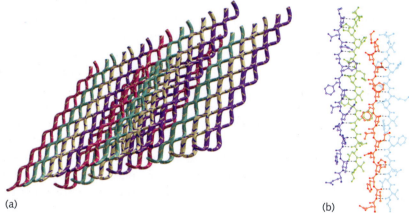

(a)   (b)

Figure 6.4 Aβ fragments of the amyloid precursor protein form deposits in the brain, associated with Alzheimer's disease. This structure of the smaller fragment of Aβ (residues 11–25) forms a cross-β structure similar to that of classic amyloid protein aggregates. (a) The fragment forms extended stacks of antiparallel β-sheets. (Each set of four strands with the same four colours corresponds to one unit cell of the crystal.) The axis of the corresponding fibre is in the plane of one of the sheets, perpendicular to the strand direction—not very far from parallel to the view direction. These pictures show repeating units. In the crystal or fibre, the sheets contain more strands, and the stacks contain more sheets. Aβ fragments containing residues 1–40 (very close to the molecules in deposits isolated from the brains of Alzheimer patients) form a fibre of diameter ~70 Å. The 15-residue fragment in this crystal structure also forms fibres, with diameter ~50 Å. The interresidue distance along the strands is ~3.3 Å. Therefore, the 15-residue extended strand is close to 50 Å long; that is, the length of the strands is close to the diameter of the fibre. The distance between the stacked sheets is ~10.6 Å. As a rough estimate, the fibre contains four or five sheets stacked as shown in this picture. These sheets are extended to macroscopic dimensions, in a direction in the plane of any of the sheets and perpendicular to the strand direction. (b) The atomic structure of one set of four strands, showing both the interstrand hydrogen bonding and the stacking of the sheets. This figure shows two pairs of hydrogen-bonded strands. In their packing together there has been a displacement of two residues in the strand direction—perpendicular to the corresponding fibre axis. Formation of such a structure from longer protein chains could be possible by connecting the hydrogen-bonding strands into a succession of hairpins. (I thank Drs L. Serpell, E.D.T. Atkins, and P. Sikorski for the coordinates of their structure.)

2. **neurofibrillary tangles**, twisted fibres inside neurons, containing microtubule-associated protein tau. There is some evidence that amyloid deposits promote tangle formation.

Abnormalities in tau appear in other neurodegenerative diseases, the **tauopathies**.

Prion diseases—spongiform encephalopathies

Prion diseases are a set of neurodegenerative conditions of animals and humans, associated with deposition of protein aggregates in the brain. A characteristic sponge-like appearance of the brains of affected individuals is observed in post-mortem investigation.

Prion diseases are unusual among protein deposition diseases in that they are transmissible, and unusual among transmissible diseases in that the infectious agent is a protein. Moreover—another unusual feature—some prion diseases are hereditary, for example **familial Creutzfeld–Jacob disease (CJD)**. (Distinguish between a disease transmitted from mother to baby perinatally by passage of an infectious agent—as in many AIDS cases—with a truly hereditary disease depending on parental genotype.) All hereditary human prion diseases involve mutations in the same gene, one that encodes the protein found in the aggregates deposited in the brains of sufferers of *both* hereditary and infectious prion diseases (see Box 6.2).

We mentioned, in connection with circular dichroism (see Chapter 1), that the prion protein can exist in two forms: the normal PrP^C and the dangerous PrP^{Sc}. PrP^{Sc}, but not PrP^C, can (1) form aggregates, (2) catalyse the conversion of additional PrP^C to PrP^{Sc} within the brain of an individual person or animal, and (3) infect other individuals, by various routes, including ingestion of nervous tissue from an affected animal (or person, in the case of kuru).

> PrP^C = prion protein – *Cellular*; PrP^{Sc} = prion protein – *Scrapie*

Prion disease presents widespread health problems for humans and animals. In 2001 a serious epidemic of **bovine spongiform encephalopathy** (BSE, colloquially, '**mad cow disease**') devastated the United Kingdom countryside. There was an apparent association with the appearance of human cases of **variant Creutzfeld–Jacob disease (vCJD)**. In the hereditary disease familial CJD, symptoms begin to appear in people aged 55–75. Variant CJD affected people in their 20s. It is hypothesized that these outbreaks were associated with transmission of prion protein infections across species barriers: sheep to cows for BSE, and cows to humans for vCJD.

Prion proteins form a family of homologous proteins in many species of animals, and also in yeast, but apparently not in *C. elegans* or *Drosophila*. Normal human prion protein is synthesized as a 253-residue polypeptide. This comprises an N-terminal signal peptide, followed by a domain containing ~five tandem repeats of the octapeptide PHGGGWGQ (in mammals), a conserved 140-residue domain, and a C-terminal hydrophobic domain. The signal

---

BOX 6.2   Some diseases associated with prion proteins

Disease	Species affected	Symptoms
scrapie	sheep	hypersensitivity, unusual gait, tremor
bovine spongiform encephalopathy, or 'mad cow disease'	cow	similar to scrapie
kuru	human	loss of coordination, dementia
Creutzfeld–Jacob disease (CJD)	human, age 55–75	impaired vision and motor control, dementia
variant CJD	human, age 20–30	psychiatric and sensory anomalies preceding dementia
Gerstmann–Straüssler–Scheinker syndrome	human	dementia
fatal familial insomnia	human	sleep disorder

domain and the C-terminal domain are cleaved off, and the protein is anchored to the extracellular side of the cell membrane of neurons by a GPI (glycosylphosphatidylinositol) group bound to the C-terminal residue Ser231 of the mature protein.

The normal role of PrP^C is not clear. Mice in which PrP^C has been knocked out develop normally for a time, and eventually die of apparently unrelated developmental defects. In fact, PrP^C–knockout mice are not susceptible to infection with PrP^{Sc}, an observation important in proving the mechanism of the disease.

The nature of the conformational change is still not entirely clear. The change from α to β structure shown by the circular dichroism is one clue (see Box 1.12). In principle, prion proteins show multiple structures from one polypeptide sequence. However, differences in glycosylation patterns between PrP^C and PrP^{Sc} have been reported; these may play a role in defining the conformation.

The mechanism by which PrP^{Sc} catalyses the transformation of additional PrP^C to PrP^{Sc} is also not clear. Inherited prion diseases are associated with mutants, presumably increasing the tendency for conformational mobility. A related question concerns the kinetics of the process—what governs the rate of accumulation of aggregates that causes many prion diseases to appear only among the elderly?

Many diseases arise from formation of protein aggregates, including:

- sickle-cell anaemia;
- amyloidoses;
- Alzheimer's disease;
- Huntington's disease;
- familial and variant Creutzfeld–Jacob disease;
- prion diseases.

Most of these are genetic, some are infectious.

The immune system

The vertebrate immune system has the job of identifying foreign pathogens, and defending the body against them. It involves the integrated activities of a number of types of multisubunit proteins, and of complexes formed among them, and with **antigens** or fragments of antigens. The system:

- must recognize foreign molecules, and accurately distinguish between 'self' and 'non-self';
- destroy the invaders: molecules, viruses, or bacterial cells;
- remember foreign substances previously encountered, and mobilize more rapid and specific responses to subsequent encounters.

The immune system is one of many biological examples of molecular recognition. Individual antibody–antigen complexes fit the classical ideas of lock-and-key complementarity and induced fit; but the system as a whole is more complex than, for example, enzymes, in which the lock and key are in most cases unchanging within the lifetime of the organism. In contrast, in immunology there is a primary response—to antigens not previously encountered—followed in a matter of days by a secondary response of greater specificity and affinity. In the primary response, the immune system achieves affinity for many keys by providing many locks, but achievement of the spectacular affinity of the secondary response requires a mechanism that perturbs and 'tunes' the locks for better fit. It is the integration of the immune response over the system as a whole, involving both genetic and structural diversity, that adds dimensions of complexity.

The immune system has evolved specific weapons at the genetic, protein, and cellular level:

- In **cellular immunity,** a macrophage ingests an antigen, digests it, and displays fragments on its surface, in complex with proteins of the **major histocompatibility complex (MHC)**. T-cells recognizing these complexes are induced to mature and proliferate. **Cytotoxic T-cells** will destroy cells that bear their target antigen.
- In **humoural immunity,** a B-cell will engulf and digest molecules that bind to the immunoglobulin on

its surface. It will display the fragments produced, on its surface in complex with MHC proteins. These complexes are recognized by **T-cell receptors** on T-helper cells, leading to activation of the B-cell to proliferate and differentiate into a factory to produce and excrete the specific antibody elicited. Some B-cells become **memory cells** that remain poised to respond to subsequent challenges from the same antigen by copious synthesis of secondary antibodies. This is the mechanism that keeps us from getting certain 'childhood' diseases more than once, and why second and subsequent pregnancies involving **Rh incompatibility** are more dangerous than the first.

> • The immune system recognizes foreign proteins, and defends against pathogens. Autoimmune diseases may result from failure to distinguish accurately between self and non-self.

Antibody structure

The immune system has evolved to be able to recognize the entire organic world. It is designed to generate diversity: during our lifetimes, we synthesize about 10^{10} different antibodies.

Antibodies contain multiple polypeptide chains, distinguished by size into **light (L) chains** (r.m.m. about 23 000) and **heavy (H) chains** (r.m.m. about 50 000–70 000). Each chain is itself modular, containing a series of homologous domains, each with a characteristic double-β-sheet structure. Light chains contain two domains, and heavy chains contain four or five domains. Light chains are distinguished into κ and λ classes or isotypes. In the human, κ and λ light chains are present in comparable proportions; in the mouse κ light chains predominate.

Different classes of immunoglobulins—IgG, IgA, IgM, and IgE—differ in the assembly of their chains and domains (see Fig. 6.5). IgGs usually have two heavy chains each containing four domains and two light chains each containing two domains (see Fig. 6.6). Figure 6.7 shows the structure of a complete IgG.

How is antibody diversity created? The genes for antibodies are assembled from a combination of DNA segments. The Ig loci in vertebrate genomes contain tandem segments of genes that join in different combinations to generate immunoglobulin genes (see Table 6.3). Each light-chain gene contains a V segment, a J segment, and a C segment. Each heavy-chain gene contains a V segment, a D segment, a J segment, and several C segments. Additional variability is introduced by imprecise joining of the V, D, and J segments. This splicing occurs at the DNA level. It is entirely independent of the exon/intron splicing that creates mature messenger RNA.

Upon challenge by a pathogen, the products of these genes mount the initial **primary immune response**, mobilizing antibodies with dissociation constants in the micromolar range. Subsequently, **somatic mutation** tunes the affinity and specificity of the antibodies, typically achieving nanomolar dissociation constants. Artificial selection of antibodies, using phage display techniques, can also routinely produce molecules of comparable affinity to those arising naturally from somatic mutation.

> • Antibodies close to those encoded in the genome produce the primary immune response. In response to an antigen, antibodies undergo somatic mutation, a process of tuning for enhanced affinity and specificity.

Molecules related to antibodies appear on cell surfaces, to mediate cell–cell recognition and signalling processes. In the immune system itself, receptors trigger proliferation of particular cells in response to antigenic challenge. Many related proteins are known; antibodies are members of a large superfamily of proteins, most but not all of which are involved in molecular recognition. Some are active in development; participating, for example, in neuronal targeting.

The constant, the variable, and the hypervariable

Patterns in multiple sequence alignments distinguished two types of domains, **variable (V)** and **constant (C)**, on the basis of the extent of sequence conservation. (A third type of domain, the I type, occurs in other molecules of the immunoglobulin superfamily.)

Within the variable domains are regions of still higher variability that determine antibody specificity.

shows affinity and requires only 'fine tuning'. Consider the alternative: if most mutations completely altered the mainchain conformation of the loops, the effect would be to produce a succession of independent primary responses, rather than a fine-tuned response with structures perturbed in only minor (but crucially important) ways from a set of already selected primary antibodies. The reader must never underestimate the power of very tiny structural changes to produce very large effects on binding affinity.

Greater variability in the H3 loop

H3, the third hypervariable region of the heavy chain, is far more variable in length, sequence, and structure than the other antigen-binding loops. It cannot therefore be included in the canonical-structure description of the conformational repertoire of the three hypervariable regions of V_L chains and the first two of V_H chains. Because the H3 loop falls in the region of the V–D–J join in the assembly of the immunoglobulin heavy-chain gene, several mechanisms contribute to the generation of its diversity, including the combinatorial choice of V_H, D, and J gene segments, and alternative splicing patterns at their junctions.

In expressed antibodies, H3 appears prominently at the centre of the antigen-binding site (see Fig. 6.9). Given this central position, H3 makes significant interactions—with other loops, with the framework, with the light-chain partner, and with ligands—that influence its conformation. Thus H3, in contrast to the other five antigen-binding loops, has a conformation that depends strongly on its molecular environment. Indeed, structures containing the V_H domain of antibody B1–8 combined with two different V_L domains show two very different conformations of H3. This important observation implies that (unlike the other five antigen-binding loops) general rules governing the conformation of H3 *must* involve nonconserved interactions outside its local region in the sequence.

Some antibodies in cows have a very unusual V_H CDR3 region. They contain a stalk protruding from the antibody, with a domain at the end. The stalk consists of a pair of β-strands 10 residues long, so that the extra domain does not pack against the rest of the antibody but stands proud of it. It is possible to use this structure as a carrier, by substituting other domains for the natural one. This produces

molecules with favourable pharmacokinetic properties, as antibodies tend to be relatively long-lived in the bloodstream. It has also been possible to reproduce this structure in a human antibody, enhancing the possiblities of therapeutic applications.

Spatial distribution of somatic mutations in the maturation of the antibody response

Somatic mutation is the process whereby antibodies active in the primary response are tuned in affinity and specificity by tinkering with their sequences to produce the 'mature' antibodies of the secondary response. Whereas the diversity in the primary antibodies is highest in H3 and L3, around the centre of the antigen-binding site, somatic mutations spread the diversity to its periphery. Usually, somatic mutations are isolated point substitutions, and not insertions or deletions.

Antibody maturation

G.J. Wedemayer, P.A. Patten, L.H. Wang, and P.G. Schultz studied maturation of an anti-nitrophenyl phosphonate catalytic antibody by X-ray crystallography. They solved the structures of the primary (germ-line) and secondary, somatically mutated, Fab fragments, each with and without ligand. Their results reveal both the sites and the structural effects of mutations. There are nine amino acid sequence changes between germ-line and mature antibody: three in the light chain and six in the heavy chain. One of the light-chain mutants appears within the L1 loop; the other two are in positions in the sequence near L1 and L2. One of the heavy-chain mutants appears in H2, three are in regions adjacent to or in contact with the antigen-binding loops, and two are surface residues at the opposite ends of the domains, and would appear to have little effect on the antigen-binding site. No mutant is at a position directly in contact with the antigen. No mutant appears in H3, even though there are extensive contacts between H3 and the antigen. The conformations of two of the antigen-binding loops (H1 and H2) differ between germ-line and mature antibody.

The antigen-binding site of the mature antibody has a similar conformation in ligated and unligated states; a fixed conformation complementary to the antigen. This structure is shared by the ligated state

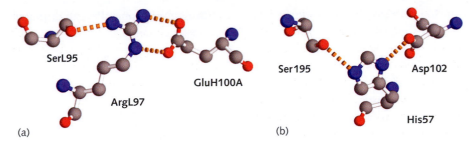

Figure 6.12 (a) Residues Ser95L, Arg97L, and Glu100aH from an antibody with proteolytic activity, [2$_{AGJ}$], compared with (b) the normal serine protease catalytic triad of bovine trypsin [1$_{BJU}$].

of the germ-line antibody, but the unligated state of the germ-line antibody shows differences in conformation. That is, the germ-line antibody *can* adopt the antigen-binding conformation, and is *induced* to do so by the ligand; the mature antibody adopts this conformation even in the absence of ligand. In fact, some of the mutations acting to 'freeze' the structure of secondary antibodies are not in the antigen-binding site at all. Their effect is similar to that of placing a matchbook under a leg of an unsteady table to select and fix a unique conformation.

Catalytic antibodies—'abzymes'

Enzymatic catalysis requires: (1) specific binding of substrate and (2) juxtaposition of bound substrate with residues that effect catalysis by interactions that stabilize the transition state. Antibodies are a suitable system to use for creating artificial enzymes because they already provide specific binding. Catalytic antibodies are called **abzymes**.

There are natural catalytic antibodies. Figure 6.12, from an antibody with natural proteolytic activity, shows three interacting residues that resemble the catalytic triad of chymotrypsin-like serine proteases. In disease, patients with certain autoimmune diseases, such as systemic lupus erythematosus, produce IgGs with DNase and RNase activities. In health, these nuclease and other enzymatic activities appearing normally in catalytic antibodies in breast milk are believed to protect the neonate against pathogens.

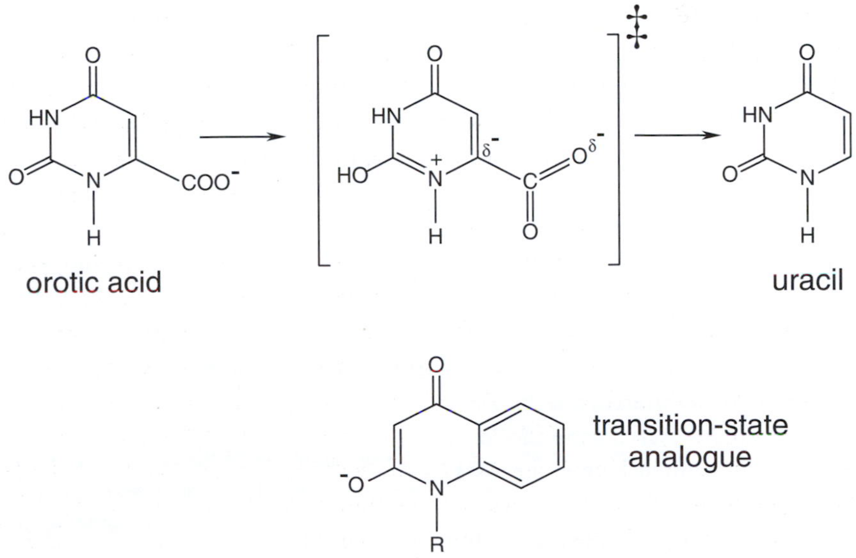

orotic acid   uracil

transition-state analogue

Figure 6.13 Hypothesized **zwitterionic transition** state for decarboxylation of orotic acid to uracil. For purposes of immunization in the work of Smiley and Benkovic, R = (CH$_2$)$_3$COOH. This hapten was conjugated to the carrier protein keyhole limpet haemocyanin.

How would one logically go about creating a catalytic antibody? Given the goal of strong binding to the transition state, one can raise natural antibodies to a transition-state analogue. This idea was suggested first by W.P. Jencks in 1969.

This approach first bore fruit in 1986, with groups at Scripps Research Institute and the University of California at Berkeley reporting antibody catalysis of the hydrolysis of aryl esters and carbonates, respectively.

Selection can choose a catalytically effective antibody from those elicited in response to a transition-state analogue. J. Smiley and S. Benkovic developed a catalytic antibody that decarboxylates orotic acid to uracil (see Fig. 6.13). The transition-state analogue used was N-carboxypropyl-2,4-dihydroxyquinoline. The intent was to elicit antibodies with a proton-donating group adjacent to the O$^-$, and a hydrophobic cavity adjacent to the fused phenyl ring to destabilize the charged carboxylate.

To select active abzymes, Smiley and Benkovic assembled antibodies produced in response to the transition-state-analogue hapten into a phage library. The phage infected a strain of *E. coli* unable to synthesize uracil. Active enzyme brought in by the phage remediated this defect, conferring a growth advantage, and allowed for selection of bacteria containing active abzyme.

An abzyme isolated from this procedure catalysed the decarboxylation of orotic acid at a rate approximately 10^8 times that of the uncatalysed reaction. Admittedly, the rate is only ~10^{-7} times that of the natural enzyme yeast OMP decarboxylase, but one must recognize that this enzyme is one of the most proficient known! (see Fig. 5.1).

There is an interesting postscript to this work. Although Smiley and Benkovic designed their transition-state analogue on the basis of the mechanism then believed to be the correct one, subsequent crystal structures of natural orotidine-monophosphate decarboxylases suggest a very different mechanism. It is likely that the abzyme-catalysed reaction follows an unnatural mechanism imprinted by the chemists' choice of transition-state analogue (see Problem 6.3).

The field is very active, and maturing. Some abzymes now rival natural enzymes in both their specificity and catalytic efficiency. Catalytic antibodies offer much promise for industrial processes and clinical applications. Potential therapeutic uses of abzymes include:

- Clearance of toxic substances from blood by specific hydrolysis.

- Abzyme-catalysed activation of a **prodrug**. This approach makes it possible to deliver active drugs to specific target locations. A prodrug-activating abzyme conjugated with a passive antibody that binds specifically to a tumour will release active drug at the site of the tumour.

- S. Planque, S. Paul, and Y. Nishiyama have described an engineered antibody that cleaves a region of the gp120 CD4 binding site of the HIV–1 virus. In contrast to a therapeutic agent that binds permanently to a single target molecule, this antibody can cleave and release many substrate particles.

- Abzymes are catalytic antibodies. Some natural catalytic antibodies are known. Others arise by raising antibodies to a hapten resembling the transition state of a reaction.

Proteins of the major histocompatibility complex

Surgical patients, if not immunosuppressed by drugs, will reject transplanted organs—unless the donor is an identical sibling—because the transplant is recognized as foreign. The immunological distinction between 'self' and 'non-self' resides in the proteins of the major histocompatibility complex (MHC) and their interaction with **T-cell receptors**. MHC proteins bind intracellularly produced peptides and present them on cell surfaces. The triggering event in alerting the immune system to the presence of a foreign protein is the recognition, by a T-cell receptor, of a complex between an MHC protein and a peptide derived from the foreign protein.

- Recognition of the distinction between self and non-self depends on the interaction between T-cell receptors and peptide-presenting MHC proteins.

MHC proteins fall into two classes, with related but different structure and function. The two classes

Class I

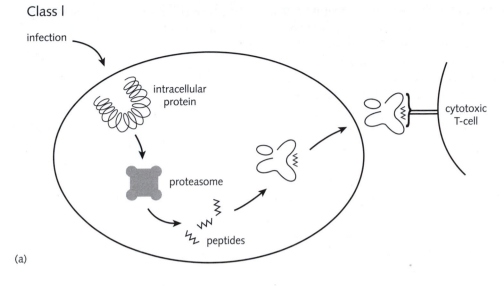

(a)

Class II

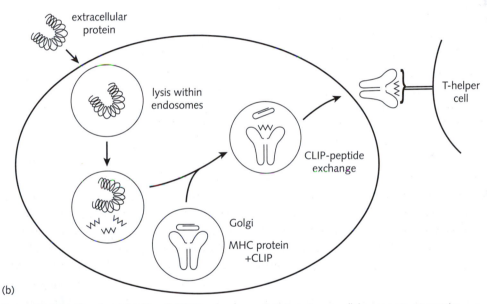

(b)

Figure 6.14 (a) Class I and (b) Class II MHC molecules participate in two parallel systems to trigger the immune response to foreign proteins originating inside and outside cells. Peptides derived from foreign proteins are loaded intracellularly and transported to the surface, where they are presented to T-cells. The representation of the class II invariant peptide (CLIP) is an icon, not a drawing of its structure.

function as parallel systems, to produce different immune responses appropriate to intracellular and extracellular pathogens, respectively (see Fig. 6.14).

- **Class I MHC molecules** appear on the surfaces of most cells of the body, and present peptides derived from proteins degraded in the cytosol.

These peptide–MHC complexes alert the body to intracellular pathogens. They interact with **cytotoxic T-cells**, and direct the immune response to the presenting cell and those in its vicinity.

- **Class II MHC molecules** appear on the surfaces of specialized cells of the immune system:

B-lymphocytes and antigen-presenting **macrophages**. They present oligopeptides derived from exogenous antigens (which have been endocytosed and chopped into peptides). These peptide–MHC complexes interact with **helper T-cells**, mediating the proliferation of cells synthesizing antibodies that circulate in the blood, and the activation of macrophages.

In addition to triggering immune responses in mature individuals, MHC–peptide complexes are also involved in the removal of self-complementary T-cells in the thymus during development, at the stage when the distinction between self and non-self is 'learnt'.

Each individual in vertebrate species expresses a set of MHC proteins selected from a diverse genetic repertoire in the species. In humans the MHC complex is a set of linked genes on chromosome 6. The system is highly polymorphic, with 50–150 alleles per locus, showing greater sequence variation than most polymorphic proteins. Each of us produces six class I molecules and a somewhat higher complement of class II. Each MHC protein must therefore be able to bind many peptides, if ~30 MHC proteins are to present the very large number of possible antigens.

The set of MHC proteins expressed defines the **haplotype** of an individual. The number of possible haplotypes has been estimated to be of the order of 10^{12}, although, because of linkage, the combinations are non-random, and because of selective pressure fewer combinations appear than expected. Like fingerprints and restriction fragment length polymorphisms, our haplotypes are a personal identification code. MHC haplotypes determine eligible donor-recipient pairs for organ transplants. In addition, MRC haplotypes have been used extensively in anthropology in measuring quantitatively the relationships between human populations, and in tracing routes of migration. There is even some evidence for a role in determining sexual attractiveness in pairs of humans.

Structures of MHC proteins

MHC proteins are modular, containing several characteristic domains. These include peptide-binding domains with folds special to the MHC system, and immunoglobulin-like domains (see Fig. 6.15).

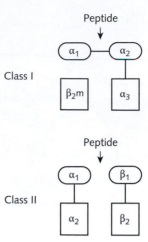

Figure 6.15 MHC proteins are modular, containing two peptide-binding domains (round cartouches) and two immunoglobulin-like domains (rectangles). Peptides bind in a groove created by the α_1/α_2 domains in class I MHC proteins and by the α_1/β_1 domains in class II MHC proteins.

Class I MHC proteins contain two polypeptide chains (see Fig. 6.16). The longer chain (r.m.m. 44 000 in humans; 47 000 in mice) has a modular structure of the form $\alpha_1–\alpha_2–\alpha_3$ followed by a short hydrophobic membrane-spanning segment and a 30-residue cytoplasmic tail. The α domains are approximately 90 residues in length. The second chain is β_2-**microglobulin**, a non-polymorphic structure (that is, constant within the species), the gene for which is unlinked from the MHC complex.

The α_1 and α_2 domains of class I MHC proteins have a common fold, and interact to form a

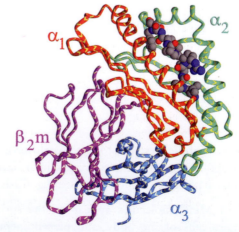

Figure 6.16 The structure of a class I MHC protein, B35, binding the peptide VPLRPMTY from the nef protein of HIV-1 [1A1N].

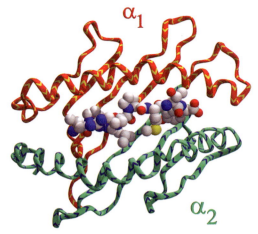

Figure 6.17 The peptide-binding domains and ligand from class I MHC protein B35 [1A1N].

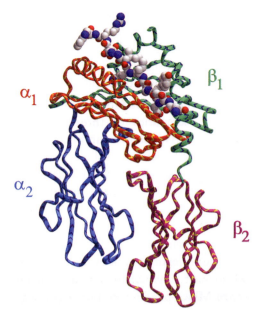

Figure 6.18 The structure of a class II MHC protein, I-Ak, binding the peptide STDYGILQINSRW from hen egg-white lysozyme [1IAK].

symmetric combined structure. They bind peptides in a groove between them, created by two long curved α-helices (see Fig. 6.17). The variability in the amino acid sequence of MHC proteins is high in the regions that surround the groove, to create variety in specificity. The α_3 domain and β_2-microglobulin do not interact with bound peptides. They are double β-sheet proteins with topologies of the immunoglobulin superfamily.

The α_1 and α_2 domains of class I MHC proteins, and the α_1 and β_1 domains of class II, have a common folding pattern. Each has a four-stranded β-sheet at the N-terminus, followed by a short bridging helix and then a long C-terminal helix that lies across the β-sheet. The strands from each domain interact laterally to form an eight-stranded β-sheet, positioning the C-terminal helices from the two domains to form the sides of a peptide-binding cleft. Each helix has a pronounced curvature. Because of the approximate dyad symmetry of the α_1–α_2 unit, the long C-terminal helices run antiparallel to each other. Peptides bind in an orientation parallel to that of the α_1 helix.

Class II MHC proteins contain an α-chain (r.m.m. 34 000) containing two domains, $\alpha_1 + \alpha_2$, a homologous β-chain (r.m.m. 29 000) containing two domains, $\beta_1 + \beta_2$, and an invariant chain I*i* (r.m.m. 31 000) (see Fig. 6.18; the invariant chain does not appear). The α_1 and β_1 domains pack together to make a structure similar to that formed by the α_1 and α_2 domains of class I MHC proteins, and they bind peptides in a similar mode (see Fig. 6.19).

Class I molecules can bind peptides of limited length—from about 8 to 11 residues—but most commonly 8 to 9. Two factors impose the limits: the closure of the cleft at either end, and a salt bridge to the C-terminal carboxyl group of the peptide. The cleft will accommodate nine-residue peptides in a nearly extended conformation; longer peptides can bulge out or zig-zag or their C-termini can extend out beyond the end of the pocket. In class II the cleft is also closed at the right, but open at the left.

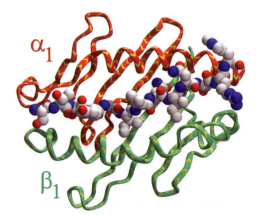

Figure 6.19 The peptide-binding domains and ligand from a class II MHC protein (I-Ak) [1IAK].

- Viruses contain nucleic acid packed inside a coat formed by a regular array of protein. Two basic architectures are: helical viruses, such as tobacco mosaic virus, and spherical or polyhedral viruses, such as tomato bushy stunt virus.

- To create a polyhedral coat with more than 60 copies of a single protein, it is not possible for all proteins to have identical environments. However, they can occupy quasi-equivalent positions, leading to small conformational deformations of proteins in response to the differences in surroundings.

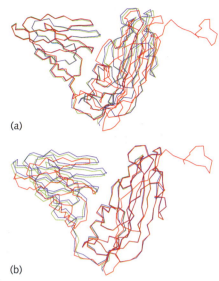

(a)

(b)

Figure 6.29 Superpositions of the three subunits of the coat protein of tomato bushy stunt virus [2TBV]. The colours correspond to those in Fig. 6.27: A = green, B = blue, C = red. (a) Structures superposed on the P domain, at the left of the picture. The red and black copies of the S domain also superpose well, but the S domain of the blue chain does not. (b) Structures superposed on the S domain, at the right of the picture. The red and black copies of the P domain also superpose well, but the P domain of the blue chain does not. The conclusion is that the red and black molecules have similar mainchain conformation. The blue molecule differs, by a hinge motion between the domains.

does not appear in virus crystal structures. The electron-density map shows a symmetry-averaged superposition of the capsid contents, and is not interpretable in atomic detail.

Tomato bushy stunt virus

Tomato bushy stunt virus (TBSV) is an **icosahedral virus** with a capsid formed from 180 copies of a 386-residue protein. The capsid is approximately 175 Å in radius, enclosing a single-stranded RNA genome, 4776 nucleotides long, which encodes five proteins.

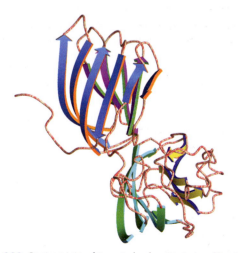

Figure 6.28 Coat protein of tomato bushy stunt virus [2TBV]. P domain above, S domain below. The R domain is not shown. The general combination of two double-β-sheet domains appears in many virus capsid proteins.

The capsid protein contains three domains (see Fig. 6.28). The N-terminal R domain projects into the capsid and interacts with the RNA. (Figure 6.28 does not show the R domains and the RNA.) The central, S, domain, forms the shell. The C-terminal domain, the P or protruding domain, creates prominent surface features. Both S and P domains have double-β-sheet structures.

A short hinge region between the S and P domain mediates the conformational change between the subunits. Tomato bushy stunt virus is a T = 3 structure. The coat proteins have three quasi-equivalent environments, and—despite their identical sequences—differ slightly in structure (see Fig. 6.29).

Bacteriophage HK97: protein chain-mail

Bacteriophage HK97 is an icosahedral virus with a double-stranded DNA genome. The capsid is 55 nm in diameter, formed of 420 copies of a 281-residue protein. In most viruses, the capsid proteins interact by familiar non-covalent interactions—Van der Waals forces, hydrogen bonding, and salt bridges. In contrast, HK97 forms covalent bonds between the subunits of its capsid.

The sidechains of Lys169 and Asn356, in adjacent subunits, form a peptide-like bond:

$$\cdots C\alpha CH_2CH_2CH_2CH_2NH_3^+ +$$
$$O=C(NH_2)CH_2C\alpha \cdots \rightarrow$$

$$\cdots C\alpha CH_2CH_2CH_2CH_2NH-COCH_2C\alpha\cdots + NH_4^+.$$

These bonds join sets of monomers into large closed rings. These rings contain five or six subunits, suitable for placement on the fivefold, threefold or twofold axes of the icosahedral symmetry.

Still more bizarre, the rings are threaded through one another, to form a structure akin to chain-mail! (see Fig. 6.30).

This section is for those readers who think they've seen everything.

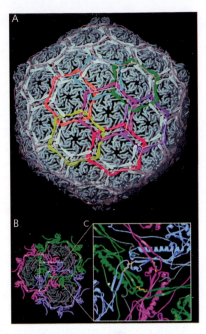

Figure 6.30 (A) Interlocking chains of subunits in the bacteriophage HK97 capsid. Six hexameric rings are shown in brighter colours. Pentagonal rings are also visible. (B, C) Details of the linkage between adjacent subunits meeting at a threefold axis.

[From: Wikoff, W.R., Liljas, L., Duda, R.L., Tsuruta, H., Hendrix, R.W. and Johnson, J.E. (2000). Topologically linked protein rings in the bacteriophage HK97 capsid. *Science*, 289, 2129–2133. Reproduced with permission from AAAS.]

Photosynthetic reaction centres

Photosynthetic reaction centres are multiprotein complexes that capture light energy. They require a precise large-scale structure in order to create the scaffolding for precise spatial distribution of the cofactors, to control the flow of excitation energy and electrons.

The reaction centre from the purple bacterium *Rhodopseudomonas viridis* is the site of the initial step in the capture of light energy in photosynthesis. The reaction centre is a membrane-bound complex of four proteins, binding 14 low molecular weight cofactors. These cofactors include the chromophores that absorb the excitation energy that is converted to electrochemical potential—or redox—energy across the membrane.

Figure 6.31 shows a representation of all atoms of the reaction centre except hydrogens, in front and side views. The assembly as a whole has dimensions 72 Å × 72 Å × 133 Å. The chromophores are embedded in the two central subunits and are only partly visible.

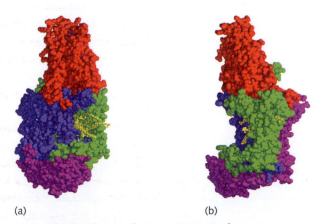

(a)   (b)

Figure 6.31 The photosynthetic reaction centre from *Rhodopseudomonas viridis* [1PRC]. This figure shows an 'all-atom' representation. Parts (a) and (b) show two orientations, at right angles to each other. The four proteins are distinguished by colour: *light*, blue; *medium*, magenta; *heavy*, green; *cytochrome*, orange; chlorophylls, yellow.

Figure 6.32 The reaction centre coloured by residue charge (positive, red; negative, blue; neutral, green). The 'green belt' around the waist of the molecule corresponds to the membrane-spanning segment [1PRC].

The four proteins are the light (L, in blue), medium (M, in purple), and heavy (H, in green) subunits, and a cytochrome (orange).

Figure 6.32 shows the structure with an alternative colour coding. Positively charged residues are coloured blue, negatively charged residues red, the others green. The wide horizontal swathe across the centre of the molecule that is entirely green (except for the cofactors) corresponds to the region that traverses the membrane. The proteins present surfaces of different character to the aqueous environment, at top and bottom, and to the membrane. The surfaces of the H subunit and the cytochrome are typical of proteins in aqueous environments: a tossed salad of charged, polar, and uncharged residues. The surfaces presented to the lipid environment of the membrane are devoid of charged residues. These membrane-exposed surfaces resemble the interiors of soluble proteins in physicochemical character.

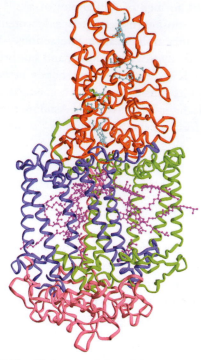

Figure 6.33 Simplified representation of the reaction centre. The cofactors are shown in all-atom detail [1PRC]. The 'special pair' of chlorophylls, the initial site of the ionization, are at the centre of the picture, near the top of the blue and green subunits.

The schematic diagrams in Fig. 6.33 show that the complex contains transmembrane helices, a structural feature common to many membrane proteins.

> • Photosynthetic reaction centres are large protein complexes sitting in membranes that contain chromophores. Reaction centres capture light energy in photosynthesis.

Protein–DNA interactions

> In this section we recognize that DNA is made of atoms, not character strings. 'One writes [poems] not with ideas, but with words.'—Mallarmé.

Protein–DNA complexes mediate several types of processes:

• replication, including repair and recombination;
• transcription;
• regulation of gene expression;
• DNA packaging, including nucleosomes and viral capsids.

Enzymes that replicate and transcribe DNA sequences are of necessity complex structures. Compare their operation with that of a typical metabolic enzyme that catalyses a single reaction of a small substrate and releases the product. In contrast, a DNA polymerase must carry out multiple manipulations of its substrate, some chemical and others mechanical. Addition of a nucleotide is a chemical step. Successive additions of nucleotides are mechanically linked, as the product is not released but translocated.

In adding different successive nucleotides, a DNA polymerase must evince a versatile but nevertheless exact specificity. It must be able to catalyse the addition of four different bases, and must select the proper one each time. This is different from promiscuous specificity, shown, for instance, by an exonuclease that can accept *any* nucleotide as a substrate at any step. Moreover, errors are unacceptable, as they would create mutations. In addition to the specificity (aimed at getting it right the first time) the enzyme does proofreading and error correction (just in case it didn't). To accomplish this, the enzyme has two active sites, one for addition of a base and an exonuclease site to remove an incorrectly incorporated one.

Other types of DNA–protein interactions require different degrees of DNA-sequence specificity. Differences in DNA-sequence specificity impose different requirements on the structures of the proteins that participate in different processes (see Box 6.6).

> • There are a great variety of protein–nucleic acid complexes. Some are permanent, others are transient. Functions include, but are not limited to, catalysis of replication, transcription, translation, and regulation of transcription.

Structural themes in protein–DNA binding and sequence recognition

What does a protein looking at a stretch of DNA in the standard B conformation see? (see Fig. 6.34) What could it hope to grab hold of? Prominent general features are the sugar–phosphate backbone, including charged phosphates suitable for salt bridges and potential hydrogen-bond partners in the sugar hydroxyl groups. Contact with the bases is accessible through the major and minor grooves, although unless the DNA is distorted the bases are visible only 'edge-on'. Hydrogen-bonding patterns between bases in the grooves and particular amino acids account for some of the DNA-sequence specificity in binding (see Fig. 6.35). However, many protein–DNA hydrogen bonds are mediated by intervening water molecules, an effect that tends to *reduce* the specificity.

BOX 6.6 DNA-binding proteins show varying degrees of DNA-sequence specificity

• Some DNA-binding proteins are relatively non-specific with respect to nucleotide sequence, including DNA replication enzymes and histones.

• Some, for instance the **restriction endonuclease** EcoRV, *bind* to DNA with low specificity, but cleave only at GATATC. This combination permits a mechanism of finding the target sequence by initial non-specific binding followed by diffusion in one dimension along the DNA.

• Some recognize specific nucleotide sequences. For example, the restriction endonuclease EcoRI binds specifically to GAATCC sequences with almost absolute specificity. It is a homodimer that recognizes palindromic sequences.

• Some DNA-binding proteins recognize consensus sequences. For example, the phage Mu transposase and repressor proteins bind 11 base-pair sequences of the form CTTT[T|A]PyNPu[A|T]A[A|T] (where [A|T] = A or T,

Py = either pyrimidine = C or T, Pu = either purine = A or G, and N = any of the four bases).

• Some recognize nucleotide sequences indirectly, *via* modulations of local DNA structure. For example, the TATA-box binding protein takes advantage of the greater flexibility of AT-rich sequences to form complexes in which the DNA is very strongly bent. The distinction between sequence specificity achieved through direct interaction with bases or through recognition of local structure has been termed 'digital *versus* analogue readout'.

• Some recognize general structural features of DNA, such as mismatched bases or supercoiling. DNA topoisomerase III is an example (see Chapter 5).

• Some DNA-binding proteins form an initial complex with high DNA-sequence specificity, followed by recruitment of other proteins of low specificity, to enhance overall binding affinity.

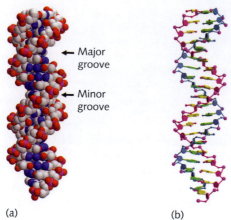

(a)   (b)

Figure 6.34 DNA in the standard B conformation. (a) All-atom representation. (b) Schematic representation. This is the conformation of DNA under physiological conditions. At lower ionic strength it forms the A conformation, with a narrower major groove and a wider and shallower minor groove. Certain sequences, rich in GC repeats, can form a left-handed double-helical Z conformation. The grooves are outlined by charged phosphate groups (P is in magenta). Bases are clearly more accessible through the major groove than through the minor groove.

The idea that an α-helix has the right size and shape to fit into the major groove of DNA was noted in the 1950s. The structures of the first protein–DNA complexes confirmed this prediction. It became the paradigm for protein–DNA interactions. Indeed, when a student solving the structure of the Met repressor–DNA complex told his supervisor that in the electron-density map he was interpreting a β-sheet

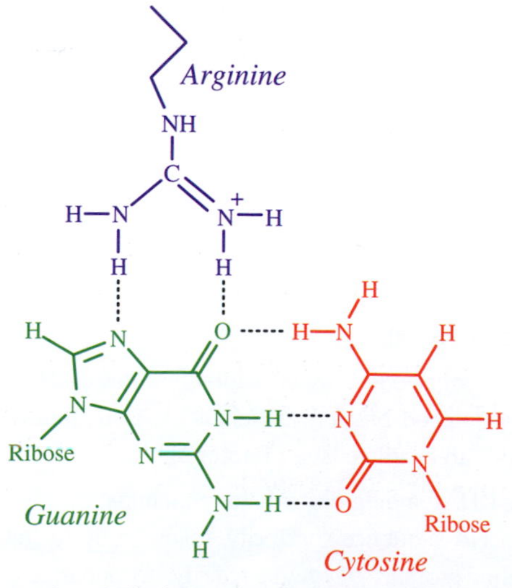

Figure 6.35 Specific hydrogen-bonding pattern between an arginine sidechain of a protein and a guanine of DNA.

appeared to bind in the major groove, he was advised, with patience strongly tinged with condescension, to go back and look for the helix.

We now recognize great structural variety in DNA–protein interactions. A few themes include:

- **Helix-turn-helix domains.** These appear in prokaryotic proteins that regulate gene expression, eukaryotic homeodomains involved in developmental control, and histones that package DNA in chromosomes.

- **Zinc fingers,** including eukaryotic transcription factors, and steroid and hormone receptors. Proteins with β-sheets that interact with DNA, for instance the gene-regulatory proteins, Met and Arc repressors, and the TATA-box binding protein. **Leucine zippers** that act as eukaryotic transcriptional regulators.

- The **high-mobility group** in eukaryotes and the prokaryotic protein HU, which bind sequences non-specifically and bend DNA.

- Enzymes that interact with DNA, involved in replication, translation, repair, and uncoiling. Some are relatively small; others are large multiprotein complexes. They show many different types of folding patterns. Many distort the DNA structure in order to get access to the bases that are the target of their activity.

- Viral capsid proteins form compact shells enclosing nucleic acid.

These examples are an anecdotal list, not a classification.

RNA-binding proteins have a separate variety. Some resemble DNA-binding proteins. Others bind to RNA molecules of defined structure; for instance, proteins of the spliceosome, responsible for excising introns from eukaryotic pre-mRNA, and enzymes that interact with tRNA, including but not limited to aminoacyl tRNA synthases, and the ribosome itself.

Bacteriophage T7 DNA polymerase

Bacteriophage T7 is a virus infecting many bacteria, including *E. coli*. It has its own DNA polymerase, which acts in conjunction with other proteins, including enzymes that unwind and prime the DNA, and which permits translocation of the primer and template into position for the next catalytic cycle, and a closed catalytically active form.

Figure 6.36 shows T7 DNA polymerase in the act of adding a residue to the primer. The individual reaction step uses up one deoxynucleoside triphosphate, adds a nucleotide to the growing strand, and releases pyrophosphate.

The enzyme takes advantage of the stereochemical compatibility of different Watson–Crick base pairs to enhance fidelity of replication. A hydrophobic pocket that closely surrounds the nascent base pair selects against mismatches, in part by size discrimination and in part by an energetic penalty against unsatisfied hydrogen bonds in a structure sequestered away from water.

In case of incorporation errors, the speed of synthesis slows down, and the DNA shifts from the polymerase site to the exonuclease site to cleave off the incorrect base.

Some protein–DNA complexes that regulate gene transcription

Proteins involved in regulation of gene expression show great structural variety, reflecting the complexity and diversity of the network of transcription control mechanisms.

λ *cro*

Bacteriophage λ is a virus containing a double-stranded DNA **genome** of 48 502 base pairs. A λ phage infecting an *E. coli* cell chooses—depending on which genes are active—between **lysis** or **lysogeny**. λ can replicate, and *lyse* the cell, releasing ~100 progeny. Alternatively, it can integrate its DNA into the host genome. The phage in such a *lysogenized* cell is dormant, and can be released by stimuli that switch from the lysogenic to the lytic state.

λ cro is a transcription regulator involved in the lytic–lysogenic switch. It binds to DNA as a symmetrical dimer (see Fig. 6.37). Its target sequence is approximately palindromic:

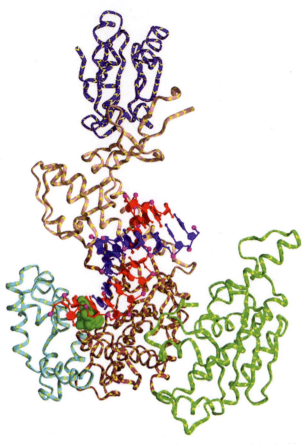

Figure 6.36 T7 DNA polymerase, binding primer-template double helix, with nucleoside triphosphate (green spheres) about to be incorporated. Green ribbon = exonuclease domain; blue, at top, thioredoxin.

that bind the unwound single-stranded DNA. The phage also recruits *E. coli* thioredoxin to keep the polymerase-template complex intact—in some cases permitting replication of the entire 39 937-base-pair T7 genome in one go.

S. Doublié, S. Tabor, A.M. Long, C.C. Richardson, and T. Ellenberger solved the structure of the T7 DNA polymerase, in complex with a primer-template double helix, and a nucleoside triphosphate about to be added to the growing strand. (To trap the structure, the 3′ end of the primer strand is a dideoxynucleotide, to block the reaction.)

As part of its mechanism, the polymerase undergoes a conformational change from an open state,

```
C T A T C A C C G C A A G G G A T A A
G A T A G T G G C G T T C C C T A T T
```

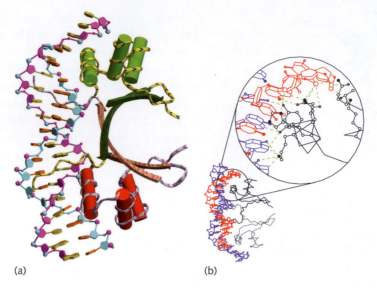

(a)   (b)

Figure 6.37 Phage λ cro repressor–DNA complex. (a) A dimer binds an approximately palindromic sequence in DNA. (b) Pattern of DNA–protein hydrogen bonds.

The bases to which the protein makes contact are shown in bold face. The protein interacts with both strands. The DNA is slightly bent.

λ cro is an example of the helix-turn-helix structural motif. Following along the chain in Fig. 6.37(a), the first secondary structure is a helix, followed by two more helices that frame the motif. The second of these two helices (the third helix in the molecule)—called the **recognition helix**—lies in the major groove and makes extensive contacts with the DNA (see Fig. 6.37(b)). The hairpin that follows is involved in dimerization of the protein. It interacts with the corresponding hairpin of the other monomer to form a four-stranded β-sheet. A long C-terminal tail wraps around the DNA, following the minor groove.

The eukaryotic homeodomain antennapedia

Homeodomains are highly conserved eukaryotic proteins, active in control of animal development. They regulate **homeotic genes**; that is, genes that specify the locations of body parts. Antennapedia is a *Drosophila* protein responsible for initiating leg development. (We saw the structure of free antennapedia in Chapter 3.) The earliest mutations found in antennapedia produced **ectopic** (= out of place) legs at the positions of, and instead of, antennae. Loss-of-function mutations convert legs into antennae.

The structure of the antennapedia–DNA complex (see Fig. 6.38(a)) resembles, in some respects,

prokaryotic helix-turn-helix proteins such as λ cro. However, the tail that wraps around into the minor groove is N-terminal to the helix-turn-helix motif in antennapedia, instead of C-terminal as in λ cro.

A comparison of the protein–DNA hydrogen bonds in the antennapedia complex (see Fig. 6.38(b)) with that of λ cro shows that in the antennapedia complex a greater fraction of DNA–protein hydrogen bonds involve phosphate oxygens rather than bases.

Leucine zippers as transcriptional regulators

Leucine zippers, which we have already met in connection with keratin, form part of another type of dimeric transcriptional regulator. The jun protein forms a homodimer consisting of an N-terminal domain with many positively charged sidechains that binds to DNA, and a C-terminal leucine zipper domain involved in dimerization.

The proteins grip the DNA as if they were picking it up with chopsticks. The α-helices bind in major grooves on opposite sides of double helix (see Fig. 6.39). This structure shares with λ cro, and many other DNA-binding proteins, the *symmetry* of the complex, which mimicks the dyad symmetry of the DNA double helix. This requires, on the part of the protein, formation of symmetrical dimers, and on the part of the DNA, an exactly or approximately palindromic target sequence. For jun dimers the target sequence is ATGACGTCAT.

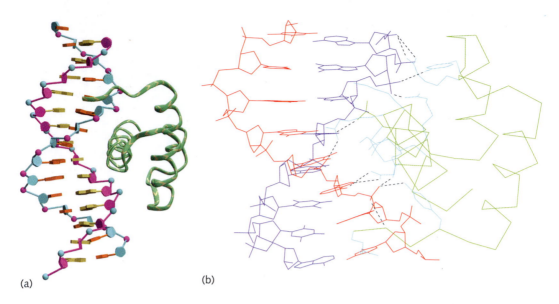

Figure 6.38 (a) Antennapedia homeodomain-DNA complex [9ANT]. (b) Details of protein–DNA interactions in antennapedia complex.

Jun can dimerize not only with itself but with other related proteins, notably fos. Different dimers have different DNA-sequence specificities and different affinities, affording subtle patterns of control.

Zinc fingers

Zinc fingers are small modules found in eukaryotic transcription regulators. Each finger recognizes a triplet of bases in DNA. Tandem arrays of fingers recognize an extended region (see Fig. 6.40(a)). Understanding the relationship between the amino acid

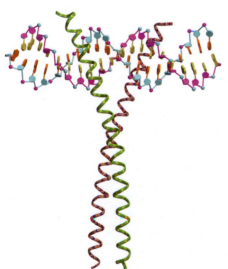

Figure 6.39 Bzip (basic DNA-binding domain)–leucine zipper transcriptional regulator jun homodimer binding to DNA [1JNM].

sequences of Zn fingers and the DNA sequences they bind would permit design of gene-specific repressors. One fairly obvious target would require only one prototype finger: If one had a Zn finger that bound preferentially to the trinucleotide CAG, a polymer of such a domain might afford an approach to treating Huntington's disease. In fact this approach works. Another established technique is the fusing of Zn-finger domains to nucleases, to create artificial restriction enzymes that target specific genes.

Three types of zinc fingers differ in structure. In class 1, one Zn^{2+} ion binds two Cys and two His residues to stabilize the packing of a β-hairpin against an α-helix (see Fig. 6.40(b)). Residues in the helix interact with three consecutive nucleotides. (In class 2 zinc fingers, one Zn^{2+} ion binds four Cys sidechains. In class 3, two Zn^{2+} ions bind six cysteines.)

The E. coli Met repressor

Like many other DNA-binding proteins, the Met repressor binds as a symmetrical dimer. In the complex, each monomer contributes one strand of two-stranded β-sheet, which sits in the major groove (the student was right!), with sidechains making hydrogen bonds to bases (see Fig. 6.41). The co-repressor, S-adenosyl methionine, increases the affinity of the complex. S-adenosyl methionine is a product of methionine metabolism, and its effect is a kind of feedback inhibition—not on the enzymatic activity directly but on its *expression*.

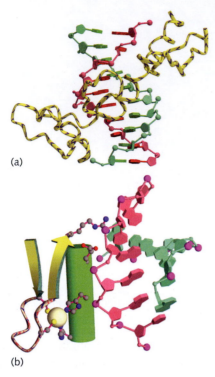

(a)

(b)

Figure 6.40 (a) Zif268, a tandem three-finger structure binding the sequence GCGTGGGCG. [1aay]. (b) Structure of a single module of a class 1 (Cys$_2$–His$_2$) zinc finger, taken from the Zif268–DNA complex. Each finger interacts with three consecutive bases. Three positions along the α-helix, non-consecutive in the amino acid sequence, contain primary determinants of the DNA-sequence specificity. One of them, a histidine, binds to both the Zn^{2+} and a phosphate of the DNA. The other three sidechains that bind the Zn^{2+} are also shown.

The TATA-box binding protein

A **TATA box** is a sequence (consensus TATA[A|T] A[A|T]) upstream of the transcriptional start site of bacterial genes. Recognition of this sequence by the

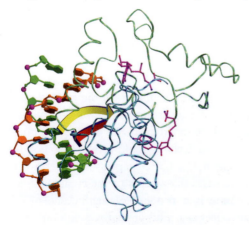

Figure 6.41 *E. coli* Met repressor–DNA complex [1CMA].

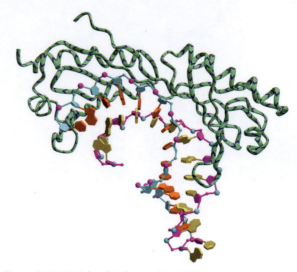

Figure 6.42 TATA-box binding protein YTBP [1YTB].

TATA-box binding protein (see Fig. 6.42) initiates the formation of the basal transcription complex, a large multiprotein particle. This is an example of initial binding of a protein to DNA followed by recruitment of other proteins to form an active complex. The most obvious feature of the complex is the very strong bending and unwinding induced in the DNA. A long curved β-sheet sits against an unusually flat surface on the DNA, the result of prying open the minor groove. Phe sidechains intercalate between the bases (see Fig. 6.43).

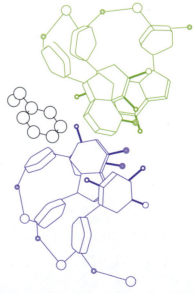

Figure 6.43 Intercalation of a Phe sidechain between bases in the distorted double helix in TATA-box binding protein–DNA complex [1YTB].

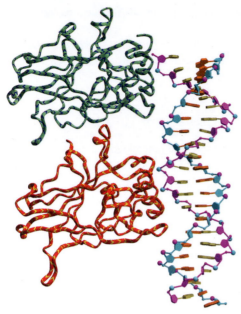

Figiure 6.44 p53 core domain in complex with DNA [1TSR].

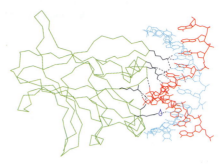

Figure 6.45 Details of protein–DNA interaction in p53 core domain–DNA complex [1TSR]. One protein–DNA contact is water mediated.

p53 is a tumour suppressor

p53 is a DNA-binding transcriptional activator. It is of great clinical importance because mutations in the p53 gene are very common in tumours.

p53 acts by surveilling for genome integrity. Damage to DNA induces enhanced expression of p53, which stalls cell-cycle progression. This gives time for DNA repair; if repair is unsuccessful, the 'fail-safe' mechanism is apoptosis.

The structure of the DNA-binding subunit of p53 shows a double-β-sheet fold (see Fig. 6.44). A helix sits in the major groove, and sidechains from loops connecting strands of the β-sheet insert into the minor groove (see Fig. 6.45).

● RECOMMENDED READING

Dobson, C.M. (2002) Protein-misfolding diseases: Getting out of shape. *Nature*, **418**, 729–30.
   Comments on the mechanism and variety of protein-aggregation diseases.

Serpell, L.C., Sunde, M. and Blake, C.C. (1997). The molecular basis of amyloidosis. *Cell Mol. Life Sci.*, **53**, 871–87.

Rambaran, Roma N and Serpell, Louise C (2008). Amyloid fibrils: abnormal protein assembly. *Prion* **2**, 112–7.

Sipe, J.D. and Cohen, A.S. (2000) Review: History of amyloid fibril. *J. Struc. Biol.*, **130**, 88–9.

Four review articles about protein aggregation and its clinical consequences.

Carrell, R.W. and Lomas, D.A. (2002). α1-antitrypsin deficiency—a model for conformational diseases. *New Engl. J. Med.*, **346**, 45–53. Discovery of clinical consequences of serpin conformational change.

Reddy, V., Natarajan, P., Okerberg, B., Li, K., Damodaran, K., Morton, R., Brooks, C. III, and Johnson, J. (2001). VIrus Particle ExploreR (VIPER), a website for virus capsid structures and their computational analyses. *J. Virol.*, **75**, 11943–7. Description of useful database of virus structures: http://viperdb.scripps.edu/

Petrey, D. and Honig, B. (2014). Structural bioinformatics of the interactome. *Annu. Rev. Biophys*. **43**, 193–210. Application of genome-wide methods to determining the nature and function of networks of interactin molecules.

Calladine, C.R. and Drew, H.R. (1997). *Understanding DNA: the molecule and how it works*, 2nd. edn. Academic Press, London.

Garvie, C.W. and Wolberger, C. (2001). Recognition of specific DNA sequences. *Mol. Cell.*, **8**, 937–46.

Tsonis, P.A. (2003). *Anatomy of gene regulation: A three-dimensional structural analysis.* Cambridge University Press, Cambridge. Three general treatments of protein–DNA interactions.

● EXERCISES AND PROBLEMS

Exercise 6.1 Find illustrations in this book of isologous and heterologous dimers.

Exercise 6.2 (a) Does the dimeric globin from *Scapharca inaequivalvis* (see Fig. 1.11(b)) appear to have an axis of twofold symmetry? One way to check is to photocopy Fig. 1.11(b) onto a transparency, rotate by 180°, and try to superpose the copy on the original. (b) Does the interface in this dimer appear to be isologous?

Exercise 6.3 (a) Why is an isologous open structure impossible? (b) Must a dimer with an isologous interface have a twofold axis of symmetry?

Exercise 6.4 On a photocopy of Fig. 6.2(a), indicate a region of intermolecular β-sheet formation.

Exercise 6.5 Knockout mice lacking the gene for prion protein cannot develop spongiform encephalopathy even if inoculated with PrP^{Sc}. How does this suggest a way to breed strains of sheep and cows that would be safe to eat?

Exercise 6.6 A complete IgG—as shown in Fig. 6.7—has an r.m.m. of about 170 000. Estimate the r.m.m of (a) a Fab fragment, (b) the F_c fragment, and (c) an F_v fragment.

Exercise 6.7 On a photocopy of Fig. 6.7, circle the antigen-binding sites.

Exercise 6.8 Estimate the minimum size of an antigen or hapten that could make contacts with all six CDRs of the antibody shown in Fig. 6.9. (The total width of the figure is 40 Å.)

Exercise 6.9 The end-to-end distance of an N-residue peptide in a nearly extended conformation is about 3.8 Å × N. The rise per turn of an α-helix is 3.6 Å. Estimate the number of turns of an α-helix needed to flank a groove designed to bind a nine-residue peptide in a nearly extended conformation. Compare with the lengths of the long helices in MHC molecule binding domains.

Exercise 6.10 On three separate photocopies of the Fig. 6.46: indicate by a '5' each of two sites related to the red dot by fivefold symmetry; indicate by a '2' each of two sites related to the red dot by twofold symmetry; indicate by a '3' each of two sites related to the red dot by threefold symmetry.

Exercise 6.11 On a photocopy of Fig. 6.30, circle the region of part A that appears in part C.

Exercise 6.12 On a photocopy of Fig. 6.30, circle two pentagonal rings.

Exercise 6.13 On a photocopy of Fig. 6.27(e), indicate the positions of (a) four additional twofold axes, (b) one additional threefold axis, and (c) three additional fivefold axes.

Figure 6.46 A point on the surface of an icosahedron. Find symmetry-related points (see Exercise 6.10).

Exercise 6.14 What fraction of a closed figure with icosahedral symmetry will be formed by attaching the broken red and black lines in Fig. 6.26(b)?

Exercise 6.15 On a photocopy of Fig. 6.34(a), indicate (a) a phosphate group and (b) a sugar. On a photocopy of Fig. 6.34(b), indicate (c) a phosphate group (shown as a single sphere), (d) a sugar, (e) a purine and (f) a pyrimidine.

Exercise 6.16 From inspection of Fig. 6.34: (a) How many base pairs are visible in the major groove? (b) Compare the accessibility of bases through the major and minor grooves.

Exercise 6.17 Draw a picture analogous to Fig. 6.35 showing what amino acid sidechain would readily form two hydrogen bonds to an A–T base pair.

Exercise 6.18 From Fig. 6.37(a), estimate the angle through which binding of λ cro bends DNA.

Exercise 6.19 On a photocopy of Fig. 6.37(a), indicate (a) the helix-turn-helix motif, (b) the four-stranded β-sheet and (c) the N-terminal tail in the minor groove.

Exercise 6.20 On a photocopy of Fig. 6.40(b), indicate (a) the three sidechains primarily responsible for recognition and (b) the four sidechains that ligate the zinc ion. (One of the sidechains does both.)

Exercise 6.21 At which positions does the sequence to which λ cro binds differ from an exact palindrome?

Exercise 6.22 On a photocopy of Fig. 6.24(a), show the approximate position of the intersection of the helix axis with the plane of the picture.

Problem 6.1 (a) Assume the following estimates: (1) stimulation of a T-cell requires approximately 100 MHC–peptide complexes per cell (as a consequence of the kinetics of dissociation of MHC–peptide complexes and of cell–cell encounters) and (2) an antigen-presenting cell expresses on the order of 10^5 MHC molecules on its surface. How many peptide species can an antigen-presenting cell effectively display at any time? (b) A typical eukaryotic cell synthesizes significant amounts of about 2000 proteins, average length 300 residues. How many eight-residue peptides can theoretically be generated from them? (c) What fraction of these peptides can a single antigen-presenting cell effectively display at any time?

Problem 6.2 The homologous antigen-binding loops of Vκ and Vλ domains of immunoglobulins have different repertoires of canonical structures. Suppose the antigen-binding loops from a Vκ antibody are 'transplanted' into the homologous positions in a Vλ domain. Would you expect affinity to be retained? Explain your reasons.

Problem 6.3 The transition state for the natural conversion of orotidine-5'-monophosphate is now believed, on the basis of crystal structures, to be as shown in Fig. 6.47. On the basis of this mechanism, suggest a transition-state analogue suitable for immunization to elicit catalytic antibodies for this reaction. (OMP = orotidine-5'-monophosphate, RP = ribose-phosphate.)

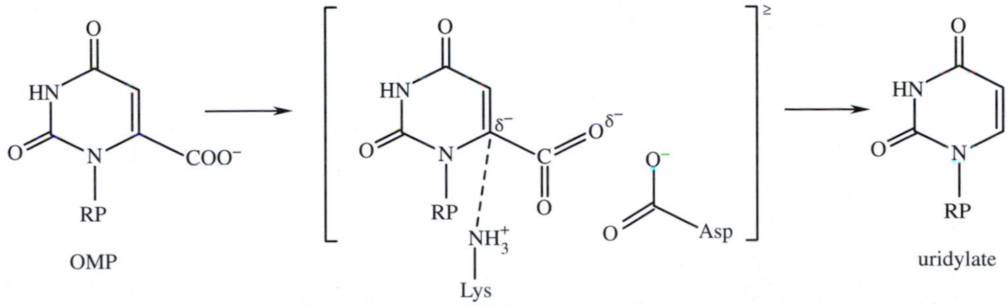

Figure 6.47 The natural reaction that produces uracil from orotidine 5'-monophosphate is now believed to proceed through the intermediate shown this figure (compare Fig. 6.13). RP = ribose-phosphate

Problem 6.4 (First do Exercise 6.13.) (a) On a photocopy of Fig. 6.27(e), draw in the positions of subunits of type A, B, and C in the rest of the area of the figure. (b) Which subunits meet at twofold axes? At threefold axes? At fivefold axes? (c) How many different types of intersubunit contacts are formed?

Problem 6.5 (a) Compare the mass ratios of nucleic acid:protein for tobacco mosaic virus and tomato bushy stunt virus. (b) Compare the capsid encoding efficiency—the ratio of the length of the capsid protein to the total genome length—for these two viruses.

Problem 6.6 The satellite tobacco necrosis virus capsid contains 60 copies of a 195-residue coat protein. How many base pairs would be required to code the capsid if it contained no repeating subunits? Estimate the volume that this nucleic acid would occupy. Would it fit in the approximately spherical particle 180 Å in diameter?

Problem 6.7 Satellite tobacco necrosis virus is an icosahedral virus with a $T = 1$ structure of inner radius 60 Å, containing one molecule of single-stranded RNA. If one ribonucleotide occupies ~600–700 Å^3 in a virion, estimate the maximum length of the RNA molecule that could fit inside the capsid, and compare with the observed genome size of 620 nucleotides.

Problem 6.8 Compare the interactions of the recognition helix in the major groove in λ cro and antennapedia. In each structure, how many hydrogen bonds are formed to (a) the sugar–phosphate backbone of DNA and (b) the bases of DNA?

CHAPTER 7

Evolution of protein structure and function

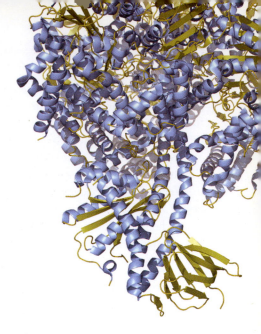

LEARNING GOALS

- *Understanding the basis for the relationships among protein folding patterns.* Distinguishing between structural similarities based on true homology—descent from a common ancestor— and general similarities arising from structural themes common to many unrelated proteins.

- *Appreciating the distinction between two types of homologues: orthologues and paralogues.* Orthologues are homologous proteins in different species, descended from a single ancestral protein. Paralogues are homologues in the same species arising from gene duplication, and their descendants.

- *Understanding the relationship between* divergence of sequence and divergence of structure in homologous proteins.

- *Being able to analyse patterns of conservation* in multiple sequence alignment tables.

- *Being familiar with evolutionary variations in protein families,* including globins, NAD-binding domains, serine proteases, and opsins.

- *Understanding domain swapping* and its appearance as a general mechanism of formation of oligomeric proteins.

- *Knowing how proteins can develop new functions during evolution:* the mechanisms, pathways, and limitations.

Introduction

Protein evolution is the exploration by a set of genomes of the space of amino acid sequences in search of selectively advantageous variants. Evolution acts at the level of protein functions, in a feedback cycle that selects gene sequences.

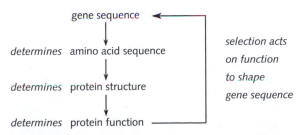

Observed evolutionary relationships among protein structures illuminate **sequence–structure–function** relationships. How are changes in sequence reflected by changes in structure? What mutations are acceptable? How do different sequences create similar structures? What is the topography of evolutionary sequence space, what pathways of evolutionary change are allowed, and how does evolution explore them? In summary: starting with one protein, what others are accessible to evolution?

Proteins evolve at several scales:

1. Individual domains are subject to amino acid substitutions, insertions, and deletions. Many, but not all, such mutations produce minimal changes in structure, and minor modifications of function. However, the cumulative effect of a succession of such local mutations can produce large changes in domain structure, and development of novel functions.

2. At a higher structural level, proteins can evolve by 'mixing and matching' entire domains (see Box 7.1). Most proteins contain more than one domain and/or more than one polypeptide chain. Evolution has recombined a relatively small repertoire of domains into many different partnerships. Sometimes individual domains contribute specialized functional components to the overall structure. For example, the **NAD-binding domain** appears in many dehydrogenases that share that cofactor, in combination with catalytic domains of very different structure and function, specific for different substrates. In other cases, the function of the protein as a whole varies more widely with the combination of domains.

> Protein evolution occurs through:
> - modification of individual domains by point mutations and insertions/deletions;
> - forming different proteins as different constellations of domains.

The challenges in studying protein evolution include several related goals:

1. A reliable method to decide which proteins are homologous.

2. An understanding of the allowable evolutionary pathways that connect homologous proteins.

BOX 7.1   Domains, modules, and modular proteins

Domains. It is difficult to give a formal definition of a domain, although most people 'know one when they see it'. What they see is that many proteins contain compact units within the folding pattern of a single chain. These units look as if they should have independent stability, although this is rarely demonstrated experimentally. In the hierarchy of protein structure analysis, domains fall between supersecondary structures and the tertiary structure of a complete monomer.

What is the difference between a module and a domain? The terms are nearly synonymous, and usage is not consistent.

- A *domain* is a compact subunit of the structure of an individual polypeptide chain in a protein. Phosphoglycerate kinase is an enzyme with its active site in a cleft between two domains (see Fig. 7.1(a)). Aspartate carbamoyltransferase is composed of subunits containing a catalytic and a regulatory chain, each composed of two domains (see Fig. 7.1(b)).

- A *module* is a protein substructure that stays together as a unit during evolution, appearing in different contexts. Serine proteases contain two homologous domains, but, because no individual serine protease domain has appeared in another structural context, the individual domains are not modules.

Modular proteins often contain many copies of closely-related domains (recall Fig. 1.13). The domains can appear in different structural contexts; that is, proteins can 'mix and match' sets of domains. For example, fibronectin, a large extracellular protein involved in cell adhesion and migration, contains 29 domains, including multiple tandem repeats of three types of domains, called F1, F2, and F3. It is a linear array of the form: $(F1)_6(F2)_2(F1)_3(F3)_{15}(F1)_3$. Fibronectin domains also appear in other modular proteins. Recombination of modules is an important mode of protein evolution.

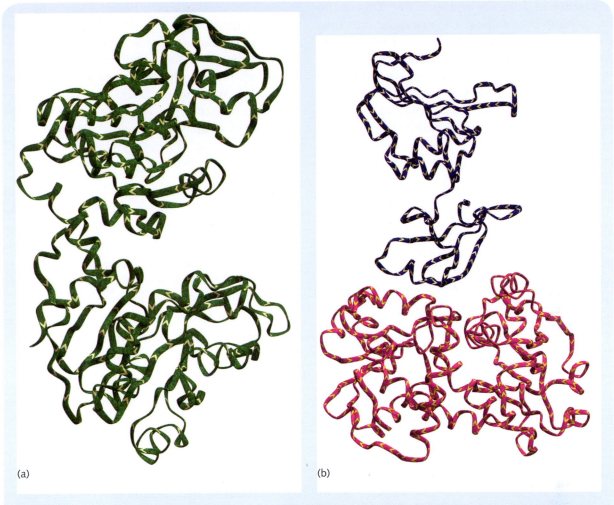

(a)

(b)

Figure 7.1 Examples of proteins composed of domains. (a) Phosphoglycerate kinase, a two-domain enzyme with the binding site in the cleft between domains. This molecule changes conformation upon binding substrate [3PGK]. (b) Two subunits of *E. coli* aspartate carbamoyltransferase, each comprising two domains. This figure contains one regulatory subunit (blue) and one catalytic subunit (red). The full molecule contains six catalytic and six regulatory subunits.

3. An understanding of what constrains the divergence of protein sequences, structures, and functions.

Two or more proteins are homologous if they are descended from a common ancestor. Because only rarely can we observe homology directly, we must infer homology from similarity in sequence, structure, and function. This has two implications:

• We must calibrate our measures of similarity to provide thresholds for confident conclusion about homology. To exclude non-homologues that, nevertheless, show some degree of similarity, we must set the criteria fairly tight.

• It would be very useful to know how far evolutionary divergence can proceed. This would allow us to assess whether the choice of strict criteria for inference from similarity to homology leads us to *miss* many genuine homologues that are highly diverged. One method for showing that two very dissimilar proteins A and B are homologues would be to exhibit a series of intermediate proteins that form a connected path between A and B, such that each successive pair on the path has high similarity. This is not always possible.

In the evolution of individual protein domains, a succession of mutations leads to progressive divergence of sequence and structure. Criteria for recognizing homologues are extrapolations from comparisons of proteins sufficiently closely related in structure and function that homology is not in question. For example, sperm-whale and seal myoglobin have 127/153 = 83% identical residues in an optimal pairwise amino acid sequence alignment, have very similar structures, including the same cofactor bound by corresponding residues, and have identical functions. These are certainly homologues.

The alignment of sequences of sperm-whale myoglobin and the phytoglobin from Yellow lupin has only 22/153 = 17% identical residues. But they are also homologues. How do we know?

Following evolution in globins, and other families, from pairs as similar as sperm-whale and seal myoglobins through progressive divergence, has shown that the structures diverge more slowly than the sequences. For distant relatives, then, homology is recognizable more confidently in structure than in sequence. In most cases the basic folding pattern of the family remains the same. Usually there is at least a small set of residues conserved at least in physico-chemical character, although it requires knowledge of the structure to appreciate their role in creating the common architecture of the fold. This, together with functional cues, gets most of the job done—the job being to distinguish homologous from non-homologous proteins.

What continues to create difficulty are:

- cases in which folds are tantalizingly similar in the absence of any sequence similarity above the noise level, and no analogy in function. In many cases it is hard to decide whether such similarities reflect evolutionary relationship or not. The case of phycocyanin and the globins is an example (see Phycocyanins and the globins section).

- cases of similar sequences with different folding patterns. These, fortunately, are quite rare.

Protein structure classification

A *family* of proteins is a set for which similarities of sequence and/or structure provide evidence that they are related by evolution.

Homologues can appear within the same species or in different species (see Box 7.2).

- Homologous proteins may be orthologues or paralogues. Orthologues appear in the same species, having arisen by gene duplication and divergence. Paralogues appear in different species, descended from a common ancestral protein.

Some features of protein structures are what they are because the laws of physics and chemistry would not allow them to be otherwise. Some are imposed by the mechanism of evolution. In addition, historical accident has played a large role in creating the roster of folding patterns observed in Nature. It is by no means easy to sort out these effects. A creative tension among them pervades and animates this field of protein science.

Although proteins from the same family have similar structures, proteins from different families often contain recurrent structural themes. Folding patterns of protein domains are built by combining local structures, such as α-helices and β-sheets. Sometimes these create structural similarities.

Nevertheless, we can compare and classify the conformations of apparently unrelated proteins, on the basis of secondary structures and their folding patterns. Several websites, such as SCOP and CATH, offer such classifications (see Chapter 2). They offer classifications encompassing all known protein structures, a useful thing to have. However, within the hierarchy of such a classification, only the relationships among classes of proteins *within* the same family necessarily reflect evolutionary divergence. Higher levels of the classification are based purely on architectural similarity, independent of provable evolutionary history and relationship. There may well be justifiable suspicion of homology—enough evidence to indict but not to convict.

BOX 7.2

Evolutionary relationships among proteins: homologues, orthologues, and paralogues

- Proteins are homologous *if and only if* they are descended from a common ancestor.

- Homologues in different species, descended from a single ancestral protein, are **orthologues**.

- Homologues in the same species, arising from gene duplication, are **paralogues**. Their descendants are also paralogues. After gene duplication, one of the resulting pairs of proteins can continue to provide its customary function, releasing the other to diverge, to develop new functions. Therefore, inferences of function from homology are more secure for orthologues than for paralogues.

The globin family contains both orthologues and paralogues. An ancestral monomeric haemoglobin gene duplicated about 450–500 million years ago to form **paralogous** α- and β-chains. Subsequent divergence produced orthologous α-chains and orthologous β-chains in horse and human (and many other species). Differential gene loss can make it difficult to distinguish orthology from paralogy. What if humans had lost their α-chains and horses their β-chains? The remaining human β- and horse α-chains would appear to be orthologues. Full-genome information can often resolve such ambiguities.

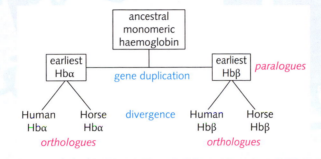

- Evolution generates homologues that progressively diverge in sequence, structure, and function. It is difficult to recognize homology with confidence when the divergence is extreme.

In practice, the finest levels of the classification are the easiest. Given a set of closely-related homologous proteins, computer programs can detect, and measure quantitatively, their degrees of similarity. In most cases, the divergence of the amino acid sequences of orthologues correlates well with the divergence of the corresponding structures, and with the classifications of species of origin established by classical methods.

Extending the classification scheme to more-distant relatives is more difficult. In many cases, alignment methods based solely on sequences are uninformative, because the sequences have diverged too far. For distantly-related proteins it is some-

times possible to align the sequences via a structural alignment.

The results of a structural alignment are:

- The alignment itself—that is, the set of residue–residue correspondences. In some cases, parts of the structure may have changed so much that no residue–residue correspondence can be made (see Fig. 7.2). In these cases, we can distinguish between *alignable* regions of the sequence and *non-alignable* regions. This distinction cannot be drawn from pairwise sequence alignments.

- A measure of structural similarity—the average (root-mean-square) deviation of the alignable atoms (see 'Calculation of optimal superposition of aligned sets of atoms' in Chapter 4).

Although structural analysis can extend recognition of homology to more-distant relationships, there are limits beyond which one enters a desert of ignorance.

Changes in proteins during evolution give clues to the roles of residues at different positions

Residues at different positions in proteins play different roles in structure and function. Some residues participate directly in function. Some are essential for creating the structure. Some are not so important to the structure of the domain in which they appear, but mediate intramolecular or intermolecular interactions. Others are not subject to any obvious constraints.

If we were smart enough, we could look at the sequence and structure of a single protein, and be able to assign the roles of the different residues. But that's very difficult.

It is much easier to let Nature do the work for us. Residues essential for structure or function will be conserved. Those that are not are dispensable. Therefore, align the amino acid sequences, and superpose the structures, of proteins from a family. Patterns of conservation will declare themselves in the sequences and the structures. These patterns of conservation provide important clues to the roles of the residues retained.

> We discussed the example of thioredoxin in Chapter 4.

Many residues involved directly in function are tightly conserved. Many residues buried in protein interiors are largely restricted to hydrophobic amino acids. Many positions on the surface are relatively free to vary. Many surface loops readily tolerate insertions and deletions. However, looking at individual positions provides much valuable information but does not tell the whole story. Subtle linkages among interresidue interactions restrict the *combinations* of amino acids that can form a viable sequence of a protein in a family. Moreover, our analysis is limited by the range of proteins in the family that are known. Be warned that inferences from a set containing only closely-related proteins are often misleading.

- Patterns of conservation provide clues to the role and importance of residues at particular positions.

To what constraints are pathways of protein evolution subject?

Although protein evolution can unquestionably create very divergent homologues, the *process* is not entirely free-wheeling. As we shall see when we examine evolution in specific protein families, a variety of constraints restricts the exploration of sequence space. Some of these constraints are general; others apply to particular families.

- The requirement for stability during the course of evolution constrains the changes in the amino acid sequence. The tertiary structural interactions—the packing of the interior—can accommodate extensive sequence changes provided that they are made one by one, but a completely novel well-packed interface cannot form in a single step. In consequence, (1) insertions and deletions in secondary structural elements tend to be confined to their ends, and (2) the pattern of residue contacts at interfaces between helices and sheets tends to be conserved.

- These considerations apply to natural proteins, and not necessarily to engineered proteins. To any two natural homologous proteins, there must have been continuous evolutionary paths to each from their common ancestor, over which all intermediates were stable and functional. This constraint does not apply to engineered proteins.

- Retention of function, or at least of binding a common ligand, constrains evolution in many families. The constraints on the structure appear to apply at the *global* level. Figure 7.6 shows a superposition of the B, E and G helices from horse methaemoglobin, α-chain, and lupin leghaemoglobin. The B and E helices and B and G helices are in contact, but the E and G helices are not. The figure shows that there have been large rotations of the helices at the B–E and B–G contacts. But these rotations have been *coupled* to leave the E and G helices in

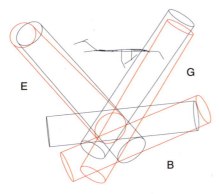

Figure 7.6 The B, E and G helices surrounding the haem group in human oxyhaemoglobin, α-chain [1HHO], and lupin leghaemoglobin [2LH7]. The E helix is closest to horizontal.

the same relative spatial position, in order to form the haem pocket.

> In most cases, the demands of function are relatively rigid. The demands of structure *per se*, less so.

- A special and unusual type of constraint applies to proteins containing β-barrels. A.D. McLachlan first analysed the possible structures of β-barrels, showing that they can form in only a *discrete* set of ways. (We shall discuss these results in the next section.) It is unlikely that evolution can jump from one discrete barrel topology to another. Therefore, retention of barrel topology constrains evolution in this family.

This is one of the few cases in which the possible structures corresponding to a type of protein architecture can be analysed on the basis of purely geometric principles.

Closed β-barrel structures

A.D. McLachlan classified β-barrel topologies. Two integral quantities, the **number of strands** and the **shear** (= the stagger, along the sequence, of the residues forced to correspond when closing the barrel), determine the tilt of the strands to the barrel axis, the twist of the strands (that is, the average angle between adjacent strands), and the radius of the barrel. McLachlan's classification of β-barrel topologies by *discrete* indices—the number of strands and the shear number—permits writing down a complete 'periodic table' of possible β-barrel folds. Quantitative predictions of β-barrel geometries agree very well with experiment.

By far the most common β-barrel structure is the $(\beta\alpha)_8$ **TIM barrel**. It has eight strands, and shear number 8. Another type of β-barrel appears in interleukin-1α and β, and certain fibroblast growth factors. This barrel has six strands and shear number 12. The serine proteinase domain contains still another type of β-barrel. This will be discussed later in this chapter.

> - β-barrels are classifiable according to two numbers: the number of strands, n, and the shear, S. For TIM barrels, $n = 8$, $S = 8$. For interleukin-1α, $n = 6$, $S = 12$. For serine protease domains, $n = 6$, $S = 8$.

The TIM barrel

The enzyme glycolate oxidase is typical of a large number of structures that contain eight β–α units in which the strands form a sheet wrapped around into a closed structure, cylindrical in topology. The helices are on the outside of the sheet. (see Fig. 7.7; see also Fig. 2.12(d)). Chicken triose phosphate isomerase (TIM), first solved in 1975, was for a long time the only example, but now very many enzymes containing TIM-like barrels are known.

In most TIM-like barrels the active site is at the end of the barrel that corresponds to the C-termini of the strands of sheet, just as in the open β–α–β structures. Although they show similar folding patterns, TIM-barrel enzymes catalyse a variety of different reactions, and the amino acid sequences of enzymes that share this fold but differ in function are very dissimilar: we cannot prove that all are diverged from a common ancestor, indeed many people believe that at least some of the structures arose by convergence.

With many structures available, it is possible to adduce general features of this folding pattern:

1. Looking from the outside, the sheet is formed by eight parallel strands, tipped by approximately

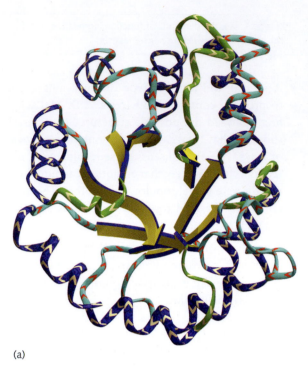

(a)

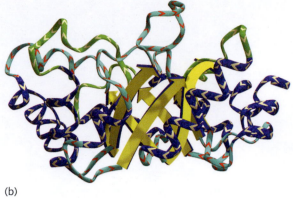

(b)

Figure 7.7 The β-barrel from spinach glycolate oxidase [1GOX]. First discovered in triose phosphate isomerase, these are known as TIM barrels. (a) View along the barrel radius. (b) View perpendicular to the barrel radius.

36° to the barrel axis (see Fig. 7.7(b)). The helices, outside the sheet, are approximately parallel to the strands, typical of α/β structures. The chain proceeds around the barrel in a consistently counterclockwise direction, viewed from the C-termini of the strands of sheet: locally the chain proceeds up a strand, down a helix, up the next strand, etc. Some exceptional cases are known, in which a helix is missing (muconate lactonizing enzyme) or a strand is inverted (enolase).

2. A perspicuous view of the inside of the structure of the sheet is afforded by 'rolling out the barrel' (see Fig. 7.8). The leftmost (the N-terminal) strand is repeated at the right, in red. To recover the three-dimensional eight-stranded barrel, this diagram must be folded over, and the two images of the first strand glued over each other: superposing A and B (black, at left) onto A and B (red, at right) in Fig. 7.8.

The tipping of the strands to the barrel axis (vertical in Fig. 7.8) produces a layered structure. Note that the sidechains of each strand point alternately into and out of the barrel; this is an important feature of the packing inside the barrel. The

individual strands vary in length from protein to protein. (The strand lengths in Fig. 7.8 are those appearing in chicken triose phosphate isomerase.) However, in all TIM barrels all strands contain three residues at the same height, forming a continuous hydrogen-bonded net girdling the barrel. Sidechains from these levels pack in three layers in the interior of the barrel.

3. The different barrels are similar in topology, having the same values of McLachlan's strand number and shear number. All TIM barrels have eight strands, shear = 8, a tilt of the strands to the barrel axis of approximately 36°, and radii of 6.5–7.5 Å, depending on the eccentricity of the cross-section.

4. The packing of residues inside the sheet shows a common structural pattern.

Figure 7.9(a) shows the hydrogen-bonding net of the β-sheet of spinach glycolate oxidase (GAO). On each strand of sheet, alternate sidechains point towards the region inside the sheet and out towards the helices. The 12 residues with identifying letters have sidechains pointing inwards. The packing inside the barrel is formed by the interactions of

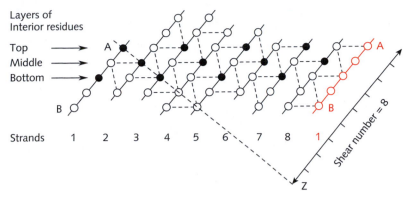

Figure 7.8 Schematic diagram of the hydrogen-bonding nets obtained by 'rolling out the TIM barrel'—a closed β-sheet with eight strands and shear number 8, which appears in triose phosphate isomerase, glycolate oxidase and many other proteins. Each circle represents a residue; short broken lines represent hydrogen bonds. Filled circles represent residues in three layers that point into the barrel interior; note their pattern of alternation. This diagram contains nine strands, because the first strand is duplicated at left (in black) and right (in red) edges. To recover the three-dimensional eight-stranded barrel, the leftmost strand must be superposed on the rightmost by folding the paper into a cylinder and glueing residues labelled A and B in black onto residues labelled A and B in red. This produces the barrel as it occurs in the proteins, with the strands tipped by 36° to the barrel axis. To form an eight-stranded barrel with strands *parallel* to the axis, residue A would have to be glued onto point Z. A.D. McLachlan defined the shear, S, as a measure of the stagger of the strands. It is the number of residues by which the residue on which A is actually superposed is displaced from Z—the residue on which A would be superposed in a barrel with strands parallel to the axis. Here S = 8. At the centre of the barrel, 12 inward-pointing sidechains pack together. These correspond to the filled circles. Note their arrangement in three parallel layers, in planes lying at the same height relative to the barrel axis. The symmetry of this pattern could accommodate the change of a strand from parallel to antiparallel, as observed in enolase.

these 12 residues. The first, third, fifth and seventh strands each contribute one sidechain, and the second, fourth, sixth and eighth strands each contribute two sidechains. Note that residues at the same height along the axis of the sheet (vertical in Figs. 7.8 and 7.9(a)) are not nearest neighbours on adjacent strands because of the tilt of the strands with respect to the barrel axis.

The packing of these sidechains in the barrel interior is shown in Figs. 7.9(b)–(e). Part (b) is a side view of the sheet of GAO, pruned to three residues per strand. The sidechains occupy three tiers or layers with almost perfect segregation.

The packing of these residues is seen in Figs. 7.9(c)–(e), which show serial sections through the three layers of GAO. In the first and third layers, atoms from the four packed sidechains are drawn in black. These belong to odd-numbered strands. In the central layer, atoms from the four packed sidechains are drawn in red. These belong to even-numbered strands. Van der Waals envelopes are drawn in green.

These results suggest a simple description of the packing of residues inside the sheet of glycolate oxidase and other TIM barrels. There is a three-tiered arrangement, involving a double alternation, a three-dimensional chessboard pattern, in which alternate strands contribute sidechains to alternate layers:

- The tilt of the strands relative to the axis of the sheet, and the twist of the sheet, place the inward-pointing sidechains in layers. Each layer contains four sidechains from alternate strands. The sidechains that point 'in' are on odd-numbered levels on odd-numbered strands, and on even-numbered levels on even-numbered strands. The central region of the barrel is filled by 12 sidechains from three layers.

- Qualitatively, the packing is a layered A–B–A type structure. (That is, A indicates the layout of units

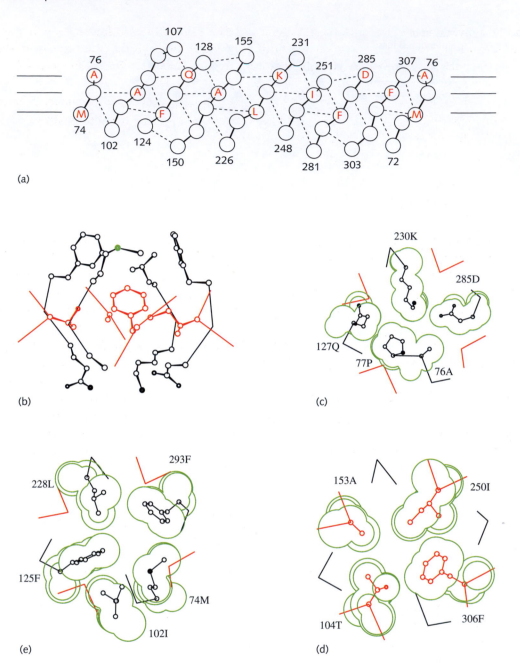

Figure 7.9 The packing of residues inside the barrel of spinach glycolate oxidase [1GOX]. (a) The β-sheet of glycolate oxidase, unrolled. Each circle represents a residue; one-letter codes identify residues the sidechains of which pack inside the barrel. Broken lines represent hydrogen bonds. Numerals indicate residue numbers. Nine strands are shown: the first strand (residues 72–76) is duplicated. This picture is a cylindrical projection drawn from atomic coordinates, and it gives an accurate picture of the positions of the residues. The previous figure, Fig. 7.8, is idealized. (b) The inward-pointing sidechains form three layers inside the barrel. This drawing shows the eight strands of the β-barrel, pruned to the three residues per strand, and the inward-pointing sidechains. The view is perpendicular to the barrel axis. (c, d, e) Serial sections cut through a space-filling model (Van der Waals slices) of the three layers of residues packing inside the barrel of spinach glycolate oxidase. Here the view is parallel to the barrel axis. In each drawing three slices separated by 1 Å are shown. Atoms from alternate strands are shown in black and red. The reader is urged to follow each of the 12 residues from part (a) to part (b) to part (c), (d) or (e) of this figure.

on the first layer. On the second level, the units have different layout, B. But the units on the third layer lie above those of the first layer, with layout A again. Such structures are common in simple inorganic crystals.) Successive layers are related by a rotation by 45° around the barrel axis and a translation along the axis by approximately 3 Å.

- The formation of a fourth layer is prevented by the protrusion of the sidechains from the top and bottom layers (see Figs. 7.9(c) and (e)).

We next examine the evolutionary relationships in several protein families of different fold types: all-α, α/β, and all-β.

Evolution of the globins

Globins are an ancient family of proteins, appearing in prokaryotes, animals, including vertebrates and invertebrates, and in plants.

Haemoglobin in our blood transports oxygen and carbon dioxide, delivering oxygen to myoglobin in other tissues. To promote this process, tetrameric haemoglobin arose by gene duplication and divergence from monomeric globins. We discussed the allosteric change in haemoglobin in Chapter 5.

The sequencing of the human genome has turned up additional globins. For example, neuroglobin is expressed in the brain. Its precise function remains obscure.

Sperm-whale myoglobin, the first protein structure to be solved by X-ray crystallography, shows a characteristic globin folding pattern (see Fig. 7.10) containing eight helices enfolding a haem group (see Fig. 7.11).

Historically, analysis of the globin structures began with the mammalian structures, the first to be solved. One then went on to treat a wide range of globins from eukaryotes, which had diverged in sequence but retained a generally similar folding pattern. Many years later, structures of a new class of **truncated globins** emerged—substantially shorter than those previously considered. This reopened questions about globin evolution. The globins first studied, with lengths around 150 residues, are now called **full-length globins**. Truncated globins have lengths around 110–120 residues.

- Full-length globins, including human haemoglobin, are ~150 residues long. Truncated globins are ~110–120 residues long.

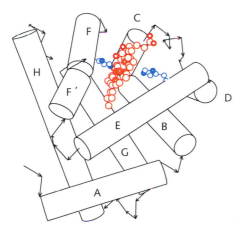

Figure 7.10 The 'globin fold' consists of a characteristic arrangement of helices, surrounding a haem group. The helices are denoted A, B, C (a 3_{10} helix), D, E, F (broken in mammalian globins into two consecutive helices, F' and F), G, H. Many globins lack the D helix. In most globins, the closest neighbours of the iron are the four pyrrole rings of the haem group and two histidine sidechains, shown here in blue, on the E and F helices. Until recently, it was thought that this overall architecture characterized the entire globin family. The discovery of *truncated globins*—about 120 residues long instead of the ~150 residues characteristic of full-length globins—overturned this idea.

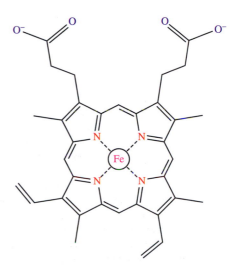

Figure 7.11 The haem group contains an iron at the centre of a porphyrin ring.

```
                          AAAAAAAAAAAAAAAAA         BBBBBBBBBBBBBBBBBBBCCCCCCC                    DD
Sperm whale myoglobin     VLSEGEWQLVLHVWAKVEA---DVAGHGQDILIRLFKSHPETLEKFDRFKHLKTE
P. caudatum                         SLFEQLGGQAAVQAVTAQFYANIQADA-TVATFFN--------
C. eugametos                        SLFAKLGGREAVEAAVDKFYNKIVADP-TVSTYFS--------
M. tuberculosis           GLLSRLRKREPISIYDKIGGHEAIEVVVEDFYVRVLADD-QLSAFFS--------
                                                                                                  F

                          DDDDDEEEEEEEEEEEEEEEEEEEEE              FFFFFFFFFF                GGGGGG
Sperm whale myoglobin     AEMKASEDLKKHGVTVLTALGAILKKKGHHEAELKPLAQSHATKH--KIPIKYLE
P. caudatum               -----GIDMPNQTNKTAAFLCAALGGPNAWTGR--NLKEVHA-NM--GVSNAQFT
C. eugametos              -----NTDMKVQRSKQFAFLAYALGGASEWGK--DMRTAHK-DLVPHLSDVHFQ
M. tuberculosis           -----GTNMSRLKGKQVEFFAAALGGPEPYTGA--PMKQVHQ-GR--GITMHHFS
                                                                                                  H

                          GGGGGGGGGGGGGG        HHHHHHHHHHHHHHHHHHHHHHHHHHH
Sperm whale myoglobin     FISEAIIHVLHSRHPGDFGADAQGAMNKALELFRKDIAAKYKELGYQG
P. caudatum               TVIGHLRSALTGAGV---AAALVEQTVAVAETVRGDVVTV
C. eugametos              AVARHLSDTLTELGV---PPEDITDAMAVVASTRTEVLNMPQQ
M. tuberculosis           LVAGHLADALTAAGV---PSETITEILGVIAPLAVDVTS
                          L
```

Figure 7.17 Alignment of the sequences of sperm-whale myoglobin and truncated globins from *Paramecium caudatum, Chlamydomonas eugametos,* and *Mycobacterium tuberculosis.* Letters on the top line indicate the extents of the helices in the sperm-whale myoglobin structure.

for the active site to be more tightly conserved than other parts of the structure.

> • Point mutations in protein interiors cause shifts and rotations of secondary structure elements, coupled to maintain function. The contact pattern of residues tends to be maintained.

These conclusions, first proposed over 35 years ago, were challenged recently when the structures of a class of shortened globin structures appeared. These may be as small as 109 residues.

Truncated globins

Truncated globins are short proteins, occurring in prokaryotes and eukaryotes, that maintain a recognizable globin fold despite typically containing only ~120 residues, substantially smaller than the ~150 residues of typical full-length globins. They have been implicated in diverse functions, including **detoxification** of NO and photosynthesis.

Which residues of full-length globins, and which structural elements, are sacrificed? Truncated globins retain most but not all of the helices of the standard globin fold with the notable exception of the loss of the F-helix, which contains the iron-linked histidine. They show a shortening of the A helix and of the

CD region. Of the 59 sites involved in conserved helix-to-helix or helix-to-haem contacts in full-length globins, 41 of them appear, with conserved contacts, in truncated globins. The helix/helix interfaces have ridge-groove packing patterns similar to those of the full-length globins, with the exception of the B/E contact, which has an unusual crossed-ridge structure in full-length globins but is normal in truncated globins.

Despite the differences, it is possible to align the truncated and full-length globins (see Figs. 7.17 and 7.18).

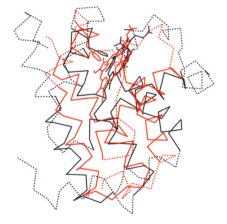

Figure 7.18 Superposition of the structures of sperm-whale myoglobin [1MBO] and truncated globin from *Paramecium caudatum* [1DLW]. Solid lines indicate the parts of the structure common to full-length and truncated globins. Broken lines indicate the parts of the sperm-whale structure absent from truncated globins or changed in conformation.

Expansion of the globin family

So far, our discussion of the globins has focused strongly on structure. This was appropriate, because globins have contributed to discoveries of so many basic principles of protein structure. For a long time, in fact, structure determinations of globins kept pace with sequence determinations, provided that one accepts that a single structure could allow rationalization of the effects of mutations in closely-related proteins. In fact, as hard as it may be to believe today, the structure of myoglobin was first solved *before* the amino acid sequence was determined. (Remember that this was long before DNA sequencing.)

Globins are now a more typical family, in that we know many more distantly-related sequences than structures. What can we say then about the evolution of the globins, from the amino acid sequences' point of view?

Globins are a very ancient and widely-dispersed family. They appear in archaea, bacteria, and eukarya. Individual globin domains are either 'full-length' (~150 residues), or 'truncated' (~120 residues or even fewer). (See Fig. 7.18). Some authors refer to full-length globins as 3/3 globins, and truncated globins as 2/2 globins. This notation refers to a proposed interpretation of the spatial assembly of six helices in full-length globins and four helices in truncated globins.

In some cases, for instance myoglobin and haemoglobin, the globin domains form a single and complete polypeptide chain. In other cases, a globin domain is fused to one (usually) or more domains of different structure and function. A very unusual relative is androglobin, which contains an N-terminal calpain-like domain, an internal, *circularly-permuted* globin domain, and an IQ calmodulin-binding motif. Androglobin appears in a wide spectrum of metazoa. Although it comprises domains related to proteins of known function, a comprehensive picture of why they are joined in a single molecule and whether the function of the molecule as a whole is different from the sum of the parts, remains unclear.

Some globins are monomeric; others show quaternary structures including dimers, tetramers, and larger polymers (see Fig. 7.19).

Classification of the globins

A classification of the known globins (see Fig. 7.20), by S. Vinogradov and colleagues, on the basis of sequence similarity and divergence observed in multiple alignments, distinguishes:

(a) The flavohaemoglobin family comprises single-domain full-length globins, and chimaeric proteins containing an N-terminal full-length globin domain and a C-terminal flavoprotein, or other other chimaeric combinations of the globin and reductase domains. Flavohaemoglobins are widely distributed in bacteria and non-metazoan eukaryotes. Related single-domain globins appear in bacteria and metazoans, with only a few examples in non-metazoan eukaryotes.

Vertebrate globins are single-domain flavohaemoglobin relatives, unfused at the level of primary structure to any partners. In addition to myoglobin and haemoglobin, vertebrates contain neuroglobin and cytoglobin. Human neuroglobin is present at low levels in the cytoplasm of neurons, including the retina, and endocrine cells. Human cytoglobin appears at low levels in the cytoplasm and nucleus of fibroblasts and similar cells.

Globin E (in birds), globin X (in fishes and amphibians), and globin Y (in monotremes and amphibians) are additional recently-discovered vertebrate homologues (see Fig. 7.21).

(b) The globin-coupled sensor family comprises:

1. chimaeric proteins containing an N-terminal full-length globin domain (sensing oxygen or other small ligands) and one or more C-terminal domains (transducing the state of the sensor). The globin domain is longer than the canonical globin domain by some 40 residues, located in a pre-A helix (Z-helix), CE loop, and FG loop. Members of this family appear in bacteria and Euryarchaeota. They are greatly outnumbered by the flavohaemoglobins.

2. Corresponding single-domain globins appear in bacteria, Euryarchaeota, and fungi.

3. Protoglobins are non-chimaeric proteins are found in bacteria and archaea. It is likely that they are derived from an ancestral single-domain version of the globin-coupled sensors.

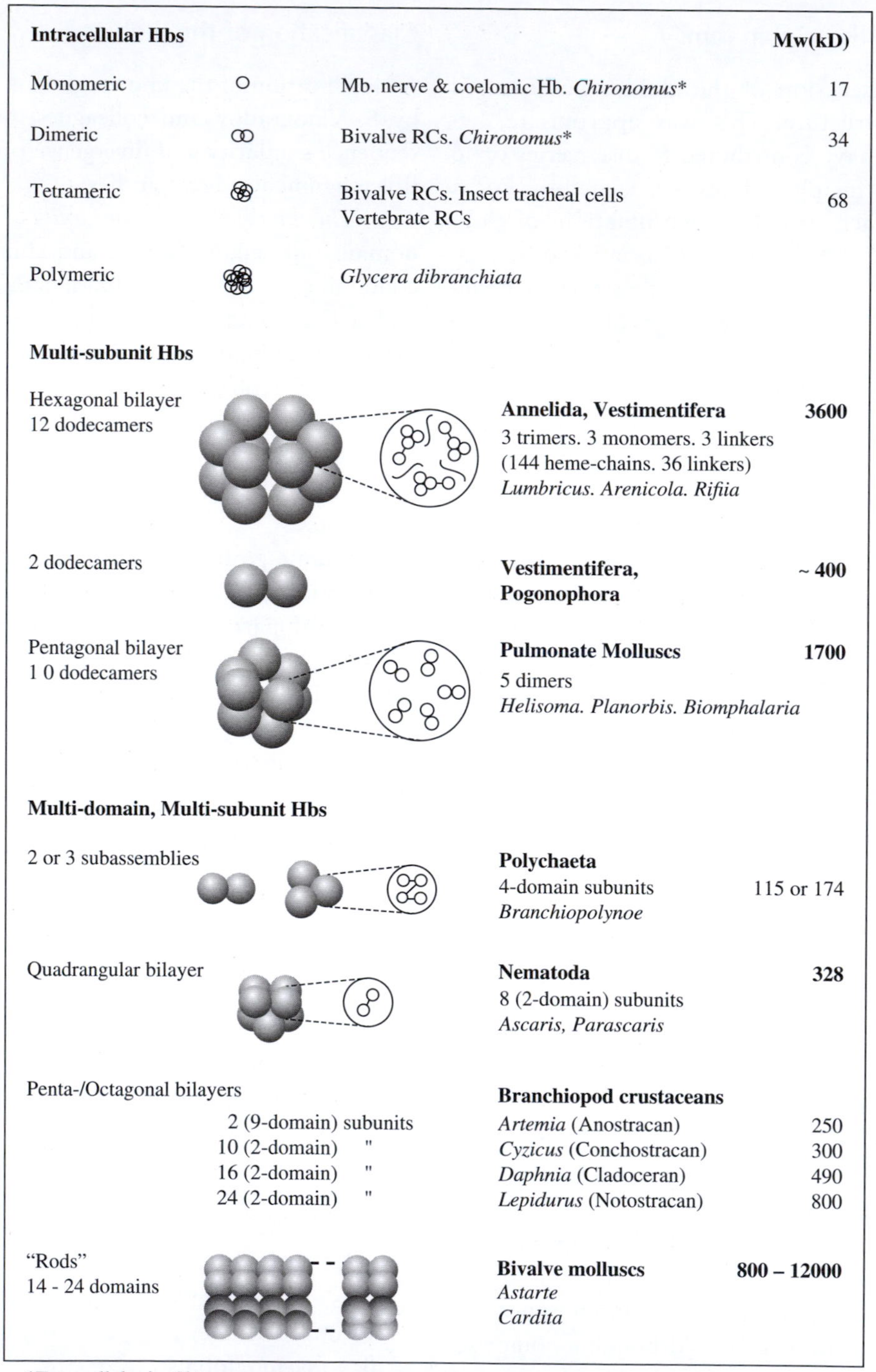

Intracellular Hbs			Mw(kD)
Monomeric	○	Mb. nerve & coelomic Hb. *Chironomus**	17
Dimeric	∞	Bivalve RCs. *Chironomus**	34
Tetrameric	⊛	Bivalve RCs. Insect tracheal cells Vertebrate RCs	68
Polymeric	⊛	*Glycera dibranchiata*	

Multi-subunit Hbs

Hexagonal bilayer 12 dodecamers		**Annelida, Vestimentifera** 3 trimers. 3 monomers. 3 linkers (144 heme-chains. 36 linkers) *Lumbricus. Arenicola. Rifiia*	3600
2 dodecamers		**Vestimentifera, Pogonophora**	~ 400
Pentagonal bilayer 1 0 dodecamers		**Pulmonate Molluscs** 5 dimers *Helisoma. Planorbis. Biomphalaria*	1700

Multi-domain, Multi-subunit Hbs

2 or 3 subassemblies		**Polychaeta** 4-domain subunits *Branchiopolynoe*	115 or 174
Quadrangular bilayer		**Nematoda** 8 (2-domain) subunits *Ascaris, Parascaris*	328

Penta-/Octagonal bilayers

	Branchiopod crustaceans	
2 (9-domain) subunits	*Artemia* (Anostracan)	250
10 (2-domain) "	*Cyzicus* (Conchostracan)	300
16 (2-domain) "	*Daphnia* (Cladoceran)	490
24 (2-domain) "	*Lepidurus* (Notostracan)	800

"Rods" 14 - 24 domains		**Bivalve molluscs** *Astarte* *Cardita*	800 – 12000

*Extracellular in *Chironomus* larvae

Figure 7.19 Taxonomic distribution of quaternary structures of globins.

From: Weber, R.E. and Vinogradov, S.N. (2001). Nonvertebrate hemoglobins: functions and molecular adaptations. Physiol. Revs. 81, 569–628.

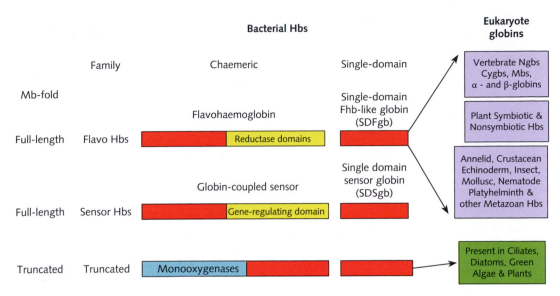

Figure 7.20 Classification of globins, and the distributions of classes. Globin domains in red. Abbreviations: Mb myoglobin; Hb, haemoglobin; FHb, flavohaemoglobin; SDFHB, single-domain flavohaemoglobin; SDSgb, single-domain sensor globin; Ngbs, neuroglobins; Cygbs, cytoglobins.

From: Vinogradov, S.N. and Moens, L. (2008). Diversity of globin function: enzymatic, transport, storage, and sensing. J. Biol. Chem. 283, 8773-8777.

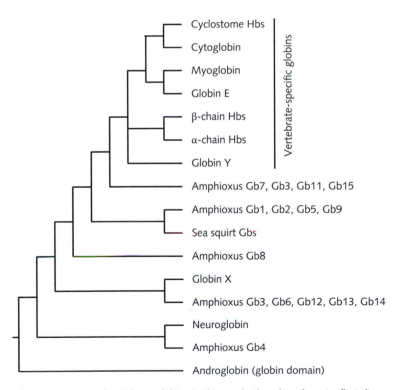

Figure 7.21 Phylogenetic tree of vertebrate globins. In this graph, the edges do *not* reflect divergence times.

(c) The truncated globin family appears in prokaryotes, protozoans, fungi, and Archaeplastida. Three paralogous groups exhibit distinctive amino acid conservation patterns, and taxonomic distributions:

Group	Distribution:
I	Only in Archaea, Cyanobacteria, Ciliates, and fungi
II	Principal truncated globins in plants
III	Largely restricted to bacteria

Only group II truncated globins form chimaeras. The added N-terminal domain is annotated as a member of the antibiotic biosynthesis monooxygenase family.

Globin functions

O_2 affinities of globins vary very widely, over four orders of magnitude. Very high affinity implies storage and sequestration; moderate affinity suggests transport and delivery. Leghaemoglobin (LegHb, where leg refers to legume) is a phytoglobin, encoded in the plant genome and expressed in the nitrogen-fixing nodules arising from infection of plant roots with rhizobia. Its role in the symbiosis is to sequester O_2: the nitrogenase complex of the rhizobia is exquisitely oxygen-sensitive.

Conversely, the role of full-length globins related to the globin domains of flavohaemoglobins of the gram-negative and obligate aerobe *Vitreoscilla* is delivery. *Vitreoscilla* Hb binds to subunit I of cytochrome *bo* oxidase and supports respiration under hypoxic conditions.

In globin-coupled sensors, the conformational or electronic perturbations pursuant to oxygen binding, trigger signalling. Aerotactic globin-coupled sensors, such as the bacterial and archaeal HemAT, sense the level of O_2 in the environment of the cell. They mediate an aerophobic or aerophilic response via a signalling domain interacting with the chemotaxis system. Gene regulatory globin-coupled sensors are fused to various C-terminal transducer domains (e.g., GAF and STAS) and associate with O_2 to report on its cellular levels.

In addition to oxygen binding, both haemoglobin and myoglobin can function as dioxygenases. They combine bound O_2 with NO to produce the relatively innocuous nitrate ion.

Escherichia coli flavohaemoglobin plays a role in cellular redox homeostasis and in the response to oxidative and nitrosative stresses. Under aerobic conditions, it shows a robust NO scavenging and dioxygenase activity.

Peroxidase activity appears to be general, but rather as a detrimental side reaction. In one contrary example, a full-length globin turned its oxidative capability into a defense mechanism: The marine polychaete *Amphitrite ornata* uses its intracellular dehaloperoxidase for detoxification as well as oxygen binding.

Some invertebrate globins are involved in sulphide transport.

An unusual adaptation appears in an extraordinary single-domain globin related to flavohaemoglobins, of the nematode *Mermis nigrescens*, concentrated in its oxy state at the tip of the female worm to shadow a photoreceptor, and direct phototaxis. Some readers may find it intriguing that the red eyes of humans who have spent too many late nights in smoke-filled rooms, and of *Mermis*, are displaying the same pigment.

There are 33 globin genes in *C. elegans*, a haem auxotroph. Most of the products of these genes have not yet been studied in detail. However, recent work provides evidence for oxygen sensing and a role in development. The properties of at least one of the full-length *C. elegans* globins suggest that it may function as an electron-transport protein. This would constitute the first example of a cytochrome-like functionality in a globin fold.

Phycocyanins and the globins

Phycocyanins are the major constituents of the **phycobilisome**, the supramolecular light-harvesting complexes of cyanobacteria and red algae. They bind a chromophore that is an open-chain tetrapyrrole structure similar to a bile pigment (the haem group is a tetrapyrrole structure *closed* into a macroring.)

When the first phycocyanin structure was solved, it showed an entirely unsuspected similarity in folding

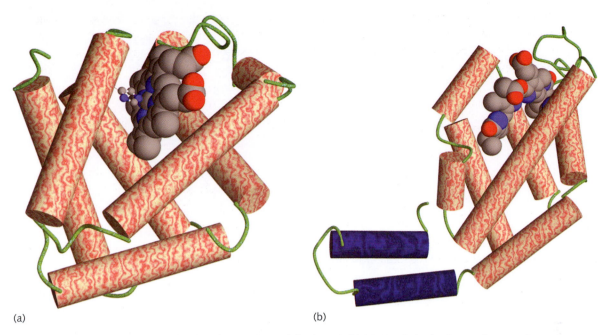

(a)   (b)

Figure 7.22 (a) Sea hare *(Aplysia limacina)* myoglobin [1MBA]. (b) *Mastigocladus laminosus* phycocyanin, α-chain.

pattern to the globins, despite the difference in prosthetic group and function, and the absence of any apparent similarity between globin and phycocyanin sequences (see Fig. 7.22). (The phycocyanin structure contains two N-terminal helices in addition to the set displaying the same structural relationships that the globin helices do. These are involved in oligomerization to form the phycobilisome, a feature that phycocyanins do not share with the globins. Ignore them.)

What is the origin of the similarity of the folds? Do globins and phycocyanins share a common ancestor, or did their very different functions somehow require the independent evolution of the same folding pattern? It is difficult to answer these questions because (1) the sequences are so different that it is not easy to align them with any confidence, and (2) some of the structural differences between globins and phycocyanins are much larger than the differences between distantly-related globin structures.

To try to distinguish between true evolutionary relationship or convergence, we can only observe the similarities of the structures and try to interpret them. The basic dilemma is the distinction between points of similarity—which may be consistent with either homology or convergence—and criteria for true relationship. W.E. Le Gros Clark, the palaeontologist,

wrote: 'While it may be broadly accepted that, as a general proposition, degrees of genetic relationship can be assessed by noting degrees of resemblance in anatomical details, it needs to be emphasized that morphological characters vary considerably in their significance for this assessment.' For us, the implication is that we must give different weight to different sorts of structural similarities, disregarding features common to many classes of proteins, and emphasizing the unusual ones. We must look for *structural similarities that are not specifically required by structure or function*. That is the principle. The problem with putting it into practice is that it is impossible to be *sure* that any feature of a protein structure is not required for structure or function.

The C-helices in globins and phycocyanins are both of the rare 3_{10} structure; the rest are all α-helices. The C-helix does not appear to play a role in the function of monomeric globins. (It is important in the allosteric change in haemoglobin, but that was surely a much later development.) If, as seems likely, a sequence compatible with an α-helix in this region could produce a viable globin or phycocyanin, then the fact that the C-helices in both globins and phycocyanins are of the unusual 3_{10} type is a structural similarity not specifically required by structure or function.

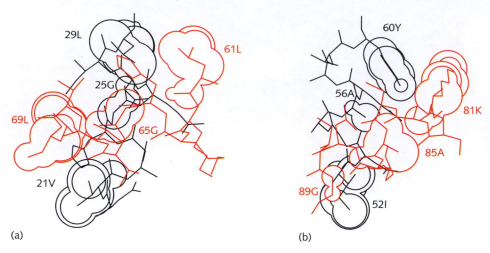

Figure 7.23 B–E contact in (a) sperm-whale myoglobin and (b) *Mastigocladus laminosus* phycocyanin, α-chain.

What about the pattern of interactions of residues at the helix interfaces? The observation of common types of ridges-into-grooves packing at interfaces between corresponding pairs of helices in globins and phycocyanins is not evidence for evolutionary relationship, as these structures appear in many unrelated proteins. More significant are the interfaces with unusual structures: the B–E and G–H packings. In full-length globins these interfaces have an unusual 'crossed-ridge' structure. It is very interesting to see a similar crossed-ridge packing in the B/E interface in phycocyanin (see Fig. 7.23). A similar crossed-ridge structure is also observed in the G–H contact in phycocyanins. But can we argue that this type of interface is not *required* by the structure? In phycocyanins the interaxial angles in the B/E interfaces are all in the region of –50° to –60° that would be expected from a *normal* '4–4' packing. Therefore, the phycocyanin structure cannot *require* the spe-

cial crossed-ridge structure to achieve an unusual interhelix-axis angle. (This argument was advanced before the discovery of the truncated globins, and it is supported by the appearance of the normal non-crossed-ridge packing in the B/E interfaces in truncated globins.)

We can never really settle the question rigorously because evolutionary events that occurred in the distant past are not directly observable. However, if we had to bet, the observation that the structural similarities between globins and phycocyanins extends down to minute structural details, apparently not required by fold or function, suggests that the structures are homologous rather than convergent.

> • Phycocyanins and globins share structure but not function. Nevertheless, they retain certain subtle structural details that suggest that they are probably homologous.

Evolution of NAD-binding domains of dehydrogenases

In 1973, knowing the structures of lactate, malate, and alcohol dehydrogenases, C.-I. Brändén, H. Eklund, B. Nordström, T. Boiwe, G. Söderlund, E. Zeppezauer, I. Ohlsson, and Å. Åkeson wrote: 'The coenzyme binding region [of horse liver alcohol dehydrogenase] has a mainchain conformation very

similar to a corresponding region in lactate and malate dehydrogenase. It is suggested that this substructure is a general one for binding of nucleotides and, in particular, the coenzyme NAD$^+$.'

Since then, many additional crystal structures have confirmed and extended this principle. The paradigm

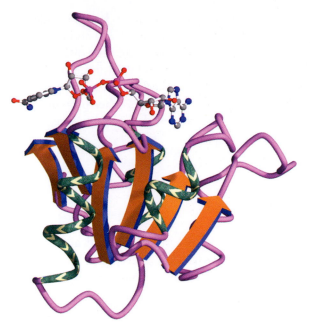

Figure 7.24 NAD-binding domain of horse liver alcohol dehydrogenase [6ADH]. The folding pattern includes a central parallel β-sheet flanked by helices. Domains with this folding pattern appear in many dehydrogenases and in other proteins that bind related molecules.

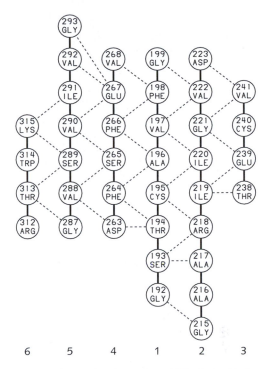

Figure 7.25 Hydrogen-bonding pattern of the β-sheet in horse liver alcohol dehydrogenase [6ADH].

nucleotide binding domain in horse liver alcohol dehydrogenase contains two sets of β–α–β–α–β units, together forming a single parallel β-sheet flanked by α-helices (see Fig. 7.24). The strands appearing from left to right in Fig. 7.24 appear in the *sequence* in the order 6–5–4–1–2–3 (see Fig. 7.25). There is a long loop, or cross-over, between strands 3 and 4. As Brändén described in 1980, this feature of the fold creates a natural cavity for binding of the adenine ring of the NAD and, in other proteins with similar supersecondary structures, for other nucleotide-containing fragments.

- NAD-binding domains of dehydrogenases have a tertiary structure based on the β–α–β supersecondary structure. They contain two sets of β–α–β–α–β subunits, connected by a cross-over loop that creates a pocket suitable for binding the adenine ring of NAD.

NAD-binding domains combine with other domains in enzymes with very different substrate specificities. They occur as oligomers of different sizes; many are tetramers. Figure 7.26 shows formate

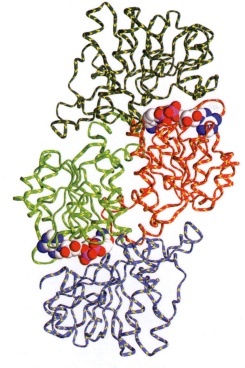

Figure 7.26 The dimeric structure of formate dehydrogenase, viewed down the axis of symmetry [2NAD]. One subunit is dark green, with its NAD-binding domain highlighted in red. The other is blue, with its NAD-binding domain highlighted in light green.

dehydrogenase, a dimer. The two domains of each monomer appear in different colours.

Proteins other than dehydrogenases bind nucleotide-containing cofactors in a manner similar to the binding of NAD to dehydrogenases, and create the binding site from domains with generally similar secondary and tertiary structure. Examples include the FMN in flavodoxin and the FAD in pyruvate oxidase. Conversely, other structures have extended the repertoire of modes of binding NAD and related ligands. Some have folding patterns very different from the NAD-binding domains of dehydrogenases. These include proteins from other general topological classes: all–β and α + β as well as other proteins in the α/β class but unrelated to the dehydrogenase NAD-binding fold.

Comparison of NAD-binding domains of dehydrogenases

Figures 7.27 to 7.32 illustrate six NAD-binding domains, showing the folding of the chain and the hydrogen-bonding net of the sheet. These give some idea of the observed structural variety.

In each of these domains, the sheets contain the canonical six strands, but are extended by additional strands. Dihydropteridine reductase has a seventh strand adjacent and parallel to the sixth strand and an eighth strand forming a hairpin with the seventh. 6-phosphogluconate dehydrogenase has a seventh strand adjacent but antiparallel to the sixth. Glyceraldehyde-3-phosphate dehydrogenase has a short stretch of antiparallel sheet between the third and fourth strand, before the cross-over. 3α, 20β-hydroxysteroid dehydrogenase and formate dehydrogenase have a seventh strand adjacent and parallel to the sixth.

The helices also differ considerably in length, and can shift relative to the sheet. In 3α,20β-hydroxysteroid dehydrogenase and dihydropteridine reductase the two helices in the C-terminal portion of the domain (between strands 4 and 5, and between strands 5 and 6) are elongated. There is also a helix in the cross-over region, which appears in many NAD-binding domains. Dihydropteridine reductase has lost the helix between strands 2 and 3.

What is the conserved core of the family? Figure 7.33 shows the superposition of the NAD-binding domains of a closely-related pair of enzymes (lactate and

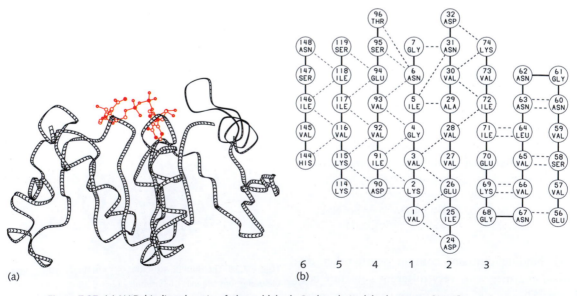

(a)
(b)

6   5   4   1   2   3

Figure 7.27 (a) NAD-binding domain of glyceraldehyde-3-phosphate dehydrogenase [1GD1].
(b) Hydrogen-bonding pattern of the β-sheet.

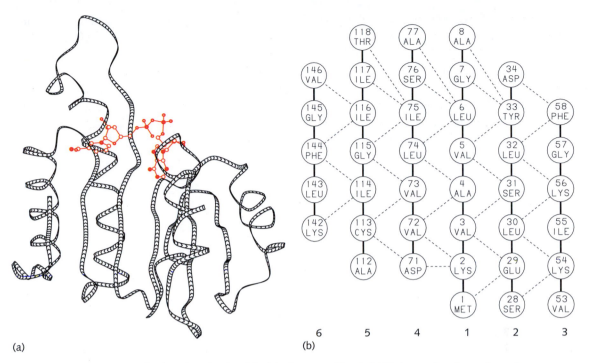

Figure 7.28 (a) NAD-binding domain of malate dehydrogenase [1EMD]. (b) Hydrogen-bonding pattern of the β-sheet.

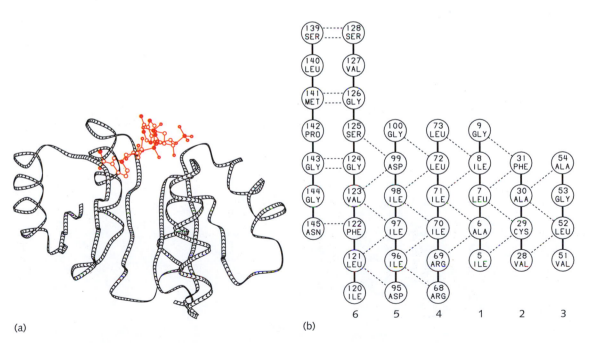

Figure 7.29 (a) NAD-binding domain of 6-phosphogluconate dehydrogenase [1PGO]. (b) Hydrogen-bonding pattern of the β-sheet.

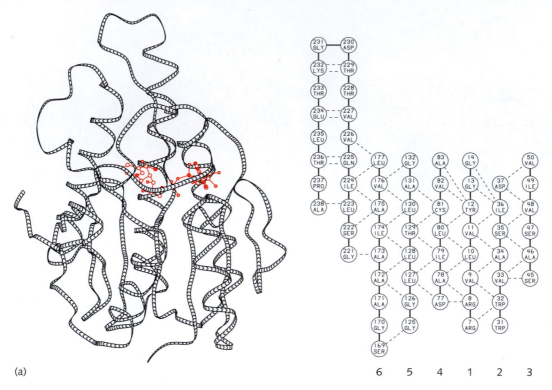

(a)

Figure 7.30 (a) NAD-binding domain of dihydropteridine reductase [1DHR]. (b) Hydrogen-bonding pattern of the β-sheet.

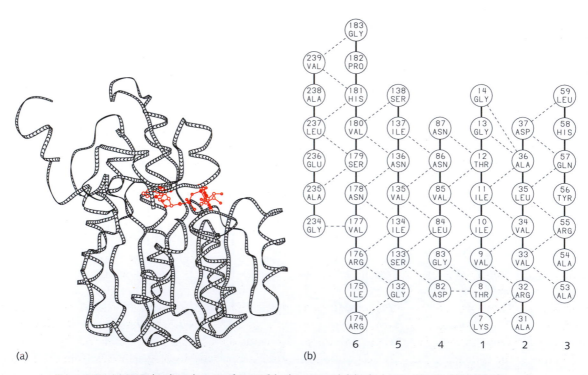

(a)   (b)

Figure 7.31 (a) NAD-binding domain of 3α,20β-hydroxysteroid dehydrogenase [2HSD]. (b) Hydrogen-bonding pattern of the β-sheet.

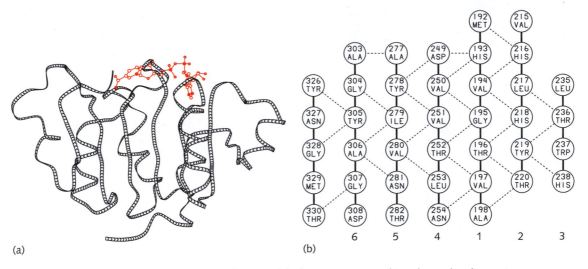

(a)

(b)

				192 MET	215 VAL
		303 ALA	277 ALA	249 ASP	193 HIS	216 HIS
326 TYR	304 GLY	278 TYR	250 VAL	194 VAL	217 LEU	235 LEU
327 ASN	305 TYR	279 ILE	251 VAL	195 GLY	218 HIS	236 THR
328 GLY	306 ALA	280 VAL	252 THR	196 THR	219 TYR	237 TRP
329 MET	307 GLY	281 ASN	253 LEU	197 VAL	220 THR	238 HIS
330 THR	308 ASP	282 THR	254 ASN	198 ALA		

6   5   4   1   2   3

Figure 7.32 (a) NAD-binding domain of formate dehydrogenase [2NAD]. (b) Hydrogen-bonding pattern of the β-sheet.

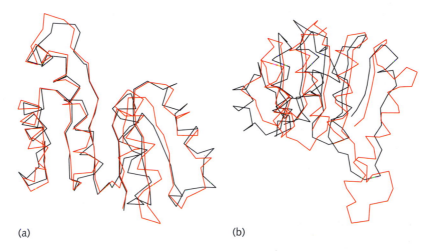

(a)

(b)

Figure 7.33 (a) Superposition of NAD-binding domains of lactate dehydrogenase [9LDT] (black) and malate dehydrogenase [1EMD] (red). The sequences of these regions have 23% identical residues upon optimal alignment. Although these molecules have developed different functions, they are still fairly closely related. (b) Superposition of NAD-binding domains of horse liver alcohol dehydrogenase [2OHX] (black) and dihydropteridine reductase [1DHR] (red). These molecules have diverged more radically. There are only 14% identical residues in an optimal alignment.

malate dehydrogenases) and a more distantly-related pair (alcohol dehydrogenase and dihydropteridine reductase). The cross-over region—the loop between strands 3 and 4—is especially variable in structure among the different enzymes. The maximal common substructure of the distantly-related pair corresponds to the core of the members of this family of domains illustrated here.

The sequence motif G*G**G

The binding of NAD to dehydrogenases involves numerous hydrogen bonds and Van der Waals contacts between the cofactor and the enzyme. In particular, there are usually hydrogen bonds from the residues in the turn between the first strand and the helix that follows it to one of the phosphate groups of the

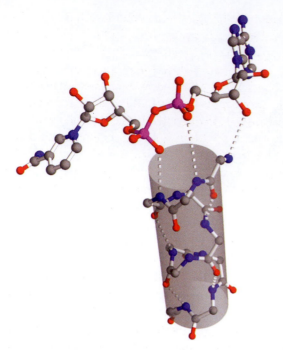

Figure 7.34 Interactions between protein and ligand in NAD-binding domains commonly include hydrogen bonding between phosphate oxygens and the N–H groups in the last turn of the first helix.

cofactor (see Fig. 7.34). These interactions give rise to a consensus sequence in this region containing the three-glycine pattern G*G**G characteristic of the first β–α–β unit in the dehydrogenase domain. (This motif signifies a hexapeptide: Gly-xxx-Gly-xxx-xxx-Gly, where xxx represents any residue.)

The first two glycines are involved in nucleotide binding and the third, which is in the helix following the first strand, is involved in the packing of the helix against the sheet. The first glycine is in the $\varphi > 0$, $\psi > 0$ region of the Sasisekharan–Ramakrishnan–Ramachandran plot, a conformation integral to the structure of the turn. A Cβ at the position of the second glycine would collide with the cofactor. In horse liver alcohol dehydrogenase, a Cβ in the residue at the position of the third glycine would clash with the carbonyl of the first glycine.

The dehydrogenases contain a well (but not absolutely) conserved aspartate approximately 20 residues C-terminal to the G*G**G motif. This aspartate appears near the C-terminus of the second strand, and forms hydrogen bonds to the ribose of the adenosine moiety of the NAD.

Structure and evolution of serine proteinases of the chymotrypsin family

The structures of serine proteinases have been of interest since the determination of the crystal structure of chymotrypsin in 1969. We discussed their mechanism of action in Chapter 5. It is common to use the term 'serine proteinase' to refer only to homologues of chymotrypsin, even though certain other families of proteinases also contain catalytic serines, and the **viral 3C proteases**, true homologues of chymotrypsin, have a catalytic cysteine in place of the serine.

What are the structural constraints to which these proteins have been subject during their evolution, and how have they explored these limits?

Each serine proteinase contains two domains of similar structure, likely to have arisen by gene duplication and divergence. (However, no protein with a single serine protease domain has yet been discovered.) The domains pack together, with the active site between them. In most members of the family, intradomain disulphide bridges help to keep the molecule intact and to maintain the structure of the active site.

The region of the substrate that includes the scissile bond binds in the cleft between the domains (see Fig. 7.35). The N-terminal domain contains the His and Asp of the catalytic triad. The C-terminal domain contains the Ser of the **catalytic triad**, the **oxyanion hole** that stabilizes the transition state, and the specificity pocket. Domain 1 is composed primarily of residues from the N-terminal portion of the molecule, and domain 2 primarily of residues from the C-terminal portion.

Structures of individual domains

Each domain contains a six-stranded antiparallel β-sheet, folded into a β-barrel. Figures 7.36 and 7.37 illustrate the domains of *S. griseus* proteinase b, one of the smaller structures. In the sheet diagram figures, residues from the first strand appear twice, at the left and—in red—at the right. It is useful, although somewhat fanciful, to think of the strands of sheet as the

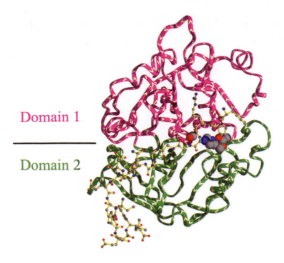

Figure 7.35 Distribution between the two domains of elements of the catalytic mechanism of serine proteinases. This figure shows human thrombin binding the artificial inhibitor hirulog–3, a 20-residue synthetic peptide related to the natural inhibitor hirudin from the leech [1ABI]. Domain 1 is above domain 2. An arginine is inserted into the specificity pocket, and the C-terminus of the molecule interacts with the anion-binding exosite.

stripes on the shirt of an American-football referee, and the long hairpins as his arms folded across his chest as in the signal for 'illegal shift' (see Fig. 7.38).

Domains from different serine proteinases differ in the lengths of the strands of β-sheet making up the common parts of the fold, and in the lengths and structures of the regions connecting the strands. The proteins vie with one another for the gaudiest decorations on their common core.

Each domain contains two repetitions of a motif consisting of three strands of antiparallel β-sheet connected by two hairpins (see Fig. 7.39). (The entire protein contains four copies, two from each domain.) The first hairpin of the motif is bent through an angle of approximately 90°. The two motifs are consecutive along the chain with a linking segment between them. This linking segment connects the only two consecutive strands that are not hydrogen bonded to each other; that is, it is the only connection between consecutive strands that is not a hairpin. (This segment corresponds to the arching line linking strands 3 and 4 in Fig. 7.37.) Each motif contains

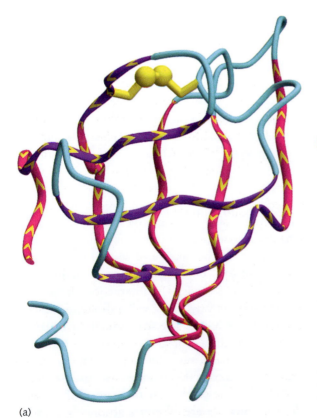

(a)

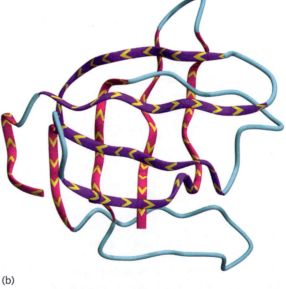

(b)

Figure 7.36 Chain tracings of the domains of *S. griseus* proteinase b [3SGB]. (a) Domain 1; (b) Domain 2. Domain 1 contains a short helix in the connection between the central vertical strand and the lower folded-over loop; a helix is present in this region in several but not all of the structures.

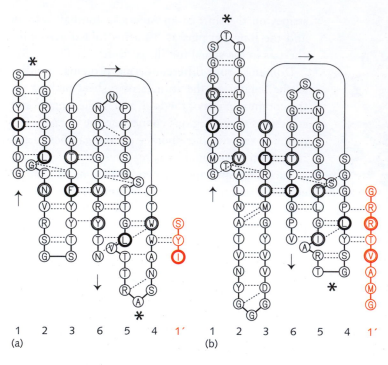

Figure 7.37 Hydrogen-bonding patterns of the β-sheets in *S. griseus* proteinase b [3sGB]. (a) Domain 1; (b) Domain 2. The structure has been unfolded to lie in a plane. Each domain contains a six-stranded β-sheet. The residues of the sheet that are packed inside the β-barrel are ringed. Strands are numbered in order of their appearance in the chain. Part of the N-terminal strand is repeated, appearing once (in black) at the left and, again (in red) at the right. To form the three-dimensional structure, strands 1 and 1' must be superimposed, and two of the hairpins—those marked by *— must be folded out of the page, and twisted to lie horizontally across the sheet. Most domains from other serine proteinases have the same general pattern but differ in the lengths of the hairpins and of the connections between them. Some of the longer connections contain α-helices. In domain 1 of Sindbis virus capsid proteinase the general pattern is incomplete.

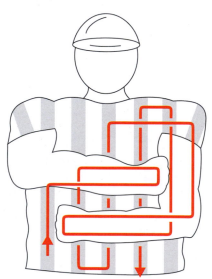

Figure 7.38 An idealized tracing of the folding pattern of the serine proteinase domain superposed onto a picture of an American football referee. This figure exaggerates the approximate symmetries within the domain. There are two axes of approximate twofold symmetry. One is perpendicular to the page and, if the direction of the chain is ignored, another is at an angle of 45° to the horizontal (compass direction northeast—southwest). In most cases, the two hairpins folded over the sheet (the referee's arms) form hydrogen bonds to each other, closing the sheet into a barrel. These hydrogen bonds appear in Fig. 7.37 between the second strand from the right and the strand at the extreme right, which is a copy of the leftmost strand. The axis of the barrel is approximately along the northeast–southwest line of Fig. 7.37.

two hairpins. The first hairpins of the two motifs form β-sheet hydrogen bonds to each other, and the second hairpins of each form β-sheet hydrogen bonds to each other; these close the barrel.

- The β-barrel of serine protease domains has $n = 6$ strands and shear number $S = 8$. Because the ratio S/n is non-integral (unlike TIM barrels, for which $S/n = 8/8 = 1$, or interleukin-1α barrels, for which $S/n = 12/6 = 2$) the sidechain packing of the interior of the barrel does not form layers.

The domain/domain interface

The domains in serine proteinases interact primarily at residues on the outside of several strands of the β-barrel of each domain, but also in the loops connecting these strands. The interactions between the domains are mostly Van der Waals contacts, with a few hydrogen bonds. The structure of the interface is characterized by a central and conserved region near the catalytic triad, surrounded by additional regions that vary among the proteinases. In mammalian proteinases, the residues at the core of the domain/domain interface are all either absolutely conserved

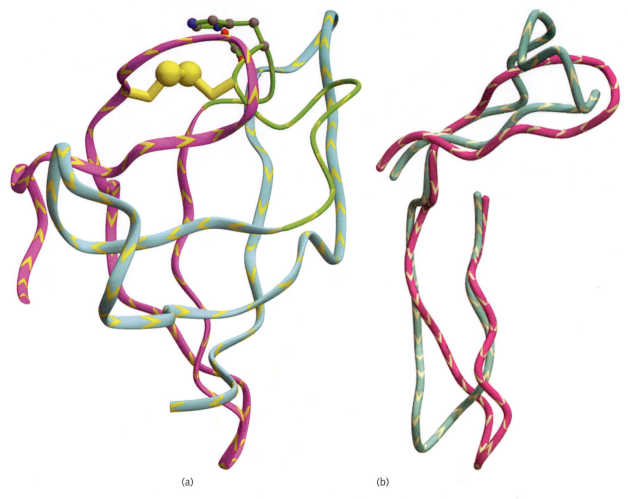

(a)   (b)

Figure 7.39 (a) Domain 1 of *S. griseus* proteinase b, showing the two similar motifs [3SGB]. The N-terminal motif (residues 1–30) is shown in magenta, S–S bridged to the linking region (residues 31–39) as a green curve, and the C-terminal moiety (residues 40–71) in cyan. Two residues of the catalytic triad, His and Asp, appear in domain 1. (b) Superposition of residues from the two motifs of domain 1 of *S. griseus* proteinase b.

or vary within fairly narrow limits, with a few exceptions. In most proteases, a Gly–Gly contact across the domain/domain interface facilitates the close approach of the domains.

The specificity pocket

The primary specificity of serine proteases is determined by the sidechain of the residue of the substrate adjacent to the scissile bond, which fits into a pocket next to the catalytic site. Figure 7.40 shows the structure of this pocket in chymotrypsin, including the

sidechain of a substrate tryptophan residue. The sides of the pocket are formed from two loops joining successive strands of the β-sheet of domain 2; they connect strands 4–5 and 5–6. The base of the pocket is occupied by Ser189 in chymotrypsin; the residue at this position contributes to the specificity. Glycines at positions 216 and 226 produce a narrow, slot-like pocket in chymotrypsin, suitable for binding a flat hydrophobic sidechain.

Comparing the specificity pockets in different proteinases, the mainchain is rather rigid, and the change in specificity is achieved primarily by mutation. To

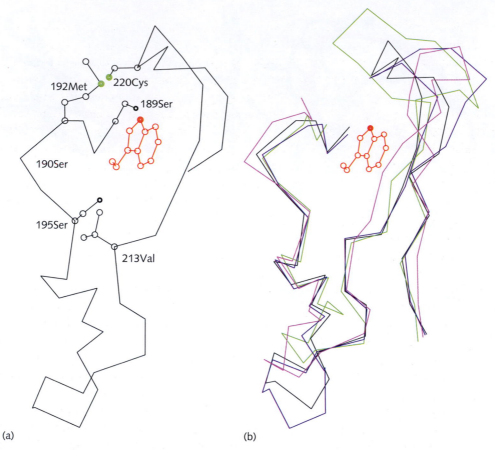

Figure 7.40 (a) Specificity pocket in chymotrypsin, occupied by a tryptophan sidechain from an autolysis product [8GCH]. (b) Superposition of the mainchains of the residues creating the specificity pocket in chymotrypsin [5CHA] (black, including Trp sidechain from autolysis product), thrombin [1PPB] (magenta), *Achromobacter* proteinase [1ARB] (green), and Sindbis virus capsid proteinase [2SNV] (blue).

the extent that this is true, it should be possible to model the structure of the specificity pocket in other proteinases, and from this to predict the specificity.

The β-barrels in serine proteinase domains and the packing of residues in their interiors

Evolution of the serine proteinases, like that of other protein families, shows a qualitative conservation of conformation at the secondary and tertiary structural level.

- The secondary and tertiary structures—β-barrels of similar topology—are conserved in mammalian and bacterial proteinases. (Greater variations appear in viral enzymes.) The loops joining the

strands of β-sheet can and do change in length and conformation.

- Mutations in and around the binding site must preserve the active site. Indeed, residues in and around the active site, including catalytic residues, many but not all residues of the specificity pocket, and residues in the interface, are well conserved.

- Preservation of the topology of the β-barrel requires that mutations be consistent with good packing of its interior. The structure of the β-barrel cannot vary continuously, because its geometry is determined by two *discrete* parameters—number of strands and shear number—which are conserved in the family. The backbone of the β-barrel is therefore relatively rigid, and the molecule can only escape this structure by giving up the closure of the barrel; this occurs in domain 1 of Sindbis virus capsid proteinase. On the other hand, the pattern of packing inside the serine protease β-barrels is more variable than those in TIM-like or β-trefoil proteins for which S/n is integral, because there is no clearly preferred arrangement of sidechains. This packing shows somewhat more consistent and regular patterns in the mammalian proteinases than in the others.

Any β-barrel topology can be characterized by two *discrete* parameters, the strand number, n, and the 'shear number', S, which measures the stagger of the strands (see Closed β-barrel structures section). The β-barrels in serine proteinases contain $n = 6$ strands and have a shear number $S = 8$. For barrels in which S/n is integral, as in the triose phosphate isomerase fold ($S = n = 8$, $S/n = 1$) or in interleukin-1β ($S = 12$, $n = 6$; $S/n = 2$), there is a regular packing of sidechains in the interior of the barrels to form natural layers, as a consequence of the geometry of the assembly of the strands. In chymotrypsin-like serine proteinases, the non-integral value of S/n produces a less regular structure in the sidechains packed inside the barrel.

Figure 7.41(a) shows the hydrogen-bonding pattern in the β-barrel of domain 1 of elastase. The first strand is repeated, in red, and appears at both the right and the left; the actual structure contains only six strands. Note the catalytic His and Asp at the top of the third and fourth strands from the left. Although the ratio S/n is non-integral, the residues packed in the interior of the barrel fall *approximately* into layers, each containing four sidechains (see Fig. 7.41(b)). Figures 7.41(c)–(e) show serial sections, cut perpendicular to the barrel axis.

The contributions of different strands to the layers is constrained by the topology of the β-barrel. The sidechains that pack in the central layer (see Fig. 7.41(d)) are surrounded by thick rings in Fig. 7.41(a), and are marked as I♠ (two copies), V♡, L◊ and V♣, in strands 1, 3, 6, 4 and 1′. Consider the relative positions of these residues in Fig. 7.41(a): start from I♠ in the leftmost strand. To get from I♠ to V♡ one must move two strands to the right and two residues down. The residue between I♠ and V♡ (the Thr in strand 2, *one*

column right and *one* residue down) has its sidechain pointing out of the sheet and does not contribute to the packing of the interior. If this pattern were repeated three additional times, four sidechains from alternate strands and alternate levels would pack together on a single layer in the interior of the barrel. The final result of moving two strands right and two residues down, four times, would be a position eight residues down, precisely the vertical displacement of the two copies of I♠, *but with a horizontal displacement of eight strands rather than six*. This pattern is observed in β-barrels with the regular triose phosphate isomerase topology, for which $S = 8$, $n = 8$, and $S/n = 1$.

Consider in contrast the relationship between the next pair of residues in the central layer: Val♡ and Leu◊. The relationship between these two can be described as 'one over, two down'. If the 'one over, two down' pattern were repeated five more times, *six* inward-pointing sidechains, from adjacent strands but alternate levels, would pack in a layer in the interior of the barrel. The final result of moving one strand to the right and two residues down, six times, would be a position six strands over, precisely the horizontal displacement of the two copies of the first strand, *but with a vertical displacement of 12 residues rather than 8*. This pattern is observed in β-barrels with the β-trefoil topology of interleukin-1β, for which $S = 12$, $n = 6$, and $S/n = 2$.

The problem faced by serine proteinases in forming their barrels is that moving two strands right and two residues down, four times, and ending up at the repeated copy of the residue from which one started, requires eight strands, but only six are present. Alternatively, moving one strand right and two residues down, six times, gives a shear number of 12, but the actual shear number is 8. The non-integral ratio of the strand and shear numbers means that an irregular pattern is required, and what the molecule does is to mix—indeed, in the case of the central layer of domain 1 of elastase, *to alternate*—the patterns of two types of regular β-barrels: $S/n = 1$ and $S/n = 2$.

A useful shorthand suggests itself if one thinks of the residues as laid out on a chessboard. The 'two over, two down' pattern connecting the positions of Ile♠ and Val♡ in strands 1 and 3 is like a bishop moving two squares along a diagonal. The 'one over,

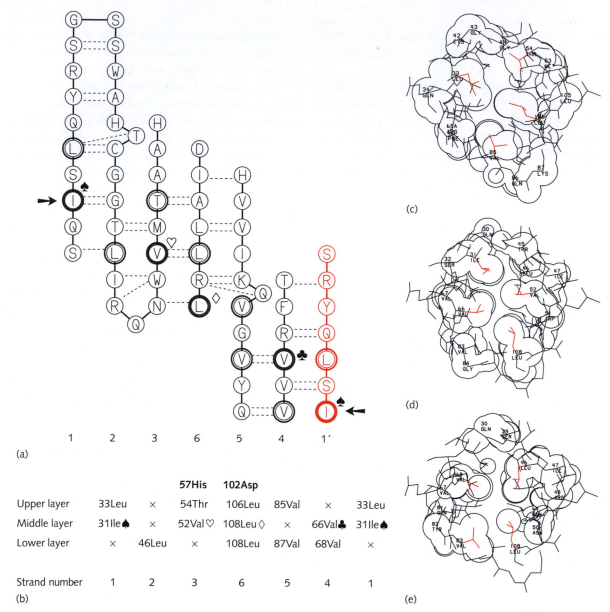

(a)

	1		57His	102Asp			
Upper layer	33Leu	×	54Thr	106Leu	85Val	×	33Leu
Middle layer	31Ile♠	×	52Val♡	108Leu◇	×	66Val♣	31Ile♠
Lower layer	×	46Leu	×	108Leu	87Val	68Val	×
Strand number	1	2	3	6	5	4	1

(b)

(c)

(d)

(e)

Figure 7.41 Packing inside the β-barrel of domain 1 of elastase [3EST]. (a) Hydrogen-bonding net of domain 1 of elastase. The arrows indicate the two copies of an Ile residue in the N-terminal strand. The axis of the barrel is orientated at approximately 45° to the edges of the page in a southwest–northeast direction. Residues that protrude from strands correspond to β-bulges. Consider the relative positions of the two copies of any residue in the repeated edge strand. For instance, two copies of an Ile residue, marked by arrows in the figure, appear in the leftmost and rightmost strands (strands 1 and 1′). To get from the marked Ile at the left to its copy at the right one must move six columns right and eight rows down. These values correspond to the strand number n and shear number S, respectively. (b) Assignment of residues to layers in the β-barrel of domain 1 of elastase. The positions of the catalytic His and Asp are also shown. (c)–(e) Serial sections, viewed along the axis of the barrel and cut perpendicular to the viewing direction. Each part shows three sections separated by 1 Å; (c) is closest to the interdomain interface and (e) is farthest from the interdomain interface. In the central layer, four residues pack together inside the β-barrel: Ile♠, Val♡, Leu◇, and Val♣, in order of appearance from left to right in the figure, where they are surrounded by thick rings. The relationship between the positions of Ile♠ and Val♡ in strands 1 and 3 and those of Leu◇ and Val♣ in strands 6 and 4 can be described as 'two over, two down'. The relationships between the positions of Val♡ and Leu◇ in the third and fourth strands from the left, and those of Val♠ and Ile♠ in strands 4 and 1′, can be described as 'one over, two down'. The combination of these patterns is required because exclusive use of either pattern is inconsistent with the non-integrality of S/n.

two down' pattern connecting the positions Val♡ and Leu◊ in strands 3 and 4 is a move like that of a knight. We can abbreviate the alternating sequence of bishop's move and knight's move patterns seen in the central layer of domain one of elastase as BNBN.

What patterns are possible? Denoting the relationship between the positions of residues in a net such as that appearing in Fig. 7.41(a) by $\begin{pmatrix} \Delta x_i \\ \Delta y_i \end{pmatrix}$, where the bishop's and knight's moves correspond to $B = \begin{pmatrix} 2 \\ 2 \end{pmatrix}$ and $N = \begin{pmatrix} 1 \\ 2 \end{pmatrix}$, the constraint on the residues forming a layer is:

$$\sum \begin{pmatrix} \Delta x_i \\ \Delta y_i \end{pmatrix} = \begin{pmatrix} n \\ S \end{pmatrix},$$

where the sum is taken over the inward-pointing residues in the layer, S is the shear number, n is the number of strands, Δx_i and Δy_i must be integral, and Δy_i must be even. (The requirement that Δy_i be even ensures that all sidechains are pointing towards the same side of the sheet; that is, towards the interior of the β-barrel. Note that this constraint refers to positions in the hydrogen-bonding net and not to residue positions in the sequence; the distinction arises where there occurs a type of irregularity observed in some β-sheets, a β-bulge. Exceptions to this rule can occur at the ends of strands where residues might have conformations outside the β-strand region of the Sasisekharan–Ramakrishnan–Ramachandran diagram.) For regular barrels—S/n integral—these equations are satisfied by $\Delta x_i = n/L = S/2$, where L is the number of residues in the layer (if $S/n = 1$, $L = n/2$; and if $S/n = 2$, $L = n$) and $\Delta y_i = (S/n)\Delta x_i$. For the serine proteinase domains with four residues per layer, the equations take the form:

$$\sum_{i=1}^{4} \begin{pmatrix} \Delta x_i \\ \Delta y_i \end{pmatrix} = \begin{pmatrix} 6 \\ 8 \end{pmatrix},$$

which has as its smallest integral solutions $\Delta y_i = 2$, for $i = 1, 2, 3, 4$, and $\Delta x_i = 1$ (twice) and 2 (twice), to give some combination of two 'two over, two down', and two 'one over, two down' relationships, or two bishop's moves and two knight's moves. The central layer of domain 1 of elastase has the most common pattern, BNBN. Different move orders are possible, of which BNBN and BNNB are the most common choices.

The other two layers in elastase domain 1 can be analysed in a similar way. The residues that contribute to the outer layers are marked by double rings in Fig. 7.41(a). In all cases, the residues five positions in the amino acid sequence before the catalytic His and four after the catalytic Asp contribute to the upper layer (T in strand 3 and the upper L in strand 6 in Fig. 7.41(a)). Leu◊ contributes to both the central and bottom layer (see Figs. 7.41(c) and (e)).

In most of the mammalian serine proteases, the packing in the β-barrel of domain 1 is similar to that of elastase. The β-barrel of domain 2 of the serine proteinases also contains three layers. As in domain 1, the packing in the layers shows preferred but not unique patterns among the mammalian proteinases, with greater variability and irregularity in the bacterial and viral molecules.

Evolution of visual pigments and related molecules

Visual pigments have evolved to absorb and process ambient light. Their spectroscopic properties are finely tuned. Different species occupy different **light environments**, and the spectral sensitivities of their visual pigments are adjusted to the quality and intensity of the ambient light. Humans have evolved to occupy two light environments—day and night—and our visual pigments have diverged to support both cone vision (high illumination, colour sensitive, high resolution), and rod vision (low illumination, colour insensitive, lower resolution). Our eyes are sensitive to light throughout the visible region (of course this is a tautology), but our visual pigments would react to ultraviolet light also, were it not for the photoprotective filter of the lens. We can deal with a dynamic range of 11 orders of magnitude, from 10^{10} photons $m^{-2}s^{-1}$—the dimmest stars visible in the night sky—to 10^{21} photons $m^{-2}s^{-1}$—bright sunlight.

The photoreceptor proteins in vision are **opsins**, membrane proteins about 360 residues in length with the seven-helix motif common to G protein-coupled receptors. Opsins combine with chromophores to

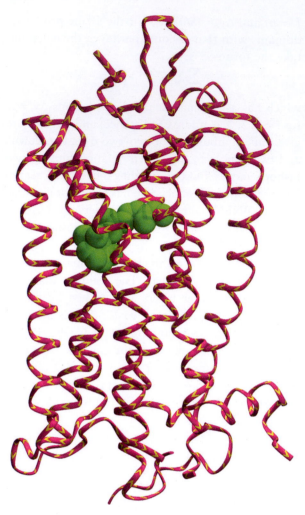

Figure 7.42 Bovine opsin, a 7-transmembrane helix G protein-coupled receptor [1F88].

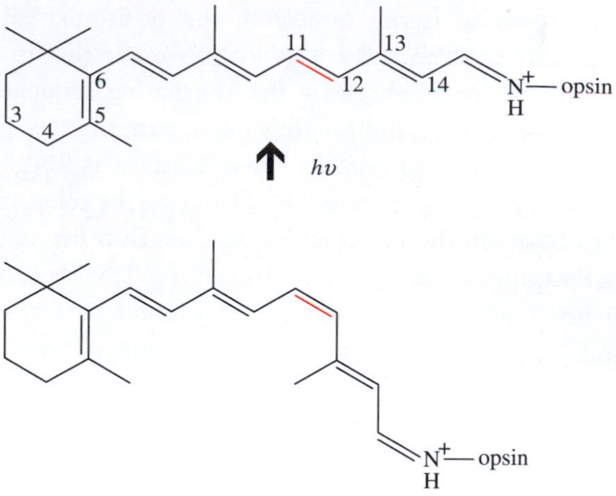

Figure 7.43 Conformational change in retinal attached to opsin upon absorption of light. One of the double bonds, indicated in red, changes from the 11-*cis* conformation in the resting state to the all-*trans* conformation. Some homologues of vertebrate opsins, including bacteriorhodopsin and the photoreceptors of green algae, bind retinal in an alternative conformation, with a different stereochemistry of the attachment of the polyene tail to the ring; the resting state is all-*trans* and the photochemical isomerization takes place at the 13–14 double bond. The visual pigments of many reptiles, amphibia, and fishes contain 3,4-dehydroretinal instead of retinal. *Drosophila* rhodopsin contains 3-hydroxyretinal.

form light-sensitive complexes (see Fig. 7.42). In vertebrate visual systems, the chromophore is 11-*cis*-retinal (see Fig. 7.43). The complex of opsin and retinal in vertebrate rods is called *rhodopsin*, because of its colour. Opsin complexes containing very similar chromophores are common to the visual systems of vertebrates, insects, and cephalopods.

• Vertebrate photoreceptors contain a protein, opsin, binding a chromophore, 11-*cis*-retinal. Absorption of light causes a conformational change in the chromophore, transmitted to the protein, which leads to generation of a nerve impulse.

Vision begins with absorption of light by retinal. Within about a picosecond, retinal isomerizes, causing a conformational change in the opsin that initiates a signal cascade, resulting ultimately in the triggering of a nerve impulse (see Fig. 7.43 and Box 7.6).

G protein-coupled receptors are common to all cells, and their specialization to photoreceptors occurred early in the history of life. Homologous proteins binding chromophores related to retinal appear in bacteria and archaea as well as eukaryotes. Figure 7.44 shows the relationships among the known homologues in higher eukaryotes. All have the same arrangement of the seven helices, and a common locus and mode of attachment of the chromophore.

Many homologues, even those in simple organisms, have functions related to primitive 'visual' sensory phenomena; for example, phototaxis in algae. In contrast, bacteriorhodopsin from *Halobacterium salinarum* is not a sensory transducer but a light-driven proton pump, converting light energy to an electrochemical gradient across the bacterial plasma membrane. This electrochemical potential

BOX 7.6   Events in the molecular mechanism of vision

- Absorption of light by retinal.

- Isomerization of retinal from 11-*cis* to all-*trans* (see Fig. 7.43). The all-*trans* isomer has a different shape, which doesn't fit the binding site well. The protein changes conformation and releases the retinal.

- The new conformation of the rhodopsin activates the G protein transducin.

- Transducin activates a phosophodiesterase.

- The phosphodiesterase hydrolyses cyclic GMP.

- Cyclic-GMP-gated cation channels close.

- Build-up of potential difference across cell membrane.

- Potential difference transmitted to adjacent neuron (in vertebrates, via a bipolar cell), triggering a nerve impulse.

- Feedback mechanisms affecting each of these steps restore the system to its resting state.

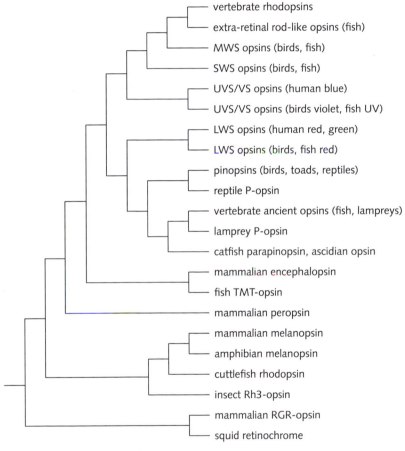

Figure 7.44 Relationships among visual opsins and other non-image-forming photoreceptors of higher eukaryotes. *Vertebrate rhodopsins*—visual pigments of rods in vertebrates; *extra-retinal rod-like opsins*—pigments related to rhodopsins but probably not involved directly in image formation; *MWS opsins*—medium-wavelength sensitive opsins in cone vision of birds and fish; *SWS opsins*—short-wavelength sensitive opsins in cone vision of birds and fish; *UVS/VS opsins*—ultraviolet- and violet-sensitive opsins in cone vision of vertebrates; human blue cone pigment is related to these; *LWS opsins*—long-wavelength sensitive opsins in cone vision of vertebrates, human red and green cone pigments are related to these; *pinopsins*—light detectors in the pineal glands of birds, reptiles and amphibia, function related to circadian rhythms; *vertebrate ancient opsin*—opsin expressed in some retinal cells and pineal gland, may function in non-image-forming light detection; *mammalian encephalopsin*—opsin expressed in mammalian brains, function may be related to circadian rhythms; *mammalian peropsin*—function unknown; *melanopsin*—probably involved in circadian rhythms; *cuttlefish rhodopsin*—visual pigment in cephalopods; *insect Rh3 opsin*—visual pigment in insects; *RGR opsins and retinochrome*—function in recovery of receptive state of visual pigments, by photoreversal of the isomerization shown in Fig. 7.43, possibly signalling also.

OSBS

NSAR

From: Brizendine, A.M., Odokonyero, D., McMillan, A.W., Zhu, M., Hull, K., Romo, D. and Glasner, M.E. (2014). Promiscuity of *Exiguobacterium* sp. AT1b o-succinylbenzoate synthase illustrates evolutionary transitions in the OSBS family. Biochem. Biophys. Res. Comm. 450, 679–684.

The OSBS reaction is part of the menaquinone (vitamin K) synthesis pathway. The NSAR activity is part of the pathway that converts D-amino acids to L-amino acids.

Homologous enzymes from different species show a range of catalytic efficiencies for the two reactions, and different ratios of OSBS and NSAR activities:

Species	k_{cat}/K_M for OSBS activity	k_{cat}/K_M for NSAR activity
Amycolatopsis sp. T-1-60c	8.4×10^4 $M^{-1}s^{-1}$	4.2×10^4 $M^{-1}s^{-1}$
Listeria innocua	2.9×10^6 $M^{-1}s^{-1}$	2.6×10^3 $M^{-1}s^{-1}$
Exiguobacterium sp. AT 1b	2.6×10^6 $M^{-1}s^{-1}$	41 $M^{-1}s^{-1}$
Staphylococcus aureus	1.1×10^6 $M^{-1}s^{-1}$	—

Data from Brizendine, A.M., *et al.*, *loc. cit.*

The *Amycolatopsis* enzyme has comparable values of k_{cat}/K_M for the two reactions. The *Listeria* enzyme has a greater catalytic efficiency for OSBS activity than for NSAR. The *Exiguobacterium* and *Staphylococcus* enzymes have efficiencies for OSBS activity approximately equal to that of *Listeria,* but a very low or undetectable NSAR activity. These numbers are consistent with, but do not prove, the idea that to raise k_{cat}/K_M for OSBS activity up to 10^6 $M^{-1}s^1$ requires some sacrifice of NSAR activity. And is it necessary to point out that the idea that selection will, if possible, always act to maximize k_{cat}/K_M is neither true nor reasonable?

How do the structures change to (a) maintain specificity and catalytic efficiency in the face of very wide sequence divergence, and (b) achieve the versatility to catalyse different or even multiple reactions? A complicating feature is that the enolase superfamily takes advantage of flexibility in the substrates as well as in the proteins. Comparisons of structures of enolase-superfamily proteins with different activities and binding different ligands reveal a variety of mechanisms *available* for evolving either to maintain a conserved function or to develop a new one:

1. Bind the same substrate in different (substrate) conformations

2. Change the relative geometry of domains around the active site

3. Accept mutations within the active site

4. Change conformation of a flexible loop

Not all these tools are necessary for every job: different homologues pick and choose. Readers who would insist on a simple general answer are doomed to disappointment. (In particular, reconformation of a flexible loop, probably high on many people's list of 'usual suspects', in involved in some cases but not in all.) In Weblem 7.5, readers are asked to superpose enolase-superfamily structures with different ligands and different catalytic activities, to analyse the structural changes in particular cases.

There is a fundamental take-home message, of which we have seen hints, in connection with the evolution of globins: what you see in closely-related proteins does not adequately prepare you for what happens when evolution really takes the gloves off.

Protein evolution at the level of domain assembly

Comparisons of protein sequences and structures show that the domain is an important unit of protein evolution. Domains appear in different proteins in different combinations. Thereby, from a relatively small roster of domain families, evolution can assemble a large number of complete proteins.

Based on known protein structures, it has been possible to define approximately 1000 domain superfamilies. Of the ~23 000 human genes, almost two-thirds contain known domains. The approximately 1000 domain superfamilies account for approximately 30 000 matches in the human genes.

The population of domains encoded by known genes is unevenly distributed. Nine domain superfamilies account for 20% of the matched domains in human genes (see Table 7.3).

Similar results apply to other eukaryotic genomes: Fugu, *D. melanogaster* and *C. elegans*, although the rank order is not the same.

The distribution of domains depends on the functional class of the protein. The number of proteins in a given functional class scales exponentially with the size of the genome; for example:

number of proteins involved in transcription regulation = constant × genome size$^{1.9}$

number of proteins involved in protein biosynthesis = constant × genome size$^{0.13}$

Domain swapping is a general mechanism for forming an oligomer from a multidomain protein

Suppose a monomer contains two domains, A and B, connected by a flexible linker, with a well-developed interface between A and B (see Fig. 7.51). A dimer, containing four domains, can be stabilized by two copies of the *same* A/B interface that appears in the monomer.

Pig odorant-binding protein is a monomer (see Fig. 7.52(a)). The C-terminal segment contains a helix and strand of sheet (red in Fig. 7.52(a)). In contrast, cow odorant-binding protein is a dimer (see Fig. 7.52(b)). In this dimer, the C-terminal helix and strand of each monomer flip over to interact with the partner. The other monomer provides these residues with an equivalent environment and interactions (see Fig. 7.52(c)).

- Domain swapping is a means of creating multidomain proteins. Domain swapping can also lead to unwanted protein aggregation, as in the Z-mutant of α_1-antitrypsin.

Table 7.3 Common domains in the human genome

Most common domains	Number of matches in human genome
C2H2 and C2HC zinc fingers	3693
Immunoglobulin	1778
P-loop nucleoside triphosphate hydrolase	1024
G-protein-coupled receptors: family A	824
Fibronectin type III	802
EGF/laminin	697
Cadherins	686
Protein kinases	539
PH domains	491

From: Chothia, C. and Gough, J. (2009). Genomic and structural aspects of protein evolution. *Biochem. J.*, **419**, 15–28. (C2H2 = two cysteines, two histidines, *not* acetylene.)

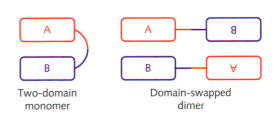

Two-domain monomer   Domain-swapped dimer

Figure 7.51 Domain swapping. The reader should verify that the two A/B interfaces in the dimers are formed from domains in the same relative position and orientation as the single interface in the monomer.

Table 8.3 Folding of mutants of Arg96 in T4 lysozyme. Negative values of ΔΔGo– imply that the mutants destabilize the structure.

Protein	ΔT_m (°C)	$\Delta\Delta G^{\oplus}$ (kJ mol^{-1})	Crystal structure?
WT	0.0	0.0	Yes
Arg96Lys	−0.2	0.0	Yes
Arg96Gln	−1.4	−1.3	Yes
Arg96Ala	−5.1	−8.4	Yes
Arg96Val	−6.4	−10.0	Yes
Arg96Ser	−7.0	−10.9	Yes
Arg96Glu	−7.0	−10.5	Yes
Arg96Gly	−7.1	−10.9	Yes
Arg96Met	−7.1	−11.3	Yes
Arg96Thr	−7.6	−11.7	No
Arg96Cys	−7.7	−12.1	No
Arg96Ile	−7.9	−12.1	No
Arg96Asn	−8.0	−12.6	Yes
Arg96His	−8.3	−13.0	Yes
Arg96Leu	−8.6	−13.4	No
Arg96Asp	−9.5	−14.6	Yes
Arg96Phe	−11.5	−17.6	No
Arg96Trp	−12.8	−18.8	Yes
Arg96Tyr	−13.2	−19.7	Yes
Arg96Pro	−15.5	−23.0	Yes

From: Mooers, H.M., Baase, W.A., Wray, J.W., and Matthews, B.W. (2009). Contributions of all 20 amino acids at site 96 to the stability and structure of T4 lysozyme. *Prot. Sci.*, **18**, 871–880.

Some of the substitutions—notably Arg96Lys and Arg96Gln—had little or no effect on the stability of the protein, or on the structure. Lysine and glutamine have long aliphatic proximal chains that terminate, like that of arginine, with polar atoms that can form hydrogen bonds. Although, in the structures, the hydrogen-bonding partners of lysine and glutamine differ from those of the wild-type arginine, these mutants make similar types of interactions.

The most destabilizing substitutions replace arginine with Asp, Phe, Trp, Tyr, and Pro. Because position 96 is in a helix, it is not surprising that the proline mutant is destabilized. Proline breaks the main-chain hydrogen-bonding pattern of an α-helix. Of the other mutants, even the sidechains that contain polar distal atoms are shorter than Arg, so that polar atoms in the mutant replace non-polar atoms in the wild type. An example of the effect of such a change is that the carbonyl oxygen of the Asn96 sidechain amide group points towards the hydrophobic core of the protein and does not make a hydrogen bond.

It is interesting that the largest volume changes—from Arg to Gly or Ala—are not the most destabilizing. They have lost the sidechain hydrogen bonds, but they do not create unsatisfied hydrogen bonds, nor do they create steric clashes by virtue of a shape change to a bulkier sidechain. They do not provide stabilizing interactions similar to Arg (as Lys and Gln do), but neither do they introduce specific destabilizing interactions. They just mind their own business. As a result their effect is smaller.

Mooers, Baase, Wray, and Matthews conclude that:

• To retain stability near to that of the wild type, the mutant should have carbon atoms at Cα, Cβ, and Cγ positions, and distal polar atoms, capable of forming hydrogen bonds. It is not necessary to form the *same* hydrogen bonds as the wild-type Arg.

• A mutant that lacks these favourable features, but does no specific damage, will have an intermediate effect on stability.

• A mutant that introduces strain or that contains buried polar atoms with unsatisfied hydrogen-bonding potential will be more seriously destabilizing.

Experimental characterization of events in protein folding

When a denatured protein is placed in renaturing conditions—perhaps by diluting out denaturant—several stages in refolding are typical and characteristic:

1. *Fast collapse and secondary structure formation (typical time scale: ms).*
The **radius of gyration** (R_G) of a polymer (in fact of any set of atoms) is a measure of its spatial extent.

If the mean position of the atoms (colloquially, the centre of gravity) is at the origin, the radius of gyration is:

$$R_G = \sqrt{\frac{\sum_i m_i (x_i^2 + y_i^2 + z_i^2)}{\sum_i m_i}}$$

where m_i are the masses and (x_i, y_i, z_i) are the coordinates of the atoms.

Denatured proteins are relatively spatially extended (large radius of gyration) and native proteins are spatially compact (small radius of gyration) (see Table 8.4).

As an early event in folding, the R_G values of proteins drop from the value in the denatured state to a smaller value typically about 10% larger than in the native state. On the same timescale, secondary structure appears.

O.B. Ptitsyn gave this state an intriguing name: the **molten globule**. By 'molten' he had in mind an analogy to a liquid state of simple substances—the state between vapour (analogous to the denatured states of proteins) and solid (analogous to native). It is interesting that the molar volumes of simple liquids are typically 10% larger than the corresponding solids. (Water is of course the most famous exception.)

- The 'molten globule' is a state that is intermediate in folding. It has a relatively compact conformation, native-like secondary structure, the correct overall folding pattern, but the packing of the interior is less dense than in the native state. The properties of the molten globule also characterize some flexible regions within ordinary native states.

2. *Appearance of tertiary structure (typical timescale ~ms to s)*

Table 8.4 Radii of gyration (Å) of native and denatured proteins

Protein	Number of residues	R_G (native)	R_G (denatured)	Ratio
PI3 kinase, SH3 domain	90	18.6	27.5	0.7
Horse heart cytochrome c	104	17.8	32.6	0.5
Hen egg white lysozyme	129	20.5	34.6	0.6
Yeast triose phosphate isomerase	247	29.7	49.7	0.6

Data from: Wilkins, D.K., Grimshaw, S.B., Receveur, V., Dobson, C.M., Jones, J.A. and Smith, L.J. (1999). Hydrodynamic radii of native and denatured proteins measured by pulse field gradient NMR techniques. *Biochemistry* **38**, 16424–16431.

Interactions between nascent secondary structure elements build up mutually stabilizing interactions, on the way to the complete native state. Sidechain conformations are still mobile, as the secondary structures seek out the interdigitating patterns of sidechain packing that fix the native state.

3. *Final 'crystallization' to native state (typical timescale ≤ s)*

The locking together of buried sidechains to form the highly compact native state expels water from the interior of the structure.

- A common scenario for protein folding:
  1. fast collapse and secondary structure formation to form the molten globule state;
  2. appearance of tertiary structure;
  3. formation of dense packing characteristic of the native state.

The molten globule

En route to the native state, proteins pass through a partially structured state, the molten globule. This state can be observed under mildly denaturing conditions. Energetically, proteins differ in the stability of their molten globule states, and in the structural relationship between the molten globule and the native states. For instance, the molten globule state of chicken lysozyme is only a transient intermediate in the pathway. In contrast, the molten globule state of its homologue α-lactalbumin is a long-lived intermediate in the unfolding. The folding of α-lactalbumin is not a simple native-denatured equilibrium.

Structurally, the molten globule state has:

- A relatively compact conformation, typically 10% higher than the native state, as shown by the values of molecular size or radius of gyration by hydrodynamic measurements, small-angle X-ray scattering, and gel-filtration chromatography.

- A substantial amount of intact secondary structure, as shown from circular dichroism measurements and NMR.

- Loose packing of the hydrophobic core. Native tertiary structure has not appeared yet, as shown by the relative lack of protection of NH protons to H↔D exchange. However, the general topology of the chain is probably close to native. Residues that are buried in the native structure are accessible to water, as revealed by interactions with the hydrophobic dye ANS, or by NMR studies that measure proton-exchange rates.

ANS = 8-anilino-1- naphthalene sulphonate

- More structural flexibility than the native state, corresponding to a spectrum of rapidly interconverting conformations.

The picture of the molten globule state is that it has at least some of the correct secondary structure and folding pattern. However, the helices and sheets have not come together with their sidechains interlocking tightly as in the native state.

Folding funnels

All discussions of reaction kinetics contain conventional diagrams in which the Gibbs free energy of the system is plotted against a single, not very precisely defined, variable called the 'reaction coordinate' (see Fig. 5.2). Even for the simplest chemical systems, this is a simplification. For protein folding the simplification is so severe as to be downright misleading.

Two problems with the conventional reaction coordinate diagram, applied to protein folding, are:

1. The diagram suggests that the system follows a unique trajectory. In fact, the denatured state comprises many different conformations, and in principle there should be a different reaction coordinate describing how each one folds to the common native state.

2. Although it would be easy to show a folding pathway that passed through several intermediates, by producing a reaction coordinate diagram with several successive minima, it is not possible in the two-dimensional diagram to show 'blind alleys'—off-path or **non-productive intermediates**.

The current view is that an appropriate schematic representation of the folding process is a funnel (see Fig. 8.3). The funnel is wide at the top, to encompass all the different denatured conformations. It is possible to get from different unfolded conformations to the native state by many routes, each passing through different sets of conformations with different partial native-like features. But these different trajectories eventually coalesce, shown by the narrowing of the funnel.

Of course, one reason for coalescence of folding trajectories is that they all share the native state as their final destination. There is also a very general reason why reactions in general, including protein folding in particular, should pass through a unique transition state. This is the exponential dependence of reaction rate on the height of the energy barrier. If any alternative reaction pathways passed over a different barrier, with a significantly different height, the 'throughput' over the lowest barrier would dominate.

The notion that the transition state sits at the peak of a reaction coordinate diagram is consistent with its having only transient existence. How then can we determine its structure?

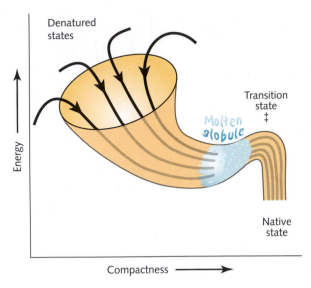

Figure 8.3 The *folding funnel*, a schematic representation of the progress from high-energy, spatially diffuse denatured conformations to the low-energy, compact native state. The process can be thought of as a kind of intramolecular crystallization. Folding proceeds by collapse of the chain to a relatively compact molten globule state, in which much of the secondary structure is formed but the sidechain packing that permits the very high compactness of the native state is not locked in. This figure generalizes the simple reaction coordinate picture (see Fig. 5.2). In recognition of the conformational heterogeneity of the unfolded state, a multiplicity of trajectories links the denatured and native states. The trajectories converge in the molten globule and transition states.

The simple reaction coordinate diagram (to speak in its defence), is intended as a description of macroscopic or thermodynamic states. In contrast, the folding-funnel idea alludes to microscopic states. Of course even this diagram is a simplification. To make it easier to understand, there is no indication of local minima, non-productive intermediates, multiple transition states, competition from aggregation, etc. Problem 8.6 invites the reader to sketch them in.

The effect of denaturants on rates of folding and unfolding: chevron plots

The kinetics of folding of mutated proteins gives clues to the structure of the transition state for folding

The usual methods of structure determination are inapplicable to unstable transition states, because their concentrations are too low and they are too short-lived. However, comparisons of the stabilities and rates of unfolding and refolding between wild-type and mutated proteins reveal the importance of each mutated site in the structure of the transition state.

Reasonable structural assumptions allow us to interpret the effects of single-site mutations on rates of folding and unfolding. Suppose that a mutated residue makes no intramolecular interactions in the denatured state, but has a destabilizing effect on the native state. If the mutated residue also makes no intramolecular interactions in the *transition state*, the activation energy for *folding*, $\Delta G^{\ddagger}_{\text{folding}} = G(\text{transition state}) - G(\text{denatured})$, will be the same in the wild type and the mutant, but the activation energy for *unfolding*, $\Delta G^{\ddagger}_{\text{unfolding}} = G(\text{transition state}) - G(\text{native})$, will be lower in the mutant (see Fig. 8.4(a)). Conversely, if the mutated residue makes the same interactions in the transition state that it does in the native state, the activation energy for unfolding $\Delta G^{\ddagger}_{\text{unfolding}} = G(\text{transition state}) - G(\text{native})$ will be the same in the wild type and the mutant, but the activation energy for folding ($G^{\ddagger}_{\text{folding}} = G(\text{transition state}) - G(\text{denatured})$ will be *higher* in the mutant (see Fig. 8.4(b)). By applying this approach to many residues spread over the sequence, a fair job of mapping out the interactions in the transition state is possible.

In many cases, the interactions of a residue in the transition state are not identical either to those in the denatured state or those in the native state. That is, many single-site mutants affect the native and transition states to different extents. The relative amount is measured by a parameter Φ:

$$\Phi = \frac{\Delta G^{\ddagger}(\text{wild type}) - \Delta G^{\ddagger}(\text{mutant})}{\Delta G_{N \to D}(\text{wild type}) - \Delta G_{N \to D}(\text{mutant})}$$

To measure Φ determine $\Delta G_{N \to D}$ for wild type and mutant either from equilibrium or rate measurements, and $\Delta \Delta G^{\ddagger} = \Delta G^{\ddagger}(\text{wild type}) - \Delta G^{\ddagger}(\text{mutant})$ from the kinetics of folding and unfolding (see Fig. 8.5).

- The Φ **value** of a residue in a protein is a number that measures the extent to which the residue makes native-like contacts in the transition state. It is determined from measurements of kinetics of folding, and thermodynamics of stabilization, of the wild-type protein, and a protein mutated at the residue in question.

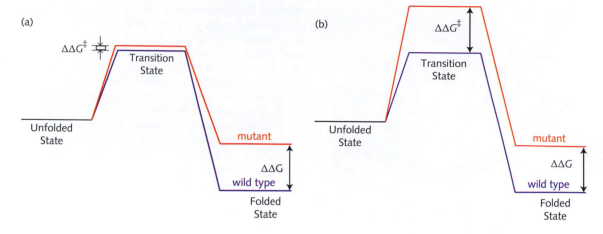

Figure 8.4 Reaction diagrams showing two possible effects of mutants on the relative destabilization of transition states and folded states relative to the unfolded state, as measured by Φ-values = ratios of destabilization of the transition state to the destabilization of the folded state. For calibration, the energies of the unfolded states for wild type and mutant are set at the same level. (a) This mutant destabilizes the native state but does *not* destabilize the transition state. In this case, Φ = 0. The implication is that the residue at the mutated position is *not* making the interactions in the transition state that it does in the native state, or that the mutated position is *not* surrounded by native-like structure in the transition state. (b) This mutant destabilizes the transition state by the same amount that it destabilizes the folded state. In this case, Φ = 1. The implication is that the residue at the mutated position is making the same interactions in the transition state as in the native state, or that the structure of the transition state in the neighbourhood of the mutated position is native-like.

Different proteins show different distributions of Φ values in the residues along the chain. In **barnase**, the Φ values are almost all nearly 0 or 1, suggesting that the stage of the folding process between the transition state and the native state involves the docking of preformed structural domains. In CI2 the values of Φ are broadly distributed, suggesting that folding forms around an extended structural nucleus.

One interesting question is whether local interactions determine the structure of the transition state. A way to test this is by studying circularly permuted proteins, in which the set of interactions local in the sequence has been altered.

Comparison of folding pathways of a natural protein and a circular permutant

S6, a 101-residue protein isolated from the small ribosomal subunit of *Thermus thermophilus*, undergoes a sharp native ⇌ denatured transition. M. Lindberg, J. Tångrot and M. Oliveberg studied the effects of mutations on the folding of S6, and of circular permutants synthesized by inserting linking peptides between the N- and C-termini, and cutting one of the natural peptide bonds, to produce new N- and C-termini.

Lindberg, M., Tångrot, J. and Oliveberg, M. (2002). Complete change of the protein folding transition state upon circular permutation. *Nat. Struct. Biol.*, **9**, 818–822.

Figure 8.7 shows the rate constants for folding and unfolding as a function of guanidinium hydrochloride concentration [GdmCl] for the wild type and a Val88Ala mutant, for the original sequence (see Fig. 8.7(a)), and for a circular permutant, cleaved between residues 13 and 14, and the original N- and C-termini linked (see Fig. 8.7(b)). Figure 8.8 shows the structure, and indicates the sites of mutation and cleavage.

For both wild type and circular permutant, $\Delta\Delta G_{D-N} > 0$; that is, the mutation destabilizes the folded states. For the wild type, $\Delta\Delta G_{D-N} = 10.5\,\text{kJ}\,\text{mol}^{-1}$. For the permutant, $\Delta\Delta G_{D-N} = 3.4\,\text{kJ}\,\text{mol}^{-1}$.

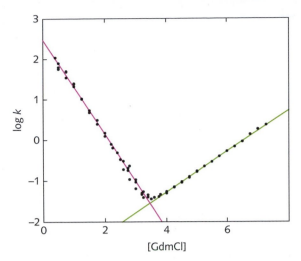

Figure 8.5 Folding and unfolding rates of ribosomal protein S6 from *Thermus thermophilus* at different concentrations of denaturant guanidinium hydrochloride. This figure shows a **chevron plot** of folding and unfolding rate constants as a function of denaturant concentration. Upon denaturation, residues buried inside the protein in the native state are exposed to solvent. The reason that high concentrations of denaturant unfold the protein is that there is a more favourable interaction of peptides with denaturant (in the unfolded state) than with other peptides (buried in the folded state) or even with water. On this basis we can explain the form of the chevron plot. Assuming that the transition state is partially unfolded, the effect of the denaturant is (see Fig. 8.6):

> the lowering of energy of the native state
> > < the lowering of energy of the transition state
> > < the lowering of energy of the unfolded state.

Don't you agree that a diagram would make this clearer? See Problem 8.2.

Therefore, as the concentration of denaturant increases, the activation energy for unfolding is reduced—and the unfolding rate increased—and the activation energy for folding is raised—and the folding rate decreased.

[Data of: Otzen, D.E., Kristensen, O., Proctor, M. and Oliveberg, M. (1999). Structural changes in the transition state of protein folding: alternative interpretations of curved chevron plots. *Biochemistry*, **38**, 6499–6511.]

In the wild type, the Val88Ala mutation has no effect on the rate of folding—$\Delta\Delta G_{D-\ddagger} \sim 0$—but a large effect on the rate of unfolding (see Fig. 8.7(a)). As $\Delta\Delta G_{D-N}(10.5\,\mathrm{kJ\,mol^{-1}}) >> \Delta\Delta G_{D-\ddagger}$, $\Phi \sim 0$. For the circular permutant, in contrast, the Val88Ala mutation has a large effect on the rate of folding and little effect on the rate of unfolding (see Fig. 8.7(b)): in this case $\Phi \sim 1$.

These data show that the transition states for folding of the wild type and circular permutant are different, at least as far as residue 88 is concerned. In the transition state of the circular permutant, this residue makes native-like contacts. In the transition state of the wild type, it remains unconstrained.

The topologies of the structures make these results reasonable. S6 contains a four-stranded β-sheet, and two helices packed against it. *In the wild type,* the order of the strands from the left in Fig. 8.8 is 2–3–1–4, with position 88 in the C-terminal, rightmost strand, and strands 2 and 3 forming an internal hairpin (light blue in Fig. 8.8). The idea that residues nearby in the sequence tend to participate in early folding events suggests that the hairpin formed by the two strands central in the sequence in the wild type might fold early, and the N- and C-terminal strands (the two rightmost strands in Fig. 8.8, containing the mutated position), might fold late, *after* passing the transition state.

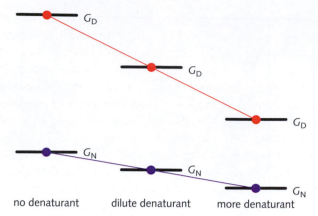

Figure 8.6 Dependence of free energy of native and denatured states on denaturant concentration. The energy of interaction of peptides with denaturant is greater than their energy of interaction with water. Therefore, increasing concentrations of denaturants stabilize both the denatured state and the native state. However, the effect on the denatured state is greater, because more peptide is exposed to solvent in the denatured state than in the native state, as a result of the compact packing in the native state. This is seen in the figure as a greater change in the level of the denatured state (G_D, red line) than the level of the native state (G_N, blue line) with increasing denaturant concentration. Increasing concentrations of denaturant reduce the stabilization of the native state with respect to the denatured state. The quantity:

$$G_{denatured\,state} - G_{native\,state} = \Delta G_{N \to D}$$

is a measure of the stability of the native state of a protein. In the absence of denaturant, it is typically in the range 20–60 kJ mol^{-1} (8–15 kcal mol^{-1}). A *linear* dependence of stability on denaturant concentration is often observed:

$$\Delta G_{N \to D} = m \times denaturant\,concentration.$$

This will certainly be true if both G_N and G_D individually depend linearly on denaturant concentration, as in this figure. Eventually the red and blue lines will cross. At this point, $G_N = G_D$ and the concentrations of denatured and native molecules will be equal. At higher denaturant concentrations, the sample will become more completely denatured (see Fig. 8.1).

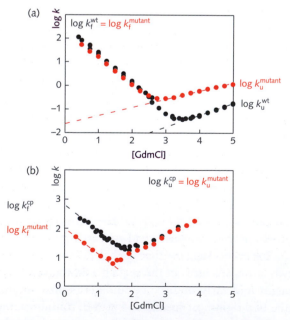

Figure 8.7 The rate constants for folding/unfolding of S6 and mutant Val88Ala, for the original sequence and for a circular permutant cleaved between residue 13 and 14. (a) Wild type. (b) Circular permutant.

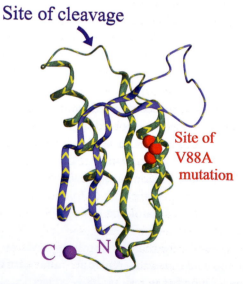

Figure 8.8 The structure of S6, showing the sites of the Val88Ala mutation and the formation of the circular permutant. To make the circular permutant, the original N- and C- termini shown here with large purple spheres are connected by a linking peptide bond, and the chain cleaved between residues 13 and 14. Residues shown in blue form an internal hairpin in this wild-type structure.

In the circular permutant, the mutated position is in a hairpin, between two strands consecutive in the sequence, and it is reasonable that they fold before the transition state.

Lindberg, Tångrot, and Oliveberg measured Φ values for additional mutants deployed throughout the molecule. For the circular permutants, the distribution of Φ values is bimodal, most values being either close to 0 or close to 1. Residues with high Φ values appeared in the newly linked strands 1 and 4; residue 88, with $\Phi \sim 0$ is an example. For the wild type, in contrast, Φ values were generally low, and distributed fairly uniformly throughout the sequence.

The results suggest that the wild type has evolved a diffuse folding pathway, by creating stronger interactions between residues distant in the sequence, which in principle have a harder time finding their correct partners than pairs of residues making local interactions. Forming the circular permutant disturbs the folding pathway. Instead, in the circular permutant, the mutated residue interacts with residues nearby in *both* the sequence and the structure, creating a localized nucleus for folding.

These S6 data show that both decentralized and locally condensed transition states are possible. Decentralized transition states are more common, and are associated with highly cooperative folding pathways, with simple $N \rightleftharpoons D$ folding equilibria. It has been suggested that decentralized transition states and highly cooperative folding transitions evolved to reduce problems caused by aggregation of partially folded proteins, and to compensate for the low thermodynamic stability of native states.

By locally condensed transition states we mean of course interactions that are local in the final native structure. This is consistent with the idea that the molten globule state, formed early in folding, has a native-like topology.

Relationship between native structure and folding

Is there a general relationship between the native structure of a protein and the structure of the transition state? Evolution has shaped both native structures and folding pathways. Should we therefore expect that homologous proteins, with similar structures, will show similar folding pathways?

One idea that seems to hold fairly generally is that it is easier to form local structure than non-local structure. That is, interactions between regions nearby in the sequence should form more easily than interactions between regions distant in the sequence. This is made quantitative in the definition of **contact order** and the observation that contact order is correlated with folding rate (see Box 8.1). Because homologous proteins have similar structures, they are expected to have similar contact orders. If the relationship between contact order and folding rate is reflecting the folding pathway, this is at least consistent with the idea that homologous proteins have similar folding pathways.

A test of the idea that homologous proteins might have similar folding pathways is to use the distribution of Φ values in a protein as a general signature of the folding pathway. Are there correlations in the Φ values of corresponding residues in homologous proteins?

For some pairs of homologous proteins, there is conservation of the Φ-value distribution. These pairs include (1) SH3 domains, (2) Ig-like domains, and (3) acylphosphatase/procarboxypeptidase. The apparent conservation of folding pathway persists even as the sequences (but not the structures) become quite dissimilar.

In other cases, such a correlation is not observed. Protein G and protein L are two small proteins from bacteria that bind to the F_c region of immunoglobulins. Structurally, both consist of a helix packed against a four-stranded β-sheet consisting of two adjacent β-hairpins (see Fig. 8.9). They are similar in native structure, indeed, they are putative homologues. However, their folding pathways differ. In each protein, the transition state contains one of the two β-hairpins. However, in the folding of the two structures, a different β-hairpin forms

BOX 8.1

The relationship between structure and kinetics of folding: contact order

Contact order is a measure of the separation in the sequence, of residues that are close by in space:

$$\text{Contact order} = \frac{1}{LN}\sum(i-j),$$

where L is the number of residues in the protein, N is the number of residues in contact, $i—j$ is the distance in the sequence between residues i and j, and the sum is taken over all pairs of residues in contact. It is observed that for proteins with simple two-state behaviour the folding rate is strongly correlated with the contact order: the greater the tendency of residues nearby in the sequence to interact in the native structure, the faster the folding tends to be.

This observation implies that the residue–residue interactions in the transition state are native-like.

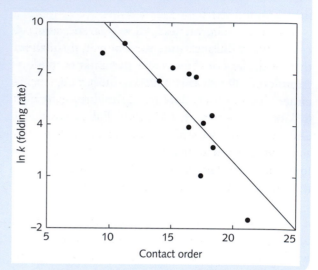

in the transition state. In the case of protein G, the transition state contains an intact second (in the sequence) β-hairpin, but the first is not formed. In the case of protein L, the transition state contains an intact first β-hairpin, but the second is not formed (see Fig. 8.9).

An engineered mutant of protein G with increased stability of the first β-turn increases the similarity of structures of the transition states. The transition state of mutant protein G better resembles the transition state of protein L than the transition state of natural protein G does. The correlation of the Φ-value distributions also increases.

The conclusion is that the folding pathway may depend on favourable local interactions. The results on the cyclically permuted form of the *Thermus thermophilus* ribosomal protein S6 support this.

- Observed folding rates correlate well with contact order, a measure of the average distance along the chain between pairs of residues that interact in space. This suggests that local interactions among residues nearby in the sequence have an important role in the folding process.

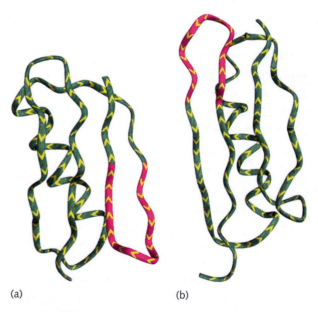

(a)   (b)

Figure 8.9 The structures of immunoglobulin-binding. (a) Protein G [2IGD]. (b) Protein L [2PTL]. These proteins have similar native structure—both contain a four-stranded β-sheet packed against a helix. However, the order of forming secondary substructures during folding is different. In the transition state for folding of each, one of the two β-hairpins is formed but the second is not formed. These regions are shown in crimson in the figure.

The hierarchical model of protein folding

Formation of the folded state requires essentially a complete domain. It is rare that a partial sequence of a protein folds into the corresponding substructure of the native state. This is certainly true of individual elements of secondary structure. Estimates of the stability of an α-helix formed by a typical isolated peptide are ≤ 10 kJ mol^{-1}. There is evidence that secondary structure elements form, but only *transiently*, on a timescale of $\sim 10^{-6}$–10^{-7} s. Is this long enough? The stability of most native states derives largely from the interaction of elements of secondary structure.

It is possible to reconcile the instability of native-like substructures of proteins with their participation in the folding process. An attractive general model of protein folding is **hierarchical condensation,** which R.L. Baldwin and G.D. Rose define as '... a process in which folding begins with structures that are local in sequence and marginal in stability; these local structures interact to produce intermediates of ever-increasing complexity and grow, ultimately, into the native conformation.'*

The hierarchical folding model can in principle apply both to proteins that show simple native $\rightleftharpoons$ denatured transitions, and those that have stable molten-globule-like intermediates. Observations that support the model include:

- Individual elements of secondary structure form, albeit transiently, in the absence of tertiary interactions.

- The signals specifying where helices and strands of sheet begin and end are present in the sequences locally, also independent of formation of tertiary structure. For instance, **helix-capping motifs** are signals that indicate the start and end positions of α-helices.

- The structures of molten-globule-like intermediates have correct folding topology. This facilitates the steps from secondary to supersecondary to tertiary structure.

*Baldwin, R.L. and Rose, G.D. (1999). Is protein folding hierarchic? I. Local structure and peptide folding. *Trends Biochem. Sci.*, **24**, 26–33.

> - The hierarchical model of protein folding is the scenario that local, marginally stable fragments of the native state form, initially transiently, and coalesce progressively.

It is consistent with the hierarchical folding model that the *order* of formation of secondary and higher-order structures may not be unique. An example in which the order of coalescence of elements of secondary structure has been analysed by simulations is a protein consisting of most of the N-terminal domain of the λ repressor (residues 6–85). This monomeric protein contains five α-helical regions and no β-sheet. The protein containing the wild-type sequence has an experimental folding time of 260 μs. A double mutant, G46A/G48A has an even shorter folding time of 12 μs. R.E. Burton, J.K. Myers, and T.G. Oas investigated the possible routes to the folded structure of wild type and the double mutant via different combinations of interacting helices. They applied a computational 'diffusion-collision' model that simulates the transient folding of individual helices, their diffusion-limited collision rate, and the probability that two helices that have collided will go on to accrete another helix before breaking up.

The model consists of 57 states, corresponding to the presence or absence of each native helix–helix contact. Burton, Myers, and Oas described the dynamics of the system in terms of the set of transitions between states, computed from the probability of helix formation and the diffusion constant.

The wild-type protein has a limited number of productive routes to folding (see Fig. 8.10). The first pair of helices to form a productive intermediate may be any of four possible combinations. One of the helices involved is either helix 1 or helix 4. There follow four possible productive states of three helices, all containing helix 4. All coalesce into a single state with four helices bound and interacting, all except helix 1. (Note that helix 3 is absent from all previous productive intermediates.)

In contrast, the double mutant has, at each level until the end, a much larger set of possible productive

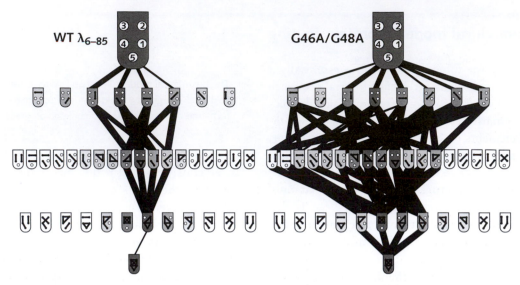

Figure 8.10 Folding of residues 6–85 of the monomeric λ repressor, (left) wild type and (right) G46A/ G48A double mutant. The states shown are a subset of all possible states containing possible native helix–helix interactions. There are five helical regions, numbered at the top, and in corresponding positions but unnumbered at lower levels. An open circle means that the corresponding helix is not formed in that state. A closed circle means that the corresponding helix is formed in that state. A bar between two helices indicates an interaction formed in that state. In any state that contains at least two helices, all helices that are present participate in some interaction. The thickness of the lines connecting the states is proportional to the rate constant of the transition between the two states. To simplify, the transitions corresponding to the lowest rate constants were suppressed.

[From: Burton, R.E., Myers, J.K. and Oas, T.G. (1998). Protein folding dynamics: quantitative comparison between theory and experiment. *Biochemistry*, **37**, 5337–5343.]

intermediates. All helices appear, in each level, in some productive intermediate. This is consistent with the idea that natural protein sequences have tended to evolve to select non-random folding pathways, or that folding funnels selected by evolution tend to become narrower sooner.

How fast could a protein fold?

Some proteins have evolved naturally to be fast folders. Others have had their folding rates increased by design and/or selection. What limits the possible speed of protein folding, and do any proteins approach this limit?

To fold, a protein must form native-like substructures, including α-helices and β-hairpins, and these substructures must diffuse together and interact. This is true no matter in which order these steps occur, or if they take place simultaneously. The data suggest that α-helices form in ~0.1–1 μs; β-hairpins in ~1–10 μs. In the absence of secondary structures, loops form much faster and can be ignored in considering rate-limiting steps in protein folding.

Putting these observations together with measurements of diffusion rates suggests a limit of ~2 μs for the folding time of a 100-residue protein. Theory suggests that this contribution to folding time should scale linearly with chain length.

The conclusion is that a protein of N residues could not fold faster than the 'speed limit' of $N/50$ μs. The fastest folders should be small α-helical proteins. This is observed: the proteins with the fastest known folding rates include the small all-α-helical proteins shown in Table 8.5.

Table 8.5 Fast-folding proteins

Protein	wwPDBentry	Number of residues	Folding time (μs)
Albumin-binding domain (K5I/K39V)	1PRB	47	1
α_3D	2A3D	73	3
Tryptophan cage	1L2Y	20	4.1
Villin headpiece subdomain	1VII	36	4.3

Protein misfolding and the GroEL–GroES chaperone protein

'Proteins fold spontaneously to their native states, based only on information contained in the amino acid sequence … but sometimes they need a little help.' After all, the dilute salt solution of the physical chemistry laboratory is one thing. The intracellular medium is quite another, with concentrations of macromolecules in the range of 300–400 mg mL^{-1}. The threat is that partially folded or misfolded proteins may form aggregates.

Cells have therefore developed molecules that catalyse protein folding, called chaperones. The name is apt: molecular chaperones supervise the states of nascent proteins, hold them to the proper pathway of folding, and keep them away from improper influences that might lead to incorrect assembly or non-specific aggregation. Heat shock or viral infection enhances the danger. These conditions induce overexpression of chaperone proteins, which is how some of them were originally discovered and why they are often called heat-shock proteins.

The participation of chaperones does not contradict the basic tenet that amino acid sequences dictate the three-dimensional structures of proteins. Chaperones act *catalytically* to speed up the process of protein folding. They do not alter the result. Indeed, the observation that a chaperone can catalyse the folding of many proteins, with very different secondary and tertiary structures, is consistent with the idea that chaperones themselves contain no information about particular folding patterns. They anneal misfolded proteins, and allow them, rather than direct them, to find the native state.

The chaperonin system GroEL–GroES of *E. coli* contains two products of the GroE operon, GroEL(L for large, r.m.m. = 58 000) and GroES (S for small, r.m.m. = 10 000). The active complex contains 14 copies of GroEL and seven copies of GroES, for a total r.m.m. of almost 10^6.

In the absence of GroES, 14 GroEL molecules form two seven-membered rings packed back-to-back (see Fig. 8.11). Each ring surrounds a cavity open at one end to receive substrate (i.e. misfolded protein). The cavity is closed at the bottom, by a wall between the two rings, so that bound protein cannot pass internally from one ring to the other. Nevertheless, the two rings are not independent; they communicate via allosteric structural changes that are an essential component of the mechanism. Mutants containing only a single GroEL ring form complexes with substrate, ATP, and GroES, but cannot release them.

With substrate in the cavity, binding of ATP and GroES enlarges the cavity, closes it off and changes its structure. The GroES subunits form another seven-membered ring that caps the GroEL ring. The GroEL ring capped by GroES is called the *cis* ring, and the non-capped ring is the *trans* ring. Formation of the GroEL–GroES complex requires a large and remarkable conformational change in the *cis* GroEL ring, changing the interior surface of the cavity from hydrophobic to hydrophilic, and breaking the symmetry between the two GroEL rings (see Fig. 8.12). An allosteric change mediates a negative cooperativity between the rings, precluding *both* rings from forming the GroES–capped structure simultaneously.

The enclosed cavity is the site of protein folding. Misfolded proteins in the cavity are given a chance to refold. After ~20 s they are expelled, either folded successfully or, if they still cannot get it right, with the chance to enter the same or another complex to try again. Misfolded proteins are treated as juvenile offenders: they are caught, incarcerated, kept in

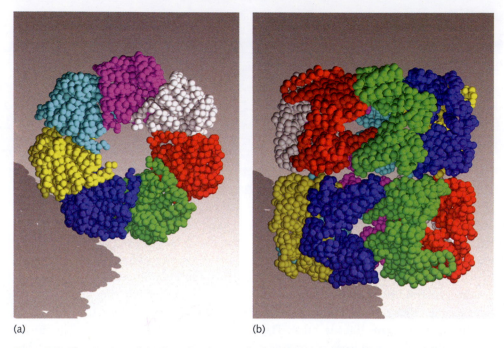

(a)   (b)

Figure 8.11 The structure of GroEL in the absence of substrate and GroES [1DER]. Fourteen subunits assemble into two seven-membered rings packed back-to-back. (a) View down axis. (b) View perpendicular to axis. What are the symmetry elements of this structure?

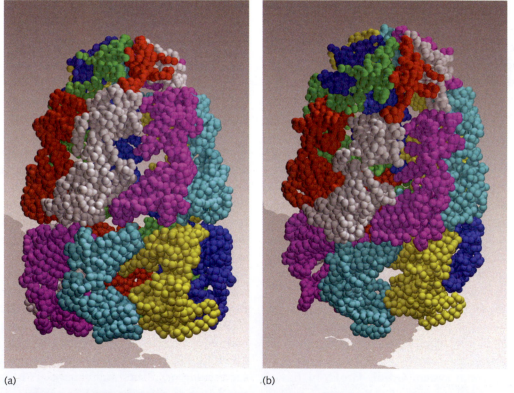

(a)   (b)

Figure 8.12 The GroEL–GroES complex [1AON]. One of the GroEL rings is 'capped' by a third seven-membered ring of GroES subunits. (a) View perpendicular to sevenfold axis. (b) View with complex tipped towards the reader to provide a better view of GroES cap.

solitary confinement, and—having been given an opportunity to reform—released.

> • Chaperones rescue misfolded proteins, protecting cells against the formation of protein aggregates.

The GroEL–GroES conformational change

The GroEL monomer contains three domains: an apical domain, a hinge domain, and an equatorial domain. Figure 8.13 compares the structures of the unbound form of GroEL and the bound form, with one subunit of GroES bound. The viewpoint is perpendicular to the sevenfold axis, that is, tangential to the ring. Figure 8.14 shows the complex, cut away to expose the cavity.

The most obvious conformational change in GroEL is the swinging up of the apical domain by a ~60° hinge motion. A second hinge, near the ATP-binding site in the equatorial domain, rotates to lock in the ATP, and to expose a residue critical for its hydrolysis. The most remarkable feature of the conformational change is the ~90° rotation of the apical domain to expose a different surface to the interior. In the unbound form the residues lining the GroEL cavity are hydrophobic. In the bound form they are hydrophilic, with the hydrophobic residues that formed the lining now taking part in intersubunit contacts.

This is inspired, but of course perfectly logical. Typical globular proteins have preferentially

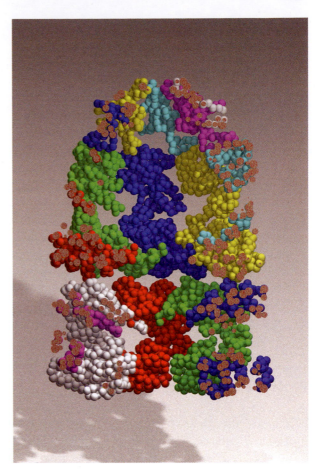

Figure 8.14 A cutaway view of the GroEL–GroES complex, showing the cavity [1AON].

hydrophobic interiors and charged/polar exteriors. The characteristic of misfolded proteins that renders them subject to non-specific aggregation is the surface exposure of hydrophobic residues that are buried in the native state. Proteins in such states bind to the open form of GroEL, with its channel lined with hydrophobic residues. What would one want a chaperone to do to such a misfolded protein, once it has it in its clutches? Altering the interior surface from hydrophobic to hydrophilic encourages the protein to turn itself right side out.

Operational cycle

The assembly functions like a two-state motor. Each of the seven-membered GroEL rings may be in one of two states: open, ready to receive misfolded proteins; or closed, containing misfolded proteins and capped by the GroES ring.

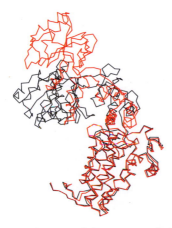

Figure 8.13 The conformational change in one GroEL subunit. The apical domain hinges up and rotates by ~90° to change the interior surface of the cavity. A second hinge motion locks in the ATP [1DER] and [1AON].

1. In the unbound state, the GroEL ring is open to allow protein to enter. The interior interface presents a flexible hydrophobic lining, suitable to bind misfolded proteins by non-specific hydrophobic and Van der Waals interactions. Indeed the binding process may even partially unfold proteins in incorrect states.

2. The binding of ATP and GroES, and the conformational change in the *cis* GroEL ring, creates the closed cavity in which the substrate protein, once released from the apical domains, can refold, sequestered away from potential aggregation partners. The conformational change more than doubles the volume of the cavity, to accommodate less-compact unfolding/refolding transition states. The interior surface changes from hydrophobic to hydrophilic, peeling the bound misfolded protein off the surface and unfolding it even further. The burial of the original interior GroEL surface in intersubunit contacts, within the GroEL–GroES complex itself, breaks the binding of the protein to the original hydrophobic internal surface. This produces a macroclathrate complex.

3. Hydrolysis of ATP in the *cis* ring weakens the structure of the *cis* ring/GroES complex. Binding of ATP (but not necessarily its hydrolysis) in the *trans* ring triggers the disassembly of the *cis* complex, and release of GroES and substrate protein, restoring the ring to its original state.

Each cycle of the engine requires hydrolysis of seven or even 14 ATP molecules. The energetic cost is much larger than the energy of unfolding of a protein, but is small compared to synthesis of the polypeptide chain, and the alternative may be the death of the cell.

Protein engineering

Protein scientists used to be like astronomers. We could observe our subjects but not influence them. Protein science has now become more invasive. We can modify amino acid sequences by genetic engineering, including introduction of point mutations, insertions, and deletions of specific peptides, and recombinations. We can introduce non-natural amino acids. Investigations of protein structure and folding have applied engineered mutants to studies aimed at clarifying different contributions to the stability of native states, or identifying intermediates or transition states in protein folding, and identifying or even modifying residues responsible for function. The same techniques have spawned the biotechnology industry.

Protein design

If we understood how amino acid sequence determines protein structure, we could create proteins unknown in Nature. The **inverse folding problem** is the challenge of starting with a structure and designing sequences that fold into that structure. A simplified version of the problem: select a few residues in a protein interior, delete them and forget what they were, and check combinations of sidechains that might pack the space vacated. Even with high-powered computers, subtle algorithms are required.

- The inverse folding problem is the challenge to develop computational methods to design an amino acid sequence that will fold into a specified structure.

ab initio design of a hyperstable variant of Streptococcal protein G, β1 domain

Streptococcal protein G is a close relative of protein G (see Fig. 8.9). In the parent 56-residue domain an α-helix packs against a four-stranded β-sheet (see Fig. 8.15). The midpoint of its thermal unfolding transition—its 'melting point'—is 83°C, with a $\Delta G°$ of stabilization of 11.7 kJ mol^{-1} at 50°C (measured by guanidinium hydrochloride denaturation).

S.L. Mayo and colleagues have redesigned its sequence to achieve remarkable thermostability.*

*Dahiyat, B.I. and Mayo, S.L. (1997). Probing the role of packing specificity in protein design. *Proc. Nat. Acad. Sci. (U.S.A.)* **94**, 10172–10177; Malakauskas, S.M. and Mayo, S.L. (1998). Design, structure and stability of a hyperthermophile protein variant. *Nat. Struct. Biol.*, **5**, 470–475.

For 11 residues buried in the core, all combinations of A, V, L, I, F, Y and W were considered, with conformations chosen from a rotamer library. The conformations of the mainchain and of other sidechains were held fixed. The seven sidechains considered have 217 possible rotamers, giving a total of $11^{217} \sim 5 \times 10^{25}$ possibilities. Optimization of an energy function reflecting Van der Waals interactions and buried surface area reproduced the wild-type residue at eight positions, and suggested the mutations Y3F, L7I, and V39I. This variant melted at 91°C. Optimization of five additional residues at the periphery of the core, allowing all residues *except* G, P, C, M, R, suggested a sequence with four additional positions changed: T16I, T18I, T25E, V29I; position 43 stayed W.

The final result differed at seven sites from the wild type, out of 16 variable positions.

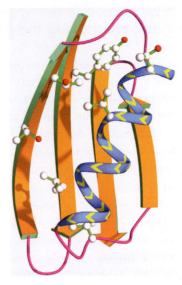

Figure 8.15 Streptococcal protein G, β1 domain [1PGA]. Sidechains shown are those mutated to form a hyperstable variant.

```
wt        MTYKLILNGKTLKGETTTEAVDAATAEKVFKQYANDNGVDGEWTYDDATKTFTVTE
          *  *        * *     *   *          *
design    MTFKLIINGKTLKGEITIEAVDAAEAEKIFKQYANDNGIDGEWTYDDATKTFTVTE
```

Its melting temperature could not be determined as the transition was incomplete at 100°C, comparable to proteins from **hyperthermophiles**. At 50°C, ΔG° of stabilization was 30 kJ mol^{-1}. This is 18 kJ mol^{-1} higher than the original.

The great change in stability did *not* require substantial structural change (see Fig. 8.16). Indeed, if it did, the calculations used in the design, based on fixed mainchain conformation, might not be adequately realistic.

It is difficult to pinpoint the source of the enhanced stability, because the structural changes are so small. The largest effects appear to be enhanced packing, reduction in sidechain strain, and additional buried surface area. Residue T25E can make additional electrostatic and hydrogen-bond interactions.

The difficulty of relating energetic effects to structural changes is a general one. There is a mismatch, unfortunate from the interpretative point of view, between the very large energies that can arise from very small structural changes. It is fortunate that computer programs can cope with these subtleties.

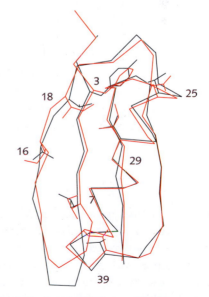

Figure 8.16 Streptococcal protein G, β1 domain; comparison of wild type (black) and hyperstable variant (red).

Expanding and contracting the genetic code

The 20 canonical amino acids give proteins a set of sidechains with a range of physicochemical properties. Most but not all proteins contain all 20. Why only 20? After all, the triplet code could accommodate many more amino acids by giving up some of its redundancy. Indeed, in some circumstances one of the Stop codons is redefined to encode an additional natural amino acid.

Conversely, several pairs of amino acids resemble each other closely: Glu/Asp and Leu/Ile, for example. Why are there as many as 20 in the canonical set? Are they really all necessary?

In the laboratory, it has been possible both to expand and to contract the repertoire of amino acids. **Solid-phase chemical synthesis of proteins** was first achieved by R.B. Merrifield in the 1960s, culminating in a full synthesis of RNAse A in 1969. With solid-phase synthesis it is possible to incorporate unnatural amino acids.

An alternative approach has been to redefine a Stop codon to specify an unnatural amino acid, and make use of a cell's protein-synthesizing machinery to incorporate the novel amino acid. Required are a dedicated toolkit, including a specialized tRNA that interacts with the redefined codon, and a corresponding aminoacyl-tRNA ligase. These must not overlap in specificity with components of the system that deal with other amino acids.

Expansion of the genetic code

The genetic code governs the 'cast of characters'—the set of amino acids from which ribosomes synthesize proteins.

The natural genetic code is almost universal. Some exceptions, as in the protein synthesis machinery of organelles, or in the nuclear genomes of certain species such as the ciliate *Blepharisma americana* (in which UGA codes for tryptophan instead of Stop), involve reassignments of codons. These reassignments do not change the roster of amino acids. Some other unusual amino acids appear in proteins through post-translational modifications.

Two natural exceptions to the restriction of codon assignments to the 20 canonical amino acids are:

1. The Stop codon UGA can be reinterpreted as encoding selenocysteine. This depends on a structural signal in the mRNA. To incorporate selenocysteine, seryl-tRNA ligase charges the special selenocysteine-tRNA with serine, which is converted on the tRNA to selenocysteine by selenocysteine synthase. The selenocysteine-charged tRNA interacts with specific elongation factors to effect incorporation. Selenocysteine-containing proteins, produced by this basic mechanism, appear in archaea, bacteria, and eukaryotes. At least 20 human proteins contain selenocysteine.

2. Some methanogenic archaea redefine the UGA Stop codon to incorporate pyrrolysine by a similar mechanism (see Fig. 8.17).

P.G. Schultz and coworkers have extended the genetic code by introducing modified tRNAs and ligases into *E. coli,* yeast, and even mammalian cells in tissue culture. Figure 8.18 shows the necessary components. It is necessary to select a codon, generally a Stop codon, to be reinterpreted to encode the unnatural amino acid. Also required are a dedicated tRNA and ligase, which do not 'cross-react': no other ligase charges the dedicated tRNA, and the dedicated ligase does not charge any other tRNA. Based on these properties, the novel tRNA and ligase are said to be 'orthogonal' to the corresponding molecules associated with the 20 natural amino acids.

The special tRNA and ligase are created by directed evolution. *M. jannaschii* tyrosyl-tRNA and the

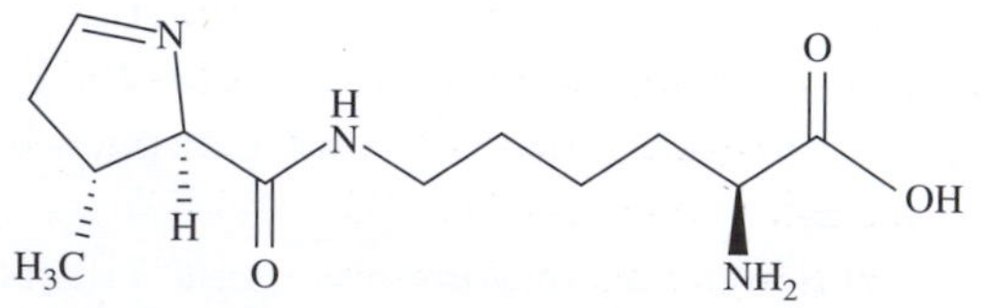

Figure 8.17 The structure of pyrrolysine, incorporated into proteins in certain methanogenic archaea, in the presence of a signal in the mRNA that directs reinterpetation of the Stop codon UGA to encode and incorporate pyrrolysine.

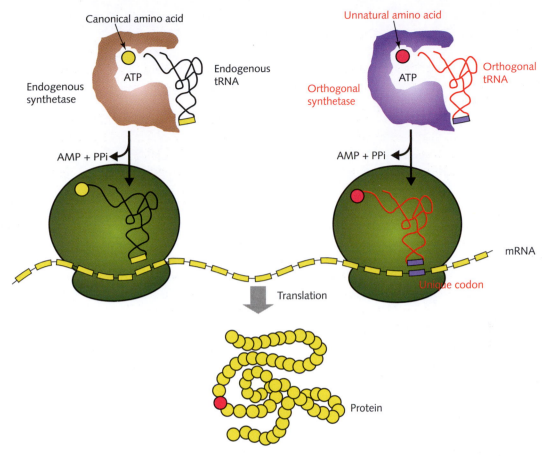

Figure 8.18 Extension of genetic code by redefining a codon. Left: incorporation of a normal amino acid from the canonical set of 20, by endogenous tRNA and the corresponding ligase. Right: incorporation of an unnatural amino acid, by a dedicated tRNA and ligase. 'Orthogonal' means that the tRNA dedicated to the unnatural amino acid must not be recognized and charged by any other ligase, and the dedicated ligase must not charge any other tRNA with the unnatural amino acid. PPi = inorganic pyrophosphate.

[From: Wang, Q., Parrish, A.R. and Wang, L. (2009). Expanding the genetic code for biological studies. *Chem. Biol.*, **16**, 323–336.]

cognate ligase are introduced into *E. coli.* The anticodon of the tRNA is mutated to CUA, so that the tRNA will recognize the Stop codon UAG. A panel of mutant tRNAs is created to contain random bases at specific sites. Selection steps (1) remove those tRNAs that cross-react with endogenous ligases, and (2) retain those with high affinity for the cognate ligase. Next, the substrate specificity of the introduced ligase is modified to charge the tRNA with the desired unnatural amino acid exclusively.

In this way, the Schultz group has introduced a wide variety of unnatural amino acids at specific positions of many proteins (see Fig. 8.19). Approximately 50 novel amino acids are now available to be introduced into proteins at specific sites. Some of the novel amino acids show designed steric or electronic properties; others contain chromophores as fluorescent reporters, or are susceptible to photocrosslinking or photochemical chain cleavage. There are glycosylated amino acids, iodine derivatives to facilitate X-ray structure determination, and sidechains containing other types of reactive groups.

- There are many natural mechanisms for post-translational modification of amino acids. In the laboratory, it is possible to incorporate unnatural amino acids into proteins by solid-phase protein synthesis or by redefining a Stop codon to encode a novel amino acid.

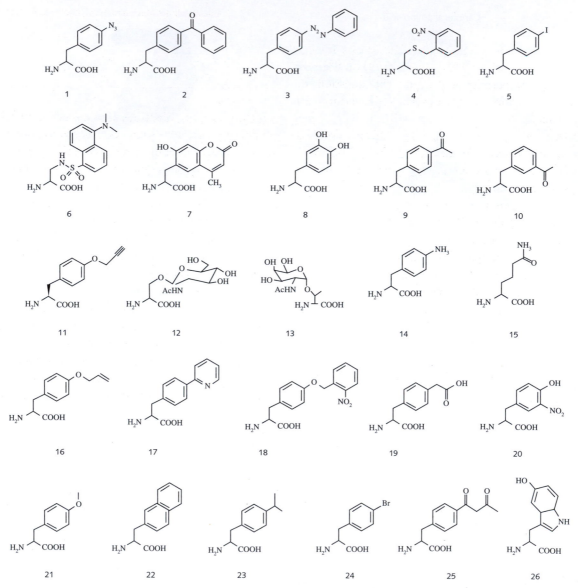

Figure 8.19 P.G. Schultz and coworkers have been able to insert a variety of unnatural amino acids into proteins, at specific positions, by redefining the amber Stop codon CUA, and supplying a dedicated tRNA and ligase. Preparation of synthetic mRNA containing a CUA at one or more positions of interest, the natural protein-synthesizing machinery of *E. coli* then incorporates the unnatural amino acid at the positions corresponding to each CUA. Examples of unnatural amino acids incorporatable into proteins include (1) *p*-azido-L-phenylalanine, (2) benzoyl-L-phenylalanine, (3) amino acids with photoisomerizable sidechains, (4) amino acid with photocleavable sidechain, (5) heavy-atom-containing p-iodophenylalanine, (6) amino acids with dansyl sidechains, (7) amino acids with 7-hydroxycoumarin sidechains, (8) the redox-active amino acid dihydroxyphenylalanine, (9) *p*-acetylphenylalanine, (10) *m*-acetylphenylalanine, (11) *p*-propargyloxyphenylalanine, (12) β–GlcNac-serine, (13) α-GalNAc-threonine, (14) *p*-aminophenylalanine, (15) homo-glutamine, (16) O-allyltyrosine, (17) a metal-binding amino acid, (18) *o*-nitrobenzyltyrosine, (19) *p*-carboxymethylphenylalanine, (20) *m*-nitrotyrosine, (21) O-methyltyrosine, (22) napthylalanine, (23) *p*-isopropylphenylalanine, (24) *p*-bromophenylalanine, (25) *p*-acetoacetylphenylalanine, (26) 5-hydroxytryptophan.

[From: Cropp, T.A. and Schultz, P.G. (2004). An expanding genetic code. *Trends Genet.*,**20**, 625–630.]

Contraction of the genetic code

Success in computational protein design, as for example in the work of Mayo (see section on 'Protein engineering'), permits addressing the question of designing proteins using subsets of the natural 20 amino acids.

K.U. Walter, K. Vamvaca, and D. Hilvert modified the *Methanococcus jannaschii* enzyme chorismate mutase.* Chorismate mutase converts chorismate to prephenate in the biosynthetic pathway of aromatic amino acids. The *Methanococcus* enzyme was introduced into a strain of *E. coli* deprived of the corresponding *E. coli* enzyme. *Methanococcus* chorismate mutase is a symmetric domain-swapped homodimer. Each monomer contains three long α-helices (see Fig. 8.20(a)). The dimer binds two copies of a transition-state analogue inhibitor 8-hydroxy-2-oxabicyclo[3.3.1]non-6-ene-3,5-dicarboxylic acid (see Fig. 8.20(b)).

*Walter, K.U., Vamvaca, K. and Hilvert, D. (2005). An active enzyme constructed from a 9-amino acid alphabet. *J. Biol.Chem.*, 280, 37742–37746.

Table 8.6 Kinetic constants of *M. jannaschii* chorismate mutase

	Wild type	**Nine-amino acid mutant**
k_{cat}	3.1 s^{-1}	0.9 s^{-1}
K_M	20 μM	830 μM

Introducing combinations of mutants that matched the polarity of residues at all but four active-site positions, produced a set of proteins containing only nine amino acids, which could be selected for function. These nine amino acids are L, I, F, M, D, E, K, R, and N. Some of the enzymes showed activity, to the extent of supporting *E. coli* growth at rates approaching that of wild type. One, chosen for detailed characterization, had kinetic constants close to those of the wild-type *Methanococcus jannaschii* enzyme (see Table 8.6).

The novel protein is not a happy camper, however. It is less stable, with a $\Delta G^{\ominus}$ for unfolding of 39.7 kJ mol^{-1} in contrast to 100 kJ mol^{-1} for the wild type. H/D exchange experiments show that the packing in the unligated state is fairly loose. (However, upon binding inhibitor it tightens up.) Thermal unfolding

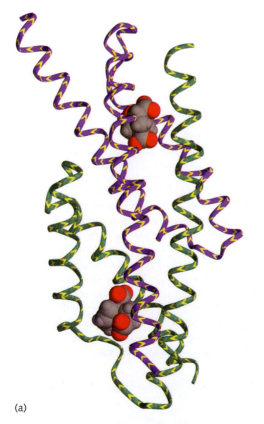

(a)

(b)

Figure 8.20 Structure of *E. coli* chorismate mutase, binding two copies of transition-state analogue inhibitor 8-hydroxy-2-oxabicyclo[3.3.1]non-6-ene-3,5-dicarboxylic acid [1ECM]. (a) Overall structure.(b) Putative structure of active site, of engineered version based on no more than nine amino acids.

[(b) from Walter, K.U., Vamvaca, K. and Hilvert, D. (2005). An active enzyme constructed from a 9-amino acid alphabet.*J. Biol. Chem.*, **280**, 37742–37746.]

is non-cooperative; that is, the change upon heating of the CD signal at 220 nm is gradual, rather than taking place over a narrow temperature range. In contrast, the wild type shows a cooperative thermal denaturation profile, with a transition temperature of 88°C.

The choice of the nine amino acids—L, I, F, M, D, E, K, R, and N—is somewhat surprising. It is likely that the constraints of retaining function dictate a choice of amino acids different from that which might be required merely to maintain the structure. The original motive for considering that reduced amino acid sets might suffice is the similarity of certain pairs of natural amino acids; for example, Leu/Ile, Asp/Glu, Lys/Arg. But for each of these pairs, both survive in the nine-amino acid enzyme! Perhaps even more surprising is that two amino acids with unique conformational properties—Gly and Pro—are dispensable. Walter, Vamvaca, and Hilvert made these observations, and conclude that 'more extensive simplification of the mutase may well be feasible'.

- It is possible to make functional proteins with many fewer than all 20 natural amino acids.

Understanding the contents and layout of the common genetic code

F.H.C. Crick once described the code as a **frozen accident**. Crick's insight must always be respected. However, given the recent results on expansion and contraction of the genetic code, can we go further?

There is now consensus that prokaryotes were and are engaged in widespread horizontal gene transfer. Leaving aside the question of what the optimal genetic code should be, this suggests that it would be to the advantage of any participating species to conform to some standard, as that would give it access to everyone else's genes. This is analogous to the observation that anyone can run any operating system on a computer that they want, but the obvious advantages of running the same system as many other people exerts pressure to conform to *some* standard. Perhaps that is at least a partial explanation of why almost all species have the *same* genetic code. On the other hand, if different species adopted different genetic codes, this might protect them against viruses jumping from other species. (Analogous considerations apply to computers!)

But why not a code with many more than 20 amino acids? Certainly, one perfectly feasible way to introduce greater versatility into the components of proteins is by expanding the genetic code. However, keeping within the general framework of a triplet code, introducing more amino acids at the expense of the redundancy of the code, threatens to reduce robustness—that is, the smoothing out of the evolutionary changes consequent to silent or conservative mutation. An alternative approach to greater versatility without this cost is to effect post-translational modifications of individual amino acids. Whether or not this reasoning is the correct explanation, post-translational modification is the choice that Nature seems largely to have made.

● RECOMMENDED READING

Baldwin, R.L. and Rose, G.D. (1999). Is protein folding hierarchic? I. Local structure and peptide folding. and II. Folding intermediates and transition states. *Trends Biochem. Sci.*, **24**, 26–33 and 77–83. A thoughtful analysis of models for folding and their observational tests.

Dill, K.A. and MacCallum, J.L. (2012). The protein-folding problem, 50 years on. Science 338, 1042–46. History and current state of the field.

Finkelstein, A.V. and Ptitsyn, O.B. (2002). *Protein physics*. Academic Press, New York. The Feynman Lectures of protein science.

Kang, T.S. and Kini, R.M. (2009). Structural determinants of protein folding. *Cell. Mol. Life Sci.*, **66**, 2341–61. A review focused on the relationships between structure and folding.

King, J. , Haase-Pettingell, C. and Gossard, D. (2002). Protein folding and misfolding. *Am. Sci.*, **90**, 445–53. A general review of both the normal folding process and what happens when it goes wrong.

Wolynes, P.G. (2001). Landscapes, funnels, glasses, and folding: From metaphor to software. *Proc. Am. Phil. Soc.* **145**, 555–63. A general statement of the physics of the folding process.

● EXERCISES AND PROBLEMS

Exercise 8.1 Consider a reaction A + B = C + D and the formula $\Delta G - \Delta G° = RT \ln K$. At equilibrium, $\Delta G = 0$. What could be inferred if $\Delta G° = 0$ also?

Exercise 8.2 A protein folds in the simplest way: N = D, no intermediates. A mutant was found that did not induce a more complicated mechanism of folding, but that destabilized the native state relative to the wild type, and left the rate of folding unaffected. Explain why the rate of unfolding must be increased.

Exercise 8.3 Calculate the Φ-value for the mutant shown in the following figure:

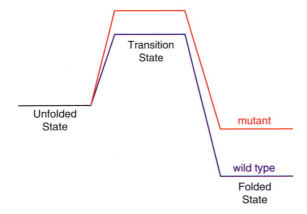

Exercise 8.4 In Fig. 8.6, does the rightmost column, labelled *more denaturant*, represent conditions in which the protein is mostly unfolded? Explain your answer.

Exercise 8.5 Calculate the contact order for the HP lattice protein shown in Fig. 1.19.

Problem 8.1 The energy difference between *cis-* and *trans*-proline conformations is about 84 kJ mol⁻¹. For an isolated proline residue in a denatured protein at 300 K, estimate the fraction that is in the *cis* conformation. [If two states differ by an energy ΔE at thermal equilibrium their populations will be in the ratio exp $(-\Delta E/RT)$ where R = the gas constant = 8.314 J mol⁻¹ K⁻¹ and *T* is the absolute temperature.] For a denatured protein containing two isolated prolines, estimate the fraction of molecules for which both prolines are in the *trans* conformation. (Proline isomerization is one of the factors governing the kinetics of refolding. Prolyl isomerases catalyse the conversion of prolines between *cis* and *trans* isomers and speed up the folding of proteins containing proline residues.)

Problem 8.2 Draw a reaction diagram showing the energetic relationships between native state, transition state, and unfolded state as described in the caption to Fig. 8.5.

Problem 8.3 From examination of Fig. 8.8, draw topology diagrams, analogous to those in the section on 'Comparison of the folding patterns of acylphosphatase and a fungal toxin' in Chapter 2, for wild-type S6 and the circular permutant.

Problem 8.4 Plot ΔS° of folding against the number of residues in the protein for the examples given in Table 8.2. Interpret the result in terms of the average number of degrees of freedom per residue.

Problem 8.5 The 56-residue amino-terminal domain of ribosomal protein L9 from *Bacillus stearo-thermophilus* shows two-state folding kinetics with $k_f = 713$ s^{-1}. (a) Calculate the contact order from the structure, 1DIV. The criterion that two residues are in contact is that at least one pair of non-hydrogen atoms from the residues are within 6 Å of each other. (b) On a photocopy of the figure in Box 8.1, add the point for this domain. (c) Does it fit the other data well?

Problem 8.6 The folding funnel shown in Fig. 8.3 is a simplification in the interests of clarity. It does not show the undoubted roughness of the energy surface that would disrupt the smoothness of the trajectories, local minima, and dead ends leading to non-productive intermediates. Modify a photocopy of Fig. 8.3 to make it more realistic.

Problem 8.7 Analyse the folding and unfolding kinetics of mutants of chicken brain α-spectrin in Table 8.7.

For each mutant:

(a) Calculate

$$\Delta\Delta G_{D-N}\Delta_{D-N}^{wt} - \Delta G_{D-N}^{mutant} = m([\text{urea}]_{50\%}^{wt} - [\text{urea}]_{50\%}^{mutant}).$$

(Take $m = 2$).

(b) Calculate $\Delta\Delta G_{D-\ddagger} = RT \ln(k_f^{wt} / k_f^{mutant})$.

(c) Calculate $\Phi_f = \Delta\Delta G_{D-\ddagger} / \Delta\Delta G_{D-N}$.

(d) Calculate $\Delta\Delta G_{\ddagger-N} = -RT \ln(k_u^{wt} / k_u^{mutant})$. (Use the data for unfolding in 5 M urea.)

(e) Calculate $\Phi_U = 1 - \Delta\Delta G_{\ddagger-N} / \Delta\Delta G_{D-N}$.

(f) To what average accuracy do the values of Φ_f and Φ_u for the same mutant agree?

(g) Divide the mutants into sets for which Φ is small ($\Phi < 0.33$), medium ($0.33 \leq \Phi < 0.67$), and large ($0.67 \leq \Phi$). Is there a correlation of Φ-values with the structure (see Fig. 8.21)? What can you say about the structure of the intermediate?

Table 8.7 Folding and unfolding kinetics of wild type and mutants of chicken brain α-spectrin

Protein	[urea]$_{50\%}$ (mol L^{-1})	k_f at [urea] = 0 (s^{-1})	k_u at [urea] = 5 M (s^{-1})
Wild type (wt)	3.0	23.3	0.80
F117L	1.9	5.4	6.00
A119G	2.5	4.0	0.89
F157L	2.3	20.5	10.0
V171A	2.4	17.1	5.62
G198A	3.4	53.8	0.39
L203	1.2	3.1	181.0

[urea]$_{50\%}$ is the concentration of urea at which [N]/[D] = 1.

All measurements were carried out at 298.15 K.

Data from S. Moran, K. Scott, and J. Clarke.

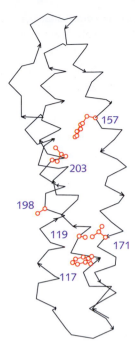

Figure 8.21 Chicken brain α-spectrin, repeat 17, showing positions of mutations for which kinetic data are given. Wild-type sidechains are shown except for 198A.

Problem 8.8 Eftink *et al.* measured the fluorescence intensity of Staphylococcal nuclease as a function of urea concentration at 20°C, see Table 8.8.

Table 8.8 Fluorescence data for Staphylococcal nuclease

[urea] (M)	Relative fluorescence intensity
0.00	69.8
0.00	74.7
0.97	70.7
1.45	72.2
1.92	68.3
1.91	64.6
2.14	56.3
2.23	52.3
2.36	40.4
2.61	30.0
2.83	22.3
3.79	15.6
4.75	15.9

[Data from: Eftink, M.R., Ghiron, C.A., Kautz, R.A. and Fox, R.O. (1991). Fluorescence and conformational stability studies of *Staphylococcus* nuclease and its mutants, including the less stable nuclease-concanavalin A hybrids. *Biochemistry*, **30**, 1193–1199.]

(a) Draw a graph of these data. Explain its general appearance. (b) Fit the data to a function of the form:

$$F([\text{urea}]) = \{1 - f_U([\text{Urea}])\}F_N + f_U([\text{Urea}])\{F_U - S_U \times [\text{Urea}]\},$$

where $F([\text{Urea}])$ is the relative fluorescence intensity and $f_U([\text{Urea}])$ is the fraction of molecules unfolded, as functions of urea concentration. F_N, F_U and S_U are constants (independent of [urea]), the values of the relative fluorescence of native and unfolded molecules *in the absence of urea*, and the slope of the relative fluorescence intensity of the unfolded state as a function of [urea] (assumed constant). The fraction of unfolded molecules as a function of urea concentration is:

$$f_U([\text{Urea}]) = \exp\{\Delta G^\circ_{UN}([\text{Urea}])/RT\} / \{1 + \exp(\Delta G^{-\circ}_{UN}([\text{Urea}])/RT\}$$

and

$$\Delta G^\circ_{UN}([\text{Urea}]) = \Delta G^\circ_{UN}([\text{Urea} = 0]) - m \times \text{Urea}.$$

Determine thereby the values of $\Delta G^\circ_{UN}([\text{Urea} = 0])$, F_N, F_u, S_u, and m. (Solution of this problem will require access to a suitable curve-fitting program.) (c) What fraction of molecules is unfolded in 2.1 M urea?

Problem 8.9 The ^{1}H NMR spectrum of the functional nine-amino acid mutant version of *Methanococcus jannaschii* chorismate mutase has fewer peaks than the spectrum of the wild-type enzyme. However, it shows a similar degree of peak dispersion as the wild type. What is the explanation of these observations?

Problem 8.10 In the nine-amino acid version of *Methanococcus jannaschii* chorismate mutase, Gln88 was changed to Glu (see Fig. 8.20). (a) What effect would you expect this mutation to have on the interaction between enzyme and inhibitor? (b) The activity of the wild-type *Methanococcus jannaschii* enzyme is independent of pH over a range from pH 5–9. What might you be able to guess about the pH optimum of the nine-amino acid modified enzyme? Explain your answer.

CHAPTER 9

Proteomics and systems biology

LEARNING GOALS

- *Gaining a sense of the discipline of systems biology* as an integrative approach to all the 'omics disciplines.
- *Knowing methods for separation and analysis of proteins,* including several techniques based on polyacrylamide gel electrophoresis (PAGE).
- *Understanding the experimental techniques and results of measurements of protein expression patterns,* including mass spectrometry and microarrays, and RNA sequencing (RNAseq).
- *Appreciating developments in metagenomics and metaproteomics:* the applications of methods from the laboratory to natural ecosystems.
- *Understanding the idea of networks* and their representation as graphs.
- *Distinguishing between static and dynamic aspects* of biological networks.
- *Knowing various types of regulatory mechanisms,* and the possible states of a network of activities that they can produce.
- *Appreciating the ideas of stability and robustness* and the mechanisms by which living systems—from cells to ecosystems—achieve them.
- *Understanding the structure, dynamics, and evolution of metabolic networks.*
- *Knowing the different ways of experimentally determining protein–protein and protein–nucleic acid interactions.*
- *Understanding the structures, dynamics, and evolution of regulatory networks.*
- *Knowing the basic features of the regulatory network of E. coli,* including the classic example of the lac operon.
- *Appreciating the adaptability of regulatory networks,* in terms of the yeast regulatory network as an example.

Introduction

The goal of contemporary biomedical science is the understanding, and beyond that the control, of life processes at the molecular and cellular level. Classical biochemistry focused on taking cells apart, purifying individual components, and studying them in isolation. It was the great achievement of biochemists to demonstrate that cells contain a large toolkit—in the form of a panoply of specialized and dedicated nucleic acids and proteins—by means of which they catalyse a large set of chemical reactions, and establish regulatory interactions to keep the traffic running smoothly.

Molecular biology undertook to put things back together, at least up to the level of macromolecular complexes. It was the great achievement of molecular biology to demonstrate the power of living things to effect the controlled manipulation of *matter, energy, and information.*

Systems biology is the study of the components of life, not in the lab, but 'on the job'—in the context of their activities. It focuses on the integration of the activities of the components; it seeks to establish connections, and to describe the physical and logical relationships that underlie these connections.

> Metabolic transformations are the 'smokestack' industries of the cell. Control networks are the 'silicon valley' industries.

Several data streams feed the systems biology enterprise. Proteomics is the study of the distribution and interactions of proteins in time and space in a cell, organism, or even an ecosystem. What are the proteins in a sample? What are their abundances? What are their stabilities and turnover rates? With what partners do they physically and/or functionally interact? What are the control mechanisms that organize the various activities?

Although almost all cells of our bodies initially contain the same genome sequence, implementation of this genomic information varies in space and time. Differentiated cells form tissues. Programmed successions of processes characterize developmental stages. The general underlying mechanism is the differential transcription and expression of genes,

under the control of **regulatory networks** composed of specialized proteins and RNAs. Even viruses and bacteria show control networks; well-studied examples include the **lytic–lysogenic switch in phage λ**, and the control of expression of genes in the **lac operon** of *E. coli.*

Systems biology tries to synthesize proteomic, genomic, and other data into an integrated picture of the structure, dynamics, logistics, and ultimately the logic of living things. A systems biologist will combine study of the proteins in a cell, their genes, the molecules that control their expression or their activities once expressed, and the set of other proteins with which they interact. A systems biologist will assemble into a metabolic network the chemical reactions catalysed by the enzymes of a cell, and assemble into control networks the mechanisms that regulate their activities and expression. Larger-scale networks describe interspecies interactions in ecosystems; for instance, predator–prey relationships, including the spread of infectious disease in human, animal, and plant populations.

High-throughput experimental methods of data collection and analysis, including mass spectrometry, microarrays, and RNAseq, are giving us a large-scale picture of the protein economy in living things.

> • The **transcriptome** is a cell's contents of RNA molecules. The **proteome** is a cell's contents of proteins. All proteins arise from mRNA molecules, but many RNAs are not translated to proteins.

We open the discussion with methods that address proteins directly: the separation, identification, and inventory of the proteome. There follows a discussion of microarrays. These began as a kind of indirect way to get at proteins through nucleic-acid-based technology. However, as the richness in variety and function of cellular RNAs has become more evident, applications of microarrays have extended to them, and are no longer limited to an interest in proteins. Indeed, for many purposes, sequencing of RNA molecules is a superior technique.

The expression patterns that these methods report are under the control of regulatory networks. With the study of regulatory networks, systems biology really comes into its own. Implementing regulatory networks are metabolite-protein, protein–protein, and protein–nucleic acid interactions (see Chapter 6). Underlying the physical interactions are the logical ones. The integration of the physical and logical forces is at the forefront of current biological research—prominently including, but not limited to, protein science.

Separation and analysis of proteins

The complete complement of a cell's proteins is a large and complex set of molecules. Metazoa contain tens of thousands of protein-encoding genes. Different splice variants multiply the number of possible proteins. Vertebrate immune systems generate billions of molecules by specialized techniques of combinatorial gene assembly.

To give some idea of the 'dynamic range' required of detection techniques, the protein inventory of a yeast cell varies from one copy per cell to one million copies per cell.

Examples of techniques for separating mixtures of proteins include gel filtration, chromatography, and electrophoresis. All methods of separating molecules require two things:

1. A difference in some physical property between the molecules to be separated.

2. A mechanism, taking advantage of that property, to set the molecules in motion; the speed differing according to the value of the property selected. This moves apart molecules with different properties.

In some separation methods, one component can stand still and the other(s) move away from it. Affinity chromatography is an example. With others, different species can all move, at different rates, and spread themselves out.

> • To measure an inventory of the proteins in a sample, the proteins must be: (1) separated, (2) identified, (3) counted.

Polyacrylamide gel electrophoresis

In polyacrylamide gel electrophoresis (PAGE), an electric field exerts force on a molecule. The force is proportional to the molecule's total or net charge. In a vacuum, the corresponding acceleration would be inversely proportional to the mass. However, counteracting the acceleration from the electric field are retarding forces from the medium through which the proteins move. Polyacrylamide gels contain networks of tunnels, with a distribution of sizes. Smaller proteins can pass through smaller tunnels as well as larger ones, and therefore move faster through the gel than larger proteins. Proteins with different mobilities move different distances during a run, spreading them out on the gel.

The mobility of a protein depends on its mass and its shape. Higher mass tends to reduce mobility; a more compact shape tends to increase it. In particular, the mobility of denatured proteins is lower than that of the corresponding native states. To achieve a separation that depends solely on relative molecular mass, denature the proteins.

Common denaturing media include urea (which competes for hydrogen bonds), and the reducing agent dithiothreitol to break S–S bridges (and iodoacetamide to prevent their reformation). Sodium dodecyl sulphate (SDS) is a negatively charged detergent that helps to denature proteins. Multiple detergent molecules bind all along the polypeptide chain. The result is a protein–detergent complex that has an extended shape, with a uniform charge density along its length.

Carrying out SDS-PAGE in one dimension spreads out a mixture of proteins into bands. Running several samples on the same gel in parallel lanes is a familiar procedure if only from sequencing gels. The results of protein gels can be made visible ('developed') by staining with Coomassie Blue or, if the samples are radioactively labelled, by autoradiography. Often markers of known molecular weight are run in a separate lane for calibration.

- Polyacrylamide gel electrophoresis (PAGE) is a common method for protein separation. Proteins migrate through a gel with different mobilities depending on their mass, shape, and charge. Proteins separated by SDS-PAGE are denatured by the detergent sodium dodecyl sulphate, creating a protein–detergent complex with a uniform layer of negative charge. Protein mobility in SDS-PAGE depends only on relative molecular mass.

Two-dimensional polyacrylamide gel electrophoresis

One-dimensional PAGE will not adequately separate a very complicated mixture of proteins. The bands in a lane on a gel will overlap and contain mixtures of proteins with similar sizes. To achieve better resolution, a two-stage procedure first separates proteins according to charge; then an SDS-PAGE step, run in a direction 90° from the original direction, separates according to size.

The charge on a protein depends on the charged residues it contains, and the pH of the medium. At different values of pH, ionizable groups on proteins have different charges. For instance, a free histidine sidechain is uncharged below pH ~5, and positively charged above pH ~7. For any protein, there is a pH at which it has a net charge of 0. This is called its **isoelectric point**.

A protein at its isoelectric point will feel no force in an electric field. It will not migrate in electrophoresis. To separate proteins according to their isoelectric points, establish a pH gradient in a medium and apply an electrophoretic field. The proteins will migrate, changing their charge as they pass through regions of different pH, until they reach their isoelectric points and then they will stop. The result, called **isoelectric focusing**, spreads proteins out according to their contents of charged sidechains.

After the proteins are spread out along a lane by isoelectric focusing, running SDS-PAGE at 90° spreads them out in two dimensions (see Fig. 9.1, 2D-PAGE). It is possible to compare the resulting patterns. Spots of interest can be eluted and identified by mass spectrometry (see next section).

Difference gel electrophoresis

A very common problem in proteomics is to compare the abundance patterns of the proteins in two or more related samples. These may be related strains of microorganisms, or, in higher organisms, different developmental stages, different tissues, or

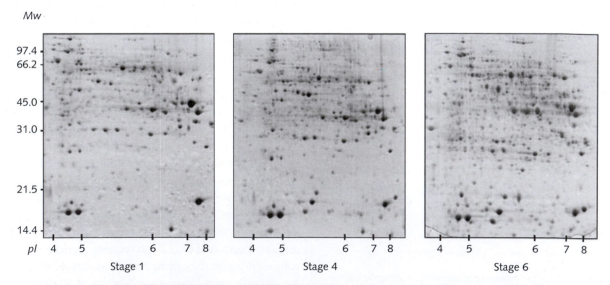

Figure 9.1 Examples of two-dimensional PAGE, showing rose petal proteins at developmental stages 1, 4 and 6. Each gel contains over 600 proteins, of which 421 are common to all three stages. About 12% of the proteins are stage specific.

[From: Dafny-Yelin, M., Guterman, I., Menda, N., Ovadis, M., Shalit, M., Pichersky, E., Zamir, D., Lewinsohn, E., Adam, Z., Weiss, D. and Vainstein, A. (2005). Flower proteome: changes in protein spectrum during the advanced stages of rose petal development. *Planta*, **222**, 37–46.]

a comparison between health and disease states. To run different samples on separate gels and then attempt to quantitate differences by comparing the gels gives unsatisfactory results, because of problems in the reproducibility of the conditions of separation.

Difference gel electrophoresis (DiGE) is a modification of 2D-PAGE in which the conditions of separation are identical for different samples. Two, or three, sets of protein mixtures are separately labelled with different coloured dyes, mixed in equal proportions, and the mixture run on a *single* 2D-PAGE gel (see Fig. 9.2). The resulting pattern contains overlapping peaks from corresponding proteins from the different sources, but these are separable by suitable coloured filters. Computer analysis of peak heights then allows quantitative comparisons of abundances.

B.J. Foth, N. Zhang, S. Mok, P.R. Preiser, and Z. Bozdech studied the dynamics of protein-abundance patterns in the schizont stage of the malarial parasite *Plasmodium falciparum.* (The schizont stage develops in the liver of the infected person 34–46 hours after infection.) They collected samples of parasite at four-hourly intervals: 34, 38, 42, and 46 h post-infection. They pooled these samples to serve as an internal reference, and labelled this mixture with Cy2. The individual samples from the four time points were separately labelled with different coloured dyes, and run by DiGE to compare them with each other and with the internal standard.

Figure 9.3(a) shows a DiGE gel overlaid with proteins from the 34-hour post-infection sample (green) and 42-hour post-infection sample (red). The proteins are spread out according to isoelectric point (horizontal axis) and molecular weight (vertical axis). Some spots correspond to proteins identified by mass spectrometry. In several cases, multiple spots correspond to the same protein appearing in different isoforms, some of which are the result of post-translational modification.

Figure 9.3(b) shows the region of the gel corresponding to the protein eIF4A (eukaryotic initiation factor 4A) or RNA helicase-1/helicase 45. This protein putatively functions in translation initation. The left-hand side of each band shows a contour map of the gel. The right-hand side shows a three-dimensional representation of the density, from which an integrated intensity can be derived to provide a quantitative measure of abundance. The top layer contains the pooled samples. The lower four layers show the results from the samples collected at 34, 38, 42, and 46 h post-infection.

RNA helicase-1 appears in five isoforms, which probably differ in state of phosphorylation. (The horizontal shift on the gel suggests a difference in charge without substantial change in molecular weight.) The different isoforms have different time courses of abundance. Isoform 1 has a low abundance in the 34-h post-infection sample, and then shows a 15.1-fold increase over the subsequent 12 h. Isoforms 4 and 5, in contrast, start high, drop off at the 38-h post-infection point, and then recover. It is interesting that the changes in isoforms 4 and 5 are opposite to those of isoforms 1, 2, and 3. The data are therefore consistent with the hypothesis that there is an interconversion of isoforms 4 and 5 and isoforms 1, 2, and 3; but this is not proven.

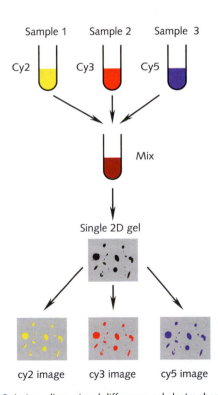

Figure 9.2 In two-dimensional difference gel electrophoresis (DiGE) samples are individually labelled by fluorescent dyes Cy2, Cy3, and Cy5 that emit at different wavelengths. A mixture of samples is run on a *single* PAGE gel, and the contributions of the different samples separated by filters.

- Mass spectrometry is a technique for identifying proteins by fragmenting them into peptides, and measuring the mass/charge ratio of the fragments very accurately. The resulting peptide mass fingerprint can be looked up in databases to find the protein that was its source.

Mass spectrometry is sensitive and fast. Peptide mass fingerprinting can identify proteins in subpicomole quantities. Measurement of fragment masses to better than 0.1 mass units is quite good enough to resolve isotopic mixtures. It is a high-throughput method, capable of processing 100 spots/day (though sample preparation time is longer). However, there are limitations. Only proteins of known sequence can be identified from peptide mass fingerprints, because only their predicted fragment masses are included in the databases. (As with other fingerprinting methods, it would be possible to show that two proteins from different samples are likely to be the same, even if no identification is possible.) Also, post-translational modifications interfere with the method because they alter the masses of the fragments.

The results shown in Fig. 9.6 are from an experiment in which the molecular masses of the ions were determined from their **time-of-flight (TOF)** over a known distance, as illustrated in Fig. 9.4. The operation of the spectrometer involves these steps:

1. Production of the sample in an ionized form in the vapour phase.

2. Acceleration of the ions in an electric field. Each ion emerges with a velocity proportional to its charge/mass ratio.

3. Passage of the ions into a field-free region, where they 'coast'.

4. Detection of the times of arrival of the ions. The TOF indicates the mass-to-charge ratio of the ions.

5. The result of the measurements is a trace showing the flux as a function of the mass-to-charge ratio of the ions detected.

Proteins, being fairly delicate objects, have been challenging to vaporize and ionize without damage. Two 'soft-ionization' methods that solve this problem are:

1. **Matrix-assisted laser desorption ionization (MALDI)**, in which the sample is introduced into the spectrometer in dry form, mixed with a substrate or matrix which moderates the delivery of energy. A laser pulse, absorbed initially by the matrix, vaporizes and ionizes the protein. The MALDI-TOF combination, which produced the results shown in Fig. 9.6, is a common experimental configuration.

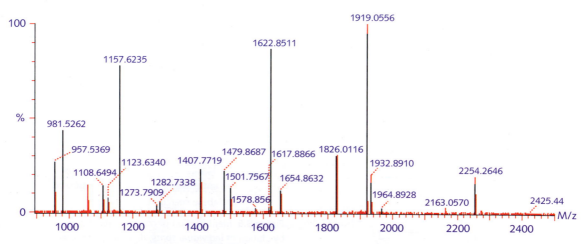

Figure 9.6 Mass spectrum of a tryptic digest. Of the 21 highest peaks (shown in black), 15 match expected tryptic peptides of the 39 kDa subunit of cow mitochondrial complex I. This easily suffices for a positive identification.

[Figure courtesy of Dr I Fearnley.]

2. **Electrospray ionization (ESI)** starts with the sample in liquid form. Spraying it through a small capillary with an electric field at the tip creates an aerosol of highly charged droplets. As the solvent evaporates, the droplets contract, bringing the charges closer together and increasing the repulsive forces between them. Eventually the droplets explode into smaller droplets, each with less total charge. This process repeats, until the ions—which may be multiply charged—are devoid of solvent. These ions are transferred into the high-vacuum region of the mass spectrometer.

Because the sample is initially in liquid form, ESI lends itself to automation in which a mixture of tryptic peptides passes through a **high-performance liquid chromatograph (HPLC)** into the mass spectrometer directly.

Protein sequencing by mass spectrometry

Fragmentation of a peptide produces a mixture of ions. Conditions under which cleavage occurs primarily at peptide bonds yield series of ions differing by the masses of single amino acids (see Fig. 9.7 and

Table 9.1 Masses of amino acid residues, standard isotopes

Gly	57.02146	Ala	71.03711	Ser	87.03203
Pro	97.05276	Val	99.06841	Thr	101.04768
Cys	103.00919	Leu	113.08406	Ile	113.08406
Asn	114.04293	Asp	115.02694	Gln	128.05858
Lys	128.09496	Glu	129.04259	Met	131.04049
His	137.05891	Phe	147.06841	Arg	156.10111
Tyr	163.06333	Trp	186.07931		

Table 9.1). The amino acid sequence of the peptide is therefore deducible from analysis of the mass spectrum (see Fig. 9.8), subject to ambiguities: Leu and Ile have the same mass and cannot be distinguished, and Lys and Gln have almost the same mass and usually cannot be distinguished. Discrepancies from the masses of standard amino acids signal posttranslational modifications. In practice, the sequence of about 9–10 amino acids can be determined from a peptide of length <20–30 residues.

In current practice, the fragments are produced *in situ*: first, the peptide is vaporized, then it is fragmented by **collision-induced dissociation (CID)** with argon gas. This approach requires two mass analysers, operating in tandem in the same instrument (called **MS/MS**). The vaporized sample first passes through one mass analyser, to separate an ion of interest. The selected ion passes into the collision cell where impact with argon atoms excites and fragments it. By keeping the energy of impact low, the fragmentation can be limited largely to peptide bond breakage (see Fig. 9.7). The second mass analyser determines the masses of the fragments.

Quantitative analysis of relative abundance

Variation in yield in the preparation steps makes it difficult to use the intensity of the peaks in a mass spectrum to determine absolute abundances. An accurate method to determine *relative* abundances of the same protein in two samples is the **SILAC technique (stable isotope labelling with amino acids in cell culture)** (see Fig. 9.9).

Suppose one wants to compare two strains of a bacterium. It is required that the two strains be unable to synthesize lysine. Cells from the two strains

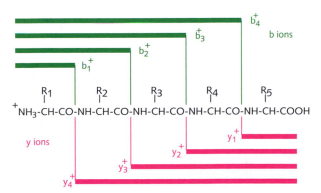

Figure 9.7 Fragments produced by peptide bond cleavage of a short peptide. 'b' ions contain the N-terminus; 'y' ions contain the C-terminus. The difference in mass between successive 'b' ions or successive 'y' ions is the mass of a single residue, from which the peptide sequence can be determined. Two ambiguities remain: Leu and Ile have the same mass and cannot be distinguished, and Lys and Gln have almost the same mass and usually cannot be distinguished. In collision-induced dissociation, bond breakage can be largely limited to peptide linkages by keeping to low-energy impacts. Higher-energy collisions can fragment sidechains, occasionally useful to distinguish Leu/Ile and Lys/Gln.

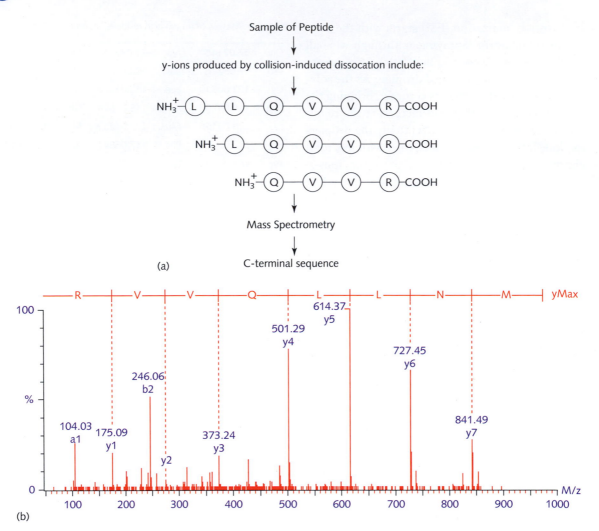

Figure 9.8 Peptide sequencing by mass spectrometry. Collision-induced dissociation (CID) produces a mixture of ions. (a) The mixture contains a series of ions, differing by the masses of successive amino acids in the sequence. In CID the ions are not *produced* in sequence as suggested by this list, but the mass-spectral measurement automatically sorts them in order of their mass/charge ratio. (b) Mass spectrum of fragments suitable for C-terminal sequence determination. The greater stability of 'y' ions over 'b' ions in fragments produced from tryptic digests simplifies the interpretation of the spectrum. The mass differences between successive 'y' ion peaks are equal to the individual residue masses of successive amino acids in the sequence. Because 'y' ions contain the C-terminus, the 'y' ion peak of smallest mass contains the C-terminal residue, etc., and therefore the sequence comes out 'in reverse'. The two leucine residues in this sequence could not be distinguished from isoleucine in this experiment.

[From Carroll, J., Fearnley, I.M., Shannon, R.J., Hirst, J. and Walker, J.E. (2003). Analysis of the subunit composition of complex I from bovine heart mitochondria. *Mol. Cell, Proteom.* **2**, 115–126 (Supplementary figure S138).]

in question are grown in media containing unlabelled ('light') lysine and $^{13}C/^{15}N$-labelled ('heavy') lysine. Mix equal amounts of 'light' and 'heavy' cells. Then, cleavage by a lysine-specific endopeptidase produces fragments containing exactly one lysine residue. The masses of the corresponding fragments from 'light' and 'heavy' cells will differ by a constant amount

(equal to the number of carbons and nitrogens in a lysine residue). The ratio of the peak heights from corresponding fragments gives the relative abundances of the protein in the two samples.

Admixture of a known amount of selectively labelled recombinant proteins as a standard allows absolute quantitation.

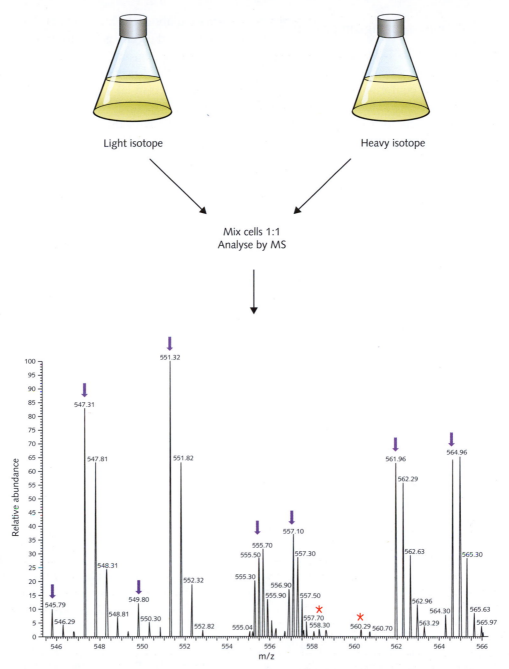

Figure 9.9 Stable isotope labelling with amino acids in cell culture (SILAC) technique to determine relative abundances of proteins in two different strains. Using cells that cannot synthesize lysine, one strain is grown in a medium containing unlabelled (^{12}C) lysine and the other is grown on a medium containing heavy lysine (^{13}C and ^{15}N). Mixing equal amounts of 'light' and 'heavy' cells, followed by digestion and analysis gives a spectrum with pairs of peaks corresponding to the corresponding proteins. If the digestion is carried out with LysC, an endopeptidase that cleaves at lysine residues, each fragment will contain only one lysine, and the mass difference between corresponding peptides will be a constant. The relative heights of the corresponding peaks give the relative abundances of the protein in the two strains. Two peaks marked by red stars have no partners; they are likely to be contaminants.

[From: de Godoy, L.M., Olsen, J.V., de Souza, G.A., Li, G., Mortensen, P. and Mann, M. (2006). Status of complete proteome analysis by mass spectrometry: SILAC labelled yeast as a model system. *Genome Biol.*, **7**, R50. Reproduced via Creative Commons Attribution License 2.0]

Measuring deuterium exchange in proteins

If a protein is exposed to D_2O, mobile hydrogen atoms will exchange with deuterium at rates dependent on the protein conformation. By exposing proteins to D_2O for variable amounts of time, mass spectrometry can give a conformational map of the protein. Applied to native proteins, the results give information about the structure. Using pulses of exposure the method can give information about intermediates in folding.

The methods of proteomics were originally developed for projects in laboratories. Recently, they have been applied to natural environments also. The challenge is to deal with the more complex mixtures of samples encountered.

'Ome, 'ome, on the range—environmental genomics and proteomics

No one doubts the immense variety of microbial life distributed around the Earth. Until recently no one believed that we had adequate tools to circumscribe it. A millilitre of ocean water may contain 100–200 species. A gram of soil may contain 4000.

Classically, microbiologists studied prokaryotes by growing them in isolated cultures, producing pure strains for detailed study. Powerful as the methods were, and important as they are for clinical applications, they were also blinders that prevented full appreciation of the variety and interactions of species in natural environments. Most microorganisms—estimates range from 80 to 99%—cannot be cultured in the laboratory. And the minority that can be cultured may well have different properties in isolation than they do in the context of their natural communities. Even without culturing individual strains, however, it is possible to amplify and determine sequences directly from natural samples containing complex mixtures of organisms.

Metagenomics

Metagenomics, the large-scale sequencing of DNA from an environmental sample, has made it possible to:

- clarify evolutionary relationships;
- use high-throughput sequencing methods to study a cross-section of the life in a natural sample;
- study the majority of microbial strains that are difficult to grow in culture;
- appreciate the relationships and interactions among different species that share an ecosystem.

An example is the sequencing of 16S rRNAs from ocean water from the Sargasso Sea. A group led by J.C. Venter sequenced 10^9 non-redundant regions.

GOLD (http://www.genomesonline.org/) lists over 550 metagenomic projects, including microbes and viruses from hot springs in Yellowstone National Park in Wyoming, U.S.A; flora of the mouth and distal gut of humans and digestive tracts of parasites; microbial communities in drinking-water sources; acid mine drainage systems; ocean waters both coastal and pelagic; and soils, including samples from the rhizospheres—the surroundings of roots—of many different types of plants.

Metagenomics reports sequences from viruses as well as from cells. Viruses are the most abundant and variable life forms on Earth. However, although human pathogens of importance in medicine and the bacteriophages of importance in the molecular biology laboratory have received intensive study, little is known about the rest. Viruses in soils and oceans remain the 'dark matter' of the biosphere.

Metaproteomics

P. Wilmes and P.L. Bond defined *metaproteomics*, analogous to metagenomics, as the large-scale characterization of the entire protein complement of environmental microbiota at a particular point in space and time. Measurements include gene sequences, transcript and protein levels, and metabolite levels and fluxes. These data permit inferences about the economy of the site, such as the cycling of living matter, nutrients, and energy, and more generally the spatial variation—with ocean depth, for example; temporal variation—diurnal or seasonal; diversity and sharing of genes; robustness to changes in conditions; interactions between species, including competition, coevolution, predation, and communication—for instance quorum sensing (see Box 9.3).

- Metagenomics and metaproteomics are extensions of laboratory techniques to natural environments. This allows study of interacting organisms and species, at the cost of greater complexity of the samples.

An example of species interaction that could not be illuminated by culturing individual strains is the complementation in competence appearing in the symbiosis of insects that are host to multiple microorganisms.

Sharpshooters are a group of insects in the family *Cicadellidae*. They are the vector of Pierce's disease, a major threat to California grapevines. The glassy-winged sharpshooter (*Homalodisca coagulata*) harbours two microbial parasites: *Sulcia muelleri*, a member of the Bacteriodetes, and *Baumannia cicadellinicola*, of the γ-proteobacteria. The three members of this tripartite symbiosis share responsibility for providing essential metabolic enzymes. The bacteria show the reduction in genome size and metabolic competence that appears frequently in parasites. The xylem sap eaten by the insect contains Asp, Asn, Glu,

and Gln. The parasites divide up amino acid biosynthesis and other pathways: *Sulcia* can synthesize Leu, Ile, Lys, Arg, Phe, Trp, Thr and Val. *Baumannia* can synthesize Met and His, and contributes to the synthesis of vitamins and cofactors (see Fig. 9.10).

Dynamic proteomics of the response to cadmium challenge

C.M.R. Lacerda, L.H. Choe, and K.F. Reardon studied the response to cadmium stress of a community of microorganisms inhabiting a wastewater treatment bioreactor in Fort Collins, Colorado, U.S.A. Cultures grown up from aliquots contained approximately 50–100 strains, mostly bacteria and a few archaea. These cultures were challenged with 10 mg L^{-1} Cd^{2+}. A comparison of proteomes of cultures with and without cadmium showed the response to the toxic heavy metal.

The levels of cadmium were chosen so that the cultures continued to grow, but at lower rates. Spreading out the proteins by 2D-PAGE showed the time course of the response (see Fig. 9.11). Selected spots

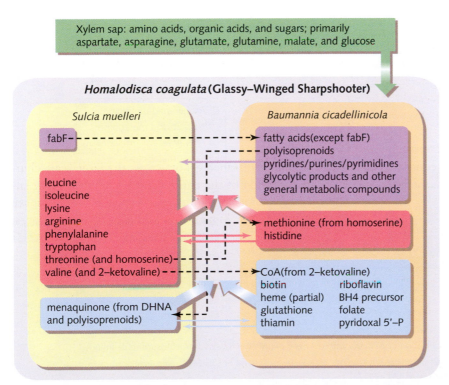

Figure 9.10 Complementarity in metabolic competence between coparasites of the glassy-winged sharpshooter (*Homalodisca coagulata*) *Sulcia muelleri*, and *Baumannia cicadellinicola*. Large coloured arrows: compounds needed by the host that the bacterial symbionts produce. Small coloured arrows (and dashed arrows): compounds (believed to be) shared between the bacterial symbionts. Red: compounds, processes, or genes involved in biosynthesis of essential amino acids. Light blue: vitamin or cofactor biosynthesis. Purple: other metabolic functions.

[From: McCutcheon, J.P. and Moran, N.A. (2007). Parallel genomic evolution and metabolic interdependence in an ancient symbiosis. *Proc. Nat. Acad. Sci. USA.*, **104**, 19392–19397. Copyright (2007) National Academy of Sciences, U.S.A.]

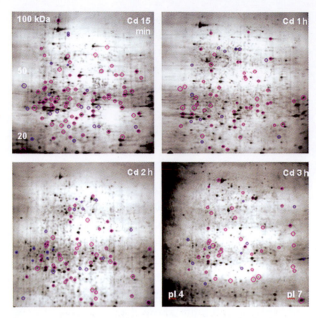

Figure 9.11 Images of gels at 15 min, 1 h, 2 h, and 3 h after addition of 10 mg L⁻¹ cadmium. Pink, upregulated proteins; blue, downregulated proteins.

[From: Lacerda, C.M., Choe, L.H. and Reardon, K.F. (2007). Metaproteomic analysis of a bacterial community response to cadmium exposure. *J. Proteome Res, 6*, 1145–1152.]

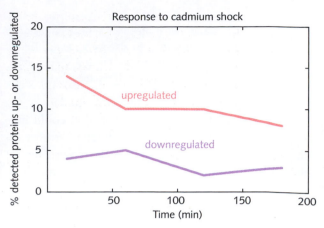

Figure 9.12 Percentage of proteins detected that are up- or downregulated during 3-h period after exposure to 10 mg L⁻¹ Cd²⁺.

were digested and identified by mass spectrometry, using database lookup of peptide mass fingerprints and *de novo* peptide sequencing.

Within 15 min after addition of cadmium, 14% of the proteins were upregulated and 4% downregulated (see Fig. 9.12). A factor of ≥2 was taken as the threshold of significance. Over 100 proteins were fast responders. Some of these returned to normal levels within the course of the experiment ('short-term only'); others remained at changed levels ('short and long term'). Other proteins showed expression changes only after a delay ('long-term only').

Which proteins showed the effect of cadmium? Figure 9.13 shows the functional classification of the proteins with changed expression levels. Cadmium has two mechanisms of toxicity. It binds to the sulphydryl groups of cysteines, and can thereby inhibit protein function. It also causes DNA damage; for

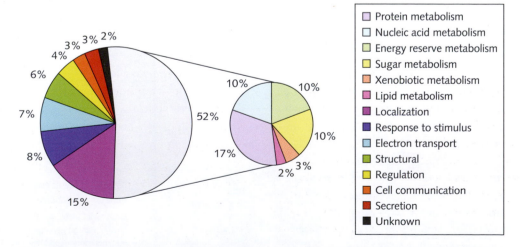

Figure 9.13 Functional classification of proteins with changed expression levels after exposure to cadmium.

[From: Lacerda, C.M., Choe, L.H. and Reardon, K.F. (2007). Metaproteomic analysis of a bacterial community response to cadmium exposure. *J. Proteome Res., 6*, 1145–1152.]

instance, cadmium can induce DNA single-strand breakage. The response to cadmium shock addresses both of these mechanisms. A major component of the response is the upregulation of proteins that pump cadmium out of the cell. This includes certain ATP-ases that specifically detoxify metal–thiolate com-plexes. Another component is the upregulation of DNA repair proteins. Upregulation of certain ribo-somal proteins may be a response to an increased de-mand for synthesis of new proteins to replace dam-aged ones.

Microarrays

Microarrays provide a link between the static gen-ome and the dynamic proteome. We use microarrays to: (1) analyse the mRNAs in a cell, to reveal the transcription patterns of genes and (2) analyse gen-omic DNA sequences, to reveal absent or mutated genes.

For an integrated characterization of cellular activity, we want to determine what RNAs and proteins are present, where, and in what amounts. Hybridization is an accurate and sensitive way to detect whether any particular nucleic acid sequence is present. Microarrays achieve high-throughput analysis by running many hybridization experi-ments in parallel (see Box 9.1). We infer protein ex-pression patterns from measurements of the relative amounts of the corresponding mRNAs. However, because of differences in efficiencies of labelling and of hybridization, microarray data do not permit measurement of absolute abundances of different mRNAs. Even if they could, relative levels of mR-NAs do not always accurately reflect relative levels of expressed proteins.

Expression patterns can help identify genes that underlie diseases. Some diseases, such as cystic fibro-sis, arise from mutations in single genes. For these, isolating a region by genetic mapping can lead to pin-pointing the genomic lesion. Other diseases, such as asthma, depend on interactions among many genes and with environmental factors. To understand the aetiology of **multifactorial diseases** requires the abil-ity to determine and analyse expression patterns of many genes, which may be distributed around dif-ferent chromosomes. To deal with environmental factors, it is necessary to compare the expression pat-terns in different external conditions.

Microarrays are also used to screen for mutations and polymorphisms. Microarrays containing many sequence variants of a single gene can detect differ-ences from a standard reference sequence.

> • Microarrays allow parallel detection of nucleic acids com-plementary to a very large set of oligonucleotides. They permit measurement of inventories of RNAs in a sample, or detection of mutations in genomic sequences.

Different types of chips support different investigations:

- In an *expression chip,* the immobilized oligos are cDNA samples, typically 20–80 bp long, derived from mRNAs of known genes. The target sample might contain mRNAs from normal or diseased tis-sue, for comparison. (The immobilized material on the chip is the *probe.* The sample tested is the *target.*)

- *Mutation or polymorphism microarray analysis* is the search for patterns of single-nucleotide poly-morphisms (SNPs). The oligos on the chip are selected from genomic data from a reference se-quence. They correspond to many known variants of individual genes.

- *Protein microarrays* are arrays of protein detect-ors—usually antibodies—that detect protein–protein interactions.

- *Tissue microarrays* collect and assemble micro-scopic samples of tissue. They permit comparative analysis of the molecular biology and immunohis-tochemistry of the samples.

Microarray data are semiquantitative

Microarrays are capable of comparing concentrations of target oligos. This allows investigation of responses

BOX 9.1   The basic innovation of microarrays is parallel processing

Compare the following types of measurements:

- **'one-to-one':** To detect whether one oligonucleotide has a particular sequence, one could test whether it can hybridize with the oligo of the complementary sequence.

- **'many-to-one':** To detect the presence or absence of a query oligo in a mixture, one could spread the mixture out, and test each component of the mixture for binding to the oligo complementary to the query. This is a **northern** or **Southern blot**.

- **'many-to-many':** To detect the presence or absence of *many* oligos in a mixture, synthesize a set of oligos, one complementary to each sequence of the query list, and test each component of the mixture for binding to each member of the set of complementary oligos. Microarrays provide an efficient, high-throughput way of carrying out these tests in parallel.

To achieve parallel hybridization analysis, affix a large number of DNA oligomers to known locations on a rigid support, in a regular two-dimensional array. The mixture to be analysed is prepared with fluorescent tags, to permit detection of the hybrids. Expose the array to the mixture. Some components of the mixture bind to some elements of the array. These elements now show the fluorescent tags. Because we know the sequence of the oligomeric probe in each spot in the array, measurement of the *positions* of the probes identifies their sequences. This identifies the components present in the sample (see Fig. 9.14).

Such a DNA microarray is based on a small wafer of glass or nylon, typically 2-cm square. Oligonucleotides are attached to the chip in a dense array. The spot size may be as small as ~5 µm. The grid is typically a few cm across. A *yeast chip* contains over 6000 oligos, covering all known genes of *Saccharomyces cerevisiae*. A DNA array, or DNA chip, may contain 400 000 probe oligomers. Note that this is larger than the total number of *genes* even in higher organisms. However, the technique requires duplicates and controls, reducing the number of different genes that can be studied simultaneously. Nevertheless, it is possible to buy a single chip containing all known human genes. A human exon array contains 5.5 million probes, four probes for every exon in the human genome. A set of seven **tiling arrays** contains a total of 4.5 million 25 bp probes, which cover the human genome at an average centre-to-centre spacing of 35 bp. Smaller genomes, such as yeast, have been tiled with overlapping probes on a single array.

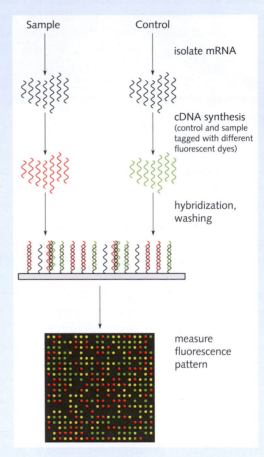

Figure 9.14 Schematic diagram of microarray experiment. A sample to be tested is compared with a control of known properties. From each source, isolate mRNA. Convert to cDNA, using reagents bearing a fluorescent tag, with different colours for control and sample. After hybridizing to the microarrays, and washing away unbound material, the bound target oligos appear at specific positions. A red spot indicates binding of oligos from the sample. A green spot indicates binding of oligos from the control. A yellow spot indicates binding of both. Each probe, represented here by a wavy black line affixed to the support, really contains *many copies of a single oligo*. Indeed, for accurate measurement the concentration of target must greatly exceed the concentration of the probe. If both red and green tagged targets are complementary to the oligo probe at one spot, both can bind to *different* probe molecules within the same spot.

Chromatin immunoprecipitation (see Fig. 9.20) is a typical application of tiling arrays.

To analyse a mixture, expose it to the microarray under conditions that promote hybridization, then wash away any unbound oligos. Scanning the array collects the data in computer-readable form.

To compare material from different sources, it was formerly common practice to tag the samples with differently coloured **fluorophores**, as in Fig. 9.14. Now, because of advances in the technology, it is more common to use a single fluorophore, and run different samples on separate arrays. Currently, the variability in labelling efficiency with different dyes is larger than the variance arising from lack of reproducibility of arrays. The cumulative technical variance of all steps of the assay is estimated to be about an order of magnitude lower than the actual biological variance that the experiment is designed to detect.

to changed conditions. Unfortunately, the precision is low. Moreover, mRNA levels, detected by the array, do not always quantitatively reflect protein levels. In addition, usually mRNAs are reverse transcribed into more stable cDNAs for microarray analysis; the yields in this step may also be non-uniform. Microarray data are therefore semiquantitative: although the distinction between presence and absence is possible, determination of relative levels of expression in a controlled experiment is more difficult, and measurement of absolute expression levels is beyond the capability of current microarray techniques. A change in expression levels of a gene between two samples by a factor ≥1.5–2 is generally considered a significant difference.

Applications of DNA microarrays

- *Investigating cellular states and processes:* For example, profiles of gene expression that change with cellular state or growth conditions can give clues to the mechanism of sporulation, or to the change from aerobic to anaerobic metabolism. In higher organisms, variations in expression patterns among different tissues, or different physiological or developmental states, illuminate the underlying biological processes.

- *Comparison of related species:* The very great similarity in genome sequence between humans and chimpanzees suggested that the profound differences must arise at the level of regulation, and patterns of protein and RNA expression, rather than in the few differences between the amino acid sequences of the proteins themselves. Microarrays are an appropriate technique for exploring this idea.

- *Diagnosis of genetic disease:* Testing for the presence of mutations can confirm the diagnosis of a suspected genetic disease. Detection of carriers can help in counselling prospective parents. This is now being done by sequencing of the appropriate regions of the genome.

- *Genetic warning signs:* Some diseases are not determined entirely and irrevocably by genotype, but the probability of their development is correlated with genes or their expression patterns. Microarray profiling can warn of enhanced risk.

- *Precise diagnosis of disease:* Different related types of leukaemia can be distinguished, from signature patterns of gene expression. Knowing the exact type of the disease is important for prognosis and for selecting optimal treatment.

- *Drug selection:* Detection of genetic factors that govern responses to drugs, which in some patients render treatment ineffective and in others cause unusually serious adverse reactions.

- *Clues to gene function:* A gene with an expression pattern similar to genes in a metabolic pathway is also likely to participate in the pathway.

- *Target selection for drug design:* Proteins showing enhanced transcription in particular disease states might be candidates for attempts at pharmacological intervention.

- *Pathogen resistance:* Comparisons of genotypes or expression patterns, between bacterial strains susceptible and resistant to an antibiotic, point to the proteins involved in the mechanism of resistance.

- *Following temporal variations in protein expression* permits timing the course of (1) responses to pathogen infection, (2) responses to environmental change, (3) changes during the cell cycle, and (4) developmental shifts in expression patterns.

High-throughput sequencing is challenging microarrays as the preferred technique for addressing many of these questions.

Analysis of microarray data

The raw data of a microarray experiment are contained in an image, in which the colour and intensity of the fluorescence reflect the extent of hybridization to alternative probes (see Fig. 9.14). In this experiment, the two sets of targets are tagged with red and green fluorophores. If only one target hybridizes, the spot appears green; if only the other target hybridizes, the spot appears red. If both hybridize, the colour of the corresponding spot appears yellow.

Extraction of reliable biological information from a microarray experiment is not straightforward. Despite extensive internal controls, there is considerable noise in the experimental technique. In many cases, variability is inherent within the samples themselves. Microorganisms can be cloned; animals can be inbred to a comparable degree of homogeneity. However, experiments using RNA from human sources—for example, a set of patients suffering from a disease and a corresponding set of healthy controls—are at the mercy of the large individual variations that unrelated humans present. Indeed, inbred animals, and even apparently identical eukaryotic tissue-culture samples, show extensive variability.

Data reduction involves many technical details of image processing, checking of internal controls, dealing with missing data, selecting reliable measurements, and putting the results of different arrays on consistent scales. There is extensive redundancy in a microarray—each sequence may be represented by several spots; for instance, they may correspond to different regions of a gene. Typically one gene may correspond to ~30–40 spots.

The initial goal of data processing is a **gene expression table**. This is a matrix containing *relative* expression levels, *derived* from the raw data. The rows of the matrix correspond to different genes, and the columns to different sources of material. Of course, the gene expression table is not a simple 'replica plate' of the microarray itself. The microarray fluorescence pattern contains the raw data from which the gene expression table must be extracted. Data from many spots on the microarray will contribute to the calculation of the relative expression level of each gene.

A typical experiment *compares* expression patterns in material from two sources: perhaps a control, of known properties, and a sample to be tested. We may wish to compare organisms growing under different experimental conditions and/or physiological states, or DNA from different individuals, or different tissues, or a series of developmental stages.

Two general approaches to the analysis of a gene expression matrix involve: (1) *comparisons focused on the genes,* that is comparing distributions of expression patterns of different genes by comparing *rows* in the expression matrix; or (2) *comparisons focused on samples,* that is comparing expression profiles of different samples by comparing *columns* of the expression matrix.

1. *Comparisons focused on genes: How do gene expression patterns vary among the different samples?* Suppose a gene is known to be involved in a disease, or to a change in physiological state in response to changed conditions. Other genes co-expressed with the known gene may participate in related processes, contributing to the disease or change in state. More generally, if two rows (two genes) of the gene expression matrix show similar expression patterns across the samples, this suggests a common pattern of regulation and some relationship between their functions, possibly involving a direct physical interaction.

2. *Comparisons focused on samples: How do samples differ in their gene expression patterns?* A consistent set of differences among the samples may distinguish and characterize the classes from which the samples originate. If the samples are from different controlled sources (for instance, diseased and healthy animals), do samples from different groups show consistently different expression patterns? If so, given a novel sample, we could assign it to its proper class on the basis of its observed gene expression pattern.

How can we measure the similarity of different rows or columns? Each row or column of the expression matrix can be considered as a vector, in a space of many dimensions. In the row-vectors, or *gene-vectors* (a row corresponds to a gene), each position refers to the same gene in different samples. The gene vector has as many elements as there are samples. In the column-vectors, or *sample-vectors* (a column corresponds to a sample), each entry refers to a different gene in a single sample. The sample-vector has as many elements as there are genes reported. It is possible to calculate the 'angle' between different gene-vectors, or between different sample-vectors, to provide a measure of their similarities.

The gene-vectors and sample-vectors correspond, separately, to points in spaces of many dimensions. The number of dimensions is either the number of genes or the number of samples. We may not be able to visualize easily points in a space with more than three dimensions, but all our intuition about geometry works fine. For instance, it is natural to ask whether subsets of the points form natural clusters—points with high mutual similarity—characterizing either sets of genes or sets of samples.

After finding clusters we can bring similar genes and samples together. This amounts to reordering the rows and columns of the gene expression matrix. The results are often displayed as a chart, coloured according to the difference in expression pattern. Figure 9.15 contains an example showing the identification of genes differentially expressed in male and female adult rats in the zona glomerulosa (ZG) of the adrenal gland, the outermost layer of the adrenal cortex. (This is only one result from an extensive study.)

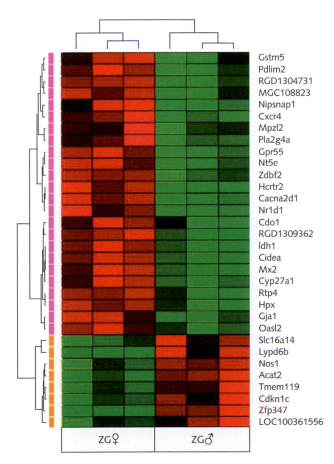

Figure 9.15 The adrenal glands of adult rats differ in size between males and females, a feature associated with functional differences in hormone secretion levels. Trejter and colleagues performed whole-transcriptome analysis of ~27000 genes, on different tissues from within the adrenal glands. This figure shows the identification of genes, differentially expressed in males and females, from the outer layer of the adrenal cortex, the zona glomerulosa. Green corresponds to increased and red to reduced relative expression. Microarrays make it possible to pick out the relatively few differentially-expressed genes from the full complement of ~27000.

In the figure, the data are clustered based on differential expression level (gene vector, according to rows), and *also* based on expression in male and female animals (sample-vector, according to columns). Note that the clusterings by gene and by male/female are independent: it would be possible to change the arrangement of the columns without altering the arrangement of the rows, and *vice versa*. The trees at the top and at the left indicate the similarities among the results, according to sample vector and gene vector, respectively. The sample-vector tree at the top separates the samples from male and female rats. The gene vector shows three classes of differently-expressed genes. Clusters 1 and 2 (nearer the top of the chart) involve genes responsible for ion transport. Genes in cluster 3 are involved in hormone response.

[From: Trejter, M., Hochol, A., Tyczewska, M., Ziolkowska, A., Jopek, K., Szyszka, M., Malendowicz, L.K. and Rucinski M. (2015). Sex-related gene expression profiles in the adrenal cortex in the mature rat: microarray analysis with emphasis on genes involved in steroidogenesis. Int. J. Mol. Med. 35, 702–714. Reproduced via Creative Commons Attribution License]

Depending on the origin of the samples, what is already known about them, and what we want to learn, data analysis can proceed in different directions.

1. The simplest case is a carefully controlled study using two different sets of samples *of known characteristics*. For instance, the samples might be taken from bacteria grown in the presence or absence of a drug, from juvenile or adult fruit flies, or from healthy humans and patients with a disease, or, in this example, male and female rats. We can focus on the question, what differences in gene expression pattern characterize the two states? Can we design a classification rule such that, given another sample, we can assign it to its proper class? Hardly necessary to identify male and female rats, but such classification algorithms are very useful in diagnosis of disease. Subject to the availability of adequate data, such an approach can be extended to systems of more than two classes.

2. In a different experimental situation, we might not be able to *pre-assign* different samples to different categories. Instead, we should hope to extract the classification of samples from the analysis. The goal is to cluster the data to *identify* classes of samples and then to investigate the differences between the genes that characterize them.

An intrinsic problem—and a severe one—in interpreting gene expression data is the fact that the number of genes is much larger than the number of samples. We are trying to understand the relationship of one space of very many variables (the genes) to another one (the phenotype), from only a few measured points (the samples). The sparsity of the observations does not provide anywhere near adequate coverage. Statistical methods bear a heavy burden in the analysis to give us confidence in the significance of our conclusions.

- Results of microarray measurements provide a gene expression table, which permits clustering of genes with comparable expression patterns among different samples, or clustering of samples with comparable profiles of gene expression.

Expression patterns in different physiological states

A fundamental question in biology is how different components of cells smoothly integrate their activities. Measurements of expression patterns tell us part of the story. They provide an inventory of the components, but suggest only inferentially how they interact.

Comparisons of alternative physiological states of an organism offer the possibility of extracting, from an entire genome, a subset of genes that underly a particular life process. An example of shift in physiological state in microorganisms is *diauxy*. Diauxy, or double growth, is the switch in metabolic state of a microorganism, when, having exhausted a preferred nutrient, it 'retools' itself for growth on an alternative. The organism may show a biphasic growth curve, with a lag period while the changed complement of proteins is synthesized. The **diauxic shift** in yeast is the transition from fermentative to oxidative metabolism upon exhaustion of glucose as an energy source.

J. Monod discovered diauxy approximately 70 years ago, during his predoctoral work in Paris. He described his observations in his 1941 thesis. Resuming his research career after the war, he and his colleagues at the Institut Pasteur made their fundamental discoveries about the mechanism of gene regulation.

Yeast as a simple eukaryote has been the subject of very intensive study. Investigations of the diauxic shift have treated not only changes in gene expression, but the activity of the regulatory network that underlies the expression changes. The combined story is a most interesting one, recounted later in this chapter.

Expression pattern changes in development: the life cycle of *Drosophila melanogaster*

During their lifetimes, insects undergo macroscopic changes in body plan, more profound than any post-embryonic development in humans and other mammals, and even in amphibians. This allows juvenile and adult flies to occupy different ecological niches. The major stages of a fly's life are embryonic, larval, pupal, and adult. Metamorphosis occurs during the pupal stage: flies spend their 'adolescence' sequestered within a pupa (to the envy of many a parent of a human teenager).

Fly development has been intensively studied at the molecular level. Impressive understanding has been achieved of the mechanism of translation of molecular signals into macroscopic anatomy. The genesis of specific organs, notably eyes and legs, has been carefully analysed.

In this section we consider how gene expression patterns vary during the lifetime of a fly.

M.N. Arbeitman, E.E. Furlong, F. Imam, E. Johnson, B.H. Null, B.S. Baker, M.A. Krasnow, M.P. Scott, R.W. Davis, and K.P. White have examined changes in transcription patterns in *Drosophila melanogaster* during different stages of its life. When they took up the problem, it was known from earlier work that large-scale changes in gene expression occurred. Microarrays made possible a more systematic and thorough study.

cDNAs containing representatives of 4028 genes (about a third of the total estimated number in *D. melanogaster*) revealed expression patterns at 66 selected time points from embryo through to adulthood (see Fig. 9.16). Expression levels were compared with pooled mRNA from all life stages to represent a (weighted) average expression level. The interval between measurements varied from 1 h, for embryos, up to several days, for adults, until a total age after fertilized egg of up to 40 days (see Table 9.2).

Most of the genes tested—3483 out of 4028, or 86%—changed expression levels significantly at *some* stage(s) of life. Of these, 3219 varied by a factor of >4 between their maximum and minimum values.

The data show that in *Drosophila*, as in other species, genes participating in a common process often exhibit parallel expression patterns, and similar perturbations of these patterns in mutants. For instance, the expression pattern of the *eyes absent* mutant, which produces an eyeless or at least reduced eye phenotype, forms a cluster, in the analysis of expression patterns, with 33 genes. Of these 11 are already known to function in eye differentiation or phototransduction. It is likely that the other 22 also have related activities. The data provide at least hypotheses, and at best reliable clues, to the function of these genes.

Different life stages make different demands on different genes

Different genes exhibit different temporal patterns of expression. Most developmentally modulated genes are expressed in the embryonic stage, as the whole system is getting its act together. Genes expressed in

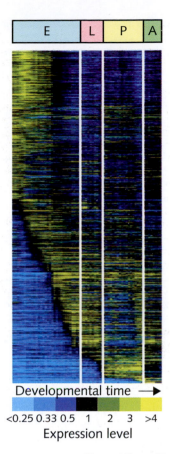

Figure 9.16 Gene expression profiles at different life stages of *Drosophila melanogaster*, ordered by time of first rise of transcript level: E = embryo, L = larva, P = pupa, A = adult. The scale of expression level, relative to that of pooled mRNA samples from all developmental stages, is shown by the intensity of colour: black, small change in expression level; dark blue→light blue, increasing downregulation relative to control; dark yellow→ light yellow, increasing upregulation relative to control. Because of the variation in measurement intervals, the developmental timescale governing the horizontal axis does not correspond to calendar time. The embryonic stage lasts ~1 day, the larval stage ~4 days, the pupal stage 5 days, and the adult stage, until cessation of data collection, 30 days. Looking at the distribution of yellow, notice that some genes are expressed at high levels at single specific stages, but that others are expressed at high levels at more than one stage. A gene expressed throughout the life of the fly, at no less than the level of the pooled sample, will appear here as a row containing only black and yellow regions, with no blue.

[From: Arbeitman, M.N., Furlong, E.E., Imam, F., Johnson, E., Null, B.H., Baker, B.S., Krasnow, M.A., Scott, M.P., Davis, R.W. and White, K.P. (2002). Gene expression during the life cycle of *Drosophila melanogaster Science*, **297**, 2270–2275. Copyright AAAS. Reproduced by permission.]

the early embryo include transcription factors, proteins involved in signalling and signal transduction, cell-adhesion molecules, channel and transport proteins, and biosynthetic enzymes. A third of these are

Table 9.2 Intervals of data collection in different life stages

Stage	Approximate interval between measurements
Embryo	1 h
Larva	5 h
Pupa	8 h
Adult	3–4 days

maternally deposited genes; many of these fall off in expression level within 6–7 h.

The genes studied include one large stage-specific class (36.3%) that shows a single major peak of expression. Some of these remain constitutively expressed (at lower levels) subsequently. Others show sharp peaks in expression level.

Another group of genes (40.3%) shows two peaks in expression level. Two patterns are common: genes with their first onset of enhanced expression early in embryogenesis generally have their second at pupation, with elevated expression levels continuing into the pupal stage; and genes with their first onset of enhanced expression late in embryogenesis generally have their second at the late pupal stage, with elevated expression levels continuing into the adult stage.

The remaining 23.4% of genes show multiple peaks in expression level.

The observation of similarities between expression patterns in embryonic stages and pupal stages, and between larval and adult stages, is interesting. Certain analogies suggest themselves. In both embryonic and pupal stages body structures are forming and physical activity is minimal. In contrast, larval and adult stages have more active lifestyles but stasis of anatomical form, although larvae but not adults grow substantially in size. Consistent with the notion of a 'go back and get it right this time' aspect to metamorphosis, there is some dedifferentiation in the pupa.

Ideas of this sort can be examined in the light of the nature of the genes that show different lifelong expression patterns. For example, maximal expression levels of most metabolic genes occur during larval and adult stages. Another set of genes involved in larval and adult muscle development has a similar two-peak expression pattern. More precise analysis is possible, showing that steps in the regulatory hierarchy for muscle development show peaks at different times, genes expressed later appearing downstream in the regulatory hierarchy. Similar time-of-onset sequences appear in both embryos and pupae. The two stages at which body plans are formed—embryo and pupa—reutilize not only the same materials, but some of the same mechanisms.

Microarrays tell us about the expression patterns of proteins, but indirectly. They measure levels of mRNA. Other methods of proteomics deal with proteins directly.

• The development of *Drosophila melanogaster* shows life-stage-dependent patterns of gene expression. Some genes show parallel expression patterns in embryonic/pupal, and in larval/adult stages.

RNAseq

The data measured in microarray experiments are also obtainable by high-throughput sequencing methods. A major advantage of sequencing is that it is 'hypothesis free': it is not necessary to design sets of probe oligos to appear on chips. What you sequence is what you get.

The transcriptome is the inventory of the RNA molecules in a cell. The proteome is the inventory of the proteins. RNAseq approaches the determination of the transcriptome by converting to DNA all the RNA extracted from a sample and then sequencng the DNA. Sequencing RNA is also a convenient way to get at the sequences of the proteins. However, although all expressed proteins correspond to some messenger RNA, by no means all RNA encodes proteins. Of course, the transcriptome is an object of very high interest in itself. And RNAseq is suited perfectly for studying divergence of RNA viruses such as hepatitis C or HIV-1, during the course of the disease in an infected patient.

What can RNA sequencing tell us about a cell's proteins that whole genome sequencing cannot? Conversely, what can RNA sequencing *not* tell us about a cell's proteins? RNA sequencing can report on splice variants, and on the results of RNA editing. By measuring the amounts of each fragment present in the original sample one can derive approximate transcription and expression profiles. Absolute quantitation, even of RNAs, is not possible, although RNAseq can identify up- and downregulated genes in controlled studies. Even worse, RNAseq can tell us only indirectly the amounts of different proteins in the cell, as these are imperfectly correlated with abundances of different mRNAs. RNAseq cannot tell us about quaternary structures of proteins, nor about post-translational modifications.

Because it is still so much easier to sequence DNA than RNA, RNAseq proceeds by converting RNAs to cDNA by reverse transcriptase. It is understood and must be accepted that this introduces additional bias into the results.

Figure 9.17 gives an outline of the procedure. Extracted RNA is converted to cDNA, then fragmented and sequenced. If a complete genome of the organism is available, mapping the sequences of the fragments onto the genome is an easy route to assembly. A pileup of fragments signifies an exon. Gaps in the mapping between the RNA and genome indicate introns: A *junction read* is a fragment that overlaps intron-exon boundaries. It will not map fully onto the reference genome (see Fig. 9.18).

Several software systems address the assembly problem if a complete genome is not available.

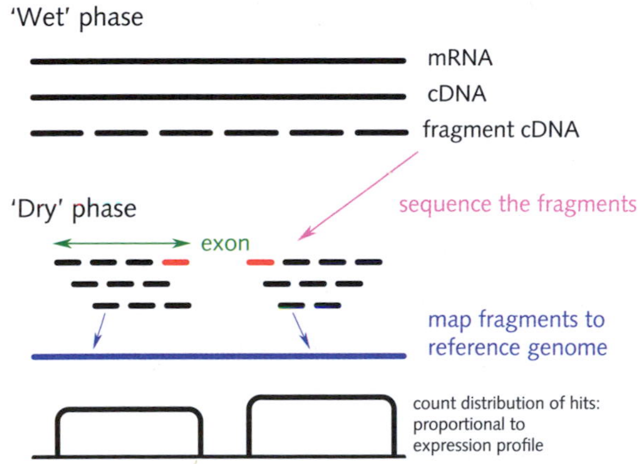

RNAseq v. microarrays

RNAseq has the following advantages over microarrays:

1. A major difference is that microarrays require specification in advance of what you will be looking for, in order to populate the chip. The array will not detect a target unprovided for on the chip: but the goal of many experiments is to detect such variants. Cancers often produce novel genetic variants to which microarrays are blind. RNAseq reports what the sample contains, with no preconceptions or limitations.

2. RNAseq has a greater dynamic range.

3. RNAseq can be tuned to look for rare transcripts. RNAseq is better than microarrays at identifying novel, low frequency RNAs associated with disease processes.

'Wet' phase

mRNA
cDNA
fragment cDNA

sequence the fragments

'Dry' phase

exon

map fragments to reference genome

count distribution of hits: proportional to expression profile

Figure 9.17 Components of the RNAseq measurement. First, the wet component: RNA is collected, and reverse transcribed. The cDNA is diced into fragments and sequenced. Next, the dry component: map the reads onto a reference genome, if available, to assemble them. The red fragments are 'junction reads' that overlap intron-exon boundaries (see Fig. 9.18). Take an inventory to give an approximate transcription profile. Identifying the genes that contributed to the RNA extracted identifies the proteins expressed. It is possible to extrapolate from transcription profiles of mRNAs to the distribution of expressed protein, but the inference is indirect and approximate.

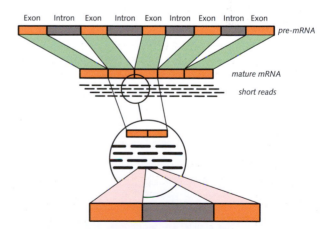

Exon Intron Exon Intron Exon Intron Exon Intron Exon

pre-mRNA

mature mRNA

short reads

Figure 9.18 A 'junction read' is a fragment that straddles an intron-exon boundary. The sequence is continuous in the mature messenger RNA but fragmented is the genomic DNA. Orange segments represent exons; grey segments represent introns. Junction reads overlap the boundary between two exons in the mature mRNA, and therefore do not map back onto a contiguous region in the pre-messenger RNA or the genomic DNA.

It is therefore not surprising to see publications entitled: 'Microarrays are dead'. They elicit the obvious response: 'Reports of the death of microarrays have been greatly exaggerated'.

Nevertheless, microarrays are perfectly appropriate for identifying *known* common allelic variants; for example, in tests for cystic fibrosis. Although over 1700 mutations in the CFTR gene have been observed, 31–72% of patients (the number varies among different populations) show a deletion of F508 in the protein. In practice, a microarray-based test can detect any of a set of mutations known to be associated with a severe disease phenotype. For this type of diagnostic goal, microarrays are not under threat from RNAseq but from genomic sequencing.

There are several databases of transcriptome data. Major institutions host some of them:

- NCBI's ENTREZ system contains gene expression omnibus http://www.ncbi.nlm.nih.gov/geo/

- EBI contains ExpressionAtlas http://www.ebi. ac.uk/rdf/services/atlas/

There are many others. Some are general, providing different 'front ends' to the major repositories, and others specialized in organisms, tissues or diseases. Study of expression profiles associated with cancer is a major industry (see Box 9.2). A number of databases specialize in cancer transcriptomes, linking the sequence data with clinical information. The Cancer Genome Atlas (TCGA) (http://cancergenome. nih.gov/), the International Cancer Genome Consortium (ICGC) (icgc.org), the Cancer Genome Project (http://www.sanger.ac.uk/genetics/CGP/), and Therapeutically Applicable Research to Generate Effective Treatments (http://target.cancer.gov/) are among the large-scale projects that curate, archive and make available the mutations observed in cancers.

The retentive reader will recognize that we have now discussed two approaches to cancer therapy:

BOX 9.2   High-throughput sequencing and cancer

Cancer imposes profound perturbations at the molecular level, with all-too-often fatal histological consequences.

DNA sequencing and RNA sequencing are complementary tools. Changes in the transcriptome are the effects of changes in the genome. Before a patient presents with cancer, DNA sequencing can identify risk factors. Detecting mutants in the BRCA1 and BRCA2 genes is a well-known example. The corresponding proteins normally enhance DNA repair processes.

Once a cancer appears, the clinical challenge begins. Understanding changes in the transcriptome and proteome is an important aid to precise diagnosis of particular types of cancer, which can guide treatment. Sequence features are also a guide to the expected course of the disease. For example, probability of recurrence of breast cancer is predictable on the basis of molecular data.

Beyond the individual patient, the challenge to the research establishment is to reveal basic biological principles that underlie the clinical consequences. High-throughput sequencing has become a basic avenue that, one hopes, will lead to this understanding, and that effective clinical applications will ensue.

Cancer is not one specific disease, but a panoply. Every patient has an individual set of molecular variants, although many features are typical and common. Indeed, the de-

velopment of resistance to drugs, which is often observed, argues for genetic divergence and change in the reproducing cancer cells. Genetic lesions in cancer include point mutations, insertions/deletions, translocations, changes in splice variants, gene fusion and changes in expression level. Knowing the set of lesions helps precisely classify the cancer and dictate treatment, provides biomarkers to assess the progress of treatment, and offers a guide to prognosis. RNAseq can identify genes that are overexpressed in cancer cells, relative to normal tissue from the same patients, and can identify splice variants. This information is not available from microarrays, nor even from genome sequencing.

One typical genome change in cancer is gene fusion, which can activate normally-quiescent **oncogenes**, often kinases. The fusions can arise through a variety of mechanisms, including translocations, insertion/deletion, and inversion. RNAseq can detect fusion events provided that the aberrant genes are transcribed. Knowing the molecular offender may provide approaches to targeted therapy. In some types of lung cancer, a chromosomal inversion creates a fusion of two genes: EML4 (echinoderm microtubule-associated protein-like 4) and ALK (anaplastic **lymphoma** kinase). The drug crizotinib specifically inhibits the kinase activity of the fusion protein, inhibiting the kinase.

drugs targeting variant proteins produced by a cancer (see Box 9.2), and the use of antibodies to enlist the body's immune system in the battle (see Chapter 7).

It has not escaped notice that these approaches could well be synergistic and perhaps their combination can provide more effective treatments.

Systems biology

Proteins are social beings, and life depends on their interactions. Because individual proteins have specialized functions, control mechanisms are required to integrate their activities. The right amount of the right protein must function in the right place at the right time. Failure of control mechanisms can lead to disease and even death.

- Systems biology focuses on the integration and control of gene and protein activity.

Under unchanging environmental conditions, an organism's biochemical systems must be stable. Under changing conditions, the system must be robust, accommodating both neutral and stressful perturbations. Over longer periods of time, processes must have their rates altered, or even be switched on and off. This regulation includes short-term adjustments, for instance in the stages of the cell cycle or responses to external stimuli such as changes in the composition or levels of nutrients or oxygen. Longer-term regulatory activities include control over developmental stages during the entire lifetime of an organism.

Metabolism is the flow of molecules and energy through pathways of chemical reactions. Many metabolic reactions involve proteins and nucleic acids, as well as small compounds such as amino acids and sugars. The full array of metabolic reactions forms complex traffic patterns. Some patterns are linear pathways, such as the multistep synthesis of tryptophan from chorismate. Others form closed loops, such as the tricarboxylic acid (Krebs) cycle. Moreover, the pathways interlock densely. The structure of the totality of metabolic pathways—its connectivity or topology—and its activity patterns, can be analysed in terms of a mathematical apparatus dealing with graphs and flows and throughputs.

To control metabolic flow patterns, *regulatory pathways* connect proteins and metabolite concentrations.

The structure and dynamics of the regulatory pathways are different from those of the metabolic pathways. Corresponding to the succession of enzymatic transformations in metabolism, a regulatory pathway is an assembly of signalling cascades.

Systems biology describes metabolic and regulatory interactions in terms of **interaction networks**.

Two parallel networks: physical and logical

In cells, the two interaction networks operate in parallel: (1) a *physical network* of protein–protein and protein–nucleic acid complexes, and (2) a *logical network* of control cascades. Metabolic pathways partake of both: many but not all metabolic pathways are mediated by physical protein–protein interactions and regulated by logical interactions.

Examples of purely physical interactions include the assembly of oligomeric proteins such as haemoglobin, or photosynthetic reaction centres; complexes of proteins and cofactors that convert light to chemical energy, ATP synthase; assemblies of collagen in connective tissue; and the ribosome. Examples of logical interactions, *not* mediated entirely by direct physical interaction between proteins, include feedback loops in which the increase in concentration of a product of a metabolic pathway inhibits an enzyme catalysing one of the early steps in the pathway, or the secretion of a small molecule as a signal to other cells ('fire-and-forget' mode). (See Box 9.3 for an example of intercell communication.) In these cases the logical interaction is transmitted by a diffusing small molecule. Many other examples appear as a very common theme in the regulation of gene expression. A transcription factor, binding to DNA, may never interact physically with the proteins the expression of which it controls.

The allosteric change in the haemoglobin is an example of simultaneous physical and logical interaction: the subunits of haemoglobin respond to changes in oxygen levels by a conformational change

BOX 9.3   Cell–cell communication in microorganisms: quorum sensing

Control mechanisms *not* involving direct protein–protein interactions mediate intercellular signalling in microorganisms. *Vibrio fischeri* is a marine bacterium that can adopt alternative physiological states in which bioluminescence is active or inactive (literally a 'light switch'). The organism can live free in sea water, or colonize the light organs of certain species of fish or squid. It is bioluminescent only when growing within the animal.

The bacteria respond to the local density of cells, a form of communication called **quorum sensing**. In *V. fischeri*, quorum sensing is mediated by secretion and detection of a small signalling molecule, N-(3-oxohexanoyl)-homoserine lactone. Related species use other N-acyl homoserine lactones, generically abbreviated AHL. AHL can diffuse freely out of the cells in which it is synthesized. Within the light organs, culture densities can reach $10^{10}–10^{11}$ cells/mL, and the AHL concentration can exceed the threshold of about 5–10 nM for triggering the physiological switch.

Bacterial genes *LuxI* and *LuxR* govern the regulation. The product of *LuxI* is involved in the synthesis of AHL. The *LuxR* gene product contains a membrane-bound domain that detects the AHL signal, and a transcriptional activator domain. *LuxR* activates an operon that includes (1) genes for synthesis of luciferase (the enzyme responsible for the bioluminescence), and (2) *LuxI*, expression of which synthesizes additional AHL, amplifying the signal and sharpening the transition.

The host also senses the bacteria: the light organs of squid grown in sterile salt water do not develop properly. This appears to be a reaction to the intensity of luminescence, rather than to the concentration of AHL. For the animal, the luminescence contributes to camouflage: disguise from predators at lower depths, by blending with illumination from the sky. The masking of shadows is a natural form of 'make-up'. (The bioluminescence also regularly surprises diners in seafood restaurants, who jump to the conclusion that their glowing dinner is of extraterrestrial origin. However, most bioluminescent bacteria are harmless, although some strains of the related *Vibrio* species, *V.cholerae*, the causative agent of cholera, are weakly bioluminescent. In fact the virulence of *V. cholerae* is also under the control of quorum sensing, by a related mechanism.)

that alters oxygen affinity. Another example is the transmission of a signal from the surface of a cell across the membrane to the interior by dimerization of a receptor. This can be the initial trigger of a process that ultimately affects gene expression. Not all links of this process need involve protein–protein interactions; some may be mediated by diffusion of small molecules such as cyclic AMP.

Even though certain protein–protein and protein–nucleic acid complexes participate in both physical and logical networks, the two networks remain distinct, and it is useful to keep the distinction in mind, *especially* when considering proteins that participate in both.

- Cells contain both physical and logical networks. Some interactions are common to both.

Networks and graphs

In the abstract, networks have the form of **graphs** (see Box 9.4).

Examples familiar to many readers are the map of the London Underground,* and maps of the tube (subway) systems of other cities. (See Box 9.5 for other examples of graphs.) Each station is a node of the graph, and edges correspond to tracks connecting the stations. The modern London Underground map shows the *topology* of the network; it does not quantitatively represent the geography of the area. An early map, from 1925, did maintain geographic accuracy[†]. This was possible when the system was

*http://www.transportforlondon.gov.uk/tfl/tube_map.shtml or http://www.afn.org/~alplatt/tube.html. Exercises 9.11 and 9.12, Problems 9.3 and 9.4, and Weblem 9.4 also make use of this map.

[†]http://www.clarksbury.com/cdl/maps.html

BOX 9.4 — The idea of a graph

- Mathematically, a graph consists of a set of vertices and a set of edges.
- Each edge links a pair of vertices.
- In a *directed graph* the edges are *ordered* pairs of vertices.
- In a *labelled graph* there is a value associated with each edge. (A directed graph is a special case of a labelled graph: consider the arrowheads as labels.)

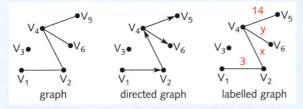

graph   directed graph   labelled graph

An undirected unlabelled graph specifies the connectivity of a network but not the distances between vertices (the topology but not the geometry, as in the modern London Underground map). Labels on the edges can indicate distances, but are unrestricted. For example, some phylogenetic trees indicate only the topology of the ancestry. Others indicate quantitatively the amount of divergence between species. Phylogenetic trees are often drawn with the lengths of the branches indicating the time since the last common ancestor. This is a pictorial device for labelling the edges.

Some graphs do not correspond to physical structures, and in any event edge labels need not reflect geometry in the usual sense. For example, the links in a network of metabolic pathways might be labelled to reflect flow patterns.

BOX 9.5 — Examples of graphs

- Sets of people who have met each other
- Electricity distribution systems
- Phylogenetic trees
- Metabolic pathways
- Chemical bonding patterns in molecules
- Citation patterns in the scientific literature
- The world-wide web

- Graphs are abstract representations of networks. They show the connectivity of the network. Labelled graphs can show physical distances between nodes or other properties of edges such as throughput capacity.

simpler than it is now. Some of the maps now posted in the Paris Métro are fairly accurate geographically. *Considered as networks, a geographically accurate map and a simplified map with the same topology correspond to the same unlabelled graph.*

The London Underground network is fully connected, in that there is a path between any two stations. Many questions familiar to commuters are shared in the analysis of biological networks; for example: What are the paths connecting Station A and Station B? Regarding different lines as subnetworks, how easy is it to transfer from one to another; that

is, what is the nature of the patterns of connectivity? In case of failure of one or more links, is the network robust—does it remain fully connected?

Robustness and redundancy

Biological systems need to be robust, both for survival of individuals under stress and for the plasticity required for evolution. In yeast, for example, single gene knockouts of over 80% of the ~6200 open reading frames are survivable injuries.

In principle, networks can achieve robustness through redundancy. The most direct mechanism is simple **substitutional redundancy**: if two proteins are each capable of doing a job, knock out one and the other takes over. In the London Underground this would correspond to a second line running over the same route. For instance, when the Circle Line is not running, passengers travelling between Paddington and King's Cross stations can travel by the District or the Hammersmith & City lines running on the same tracks.

In cells, some genes have closely related homologues resulting from gene duplication, and some of these contribute to substitutional redundancy. For example, in establishing mouse models for diabetes it was observed that mice and rats (but not humans) have two similar but non-allelic insulin genes. However, substitutional redundancy requires equivalence not only of function but of control of expression. This is the case for mouse insulin production: knocking out either insulin gene leads to compensatory increased expression of the other, resulting in a normal phenotype.

Equivalent expression patterns are more probable among duplicated genes than among unrelated ones. For an example of non-equivalent expression patterns, *E. coli* contains two fructose-1,6-bisphosphate aldolases. One, expressed only in the presence of special nutrients, is non-essential under normal growth conditions. However, the other is essential. In this case functional redundancy does *not* provide robustness. These two enzymes are probably homologous, but they are distant relatives, not the product of a recent gene duplication. One is a member of a family of fructose-1,6-bisphosphate aldolases typical of bacteria and eukarya, and the other is a member of another family that occurs in archaea. *E. coli* is unusual in containing both.

An alternative mechanism of network robustness is **distributed redundancy**: the same effect achieved through different routes. In normal *E. coli* approximately two-thirds of the NADPH produced in metabolism arises via the pentose phosphate shunt, which requires the enzyme glucose-6-phosphate dehydrogenase. Knocking out the gene for this enzyme leads to metabolic shifts, after which increased NADH produced by the tricarboxylic acid cycle is converted to NADPH by a transhydrogenase reaction. The growth rate of the knockout strain is comparable to that of the parent.

> • Biological networks need to be robust. The basic source of robustness is redundancy. Substitutional redundancy is roughly equivalent to multiple copies. (Think of 'backups'.) Distributed redundancy is roughly equivalent to alternative providers of an essential function.

Connectivity in networks

If V_A and V_Z are vertices in a graph, a *path* from V_A to V_Z is a series of vertices: V_A, V_B, V_C, ... V_Z, such that an edge in the graph connects each successive pair of vertices. The number of vertices in the chain is called the *length* of the path. For instance, in the graph in Box 9.4 there is a path of length 4 from V_1 to V_5. A *cycle* is a path of length >2 in a non-directed graph for which the initial and final endpoints are the same, but in which no intermediate link is repeated.

A graph that contains a path between any two vertices is called connected. Alternatively, a graph may split into several **connected components**. The graph in Box 9.4 contains two components, one containing five vertices and an isolated component containing one vertex. (In the extreme, a graph could contain many vertices but no edges at all.) It is often useful to determine the **shortest path** between any two nodes, and to characterize a network by the distribution of path lengths. The phrase 'six degrees of separation'—the title of a play by John Guare—refers to the assertion (attributed originally to Marconi) that if the people in the world are vertices of a graph and the graph contains an edge whenever two people know each other, then the graph is connected, and there is a path between any two vertices with length ≤6.

A **tree** is a special form of graph. A tree is a connected graph containing only one path between each

BOX 9.6   **'Small-world' networks**

Many observed networks, including biological networks, the world-wide web, and electric power distribution grids, have the characteristics of high clustering and short path lengths. They include relatively few nodes with very large numbers of connections, called **hubs**, and many that contain few connections. These combine to produce short path lengths between all nodes. From this feature they are called **'small-world networks'**. Such networks tend to be fairly robust—staying connected after failure of random nodes. Failure of a hub would be disastrous but is unlikely, because there are so few of them.

Many networks, notably the world-wide web, are continuously adding nodes. The connectivity distribution tends to remain fairly constant as the network grows. These are called **'scale-free'** networks.

pair of vertices. A hierarchy is a tree: examples include military chains of command and Linnaean taxonomy. Note that some family trees are not trees in the mathematical sense; examples are plentiful in the royal families of Europe. A tree cannot contain a cycle: if it did, there would be two paths from the initial point (= the final point) to each intermediate point. In the graph in Box 9.4, the subgraph consisting of vertices V_1, V_2, V_4, V_5, and V_6 is a tree. Adding an edge from V_1 to V_5 would create an alternative path from V_1 to V_5, and the cycle $V_1 \rightarrow V_2 \rightarrow V_4 \rightarrow V_5 \rightarrow V_1$; the graph would no longer be a tree.

> The Enzyme Commission classification of protein function is a tree; the Gene Ontology classification is not.

The **density of connections** = the mean number of edges per vertex, and characterizes the structure of a graph. A fully connected graph of N vertices has $N-1$ connections per vertex; a graph with no edges has 0. Nervous systems of higher animals achieve their power not only by containing a large number of neurons but also by high connectivities.

In some systems there are limits on numbers of connections. For many human societies, in the graph in which individuals are the vertices and edges link people married to each other, each node has connectivity 0 or 1. In hydrocarbons, the graphs in which carbon and hydrogen atoms are the vertices and edges link atoms bonded to each other, each node has ≤4 connections. In other networks, connectivities follow observable regularities (see Box 9.6). For instance, the world-wide web can be considered as a directed graph. Individual documents are the nodes, and hyperlinks are the edges. It is observed that the distribution of incoming and outgoing links follow power laws: $P(k)$ = probability of k edges is proportional to k^{-q}, where $q = 2.1$ for incoming links and $q = 2.45$ for outgoing links.

> Topological features of networks that govern their general behaviour include:
> - the density of connections;
> - the topology of the connections—for instance, a tree is a graph with no cycles;
> - the distribution of connections—for instance, small-world networks have a few nodes with large numbers of connections, called 'hubs'.

The density of connections is very important in defining the properties of a network. For instance, the interactions that spread disease among humans and/or animals form a network. Whether a disease will cause an epidemic depends not only on the ease of transmission in any particular interaction, but on the density of connections. As the density of connections—the rate of interactions—increases, the system can exhibit a *qualitative* change in behaviour, analogous to a phase change in physical chemistry, from a situation in which the disease remains under control to an epidemic spreading through an entire population. The classic approach of 'quarantine'—isolating people for 40 days—works by cutting down the degree of connectivity of the disease-transmission network. Note that a carrier who shows no symptoms—'typhoid Mary' was a classic case—serves as a hub of the transmission network.

> Mary Mallon (1869–1938) presented the following unfortunate combination of features: (1) she was infected with typhoid, (2) she did not show symptoms, and (3) she worked for many families as a cook.

Two historical epidemics associated with wars demonstrate the distinction between topology and geometry in network connectivity.

- In the early years of the Peloponnesian War, Athens suffered a severe epidemic. (From Thucydides' detailed description of the symptoms, the disease was probably bubonic plague.) A factor contributing to its transmission was the crowding of people into the city from the surrounding countryside, out of fear of greater vulnerability to military invasion.
- After World War I, an epidemic of influenza killed over 20 million people, more than died in the war itself. Long-distance travel by soldiers returning from the war helped spread the disease. Any epidemic needs an infectious agent and a high density of routes of transmission.

These examples show that the controlling factor is the density of the *connections* and not the density of the people.

Dynamics, stability, and robustness

An unlabelled, undirected graph describes the *static* structure of the topology of a network. For our molecular interaction networks, this may be an adequate description of many of the physical interactions.

For some networks, such as metabolic pathways or patterns of traffic in cities, the *dynamics* of the system depend on the transmission capacities of the individual links. These capacities can be indicated as labels of the edges of the graph. This allows modelling of patterns of flow through the network. Examples include route planning, in travel or deliveries. Note that the shortest path may well not give optimal throughput. In many cities, taxi drivers are exquisitely sensitive—and insensitively garrulous—about optimal traffic paths. In molecular biology, metabolic pathways and signal transduction cascades are networks that lend themselves to pathway and flow analysis.

Although much is known about the mechanisms of individual elements of control and signalling pathways, understanding their integration is a subject of current research. For instance, the idea that healthy cells and organisms are in stable states is certainly no more than an approximation (and in most cases a gross idealization). The description of the actual dynamic state of the metabolic and regulatory networks is a very delicate problem. Understanding *how* cells

achieve even an apparent approximation to stability is also quite tricky. It is likely that great redundancy of control processes lies at its basis. Regulation is based on the resultant of many individual control mechanisms—here a short feedback loop, there a multistep cascade. Somehow, the independent actions of all the individual signals combine to achieve an overall, integrated result. It is like the operation of the 'invisible hand' that, according to Adam Smith, coordinates individual behaviour into the regulation of national economies.

Several types of dynamic states of a network are possible (see Box 9.7):

- equilibrium;
- steady state;
- states that vary periodically;
- unfolding of developmental programs;
- chaotic states;
- runaway or divergence;
- shutdown.

- In addition to their static topology, networks have several possible dynamic states. Some of these are stable and others are unstable.

BOX 9.7   States of a network of processes

- At *equilibrium* one or more forward and reverse processes occur at compensating rates, to leave the amounts of different substances unchanging:

$$A \rightleftharpoons B$$

Chemical equilibria are generally self-adjusting upon changes in conditions, or upon addition of reactants or products.

- A **steady state** will exist if the total rate of processes that produce a substance is the same as the total rate of processes that consume it. For instance, the two-step conversion

$$A \rightarrow B \rightarrow C$$

could maintain the amount of B constant, provided that the rate of production of B (the process A → B) is the same as the rate of its consumption (the process B → C). The net effect would be to convert A to C.

A cyclic process could maintain a steady state in all its components:

A steady state in such a cyclic process with all reactions proceeding in one direction is very different from an equilibrium state. Nevertheless, in some cases it is still true that altering external conditions produces a shift to another, neighbouring, steady state.

- *States that vary periodically* appear in the regulation of the cell cycle, circadian rhythms, and seasonal changes such as annual patterns of breeding in animals and flowering in plants. Circadian and seasonal cycles have their origins in the regular progressions of the day and year, but have evolved a certain degree of internalization.

- Many equilibrium and some steady-state conditions are *stable*, in the sense that concentrations of most metabolites are changing slowly if at all, and the system is robust to small changes in external conditions. The alternative is a *chaotic state,* in which small changes in conditions can cause very large responses. Weather is a chaotic system: the meteorologist Lorenz asked, 'Does the flap of a butterfly's wings in Brazil set off a tornado in Texas?' In a carefully regulated system, chaos is usually well worth avoiding, and it is likely that life has evolved to damp down the responses to the kinds of fluctuations that might

give rise to it. Chaotic dynamics does sometimes produce approximations to stable states—these are called **strange attractors**. Understanding stability in living dynamical systems subject to changing environmental stimuli is an important topic, but beyond the scope of this book.

- *Unfolding of developmental programmes* occurs over the course of the lifetime of the cell or organism. Many developmental events are relatively independent of external conditions, and are controlled primarily by programmed regulation of gene expression patterns.

- *Runaway* or *divergence*. Breakdown in control over cellular proliferation leads to unconstrained growth in cancer.

- *Shutdown* is part of the picture. Apoptosis is the programmed death of a cell, as part of normal developmental processes, or in response to damage that could threaten the organism, such as DNA strand breaks. Breakdown of mechanisms of apoptosis—for instance, mutations in protein p53—is an important cause of cancer.

Protein complexes and aggregates

The basis of our understanding of how life within a cell is organized and regulated is the set of protein–protein and protein–nucleic acid interactions. We have already encountered several protein complexes in discussions of individual systems:

- Simple dimers or oligomers in which the monomers appear to function independently.

- Oligomers in which the monomers interact functionally, for instance ligand-induced dimerization of receptors, and allosteric proteins such as haemoglobin, phosphofructokinase, and asparate carbamoyltransferase.

- Large fibrous proteins such as actin or keratin.

- Non-fibrous structural aggregates such as viral capsids.

- Protein–nucleic acid complexes (see Chapter 6), including ribosomes, nucleosomes, splicing and repair particles, and viruses. Of particular interest in systems biology are protein–nucleic acid complexes involved in the regulation of transcription.

- Many proteins, whether monomeric or oligomeric, *function* by interacting with other proteins. These include all enzymes with protein substrates, and many antibodies, inhibitors, and regulatory proteins.

Recent developments have focused on developing methods for systematic and comprehensive identification of protein interactions.

Protein interaction networks

The units from which interaction networks are assembled are:

- for physical networks, a protein–protein or protein–nucleic acid complex;

- for logical networks, a dynamic connection in which the activity of a process is affected by a change in external conditions, or by the activity of another process.

Most experiments reveal only pairwise interactions. The challenges are to integrate pairwise interactions into a network and then to study the structure and dynamics of the system.

Many techniques detect physical interactions directly:

- *X-ray and NMR structure determinations* cannot only identify the components of the complex, but reveal how they interact, and whether conformational changes occur upon binding.

- *Two-hybrid screening systems:* transcriptional activators such as Gal4 contain a DNA-binding domain and an activation domain. Suppose these two domains are separated, and one test protein is fused to the DNA-binding domain and a second test protein is fused to the activation domain. Then a reporter gene will be transcribed only if the components of the activator are brought together by formation of a complex between two test proteins (see Fig. 9.19). High-throughput methods allow parallel screening of a 'bait' protein for interaction with a large number of potential 'prey' proteins.

- *Chemical cross-linking* fixes complexes so that they can be isolated. Subsequent proteolytic digestion and mass spectrometry permit identification of the components.

- *Coimmunoprecipitation.* An antibody raised to a 'bait' protein binds the bait together with any other 'prey' proteins that interact with it.

The interacting proteins can be purified and analysed, for instance by western blotting or mass spectrometry.

- *Chromatin immunoprecipitation* identifies DNA sequences that bind proteins (see Fig. 9.20).

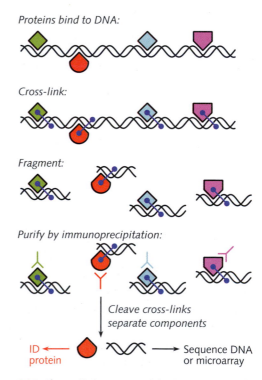

Figure 9.20 Chromatin immunoprecipitation. Treatment with formaldehyde cross-links proteins and DNA, fixing the complexes that exist within a cell. After isolation of chromatin, breaking the DNA into small fragments allows separation of proteins by binding to specific antibodies, carrying the DNA sequences along with them. Reversal of the cross-link followed by sequencing of the DNA identifies the specific DNA sequence to which each protein binds. To identify multiple sites in the genome to which the protein binds, the DNA fragments can be analysed using a microarray. To avoid the requirement for antibodies specific for each protein to be tested, the proteins can be fused to a standard epitope or to a sequence that can be biotinylated, taking advantage of the very high biotin-streptavidin affinity.

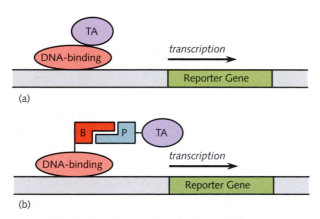

Figure 9.19 (a) A transcription activator contains two domains, a DNA-binding domain, and a transcription-activator domain (TA). Together they induce expression of a reporter gene. *lacZ*, encoding β-galactosidase (see Table 9.3) is a common choice of reporter gene because chromogenic substrates make β-galactosidase easy to detect. (b) Transcription proceeds if the DNA-binding domain and transcription-activator domain are separated in different proteins that can form a complex. A 'bait' protein B (red) is fused to the DNA-binding domain and a 'prey' protein P (cyan) is fused to the transcription-activation domain. Formation of a complex between bait and prey brings together the DNA-binding and TA domains, inducing transcription.

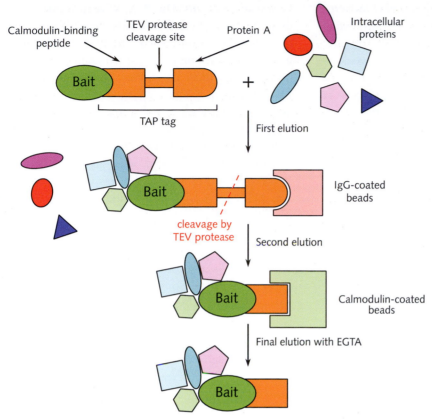

Figure 9.21 The construct and protocol in the tandem affinity purification (TAP) method for purification of complexes with a selected 'bait' protein. The fusion protein containing the bait and the two tags separated by the TEV cleavage site bind *in vivo* to proteins in the cell. A first affinity purification step binds the bait protein to a column containing IgG-coated beads that bind specifically to the first tag, protein A. After thorough washing, cleavage by TEV protease releases the bound complexes, and exposes the second tag. A second affinity purification step binds the bait protein to a column containing calmodulin-coated beads, which bind specifically to the second tag, the calmodulin-binding peptide. After washing, elution with the chelating agent ethylene glycol tetraacetic acid (EGTA) releases the purified complexes.

- *Phage display.* Genes for a large number of proteins are individually fused to the gene for a phage coat protein to create a population of phage, each of which carries copies of one of the extra proteins exposed on its surface. Affinity purification against an immobilized 'bait' protein selects phage displaying potential 'prey' proteins. DNA extracted from the interacting phages reveals the amino acid sequences of these proteins.

- *Surface plasmon resonance* analyses the reflection of light from a gold surface to which a protein has been attached. The signal changes if a ligand binds to the immobilized protein. (The method detects localized changes in the refractive index of the medium adjacent to the gold surface. This is related to the mass being immobilized.)

- *Fluorescence resonance energy transfer.* If two proteins are tagged by different chromophores, transfer of excitation energy can be observed over distances up to about 60 Å (see Chapter 1).

- *Tandem affinity purification (TAP)* allows probing cells *in vivo* for partners that bind to a selected 'bait' protein.

Fusion of the bait protein to *two* affinity tags, separated by a cleavage site, permits high efficiency in extraction of complexes. The method uses two successive, or tandem, affinity purification steps separated by a cleavage step to expose the second tag. The cleavage steps are specific and require only mild conditions in order to leave the bait–prey complex intact (see Fig. 9.21).

The fusion proteins contain an individual bait protein extended at its N or C-terminus by a calmodulin-binding peptide, a TEV protease cleavage site (TEV is a cysteine protease from tobacco etch virus), and protein A (which binds to high affinity to an available antibody). The double tag, and the two-step purification, gives superior performance relative to a single-tag technique, in yield of low-concentration complexes.

In vivo, the expressed tagged bait protein binds to a set of prey proteins. Figure 9.21 shows the purification protocol to recover the complexes from cell extracts.

Determination of a protein interaction network requires measurements of many bait proteins. Because all bait proteins carry the same fusion sequences, the purification and cleavage steps in TAP are unique. There is no need to design different purification protocols for different bait proteins.

Other methods provide complementary information:

- *Domain recombination networks.* Many eukaryotic proteins contain multiple domains. A feature of eukaryotic evolution is that a domain may appear in different proteins with different partners. In some cases proteins in a bacterial operon catalysing successive steps in a metabolic pathway are fused into a single multidomain protein in eukarya. The domains of the eukaryotic protein are individually homologous to the separate bacterial proteins. (Examples of proteins fused in eukarya and separate in prokaryotes are also known.)

It is possible to create a network by defining an interaction between two protein domains whenever homologues of the two domains appear in the same protein. This is evidence for some functional link between the domains, even in species where the domains appear in separate proteins.

- *Coexpression patterns.* Clustering of microarray or RNAseq data identifies proteins with common expression patterns. They may have the same tissue distribution, or be up- or downregulated in parallel in different physiological states. This is also suggestive evidence that they share some functional link. But this must be verified. In the response of *M. tuberculosis* to the drug isoniazid, genes for the fatty acid synthesis complex are coordinately upregulated. They are on an operon-like gene cluster, and in fact these proteins do form a physical complex. On the other hand, alkyl hydroperoxidase (AHPC) is also upregulated in response to isoniazid. AHPC acts to relieve oxidative stress. There is no evidence that it physically interacts with the fatty acid synthesis complex, or that it mediates a metabolic transformation coupled to fatty acid synthesis. It is a second, independent, component of the response to isoniazid.

- *Phylogenetic distribution patterns.* The *phylogenetic profile* of a protein is the set of organisms in which it and its homologues appear. Proteins in a common structural complex or pathway are functionally linked and expected to coevolve. Therefore proteins that share a phylogenetic profile are likely to have a functional link, or at least to have a common subcellular origin. There need be no sequence or structural similarity between the proteins that share a phylogenetic distribution pattern. A welcome feature of this method is that it derives information about the function of a protein from its relationship to *non-homologous* proteins.

BOX 9.8   **Web resource: interaction databases**

Intact: An open source molecular interaction database
http://www.ebi.ac.uk/intact/

DIP: Database of Interacting Proteins
http://dip.doe-mbi.ucla.edu/

MIPS Comprehensive Yeast Genome Database
http://mips.gsf.de/

The MIPS Mammalian Protein–Protein Interaction Database
http://mips.helmholtz-muenchen.de/proj/ppi

BIND: Biomolecular Interaction Network Database
http://www.bind.ca/

MINT: Molecular Interactions Database
http://mint.bio.uniroma2.it/mint/

APID: Agile Protein Interaction DataAnalyzer
http://bioinfow.dep.usal.es/apid/index.htm

BioGRID: Biological General Repository for Interaction Datasets
http://thebiogrid.org/

HPID: The Human Protein Interaction Database
http://www.hpid.org/

3D Complex: a hierarchical classification of protein complexes
http://www.3dcomplex.org/

Lists of links to protein-protein interaction databases:
http://biocreative.sourceforge.net/ppi_links.html
http://openwetware.org/wiki/Protein-protein_interaction_databases

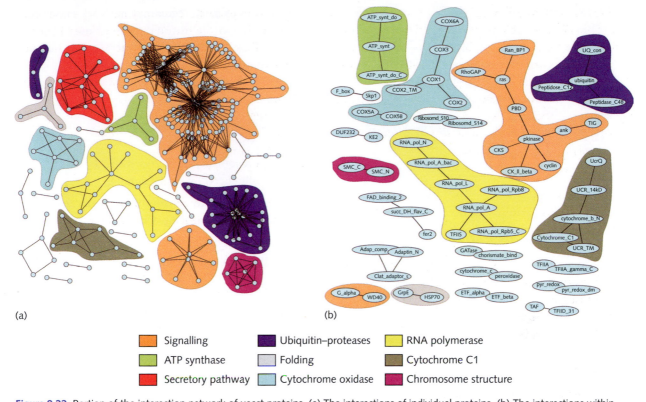

Signalling

ATP synthase

Secretory pathway

Ubiquitin–proteases

Folding

Cytochrome oxidase

RNA polymerase

Cytochrome C1

Chromosome structure

Figure 9.22 Portion of the interaction network of yeast proteins. (a) The interactions of individual proteins. (b) The interactions within a subnetwork based on representations of different protein families, in different functional categories, linked in (a). This figure is based on structural data and modelling. Each relationship implies a physical interaction between the proteins. Some of the interactions involve stable complexes (for instance, RNA polymerase II); others involve transient complexes.

[From: Aloy, P. and Russell, R. (2005). Structure-based systems biology: a zoom lens for the cell. *FEBS Lett.,* **579**, 1854–1858.]

Each of these methods provides a basis for a protein interaction network. The networks formed from each set of interactions are different, although they overlap, to a greater or lesser extent. They give different views of the kinds of relationships between proteins that exist in cells. It is possible to form a more comprehensive network by combining different types of interactions. For instance, the DIP database is a curated collection of experimentally determined protein–protein interactions, http://dip.doe-mbi.ucla.edu/. It contains data about 78 781 interactions between 27 422 proteins from 720 organisms. (For others, see Box 9.8).

Figure 9.22 shows a portion of an interaction network of yeast proteins, based on sets of proteins that have been found together in solved structures.

Regulatory networks

Regulatory networks pervade living processes. Control interactions are organized into linear signal transduction cascades and reticulated into control networks.

Any individual regulatory action requires (1) a stimulus, (2) transmission of a signal to a target, (3) a response, and (4) a 'reset' mechanism to restore the resting state (see Fig. 9.23). Many regulatory actions are mediated by protein–protein complexes. Transient complexes are common in regulation, as dissociation provides a natural reset mechanism.

Some stimuli arise from genetic programs. Some regulatory events are responses to current internal

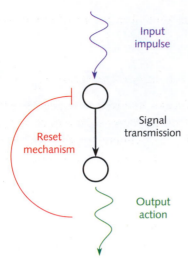

Figure 9.23 The elementary step in a regulatory network. An input impulse is received by an upstream node, which transmits a signal to a downstream node, causing an output action. This is followed by reset of the upstream node to its inactive state. Combination of such elementary diagrams gives rise to the complex regulatory networks in biology.

metabolite concentrations. Others originate outside the cell: a signal detected by surface receptors is transmitted across the membrane to an intracellular target.

Control may be exerted:

- 'in the field' by several mechanisms, such as inhibitors, dimerization, ligand-induced conformational changes, including but not limited to allosteric effects, GDP–GTP exchange or kinase-phosphorylase switches, and differential turnover rates.
- 'at headquarters' through control over gene expression.

One signal can trigger many responses. Each response may be **stimulatory**—increasing an activity— or **inhibitory**—decreasing an activity. Transmission of signals may damp out stimuli or amplify them. There are ample opportunities for complexity, and cells have taken extensive advantage of these.

The signal transduction network exerts control 'in the field' by a variety of mechanisms including inhibitors, dimerization, ligand-induced conformational changes, including but not limited to allosteric effects, GDP–GTP exchange or kinase-phosphorylase switches, and differential turnover rates. This component acts fast, on subsecond timescales. (We discussed G-protein-coupled receptors in Chapter 2.)

The transcriptional regulatory network exerts control 'at headquarters' through control over gene expression. This component is slower, acting on a timescale of minutes.

General characteristics of all control pathways include:

- A single signal can trigger a single response or many responses.
- A single response can be controlled by a single signal or influenced by many signals.
- Each response may be stimulatory—increasing an activity—or inhibitory—decreasing an activity.
- Transmission of signals may damp out stimuli or amplify them.

Structures of regulatory networks

Think of control, or regulatory networks, as assemblies of *activities*. Although mediated in part by physical assemblies of macromolecules—protein–protein and protein–nucleic acid complexes—regulatory networks:

1. *tend to be unidirectional.* A transcription activator may stimulate the expression of a metabolic enzyme, but the enzyme may not be involved directly in regulating the expression of the transcription factor.

2. *have a logical dimension.* It is not enough to describe the connectivity of a regulatory network. Any regulatory action may stimulate or repress the activity of its target. If two interactions combine to activate a target, activation may require *both* stimuli (logical 'and') or *either* stimulus may suffice (logical 'or').

3. *produce dynamic patterns.* Signals may produce combinations of effects with specified time courses. Cell-cycle regulation is a classic example.

The structure of a regulatory network can be described by a graph in which edges indicate steps in pathways of control. Regulatory networks are directed graphs: the influence of vertex A on vertex B is expressed by a directed edge connecting A and B. An edge directed from vertex A to vertex B is called *an outgoing connection* from A and *an incoming connection* to B. Conventionally, an arrow indicates a stimulatory interaction, and a T-symbol indicates an inhibitory interaction. An edge connecting a vertex to itself indicates autoregulation. A double-headed arrow indicates reciprocal stimulation of two nodes; note that this is *not* the same as an undirected edge.

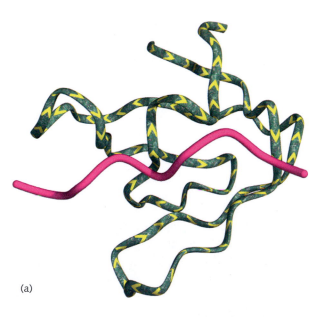

| stimulatory interaction | inhibitory interaction | autoregulatory interaction | reciprocal interaction |

In a complex network, many 'upstream nodes' may govern the activity of nodes connected to them. In addition to the topological complexity of the pattern of connectivity in the network, there is the question of the logic applied to the set of inputs to the target node.

Structural biology of regulatory networks

Many molecules involved in regulation are multidomain proteins. Each domain in a multidomain protein is relatively free to interact with other molecules. An *interaction domain* is a part of a protein that confers specificity in ligation of a partner. Regulatory proteins contain a limited number of types of interaction domains which have diverged to form large families with different individual specificities. For instance, the human genome contains 115 SH2 domains and 253 SH3 domains. (*Src-H*omology domains SH2 and

(a)

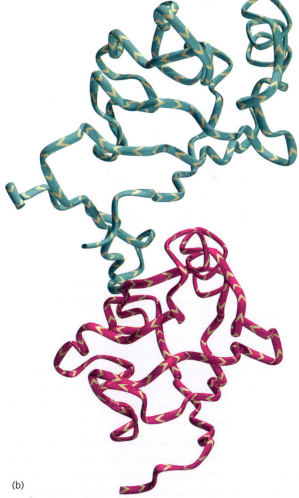

(b)

Figure 9.24 Some types of interactions involved in regulatory signalling. (a) Binding of a peptide (magenta) by an SH3 domain [1CKA]. SH3 domains are common constituents of regulatory proteins. Functions of SH3 domains include signal transduction, protein and vesicle trafficking, cytoskeletal organization, cell polarization, and organelle biosynthesis. (b) Domain–domain interaction: PDZ domains in syntrophin (magenta) and neuronal nitric oxide synthase (cyan) [1QAV].

SH3 are named for their homologies to domains of the src family of cytoplasmic tyrosine kinases.) Many individual interaction domains even interact with different partners as they participate in successive steps of a control cascade. Initial interactions may also trigger recruitment of additional proteins to form large regulatory complexes.

Figure 9.24 shows two types of interaction domain complexes with ligands, binding peptides (which may be attached to proteins), and protein–protein complexes. Other examples have appeared in earlier chapters, including extracellular dimer formation upon binding a hormone (see Fig. 2.29), and protein–nucleic acid complexes (see Chapter 6).

Many interaction domains are sensitive to the state of post-translational modification of their ligands, for instance binding preferentially to states of a ligand in which specific tyrosines, serines, or threonines are phosphorylated. These and other post-translational modifications function as switches, turning on or interrupting/resetting a signalling cascade.

> Regulatory networks:
> - tend to be unidirectional;
> - have a logical as well as a physical component;
> - produce dynamic patterns.

Gene regulation

Cells regulate the expression patterns of their genes. They sense internal cues to maintain metabolic stability, and external cues to respond to changes in the surroundings. The point of contact between genome and expression is the binding of RNA polymerase to promoter sequences, upstream of genes, to initiate transcription. This sensitive point is a juicy target for regulatory interactions (see Box 9.9).

The transcriptional regulatory network of *E. coli*

Investigation of the mechanism of transcription regulation began with the work of F. Jacob and J. Monod on the lac operon in *E. coli*. The field has burgeoned, with comprehensive studies of the *E. coli* regulatory network, together with work on other organisms, notably yeast. *E. coli* contains genes for 4398 proteins, 167 of which are recognized transcription factors. There are 2369 regulatory interactions among the transcription factors and the genes they control (see Fig. 9.25).

There are many fewer known regulatory interactions than genes. Many genes in *E. coli* are organized into operons, under coordinated control—one regulatory interaction controlling many genes. *E. coli* is estimated to contain ~2700 operons. Conclusion: more interactions are still to be discovered.

This network has been the subject of many investigations. Some focus on questions about the *static*

BOX 9.9   **Vocabulary of gene regulation**

- **operator** a control region associated with a gene
- **promoter** a region upstream of a gene, site of RNA polymerase binding to initiate transcription
- **repressor** a DNA-binding protein that blocks transcription
- **operon** a set of tandem genes in bacteria, usually catalysing consecutive steps in a metabolic pathway, under coordinated transcriptional control

- **cis-regulatory region** a segment of DNA that regulates expression of genes on the same DNA molecule. The lac repressor binding site is a *cis*-regulator of the adjacent protein-coding genes *lacZ*, *lacY*, *lacA*
- **transcription start site** position in the gene that corresponds to the first residue in the mRNA
- **constitutive mutant** a mutant defective in repression of a gene, which in consequence is expressed continuously

Combinations of the elementary motifs form modules within the network. These clusters of nodes are often dedicated to control of expression of genes with related physiological functions, such as a group of proteins responding to oxidative stress, or a group involved in aromatic amino acid biosynthesis. Such sets of genes need not be linked as a single operon.

> • Regulatory networks are directed graphs. Some simple motifs, or common small subgraphs, form the lowest level of network structure. Networks can reprogram themselves within an organism and evolve between species.

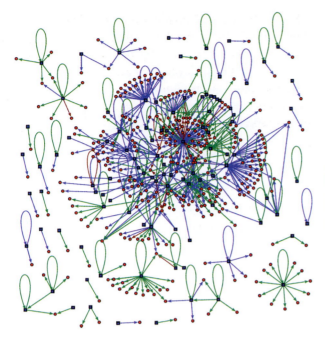

Figure 9.25 The *E. coli* transcriptional regulatory network represented as a directed graph. Colour coding of nodes: transcription factors are shown as blue squares; regulated operons are shown as red circles. Colour coding of links: blue, activators; green, repressors; brown, indeterminate.

[From: Dobrin, R., Beg, Q.K., Barabási, A.L. and Oltvai, Z.N. (2004). Aggregation of topological motifs in the *Escherichia coli* transcriptional regulatory network. *BMC Bioinformat.*, **5**, 10. This is an Open Access article: verbatim copying and redistribution of this article are permitted in all media for any purpose, provided this notice is preserved along with the article's original URL. http://www.biomedcentral.com/1471-2105/5/10/]

Other analyses address the large-scale structure of the network. The distribution of degrees of the nodes, that is, the histogram of the number of edges meeting at a node, follows a power law:

$$\text{number of nodes with } k \text{ edges} \propto 10^{-\beta k},$$

with $\beta = 0.8$. The scale-free topology means that some nodes have many connections, and form the 'hubs' of the network.

Is there substantial feedback from downstream nodes? (A social analogy: is the network hierarchical or democratic? That is, does your boss listen to you, or just give orders? Ring Lardner's classic sentence, 'Shut up, he explained', emphasizes an absence, or dysfunction, of receptors for feedback signals from lower levels back to higher ones.) Although the *E. coli* transcriptional regulatory network contains many autoregulatory interactions, it does not contain larger cycles; that is, paths in which gene A regulates gene B regulates gene C regulates … regulates gene A. Such cycles can lead to instabilities in the dynamics of a network.

Correlation with function of the topology of segments of the network suggests that short pathways, feedforward loops, and repressive autoregulatory interactions are involved in control of metabolic functions, such as a switch to alternative nutrients. This type of network topology is adapted to maintenance of homeostasis. Long hierarchical cascades, and activating autoregulatory interactions, regulate developmental processes, such as biofilm formation and flagellar development involved in mobility and chemotaxis.

topology. Some of these address the local structure of the network, deriving the common types of small subgraphs or the motifs. The fork, scatter pattern, and feedforward loop are motifs in the regulatory networks of *E. coli* and other organisms (see Box 9.10). The network contains a large number of autoregulatory connections. An autoregulatory activator amplifies responses; an autoregulatory repressor damps them out. In the *E. coli* regulatory network, links between different transcription factors are primarily activating, and autoregulatory interactions are often repressive. A high density of repressive autoregulatory interactions increases what might be thought of as the viscosity of the medium in which the network is active. The 'one-two punch', or feedforward loop, motif can also act to filter out random fluctuations, preventing the propagation of noise (see Fig. 9.26).

BOX 9.10   Common motifs in biological control networks

Within the high complexity of typical regulatory networks, certain common patterns appear frequently. In the architecture of networks, these form building blocks that contribute to higher levels of organization. Shen-Orr, Milo, Mangan, and Alon* have described examples including the *fork*, the *scatter*, and the *'one-two punch'* (a phrase from the boxing ring):

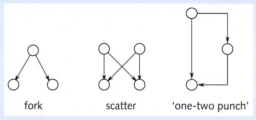

The *fork*, also called the single-input motif, transmits a single incoming signal to two outputs. Successive forks, or forks with higher branching degrees, are an effective way to activate large sets of genes from a single impulse. Generalizations of the binary fork include more downstream genes under common control (more tines to the fork), and autoregulation of the control node. Forks can achieve general mobilization. Moreover, if the regulatory genes have different thresholds for activation, the dynamics of building up the signal can produce a temporal pattern of successive initiation of the expression of different genes.

The *scatter* configuration, also called the multiple input motif, can function as a logical 'or' operation: both downstream targets become active if *either* of the input impulses is active. Generalizations of the square scatter pattern shown may contain different numbers of nodes on both layers. Note that scatter patterns are superpositions of forks.

The *'one-two punch'*, also called the 'feedforward loop', affects the output both directly through the vertical link and indirectly and subsequently through the intermediate link. This motif can show interesting temporal behaviour if activation of the target requires simultaneous input from both direct and indirect paths (logical 'and'). Because build-up of the

*Shen-Orr, S.S., Milo, R., Mangan, S. and Alon, U. (2002). Network motifs in the transcriptional regulation network of *Escherichia coli.Nat. Genet.*,**31**, 64–68.

messenger molecule requires time, the direct signal will arrive before the indirect one. Therefore a short pulsed input to the complex will not activate the output—by the time the indirect signal builds up, the direct signal is no longer active. The system can thereby filter out transient stimuli in noisy inputs (see Fig. 9.26). Conversely, the active state of the system can shut down quickly upon withdrawal of the external trigger.

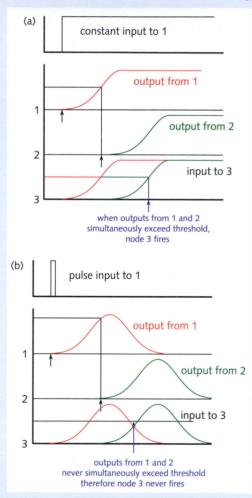

Figure 9.26 A 'one-two punch', or feedforward loop, equipped with suitable AND logic at the downstream node, can filter out transient noise. (a) Constant input. (b) Pulse input.

The dynamic properties of the network are also of interest. These include both the response of the network to changing conditions, as in the lac operon, and the comparison of regulatory networks in related organisms to understand how networks evolve.

Even a constant network can produce different outputs from different inputs. Moreover, even within an organism, networks can change their structure in response to changes in conditions. This can affect even some of the hubs of the network, the points at which changes have the most far-reaching effects. This has been examined most closely in yeast (see Box 9.11).

Similarities and differences in the regulatory interactions in related organisms illuminate how their networks evolve. The evolutionary retention of transcription factors is smaller than that of target genes. Even transcription factors that serve as hubs are not more highly conserved. Different organisms are relatively free to explore different regulatory pathways, even to regulate orthologous genes. This may well be the 'other side of the coin' of the redundancy in the networks that provides robustness.

There is evidence that larger changes in regulatory networks reflect changes in lifestyle: organisms with similar lifestyle—several species of soil bacteria, genuses *Bacillus, Corynebacterium,* and *Mycobacterium*—conserve regulatory interactions, as do intracellular parasites *Mycoplasma, rickettsiae,* and *chlamidiae.* Conversely, comparing organisms that adopt different lifestyles, individual transcription factors can be deleted.

Regulation of the lactose operon in *E. coli*

The lactose operon of *E. coli* contains structural genes and regulatory regions (see Fig. 9.27(a)). Three regions of the operon encode proteins (see Table 9.3). *lacZ, lacY, lacA* are cotranscribed into a single mRNA and separately translated.

The *lacI* gene, encoding the lac repressor protein, lies upstream of the operator. It is constitutively expressed.

The control regions of the lactose operon function as a switch, turning on and off transcription of the protein-encoding genes. Control is exerted through binding of CAP protein (catabolite activation protein) to the CAP site, and lac repressor protein to the operator site. Cyclic AMP (cAMP, produced in higher quantities if glucose is absent) and lactose analogues bind to CAP and lac repressor proteins, respectively, to control their binding affinities:

- Binding lactose induces a conformational change in the repressor, from a tightly binding form to a weakly binding one. *Lac repressor will bind promoter only in the absence of lactose.*

- The presence of glucose reduces the concentration of cyclic AMP, causing a conformational change in CAP to a weakly binding form. *CAP will bind promoter only in the absence of glucose.*

The actual molecule that binds to repressor to reduce its affinity for its site on DNA is *allolactose.* Allolactose is an isomer of lactose, produced from lactose by β-galactosidase. Alternative lactose analogues also stimulate transcription. One that is useful in the laboratory is isopropylthiogalactoside (IPTG), for two reasons: (1) IPTG enters the cell even if the *lacY*-encoded transporter is dysfunctional or not expressed, and (2) because IPTG is not metabolized, its concentration stays constant during the course of an experiment.

The switch responds to the type of sugar in medium:

- If both glucose and lactose are present, neither control protein binds to the promoter (see Fig. 9.27(a)). RNA polymerase binds only weakly. Transcription occurs at a low basal level.

- If glucose is not present, and lactose *is* present, the CAP-cAMP complex binds to the promoter. The binding of CAP-cAMP with RNA polymerase is cooperative. Interactions with CAP-cAMP increase the affinity of RNA polymerase, thereby stimulating transcription to approximately 40 times the basal level (see Fig. 9.27(b)).

- If lactose is not present, the repressor binds to the operator site. This blocks RNA polymerase and turns off transcription (see Fig. 9.27(c)).

Table 9.3 Genes in the lactose operon of *E. coli*

Gene	Enzyme	Function
lacZ	β-galactosidase	hydrolyses lactose → glucose + galactose, *and* isomerizes lactose → allolactose
lacy	β-galactoside permease	pumps lactose into cell
lacA	β-galactoside transacetylase	unknown, possibly detoxification?

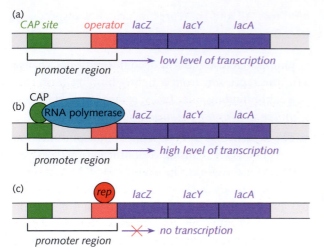

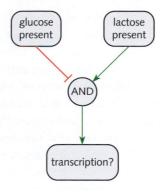

Figure 9.28 Logical diagram of lac operon. Green arrows show positive regulation. The red 'tee' shows negative regulation. The circle containing AND will pass through a positive signal only if glucose is NOT present (red 'tee') AND lactose IS present (green arrow). For many regulatory circuits, we know many of the inputs to a node but do not know the logic.

Figure 9.27 States of the lactose operon. (a) The promoter region contains regulatory sites upstream of protein-encoding genes *lacZ,lacY,lacA*. (b) Binding of CAP to its upstream site within the promoter enhances the binding affinity of RNA polymerase, turning transcription on. (c) Binding of repressor (rep) blocks binding of RNA polymerase, turning transcription off. Two additional subsidiary repressor binding sites are not shown. (Figure 9.27(c) is a simplification: lac repressor actually binds to three sites. The cover of the 1 March 1996 issue of *Science* magazine shows a model of the lac repressor, binding to two sites on a ~100-bp section of DNA, plus the CAP protein. The DNA is bent into a loop.)

The effect is to express the proteins of the lac operon only if there is only lactose in the medium. The bacteria are saying, in effect, 'We prefer to grow on glucose. If glucose is there, we don't want high expression levels of the genes for lactose transport and metabolism, *even if* lactose is present. Only if lactose is present and glucose is not present, express the genes that transport and cleave lactose.'

The lactose operon switch is an example of a 'fire-and-forget' mechanism. Once the messenger RNA is synthesized, what happens on the DNA does not affect it. However, the mRNA for the protein-coding genes of the lac operon (*lacZ, lacY, lacA*) has a half-life of ~3 min. In the absence of continuous or repeated induction, synthesis of the lactose-metabolizing enzymes will cease within minutes. This resets the switch.

Logical diagram of lac operon

Figure 9.28 represents the lac operon control logic as a network. This diagram is almost equivalent to the table showing the response to the presence and absence of glucose.

In summary (−means 'absence of'; + means 'presence of'):

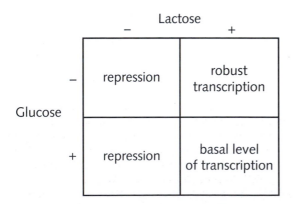

The operator site is between the CAP site and the origin of transcription. As a result, *lactose absence 'trumps' glucose absence*. That is, in the absence of lactose, the binding of repressor stops transcription whether or not glucose is absent.

- The lactose operon in *E. coli* encodes several proteins involved in the uptake and utilization of lactose. Control mechanisms activate this operon for transcription when the medium contains lactose but not glucose.

The genetic regulatory network of *Saccharomyces cerevisiae*

A recent study of transcription regulation in yeast by N.M. Luscombe, M.M. Babu, H. Yu, M.Snyder, M., S.A.Teichmann, and M. Gerstein treated a network containing 3562 genes, corresponding to approximately half the known proteome of *S. cerevisiae*. The genes included 142 that encode transcription regulators, and 3420 that encode target genes exclusive of transcription regulators. There are 7074 known regulatory interactions among these genes, including effects of regulators on one another, and of regulators on non-regulatory targets.

Analysis of the overall network architecture reveals that:

- The distribution of incoming connections to target genes has a mean value of 2.1, and is distributed exponentially. Most target genes receive direct input from about two transcriptional regulators. The probability that a gene is controlled by k transcription regulators, $k = 1,2,\ldots$, is proportional to $e^{-\alpha k}$, with $\alpha = 0.8$.

- The distribution of outgoing connections has a mean value of 49.8, and obeys a power law. The probability that a given transcriptional regulator controls k genes is proportional to $k^{-\beta}$, with $\beta = 0.6$. Power-law behaviour characterizes topologies in which a few nodes—the 'hubs'—have many connections and many nodes have few. In regulatory networks, hubs tend to be fairly far upstream, forming important foci of regulation with far-reaching control.

- The average number of intermediate nodes in a minimal path between a transcriptional regulator and a target gene is 4.7. The maximal number of intermediate nodes in a path between two nodes is 12.

- The clustering coefficient of a node is a measure of the degree of local connectivity within a network. If all neighbours of a node are connected to one another, the clustering coefficient of the node = 1. If no pairs of neighbours of a node are connected to each other, the clustering coefficient of the node = 0. The mean clustering coefficient, averaged over all nodes, is a measure of the overall density of the network. For the yeast transcriptional regulatory network, the mean clustering coefficient is 0.11.

Figure 9.29 is a cartoon-like sketch of a fragment of such a network, indicating rather loosely some of its general features. Nodes are divided into *transcriptional regulators,* shown as circles, and *target genes,* shown as squares. Target genes are distinguished by having no output connections. There is extensive interregulation among the transcription factors, to a much higher density of interconnections than can intelligibly be shown in this diagram. Think of a seething broth of transcription factors, within the shaded area, sending out signals to target genes. The shaded area indicates only the *logical* clustering of the transcriptional regulators. There is no suggestion about physical localization; indeed, transcriptional regu-

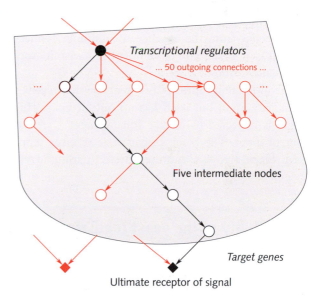

Five intermediate nodes

Target genes

Ultimate receptor of signal

Figure 9.29 Simplified sketch illustrating some features of an 'average' segment of the pathways in the yeast interaction network. Transcriptional regulators appear as circles. Target genes appear as squares. A transcriptional regulator typically has direct influence over about 50 genes, indicated by multiple connections from the filled black circle to the circles on the line below it. Roughly one in 10 of the neighbours of any node is connected to another neighbour, indicated by the horizontal arrow on the second row. The ultimate receptor of the signal lies at the end of a pathway typically containing about five intermediate nodes (shown in black). This ultimate target gene receives on average about two inputs. This diagram shows only a small fragment of a network that is in fact quite dense. The diagram also suggests that there is no 'feedback' in the form of cycles; in fact cycles are present but they are rare.

lators interact with DNA, and almost never interact physically with the proteins, the expression of which they control.

Each transcriptional regulator directly influences approximately 50 genes on average, although, as with other 'small-world' networks following power-law distributions of connectivities, the distribution is very skewed—some 'hubs' have very many output connections, but most nodes have very few. A few of the interregulatory connections between transcription factors are shown in black. In about 10% of the cases, two neighbours of the same transcription factor interact with each other. A pathway from one regulator (filled black circle) to one ultimate receptor (filled black square), through five intermediate nodes, is shown in black. The intermediate nodes are other transcriptional regulators, connected both within the path drawn in black and off this path. Even the transcription factor used as the origin of the path receives input connections. Although it is possible to identify target genes from the absence of outgoing connections, it is more difficult to identify ultimate initiators of signal cascades.

The ultimate receptor is a target gene that receives regulatory input but itself has no output links. This target is expected to receive (on average) a second control input. The black target node receives input via a black arrow, along the selected path, and via a red arrow, suggesting the second input. Of course the second input may arrive via a path that shares common nodes with the black path, including other routes from the filled black circle.

The dense forest of additional pathways, from which this fragment is extracted, is not shown. A 'back-of-the-envelope' calculation: there are ~3500 nodes, each receiving on average two input connections. There are ~140 transcription factors, making an average of 50 output connections. The number of input connections must equal the number of output connections, and indeed $3500 \times 2 = 140 \times 50 = 7000$.

The high ratio of interactions to transcription regulators implies that we cannot expect to associate individual regulatory molecules with single, dedicated, activities (as we can, for the most part, with metabolic enzymes). Instead, the activity of the network involves the coordinated activities of many individual regulatory molecules.

The yeast regulatory network involves:

- regulatory genes: the products of regulatory genes affect the transcription of other genes
- target genes: the products of target genes do not regulate other genes. The network is dense; each target gene feels the influence of a large number of regulatory genes.

Adaptability of the yeast regulatory network

The yeast regulatory network achieves versatility and responsiveness by reconfiguring its activities. This is seen by comparing the changes in the activities of networks controlling yeast gene expression patterns in different physiological regimes: cell cycle, sporulation, diauxic shift (the change from anaerobic fermentative metabolism to aerobic respiration as glucose is exhausted), DNA damage, and stress response. Cell cycling and sporulation involve the unfolding of endogenous gene expression programs; the others are responses to environmental changes (see Box 9.11).

O_2 is essential for aerobic life, yet its reduced forms include some of the most toxic substances with which cells must cope.

Different states are characterized by both similarities and differences in gene expression patterns, and by the components of the regulatory network that are active. There is considerable shift in expression of target genes. About a quarter of the target genes are specialized to individual physiological states. That is, of the total of 3420 target genes, the expression of almost half (1514) do not show major changes in the different states. Of the 1906 that show altered expression levels in different states, almost half of them (803) are specialized to a single physiological state.

In contrast, different states show much more overlap in the usage of transcriptional regulators. For instance, for cell-cycle control, 280 target genes (8%) are differentially regulated by 70 (49%) of

BOX 9.11   The diauxic shift in *Saccharomyces cerevisiae*

Yeast is capable of adapting its metabolism to a variety of environmental conditions. In the presence of glucose, *S. cerevisiae* will—even in the presence of oxygen—preferentially use the Ebden–Meyerhof fermentative pathway, reducing glucose to ethanol. Exhaustion of available glucose produces the 'diauxic shift', to oxidative metabolism, sending the products of fermentation through the Krebs (tricarboxylic acid) cycle and mitochondrial oxidative phosphorylation (see Fig. 9.30).

The shift is not merely a redirection of metabolic flux through alternative pathways using pre-existing proteins, but involves a substantial change in expression pattern. Many genes are involved, not only enzymes in the awakened metabolic pathways. Another consequence of switching to respiratory metabolism is the danger of oxidative damage, and the oxidative stress response requires enhanced expression of many other genes.

The cells are effectively sensing the level of glucose. As glucose itself acts as a repressor of expression of many genes, depletion of glucose releases this repression—'turning on' a variety of genes.

An important component of the mechanism by which high glucose levels lead to repression is the state of phosphorylation of a transcriptional regulatory protein called Mig1. The phosphorylated form of Mig1 stays in the cytoplasm. On dephosphorylation it enters the nucleus and binds to promotors of various genes, repressing their expression. The activity of kinase Snf1-Snf4, which phosphorylates Mig1, depends on the AMP/ATP concentration ratio in the cell. Growth of high levels of glucose generates high levels of ATP via glycolysis, keeping the Snf1-Snf4 kinase inactive. This leaves Mig1 in its active, dephosphorylated form, repressing glucose-sensitive genes.

In a seminal paper in 1997, J.L. DeRisi, V.R. Iyer, and P.O. Brown reported on the changes in gene expression in the diauxic shift in *S. cerevisiae*. They observed that the patterns of gene expression were stable during exponential growth in a glucose-rich medium. This justifies considering the initial anaerobically growing population as being in a *state* that can be characterized by a set of expression levels of genes. (If, alternatively, fluctuations in expression levels were large compared to the differences between anaerobic and aerobic regimes, it would not be possible to analyse the diauxic shift in terms of expression patterns. In fact, if there are fluctuations, they average out over the population.)

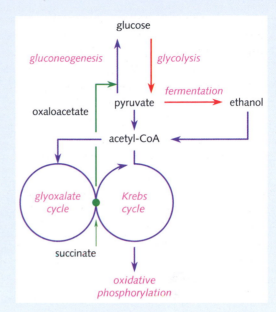

Figure 9.30 Some metabolic pathways in yeast affected by the diauxic shift. In the presence of ample glucose, yeast will adopt an anaerobic metabolic regime, converting glucose to ethanol via glycolysis and fermentation (red). On running out of glucose it will shift to an aerobic metabolic state in which ethanol is converted to CO_2 and H_2O, via the Krebs cycle and oxidative phosphorylation (blue). For the alternative pathways of energy release, pyruvate is the branch compound. However, the shift to a utilization of a different energy source creates two concomitant problems: (1) The oxidation of ethanol does not provide precursors for essential biosynthetic pathways. Oxidation of ethanol converts all carbon to CO_2. Some must be retained and converted to 3- and 4-carbon compounds, and even glucose, via the glyoxylate cycle and gluconeogenesis. Acetyl-CoA is the branch compound for this shift—it enters *both* the Krebs cycle and the glyoxylate cycle. The glyoxylate cycle shares several intermediates with the Krebs cycle, but, to conserve carbon, leaves out the decarboxylations that produce CO_2. Succinate (green dot), one of the metabolites common to the Krebs and glyoxylate cycles, is converted to oxaloacetate in mitochondria. It then feeds into numerous biosynthetic pathways (green arrow). (2) In addition to the metabolic pathways shown in this figure, yeast must also activate pathways of defence against oxidative stress. The danger is from potential chemical attack by *reactive oxygen species*, produced by partial reduction of O_2: the superoxide radical O_2^-, hydrogen peroxide H_2O_2, and the hydroxyl radical $OH^•$. In its defence, yeast increases expression of genes involved in detoxification of reactive oxygen species. The mechanism of induction of these genes depends on the transcription factor Yap1p. Increase in the concentration of reactive oxygen species leads to formation of disulphide bridges in Yap1p. This causes a conformational change that masks a nuclear export signal. The result is redistribution of the transcription factor to the nucleus, the site of its activity.

As glucose is depleted, the expression pattern changes (see Fig. 9.31, from work by M.J. Brauer, A.J. Saldanha, K. Dolinski, and D. Botstein). The changes affect a large fraction, almost 30%, of the genes:

Expression ratio in aerobic/anaerobic states	>2	<$\frac{1}{2}$	>4	<$\frac{1}{4}$
Number of genes	710	1030	183	203

The genes, differentially expressed, are associated with proteins in several functional classes. On converting from anaerobic to aerobic metabolism:

- *Metabolic pathways are rerouted.* Synthesis of pyruvate decarboxylase is shut down; synthesis of pyruvate carboxylase is enhanced. This switches the reaction of pyruvate from acetaldehyde to oxaloacetate. Enhanced synthesis of genes for fructose-1,6-bisphosphatase and phosphoenolpyruvate carboxylase changes the direction of two steps in glycolysis. Expression is enhanced for genes encoding the enzymes that carry out the new Krebs and glyoxylate cycle reactions, and oxidative phosphorylation (blue arrows in Fig. 9.30).

- *Genes related to protein synthesis show a decrease in expression level.* These include genes for ribosomal proteins, tRNA synthetases, and initiation and elongation factors. An exception is that genes encoding *mitochondrial* ribosomal molecules have generally enhanced expression.

- *Genes related to a number of biosynthetic pathways show reduced expression.* These include genes encoding enzymes of amino acid and nucleotide metabolism.

- *Genes involved in defence against oxidative stress from reactive oxygen species show enhanced expression.* Proteins encoded include catalases, peroxidases, superoxide dismutases, and glutathione S-transferases.

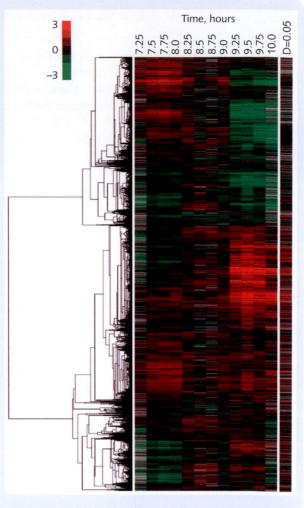

Figure 9.31 Expression patterns of yeast growing on glucose, then undergoing diauxic shift. Column headings indicate time in hours after exhaustion of glucose in the samples undergoing diauxic shift.

[From: Brauer, M.J., Saldanha, A.J, Dolinski, K. and Botstein, D. (2005). Homeostatic adjustment and metabolic remodelling in glucose-limited yeast cultures. *Mol. Biol. Cell*, **16**, 2503–2517.]

the transcription regulators. Clearly there is a much greater degree of specialization in the target genes. In general, half the transcription factors are active in at least three out of the five physiological regimes. However, in contrast with the high overlap of usage of the transcriptional regulators (the nodes), the overlap of the activities within the network (the connections) is relatively low. Different components of the interaction network organize the different gene expression patterns in different states.

Whereas different physiological states are characterized by substitutions of different sets of synthesized proteins, the regulatory network uses much of the same structure but reconfigures the pattern of activity. Think of the transcription factors as 'hardware' and the connections as reprogrammable 'software'. The molecules do not change but the interactions do: in different states, many transcription regulators change most, or a substantial part, of their interactions. In particular, the set of transcription regulators that form the hubs of the network—those with many outgoing nodes that form foci of control—are not a constant feature of the system. Some hubs are common to all states, but others step forward to take control in different physiological regimes. The result of the reconfiguration of activity is that over half of the regulatory interactions are *unique* to the different states.

The effect of the changes in the active interaction patterns is to alter the topological characteristics of the network in different states. For instance, under panic conditions—DNA damage and stress—the average number of genes under control of individual transcriptional regulators increases, the average minimal path length between regulator and target decreases, and the clustering becomes less dense (that is, there is less interregulation among transcription factors). This can be understood in terms of a need for fast and general mobilization—the equivalent of broadcasting 'Go! Go! Go!' over the radio. Normal circumstances—cell-cycle control for instance—allow for a more dignified and precise regulatory state, which permits finer control over the temporal course of expression patterns. In cell-cycle control and sporulation, there is a much denser interregulation among transcription factors, and longer minimal path lengths between transcriptional regulators and target genes.

Different physiological states also differ in their usage of the common motifs—fork, scatter, and 'one-two punch' (see Box 9.10). Scatter motifs are more used in conditions of stress, diauxic shift, and DNA damage. They are appropriate to the need for quick action. Requirements for build-up of intermediates would delay the response. Conversely, the 'one-two punch' motif is more common in cell-cycle control. This is consistent with the need for a signal from one stage to be stabilized before the cell enters the next stage.

Much of evolution proceeds towards greater specialization. The human eye is a classic example. It is an intricate and fine-tuned structure, features that were once adduced as evidence *against* Darwin's theory. Many evolutionary pathways show a trade-off between specialized adaptation and generalized adaptability.

Regulatory networks are an exception. The reconfigurability of regulatory networks allows them to respond robustly to changes in conditions by creating many different structures, specialized to the conditions that elicit them. Evolution has produced structures that are both specialized *and* versatile.

● RECOMMENDED READING

Tyers M. and Mann M. (2003). From genomics to proteomics. *Nature,* **422**, 193–7. An overview of high-throughput methods and applications.

Cutler, P. and Voshol, H. (2015) Proteomics in pharmaceutical research and development. *Proteomics Clin. Appl.,* **9**, 643–50. The history and growing role of proteomics in drug discovery and development.

Albert, R. and Barabási, A.-L. (2002). Statistical mechanics of complex networks. *Rev. Mod. Phys.,* **74**, 47–97. A technical exposition of the main models and analytic tools that govern interaction networks, not limited to those appearing in molecular biology.

Barabási, A.-L. (2003). *Linked: How everything is connected to everything else and what it means.* Plume Books, New York. A book for general readers, by important contributors to the field of systems biology.

Mutz, K.-O., Heilkenbrinker, A., Lönne, M., Walter, J.-G. and Stahl, F. (2013). Transcriptome analysis using next-generation sequencing. *Curr. Opin. Biotechnol.,* **24**, 22–30. Next-generation sequencing and applications to transcriptomics via RNAseq.

Wang, Z., Gerstein, M. and Snyder, M. (2009). RNA-Seq: a revolutionary tool for transcriptomics. *Nat. Rev. Genet.,* **10**, 57–63. Statement of the nature of the methods and results of RNAseq.

Chmielecki, J. and Meyerson, M. (2014). DNA sequencing of cancer: what have we learned? *Ann. Rev. Med.*, **65**, 63–79. Cancer genomics is a now a very heavy industry, the results of which are already guiding clinical practice.

Ozsolak, F. and Milos P.M. (2011) RNA sequencing: advances, challenges and opportunities, *Nature Rev Genet.*, **12**, 87–98. New techniques promise the ability to sequence RNA directly, without conversion to complementary DNA.

Ideker, T. (2004). A systems approach to discovering signaling and regulatory pathways—or, how to digest large interaction networks into relevant pieces. *Adv. Exp. Med. Biol.*, **547**, 21–30. Systems biology of important biological networks.

Babu, M.M., Luscombe, N.M., Aravind, L., Gerstein, M. and Teichmann, S.A. (2004). Structure and evolution of transcriptional regulatory networks. *Curr. Opin. Struct. Biol.*, **14**, 283–91. A review article including information about how networks evolve.

Bechhoefer, J. (2005). Feedback for physicists: A tutorial essay on control. *Rev. Mod. Phys.*, **77**, 783–836. Description of the theoretical background of network function.

Lacroix, V., Cottret, L., Thébault, P. and Sagot, M.-F. (2008). An introduction to metabolic networks and their structural analysis. *IEEE/ACM Trans. Comput. Biol. Bioinform.*, **5**, 594–617. An introduction and review of structural aspects of metabolic networks. Describes the concepts that have been used to analyse metabolic networks, and addresses some of the controversies that are still active.

● EXERCISES AND PROBLEMS

Exercise 9.1 Hen egg-white lysozyme has a relative molecular mass of about 14 300. If mass spectroscopy can measure mass to within 0.01%, could the following be confidently distinguished from the unmodified protein: (a) N-terminal acetylation, (b) phosphorylation of a single serine residue, (c) a single Lys→Gln substitution?

Exercise 9.2 On photocopies of Fig. 9.8, indicate the positions of the peaks if the sequence were: (a) MNLVQVR, (b) GNLQVVR, (c) MNLQVVG.

Exercise 9.3 (a) What is the sequence of the fragment y_6 in Fig. 9.8(b)? (b) To which peak in Fig. 9.8(b) does the fragment NH^{3+} LQVVR correspond?

Exercise 9.4 Oligonucleotide samples may vary by the binding of a Na^+ or K^+ ion to a phosphate, instead of a proton. (a) What is the difference in mass between an oligonucleotide binding a proton or a Na^+ ion at a single site? (b) What base change has the closest mass difference to the H^+–Na^+ mass difference? (c) Would measuring mass to within 1 D be sufficient accuracy to distinguish this base change from the binding of a Na^+ ion instead of a proton, at a single site? (d) In a mass spectrum of an oligonucleotide, what is the difference in mass between an oligonucleotide with a proton or a Mg^{2+} ion at a single site? (e) What base change has the closest mass difference to the H^+–Mg^{2+} mass difference? (f) Would measuring mass to within 1 D be sufficient accuracy to distinguish this base change from the binding of a Mg^{2+} ion instead of a proton, at a single site?

Exercise 9.5 Assuming a typical single-nucleotide polymorphism (SNP) density of 1 SNP/5 kb in a human genome, and only two possible bases observed at the position of any SNP, how many sequences could you expect to find throughout a population, within a 100-kb region, if recombination were common at every position in the region? If only three of the possible combinations of SNPs—that is, three haplotypes—are observed, what fraction of possible sequences does this represent?

Exercise 9.6 For which of the methods for determining interacting proteins (see Protein interaction networks section) (a) must one of the proteins be purified and (b) must both of the proteins be purified?

Exercise 9.7 On a photocopy of Fig. 9.11, indicate two spots that correspond to proteins that are upregulated after 15 min, but return to normal levels after 3 h.

Exercise 9.8 Cells are grown on a medium containing heavy lysine (^{13}C and ^{15}N). *All* lysine carbons are ^{13}C and both lysine nitrogens are ^{15}N. If a fragment of the protein contains a single lysine, what is the mass difference between the fragment derived from the protein from the cells grown on heavy lysine and the corresponding one derived from cells grown on unlabelled lysine?

Exercise 9.9 In the undirected, unlabelled graph in Box 9.4, (a) name two vertices such that if you add an edge between them at least one vertex has exactly four neighbours. (Note that two edges may cross without making a new vertex at their point of intersection.) (b) Name two vertices such that if you add an edge between them to the graph, the connected subgraph remains a tree. (c) Name two vertices (neither of them V_1) such that if you add an edge between them to the graph, the connected subgraph does not remain a tree. (d) Name two vertices such that if you add an edge between them to the original graph, there are alternative paths, of lengths 3 and 4, between V_1 and V_5, with no vertices repeated. (In determining the length of a path, you have to count the initial and final vertices. A path of length 3 between V_1 and V_5 contains one intermediate vertex.) (e) Name two vertices such that if you add an edge between them to the original graph, there is exactly one path between V_1 and V_5, with no vertices repeated, and it has length 4.

Exercise 9.10 Of the examples of graphs in Box 9.5, (a) Which are directed graphs? (b) Which are labelled graphs? (c) In each example, what is the set of nodes? (d) In each example, what is the set of edges?

Exercise 9.11 In the London Underground: (a) What is the shortest path between Moorgate and Embankment stations? Note that, considered as a graph, the shortest path between two nodes is the path with the fewest intervening nodes, not the path that would take the minimal time or fewest interchanges. (b) What is the shortest cycle containing Kings Cross, Holborn, and Oxford Circus stations? (c) The clustering coefficient of a node in a graph is defined as follows. Suppose the node has k neighbours. The total possible connections between the neighbours is $k(k-1)/2$. The clustering coefficient is the observed number of neighbours divided by this maximum potential number of neighbours. If the neighbours of a station are the other stations that can be reached without passing through any intervening stations, what is the clustering coefficient of the Oxford Circus station? (If necessary, see http://www.afn.org/~alplatt/tube.html.)

Exercise 9.12 In the London Underground (a) what is the maximum path length between any two stations? That is, for which two stations does the shortest trip between them involve the maximum number of intervening stops? (b) If the District Line were not active, what stations, if any, would be inaccessible by underground? (c) If the Jubilee line were not active, what stations, if any, would be inaccessible by underground?

Exercise 9.13 On a photocopy of the three common network control motifs (see Box 9.10) (a) indicate which nodes are controlled by only *one* upstream node; (b) indicate which node exerts control over only *one* downstream node.

Exercise 9.14 On a photocopy of the simplified fragment of the yeast regulatory network (see Fig. 9.29) indicate examples of the network control motifs (a) fork and (b) 'one-two punch'. (c) Add one arrow to create a scatter motif.

Exercise 9.15 (a) In the dimer between syntrophin and neuronal nitric oxide synthase (see Fig. 9.24b), is the dimer structure open or closed? (b) What secondary structure element is shared between the two domains?

Exercise 9.16 In the overall yeast transcriptional regulatory network the number of incoming connections to target genes follows an exponential distribution, that is, the probability that a gene is controlled by k transcriptional regulators is proportional to $e^{-\beta k}$ with $\beta = 0.8$, $k = 1, 2,....$ What is the ratio of the number of target genes receiving four input connections to the number receiving two input connections?

Problem 9.1 (a) How many positions in all are there in the microarray in Fig. 1.18? (b) How many are complementary to RNAs from liver? (c) How many are complementary to RNAs from brain? (d) How many are complementary to RNAs from liver and brain? (e) How many are complementary to neither?

Problem 9.2 For dissociation of a complex involving a simple equilibrium: $AB \rightleftharpoons A + B$, the equilibrium constant, $K_D = \dfrac{[A][B]}{[AB]}$, is equal to the ratio of forward and reverse rate constants: $K_D = k_{off}/k_{on}$. For avidin–biotin, $K_D = 10^{-15}$. Suppose k_{on} were as fast as the diffusion limit, $\sim 10^{-9}$ M s^{-1}. (a) What is the value of k_{off}? (b) What would be the half-life of the avidin–biotin complex? (c) Suppose k_{on} for avidin–biotin were 10^{-7} M s^{-1}. What would be the half-life of the complex?

Problem 9.3 In the map of the London Underground, what is the distribution of numbers of neighbours of vertices?

Problem 9.4 Analyse the map of the London Underground by counting the number of connections made from each station in zone 1 (the central portion). Count connections *to* stations inside and outside zone 1 as long as they originate within zone 1. Count only one connection if two stations are connected by more than one line; in other words, for each station, the question is: how many other stations can be reached without passing through any intermediate stops? (a) What is the maximum number of connections of any station? (b) For each integer k from 1 to this maximum number, how many stations have k connections? (c) Plot these data on a log–log plot. Does the relationship appear reasonably linear? (d) If so, fit a straight line to the log–log plot and determine the exponent. Results of network analysis of this sort are more significant if the data cover several orders of magnitude, but this is not possible for this example.

Problem 9.5 In the overall yeast transcriptional regulatory network the number of incoming connections to nodes follows an exponential distribution. That is, the probability P_k that a gene is controlled by k transcription regulators is given by $P_k = Ce^{-\beta k}$, $k = 1, 2,...$, with $\beta = 0.8$. (a) Determine the constant of proportionality C in terms of β, by summing the series $\sum_{k=1}^{\infty} Ce^{-\beta k} = 1$. (b) If $\beta = 0.8$, what is the maximum value of k for which at least 1% of the nodes would be expected to have at least k incoming connections? (c) If $\beta = 0.8$, plot the expected histogram for $1 \leq k \leq 7$. (d) Determine the mean value of k in terms of β. [Hint: In the solution of (a) you expressed $\sum_{k=1}^{\infty} e^{-\beta k}$ as a function $f(\beta)$. Differentiate this relationship with respect to β to produce the equation: $\sum_{k=1}^{\infty} ke^{-\beta k} = f'(\beta)$. Then the mean value of k is given by $-f'(\beta)/f(\beta)$.] (e) What is the mean value $<k>$ corresponding to $\beta = 0.8$? (f) What is the median value of k? This is the value k such that half the nodes have $\leq k$ incoming connections, and half the nodes have $\geq k$ incoming connections. Find k in terms of β. [Hint: if $\sum_{k=1}^{\infty} Ce^{-\beta k} = 1$, then $\sum_{k=\kappa+1}^{\infty} Ce^{-\beta k} = \dfrac{1}{2}$, but $\sum_{k=\kappa+1}^{\infty} Ce^{-\beta k} = e^{-\beta k} \sum_{k=1}^{\infty} Ce^{-\beta k}$. In general, this approach will provide a non-integral estimate of k, just round this result to the nearest integer.] (g) If $\beta = 0.8$, what is the median value k? How does it compare with the average value $<k>$? Are the two values approximately equal?

Problem 9.6 Indicate how to connect a selection of the three common network control motifs so that a single-input node can influence three output nodes.

Many years ago, James Watson asserted that molecular biologists were 'spoiled' by the structure of DNA: the logic of the activity of DNA was so obvious, that when protein structures came along, their complexity was frustrating in comparison. And that was when haemoglobin was people's idea of a protein with a complicated mechanism of action.

The perspective has shifted. For one thing, although unquestionably the basic logic of how DNA stores and replicates information is implicit in the double helix, the mechanisms of transcription and translation are anything but simple. One can't even blame it all on proteins: remember that the ribosome contains a ribozyme!

It is undeniable that there has been progress in understanding the activity of more and more complex combinations—larger macromolecular aggregates, with complex conformational changes; and more densely reticulated physical and regulatory networks. Again the ribosome is a prime example. But the real difficulty is that the problems lying at the heart of life processes involve dynamic patterns of interaction, in space and in time, between the components. Individual sequences and structures give us pieces of the puzzle. Knowing full genomes and thousands of protein structures denies us the excuse that we're missing too many pieces. But these are only the static and isolated data. To integrate them will require new kinds of information. These new data will be multi-dimensional: revealing networks of interaction and control, and their shifting patterns, in space and time.

For the last 150 years or so, we have been taking living things apart. Now our job is to put all the pieces back together and to see how they work.